International Federation of Automatic Control

MATHEMATICAL AND CONTROL APPLICATIONS IN AGRICULTURE AND HORTICULTURE

IFAC Workshop Series, 1991. Number 1

IFAC WORKSHOP SERIES

Editor-in-Chief

Pieter Eykhoff, University of Technology,
NL-5600 MB Eindhoven, The Netherlands

HASHIMOTO & DAY: Mathematical and Control Applications in Agriculture and Horticulture (*1991, No. 1*)

CHESTNUT *et al*: International Conflict Resolution Using System Engineering (*1990, No.1*)
SIGUERDIDJANE & BERNHARD: Control Applications of Nonlinear Programming and Optimization (*1990, No.2*)
VILLA & MURARI: Decisional Structures in Automated Manufacturing (*1990, No.3*)
RODD *et al*: Artificial Intelligence in Real Time Control (*1990, No.4*)
MOTUS & NARITA : Distributed Computer Control Systems (DCCS'89) (*1990, No.5*)
KNUTH & RODD: Distributed Databases in Real Time Control (*1990, No.6*)
LOTOTSKY: Evaluation of Adaptive Control Strategies in Industrial Applications (*1990, No.7*)
O'SHIMA & VAN RIJN: Production Control in the Process Industry (*1990, No.8*)
MOWLE & ELZER: Experience with the Management of Software Products (*1990, No.9*)

RODD: Distributed Computer Control Systems (*1989*)
CRESPO & DE LA PUENTE: Real Time Programming (*1989*)
McAVOY: Model Based Process Control (*1989*)
RODD & SUSKI: Artificial Intelligence in Real Time Control (*1989*)
BOULLART *et al*: Industrial Process Control Systems (*1989*)
SOMMER: Applied Measurements in Mineral and Metallurgical Processing (*1989*)
GOODWIN: Robust Adaptive Control (*1989*)
MILOVANOVIC & ELZER: Experience with the Management of Software Projects (*1989*)
GENSER *et al*: Safety of Computer Control Systems (SAFECOMP'89) (*1989*)

Other IFAC Publications

AUTOMATICA

the journal of IFAC, the International Federation of Automatic Control
Editor-in-Chief: G.S. Axelby, 211 Coronet Drive, North Linthicum,
Maryland 21090, USA

IFAC SYMPOSIA SERIES

Editor-in-Chief: Janos Gertler, Department of Electrical and Computer Engineering,
George Mason University, Fairfax, Virginia 22030, USA

Full list of IFAC Publications appears at the end of this volume

MATHEMATICAL AND CONTROL APPLICATIONS IN AGRICULTURE AND HORTICULTURE

Proceedings of the IFAC/ISHS Workshop,
Matsuyama, Japan, 30 September - 3 October 1991

Edited by

Y. HASHIMOTO

Department of Biomechanical Systems,
Ehime University, Matsuyama, Japan

and

W. DAY

AFRC Silsoe Research Institute,
Bedford, UK

Published for the

INTERNATIONAL FEDERATION OF AUTOMATIC CONTROL

by

PERGAMON PRESS

OXFORD · NEW YORK · SEOUL · TOKYO

UK	Pergamon Press plc, Headington Hill Hall, Oxford OX3 0BW, England
USA	Pergamon Press, Inc., 395 Saw Mill River Road, Elmsford, New York 10523, USA
KOREA	Pergamon Press Korea, KPO Box 315, Seoul 110-603, Korea
JAPAN	Pergamon Press, 8th Floor, Matsuoka Central Building, 1-7-1 Nishi-Shinjuku, Shinjuku-ku, Tokyo 160, Japan

First edition 1991

Library of Congress Cataloguing in Publication Data

Data applied for

British Library Cataloguing in Publication Data

IFAC Symposium on Mathematical and Control Applications in
Agriculture and Horticulture (1991: Matsuyama, Japan)
Mathematical and control applications in agriculture and horticulture. -
(IFAC workshop series; 1991, no.1)
I. Title   II. Hashimoto, Y.   III. Day, W.   IV. Series
338.10724

ISBN: 9780080412733

These proceedings were reproduced by means of the photo-offset process using the manuscripts supplied by the authors of the different papers. The manuscripts have been typed using different typewriters and typefaces. The lay-out, figures and tables of some papers did not agree completely with the standard requirements: consequently the reproduction does not display complete uniformity. To ensure rapid publication this discrepancy could not be changed: nor could the English be checked completely. Therefore, the readers are asked to excuse any deficiencies of this publication which may be due to the above mentioned reasons.

The Editors

Transferred to digital print 2009

Printed and bound in Great Britain by CPI Antony Rowe, Chippenham and Eastbourne

IFAC WORKSHOP ON MATHEMATICAL AND CONTROL APPLICATIONS IN AGRICULTURE AND HORTICULTURE

Sponsored by
International Federation of Automatic Control (IFAC)
  Technical Committee on Applications

Co-sponsored by
International Society for Horticultural Science (ISHS)

Organized by
The Society of Instrument and Control Engineers (SICE)
Ehime University

International Programme Committee
Y. Hashimoto (J)   (Chairman)
P.C. Austin (NZ)
H. Challa (NL)
W. Day (UK)   (Co-chairman)
I. Farkas (H)
G. Germing (NL)
B.K. Huang (USA)
L. Keviczky (H)
T. Nybrant (S)
N. Suda (J)
T. Takatsuji (J)   (Vice-Chairman)
H.J. Tantau (D)
A.J. Udink ten Cate (NL)
H. Wernstedt (D)
P.C. Young (UK)
Chr. v. Zabeltitz (D)

National Organizing Committee
Y. Asada   (Chairman)
Y. Hori
A. Ichikawa
J. Kondo
S Nakagawa
R. Sakiyama
Y. Sawaragi
J. Sugi
T. Takakura   (Vice-chairman)
M. Takatsuji
K. Yabuki
H. Yamaguchi
H. Yamasaki

CONTENTS

GREENHOUSE SYSTEMS

PLANT FACTORY SYSTEMS

POST-HARVEST TECHNOLOGY

ROBOTICS

INFORMATION, COMPUTERS AND AI

MODELS AND CONTROL IN AGRICULTURE AND HORTICULTURE

MEASUREMENTS IN CONTROLLED SYSTEMS

CONTROL IN IRRIGATION SYSTEMS

OPTIMAL CONTROL FOR PLANT PRODUCTION
IN GREENHOUSES

H.-J. Tantau

Institute for Horticultural Engineering,
University of Hannover, Hannover, Germany

Abstract. In the past, greenhouse climate control has been improved by using feedback-feedforward control in combination with adaptive control algorithms and on-line adaption of model parameter.
Controlled environment in greenhouses leads to a high level of intensity in plant production. Microcomputers give the possibilities to save labor. but also to raise the plant productivity to maximum by applying optimal environmental control for the individual cultivation. Optimal environmental control includes regulation of temperature and humidity, control of CO_2-concentration, pH- and EC-values to an optimal level, and supply of artificial lighting. This results in reduced personal expenses and decreased growing periods for maximum crop harvest, combined with less overall heat consumption. Plant growth models and economic models are indispersible for optimal control of plant production. The growth models describe the plant`s reaction on changing environmental factors. In order to increase the accuracy of optimization, a feedback from the plants is needed to realize an on-line adaption of model parameter. Several methods have been developed and tested to measure "plant growth".
Besides the growth models, heuristic knowledge is important to improve on-line optimization. Artificial intelligence and neurocomputing technique will be useful for this purpose.

Keywords. Environmental control; optimal control; computer application; control application; horticulture.

INTRODUCTION

One of the goals of horticultural production in greenhouses is to increase the sustainable income of the grower. The investment costs for greenhouses as well as labour and energy cost are much higher compared with conventional plant production. This can only be balanced out with a better utilization of the yielding potential of plants, higher labour productivity and higher energy efficiency.
Higher plant productivity and quality in combination with a reduction of energy consumption require a better control of the environment. In addition to temperature the air humidity, CO_2-concentration and light intensity are controlled in commercial greenhouses. For reducing the pollution of ground water, the control of water and nutrition supply is getting more and more important. Closed irrigation systems including soilless culture such as NFT and Rockwool, can solve the pollution problem but require an improved control of nutrition, e.g. using ion-selective sensors.
As more and more microcomputers are used in commercial greenhouses, it is possible to increase the accuracy of environmental control by highly sophisticated control algorithms.

Modern control strategies lead to dynamic control. Optimal conditions must be identified based on the knowledge of plant physiology and ecology. That means, optimal environmental conditions cannot be identified without measuring the plants` response (Hashimoto et al., 1982; Hashimoto, 1985). This concept has been defined as the speaking plant approach.
The optimizing of the horticultural production needs plant growth models for the description of the plant reaction to the environmental changes. These models consist of a series of mathematical equations and opens a large field for application of mathematics in horticulture.

CONCEPTION OF ENVIRON-
MENTAL CONTROL

Research and development have been done to improve the control accuracy by more sophisticated control algorithms and new control strategies. As new tasks have been added, the complexity of the system is increasing. For the use in commercial greenhouses the system must be as flexible as possible in order to accomplish individual demands. This can be realised by a modular system. Furthermore, dividing the system into different optimizing and control levels creates a hierarchical system. Fig. 1 shows a system with three main levels. Each level may be divided into several sublevels.

The top level is the level of production planning and management. The middle level represents the control of plant production. Decision rules and limitations for the on-line optimization are the input of this level.

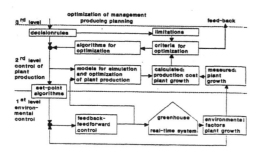

Fig. 1: Structures of environmental control.

The output gives the setpoints for the bottom level. There, the environmental control is realised using feedback-feedforward control.

In order to increase flexibility the system should be modular, that means, that each level will be divided into several modules or components which are principally independent to each other. In such a control system the modules must be able to communicate, what makes necessary a network. The principle requiries for a network are shown in fig. 2. From the bottom level (actor-sensor-level) to the top level the number of components is decreasing as well as the frequency of communication; the number of bytes for one data transfer and the life time of data is increasing. Local area networks (LAN, e.g. Ethernet) can be used on the top level, while on the lower levels other net

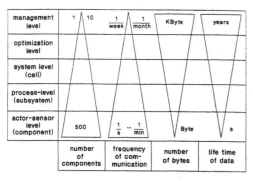

Fig. 2: Criteria for a hierarchical structure of communication

works are required which have been developed for industrial control applications. In fig. 3 some examples are shown. For the application in horticulture the maximum length of the connection link is very important, especially for the system level (Fig. 2). The UART-Bus with the physical link RS 485 is a useable solution in this level.

Bussystem	control of bus	transfer rate [KBit/s]	length [m]	update [ms]	controller
PDV-Bus	central	312.5	1000 Koax	27.2	NEB 3000
UART-Bus Profibus	central decentral	500	200	40	80 C 51 80 C 31
Bitbus	central	375	300	20-40	RUPI
CAN	central decentral	1000	30	6.4-10.8	82 C 200 82526
ABUS	central decentral	500	30	4.6-6.8	Abus-IC

Fig. 3: Bus systems for industrial control applications.

ENVIRONMENTAL CONTROL
(Bottom level, fig.1)

In the past most of research and development has been carried out to improve control on the bottom level. For environmental control commercial climate computers for climate control and substrate computers for control of water and nutrition supply (O`Flaherty, 1989; Takakura, 1989) are available.

From the engineering point of view exacte control of the climate in greenhouse is difficult, because of time delay and long time constant of the system (Veerwaaijen et al., 1985). The very thin covering material is necessary to get a high light transmissivity, but disturbances, e.g. solar radiation, outside air temperature and wind velocity, can so influence the climate inside a greenhouse very rapidly. Besides this, the heat load of a greenhouse can be reduced to 50% or even less within a few minutes by using thermal screens or other measurings for energy saving.

Several investigations have been carried out to improve the control accuracy. One possibility is the use of adaptive control algorithms (Henten, 1989; Hooper, 1985; Tantau, 1985).

For this aim two approaches are of interrest: the self-tuning regulator and the model reference adaptive system. The self-tuning controller consists of a combination of a recursive parameter estimation algorithm and a control algorithm (Udink ten Cate, 1983). Normally a parameter optimized controller is used (e.g. PI or PID). However, several problems restrict the application of parameter-adaptive-controllers to real process: the necessity to provide a number of spezification parameter in advance, unsatisfactory control behaviour during the adaption

phase, undesired effects like bursting of
parameter estimates.

niques are applicable for parameter estimation.

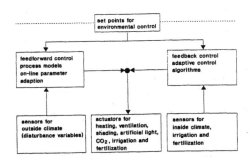

Fig. 4: Structure of feedback-feedforward
control (bottom level, fig.1)

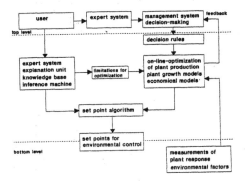

Fig. 5: Structure of the middle level
(Fig.1) using on-line optimization and expert system

Feedback control is improved by using
feed-forward control additionally.
(Tantau, 1985; Tantau, 1989). Fig. 4
shows the structure of the bottom level
(Fig. 1) using feedforward-
feedback control. Feedforward control
includes the measuring of the disturbance
variables (outside climate) and the calculation of the actuators' change by
means of process models.
Black box models or mathematical models
based on heat balances may be used. The
processes in the greenhouses are time and
place dependent and must be desribed by
partial differential equations (Bot,
1983; Bot, 1989). The system has distributed parameter. Ordinary differential
equations, describing the greenhouse as a
single compartment, working with single
energy buffers, are usual for feedforward
control. This method seems to give a reasonable result for the "average" temperature, humidity or CO_2-concentration,
even when gradients in the greenhouse occur (Udink ten Cate, Vooren, 1983; Biemond, 1989).
The use of feedforward control has several advantages compared with feedback
control (Tantau, 1990).
-The stability of the control loop will
increase
-The control system can react earlier on
changes of outdoor
weather conditions.
-Overshoots and undershoots will be
avoided.
-Movements of the actuators are reduced.

Applying feedforward control needs an on-
line process identification and an adaption of the model parameter. This requires models on a higher level of ab-
straction or the reduction of "tunable"
model parameter. On-line parameter estimation is very important for the use of
feedforward control in commercial climate
computers. An exacte technique must prevent wrong identification and "bursting"
of parameter estimates. Statistical methods and numerical optimization tech-

CONTROL OF PLANT PRODUCTION
(middle level, fig. 1)

For operation of commercial climate computers (bottom level) the possibility is
given to use about 300 parameter or setpoints for one compartment. With increasing complexity of the control system
the number of parameter will increase
too. Thus, it gets more difficult for the
grower to operate such a system. The task
of the middle level is to calculate the
setpoints for the bottom level. Two interesting approaches should be combined
at the middle level (Fig. 5).
One approach is the use of on-line optimization. Plant growth and plant development is controlled in order to achieve
the crop production which is expected at
top level. An optimal control of production should effect increasing plant productivity and e.g. decreasing energy input for maximizing economic net results
(Meurs, Stranghellini, 1989; Schapendonk
et al., 1984; Seginer et al., 1986).
The economic setpoints, decision rules
and limitations are inputs from the top
level (Vogel et al., 1989). Repeated optimization helps to compensate the
decline that may be caused by deviations
from the standard course of the climate
(Rheinisch et al., 1989).
Considering the corrections which are necessary due to the deviations from
assumed standard outdoor climate the demand for a long term strategy arises.
Knowledge about the crop reactions on a
change of the environment must be available for optimization. This can be expressed with plant growth models. Several
researchers are working in the field of
modelling plant growth (Augustin et al.,
1980; Challa et al., 1986; Enoch, 1984;
Liebig, 1989; Nederhoff et al., 1989).
Several growth models are available, especially for different vegetables. De-
scriptive models are used normally for
optimization.

One example for the use of such models is the optimized control of thermal screen (Bailey, 1988). By such a strategy the screen will be closed whenever the amount of energy exceeds the value of growth, which is lost due to light reduction. Temperature control by temperature integration is an other example (Bailey, 1985; Bailey, Seginer, 1989).

For control and optimization of plant production a feedback from the plants would be favourable. This feedback can be used for feedback control and for an online parameter adaption of the growth models.

Several methods of measurements of plant growth and plant development have been investigated (Hack, 1989; Matsui, Eguchi, 1976, 1977, Hashimoto, 1989; Schoch, 1987):

-net CO_2-uptake (net photosynthesis) using the greenhouse as a cuvette;
-remote sensing of plant temperature as an indicator of stomatal opening
-transpiration rate using lysimeter or an electronic balance;
-growth rate of leaves with image processing;
-fresh weight with an electronic balance;
-height of the plants with mechanical sensor;
-stem diameter to detect the water status;
-plant development by image processing.

These methods are applicable in experimental greenhouses, especially for development of new control strategies. The frequency for measurements of plant growth will be in the range 1/h to 1/day (Fig. 2).

This approach has been called "speaking plant-approach" (Hashimoto, 1989).

For the on-line optimization of plant production numerical optimization techniques offer opportunities to derive optimal control from complex simulation models. The performance of different methods depends largely on the nature of the problem. If little is known about the surface or the solution space, random strategies mostly perform better than directed search methods. One problem is to find the global optimum and not local optima, especially when an on-line optimization is gained.

The other approach, shown in fig. 5, is the use of expert systems or artificial intelligence in order to make heuristic knowledge available for the control system. In this case the expert system works as an interface between the user and control system. For this the explanation unit is very important, but unfortunately not included in most of the commercial expert systems.

Several researchers are working in the field of artificial intelligence and expert systems for the control of plant production (Harazono et al., 1988; Jacobsen et al., 1987; Jones et al., 1988; Kozai, 1985; Kurata, 1988; Leung et al., 1985; Hatou et al., 1990).

The combination of heuristic and procedural knowledge is of high importance for optimal control strategies on the middle level.

CONCLUSIONS

In the past most of research in the field of environmental control has been carried out on the bottom level (Fig. 1). The application of mathematics and control have led to an increase of control precision, an increase of stability of the control loops and a decrease of energy consumption. This is one base for an on-line optimization of plant production (middle level).

Another base is the knowledge of the plant response to a change of environmental condition. This must be expressed with plant growth models and rules for plant production. The lack of knowledge is still a serious restriction for the use of on-line optimization in commercial holdings. The use of expert systems is a possibility to make heuristic knowledge available for the control system and to develop an "user friendly" interface. For further research activities is the development of an explanation unit, which allows to handle "uncertain" knowledge, important. Another demand is a self-learning facility.

A problem, which must be solved is the standardization of the communication link between the different levels (Fig. 1), especially between the top level and the middle level.

REFERENCES

Augustin, P., and Schmidt, M., 1980. Model of basis for optimal control of growth factors in the cul-ture of the greenhouse cucumber. Int. Symposium on System Analysis and Simulation, 5, 420-424.

Bailey, B.J., 1985. Wind dependent control of greenhouse temperature. Acta Hort. 174, 381-386.

Bailey, B.J., 1988. Improved control strategies for greenhouse thermal screens. Acta Hort. 230, 485-492.

Bailey, B.J., Seginer, I., 1989. Optimum control of greenhouse heating. Acta Hort., 245, 512-518.

Bakker, J.C., 1985. A CO_2 control algorithm based on simulated photosynthesis and ventilation rate. Acta Hort. 174, 387-392.

Biemond, F., 1989. A glasshouse climate model as part of a bio-economic model. Acta Hort. 260, 275-284.

Bot, G.P.A., 1983. Greenhouse climate: from physical processes to a dynamic model. Thesis, Agric. Univ. Wageningen, The Netherlands.

Bot, G.P.A., 1989. Greenhouse simulation models. Acta Hort. 245, 315-325.

Challa, H., and Schapendonk, A.H.C.M., 1986. Dynamic optimization of CO_2-concentration in relation to climate control in greenhouses. In: Carbon dioxide enrichment of greenhouse crops. Florida. Vol. I, 147-160.

Enoch, H.Z., 1984. Carbon dioxide uptake efficiency in relation to crop-intercepted solar radiation. Acta Horticulturae 162, 137-147.

Gal, S., Angel, A., and Seginer, I., 1984. Optimal control of greenhouse climate: Methodology. European Journal of Operational Research, North Holland. 17, 45-56.

Hack, G., 1989. The use of image processing under greenhouse conditions for growth and climate control. Acta Hort. 245, 455-461.

Harazono, Y., Kamiya, H., Yabuke, K., 1988. A control logic based on an artificial intelligence technique and its application for controlling plant environment. Acta Hort.230, 209-214.

Hashimoto, Y., Fukuyama, T., Morimoto, T., 1982. Computer aided protected cultivation (CAPC) by monitoring "Speaking Plant". Abstracts 21st Int. Hortic. Congr. Hamburg, 1982.

Hashimoto, Y., Morimoto, T., 1985. Some speaking plant approach to the synthesis of control system in the greenhouse. Acta Hort. 174, 219-226.

Hashimoto, Y., 1989. Recent strategies of optimal growth regulations by the speaking plant concept. Acta Hort. 260, 115-122.

Hatou, K., Nishina, H., Hashimoto, Y., 1990. Computer integrated agricultural production. Reprints of 11th IFAC World Congress. Pergamon Press, Oxford, Vol.11, 306-310.

Henten, E.J. van, 1989. Model based design of optimal multivariable control systems. Acta Hort. 248, 301- 306.

Heuvelink, E., Challa, H., 1989. Dynamic optimization of artificial lighting in greenhouses. Acta Hort. 260, 401-412.

Hooper, A.W., Davis, P.F., 1985. Control of greenhouse air temperature with an adaptive control algorithm. Acta Hort. 174, 407-412.

Jacobsen, B.K., Jones, J.W., Jones, P., 1987. Tomato greenhouse environment controller: real-time, expert system supervisor. American Society of Agricultural Engineers, St. Joseph, Paper No. 87-5022.

Jones, P., Jacobsen, B.K., Jones, J.W., 1988. Applying expert system concepts to real-time greenhouse control. Acta Hort. 230, 201-208.

Kozai, T., 1985. Ideas of greenhouse climate control based on knowledge engineering techniques. Acta Horticulturae 174, 365-373.

Kurata, K., 1988. Greenhouse control by machine learning. Acta Hort. 230, 195-200

Leung, C.M., Wilson, D.R., Marquand, C.J., Tassou, S.A., 1985. A knowledge based system for the control of a greenhouse environment. Acta Hort. 174, 425-432.

Liebig, H.P., 1989. Growth and yield models as an aid for decision making in protected crop production control. Acta Hort. 260, 99-114.

Matsui, T., Eguchi, H., 1976/77. Computer control of plant growth by image processing. I. Mathematical representation of relations between growth and pattern area taken in photographs of plant. Envir. Control in Biol. 14, 1-7, 1976.
II. Pattern recognition of growth in on-line system. Envir. Control in Biol. 15, 37-45, 1977.

Meurs, W.T.M. van, Stanghellini, C., 1989. A transpiration-based climate control algorithm. Acta Hort. 245, 476-481.

Meyer, J., 1989. Evaluation of artificial light systems. Acta Hort. 245, 370-376.

Nederhoff, E.M., Schapendonk, A.H.C.M., 1985. Effects of environmental conditions on growth and production of cucumbers; comparison between empirical and simulation data. Acta Hortic. 174, 251-258.

Nederhoff, E.M., Gijzen, H., Vegter, J.G., Rijsdijh, A.A., 1989. Dynamic model for greenhouse crop photosynthesis. Validation by measurements and application for CO_2 optimization. Acta Hort. 260, 137-148.

O'Flaherty, T., 1989. Microcomputers in energy saving greenhouses. Acta Hort. 245, 416-423.

Rheinisch, K., Tümmer, C., Hopfgarten, S., 1984. Hierarchical on-line control algorithm for repetitive optimization with predicted environment and its application to water management problems. Syst. Anal. Model. Simul. 1, 4, 263-280.

Rheinisch, K., Arnold, E., Markert, A., Puta, H., 1989. Development of strategies for temperature and CO_2-control in the greenhouse production of cucumbers and tomatoes based on model-building and optimization. Acta Hort. 260, 67-76.

Schapendonk, A.H.C.M., Challa, H., Broekharst, P.W., Udink ten Cate, A.J. 1984. Dynamic climate control: an optimization study for earliness of cucumber production. Scienta Horticulturae 23, 137-150.

Schoch,H.-J., 1987, Influence du niveau d'alimentation hydrique sur les variations du diametre des tigres, du potentiel hydrique, de la resistance stomatique, de la transpiration et de la photosynthese de l'aubergine. Agric. and Forest Meteorology, 1987, 40, 89-104

Seginer, I., Angel, A., Gal, S., Kantz, 1986. Optimal CO_2-enrichment strategy for greenhouses: a simulation study. J. Agric. Engig. Res. 34, 285-304.

Takakura, T., 1989. Climate control to reduce energy inputs. Acta Hort. 245, 406-415.

Tantau, H.-J., 1985. Analysis and synthesis of climate control algorithms. Acta Hort. 174, 375-380.

Tantau, H.-J., 1985. Greenhouse climate control using mathematical models. Acta Hort. 174, 449-459.

Tantau, H.-J., 1989. Models for greenhouse climate control. Acta Hort. 245, 397-404.

Tantau,H.-J., 1990. Automatic control application in greenhouse. Reprints of 11th IFAC World Congress Pergamon Press, Oxford, Vol.11, 302-305.

Udink ten Cate, A.J., 1983. Modelling and (adaptive) control of greenhouse climates. Thesis, Agric. Univ. Wageningen.

Udink ten Cate, A.J., Vooren, J. van de, 1983. New models for greenhouse climate control. Acta Hort. 148.

Udink ten Cate, A.J., 1985. Modelling and
    simulation in greenhouse climate con-
    trol. Acta Hort. 174, 461-468.
Veerwaaijen, P.W.T., Gieling, T.H.,
    Meurs, W.T.M. van, 1985. Measurement
    and control of the climate in a
    greenhouse, where a heated concrete
    floor is used as low temperature en-
    ergy source. Acta Hort. 174, 469-475.
Vogel, G., Heißner, A., Klåring, P.,
    1989. Ways and possibilities for
    automatic control. Acta Hort. 260,
    25-66.
Vooren, J. van de, Steinbuch, F.,
    Hendricks, P., 1985. A multi-level
    control system. Acta Hort. 174,
    547-548.

PHYSICAL MODELING OF GREENHOUSE CLIMATE

G. P. A. Bot

Institute of Agricultural Engineering (IMAG-DLO), P.O. Box 43,
6700 AA Wageningen, The Netherlands

Abstract. The physical environment in a greenhouse is a modification of
the outdoor weather due to the physical properties of the enclosure and
its content and the control actions. This modification can be understood
from the various physical processes for the exchange of energy and mass
with the outdoor environment and that within the system. For a progress
in greenhouse operation and design a quantitative description of the
various processes is needed. In this paper a survey of the physical
processes in the system is given with emphasis on a qualitative description needed to understand the constraints for modeling.

Keywords. Agriculture; Computer simulation; Greenhouse climate; Modeling; Nonlinear systems; Physical processes.

INTRODUCTION

A greenhouse is a building constructed as
light transparent enclosure to improve
environmental conditions for plant production. The set of momentaneous environmental factors in the greenhouse affecting crop growth and development is referred to as the greenhouse climate. As
the meteorological term climate indicates
a long term average of environmental factors and the term weather is used for the
momentaneous set, the proper term in the
greenhouse would be greenhouse weather.
However it is common practice to use the
term greenhouse climate in this respect.
From the physical point of view a greenhouse can be considered as a bioreactor,
in which a crop produces a marketable
product from carbondioxide and water containing fertilizer with the aid of shortwave radiation. The greenhouse climate
affects the exploitation of the growing
factor light, so it is extremely important to control the greenhouse climate in
order to have a maximal effect. More and
more information is gathered on the relation between crop growth and environmental conditions (Bot and Challa, 1991).
This opens the possibility of growth optimization to replace the control of a
desired temperature and humidity level in
the greenhouse by an optimization scheme
calculating these levels by the system
(Challa and others, 1988). To do this in
a proper way, the physical behaviour of
the reactor has to be known.
Physical modelling of greenhouse climate
started already in the seventies. However, the information available was diffuse, not aimed at the before mentioned
purpose or lacking at all (Udink ten Cate, Bot and Van Dixhoorn, 1978). Bot
(1983) presented a physical model based
on in-situ measurements of the main physical processes. Various physical processes are studied in more detail (Stanghellini, 1987; De Jong, 1990) and the adap-

ted model is applied in recent optimiza-
tion studies (Houter and others, 1989;
Houter, 1990; Gijzen and Ten Cate, 1988;
Gijzen, Vegter and Nederhoff, 1990; Van
Henten and Bontsema, 1991)

GENERAL ASPECTS

The difference between greenhouse climate
and outside weather (sometimes called
greenhouse effect) is mainly caused by
two mechanisms:
- The air in the greenhouse is stagnant
  due to the enclosure, so the air exchange with the surrounding (outside)
  air is strongly decreased compared to
  that of the air without envelope and
  the local air velocities are small compared to that in the open air.
  The reduction of the air exchange (or
  ventilation) directly affects the energy and mass balances of the greenhouse
  air while the smaller local air velocities affect the transport processes
  between the greenhouse air and the
  greenhouse inventory.
- The inward shortwave radiation is decreased due to the light interception
  by the opaque and transparent components of the greenhouse and the long
  wave radiative exchange between inside
  and outside the greenhouse is influenced by the radiative properties of the
  covering materials. With glass as covering material this leads to the mouse
  trap theory: glass is (partly) transparent for the incoming shortwave radiation and opaque for the longwave radiation emitted from the interior. However
  this effect is of minor importance to
  explain the increased air temperature
  in the greenhouse. The common name
  greenhouse effect for the mouse trap
  theory is therefore misleading. Nevertheless the radiative effects are indispensable to a description of the
  greenhouse climate because they direct-

ly affect all energy balances.

The general mechanisms may be relatively simple, the impact on the greenhouse climate itself is of a more complex nature. In this paper the main physical processes are discussed seperately first and then combined to a lumped parameter physical model that can be used to simulate dynamically the greenhouse climate.

PHYSICAL PROCESSES

General overview
The various physical processes can be schematised according to fig. 1 a,b. In this scheme the major greenhouse components are considered to be the greenhouse cover, the greenhouse air, the crop and the soil. In fig. 1 a the radiative processes are outlined, in fig. 1 b the various convective and conductive exchange processes. The main processes will be discussed seperately in the following sections with the emphasis on physical understanding.

Radiative exchange

The production in a greenhouse is aimed at an optimal use of incoming direct and diffuse solar radiation as driving force for the photosynthesis process. The spectral distribution of solar irradiance at earth surface is well known for various meteorological conditions (e.g. Kondratyev (1972)). For planth growth a special part of the spectrum in the visible region is of interest, the so called photosynthetic active radiation (PAR) with wave length region of 400 to 700 nm. In this region about half of the solar energy is irradiated. Only a small part of the PAR energy is directly converted into the photosynthetis process.
The solar radiation at crop level directly contributes to the energy balance of the crop. To translate the solar radiation at earth level to that inside the greenhouse at crop level is therefore of great importance.
The interaction of the solar radiation with the greenhouse cover determines how much radiation is transmitted and available at crop level. The interaction can be calculated from the optical laws for reflection, absorbtion and transmission of transparent layers and opaque construction parts. The optical properties of the transparent materials, the angle of incoming radiation relative to the observed surface and the geometry of the construction have to be known. For the direct light, the angle between radiation and surface follows from the solar position (given by the latitude of the observed greenhouse and the time and date) and of the orientation and geometry of the surfaces. For the diffuse radiation it follows from the distribution of the radiation intensity over the hemisphere, which is different for various meteorological conditions. The most striking difference is that between a clear and cloudy sky. Bot (1983) described the considerations given above in full detail. Comparison between the theoretical calculations and measurements of transmission factors gave an agreement for both diffuse and direct radiation within a few percents of trans-

mission. Therefore this light transmission model of greenhouses is an operational part of crop production models (Gijzen and Ten Cate, 1988; Gijzen, Vegter and Nederhoff 1990) and of greenhouse climate models (Bot, 1989a: Houter, 1990).

The thermal radiation exchange from the greenhouse interior to outside is shielded by the cover. The efficiency of this shield depends on the radiative properties in the thermal wavelenght region (about 5000-50000nm). The exchange can be calculated according to the well known non-linear Stefan-Boltzmann relation, implementing emission coefficients to account for the radiative properties of the surfaces and view factors to account for the mutual interception of the radiation between the surfaces:

$$Q_{rad} = \epsilon_{12} \, F_{12} \, \{(T_1)^4 - (T_2)^4\} \quad (1)$$

with Q_{rad} energy flux density due to thermal radiation from surface 1 to 2 (Wm^{-2}), ϵ_{12} the effective emissivity between the surfaces (-), F_{12} the view factor from surface 1 to surface 2 (-) and T_1 and T_2 the temperature of surface 1 and 2 respectively (K).
The view factors for the thermal radiation exchange with the crop are of course dependent on the development of the crop and determined in a seperate study on the physical behaviour of the crop (Stanghellini, 1987). Calculations and measurements in the thermal wavelength region were in good agreeement. A special difficulty in this respect is the characterization of hemisperical thermal radiation. This is represented by defining the sky temperature as a temperature of a black hemisphere exchanging thermal radiation with the greenhouse cover according to Stefan-Boltzmann. This sky temperature is not measured as a standard meteorological factor. In literature (e.g. Wartena and others, 1973) relations are known to relate sky temperature to standard meteorological obeservations, but these relations are only applicable for an average sky temperature over a long period and only valid for regions with the same meteorological characteristics as the region in which the measurements are performed. For realistic dynamic modelling sky temperature is a dynamical boundary condition. Therefore in our experiments it is calculated from real time measurements of the thermal radiation exchange between a surface with known temperature and the hemisphere.

Ventilation exchange

The greenhouse enclosure envelopes the interior and prevents mixing of the internal with the external air. Direct exchange of air through openings in the enclosure (leaks and ventilation windows) is called ventilation flux and can be expressed in terms of volumetric flow. While this ventilation process affects the most important greenhouse effect, i.e. the enveloping of air, a proper description of the dependency of the ventilation flux on external and internal effects is a prerequisite for a description of greenhouse climate.
In general, ventilation flux is the flow of air from inside to outside and reverse

through openings. This flow has to be generated by some pressure difference between both sides of the opening. The pressure difference can be due to the effect of the outside wind (wind effect) or due to the density differences (generated mainly by temperature and to a much lesser extent by concentration differences) between internal and external air (temperature effect). The flow through the opening, with known pressure difference between both sides of the opening, is dependent on the flow resistance of the opening itself. In literature on the ventilation of buildings specific overall relations are deducted for the ventilation rate through specific openings as function of relevant parameters. Bot (1983) adopted an approach in which the wind and temperature effects and the flow characteristics of the opening are discussed seperately. Recently this aproach is more soundly based and validated in more detail (De Jong, 1990)

From this studies the ventilation volume flux Φ_v ($m^3 s^{-1}$) can be expected to be linear proportional to the outside wind speed u at reference height and the area of ventilation windows A_o for any window opening angle ß.

Full scale measurements of the ventilation flux with a tracer gas technique proved these relations and validated them for various conditions and window types (De Jong, 1990).

With the known dependency of ventilation flux on wind speed, temperature difference between in- and outdoor, window characteristics and window opening the exchanged energy q_{vent} (Js^{-1}) and mass m_{vent} (kgs^{-1}) (water vapour, CO_2, etc) between the greenhouse interior and the ambient can be calculated as respectively:

$$q_{vent} = \Phi_v \; Cap_{air} \; (T_i - T_a) \qquad (2)$$

$$m_{vent} = \Phi_v \; (C_i - C_a) \qquad (3)$$

with Cap_{air} the volumetric specific heat of the greenhouse air ($Jm^{-3}K^{-1}$), T_i and T_a the in- and exterior temperature respectively (K) and C_i an C_a the in- and exterior concentration of the gas component considered (kgm^{-3}).

So the non-linearity of ventilation exchange is in the non-linear dependency of Φ_v; at constant Φ_v the exchange is linear.

Crop transpiration

All greenhouse activities are aimed at an optimal growth and development of the crop as the supplier of greenhouse products. In the greenhouse climate model the impact of the greenhouse climate on the growth and development of the crop is out of the scope; the vegetation is considered as a well-defined body that exchanges energy, water vapour and carbondioxide with the greenhouse air. The greenhouse climate and the plants interact in this respect. Generally speaking, without plants the greenhouse climate can be marked as a desert climate. With plants is is transformed into a humid, hot climate. The physical interaction between plants and environment can explain how this transformation is accomplished.

The transport of energy and water vapour from the leaf is in general defined in the same way as the heat and mass transfer from other surfaces. Water evaporates in the internal water saturated cavities of the leaves with saturated vapour pressure, is transported to the surface of the leaves meeting a crop resistance and then transported from the surface to the greenhouse air meeting some boundary layer resistance. The crop resistance is a complex function of the environmental conditions, translating the plant-physiological reactions of the crop to the environment into a phenomenological-physical model. The boundary layer resistance depends on the local air movement and temperature differences and differs quantitatively from the extensively reported outdoor conditions. The strong non-linear relations for the mentioned resistances are reported by Stanghellini (1987).
The exchange can be described as the product of resistance and vapour-pressure difference so as a linear relation if the resistance is constant.

Exchange with the cover

The cover of a multispan greenhouse complex is a saw tooth surface. A common Dutch greenhouse type, manufactured in an industrial way, is a so called Venlo greenhouse. It has standardised measures. The cover has a span width of 3.2 m and the roof slope can vary between 22 and 26 degrees, so the geometry of the saw tooth surface is known.

The exchange with the cover at both the inside and outside are of interest. At the inside, local air velocities are low ($\approx$ 10 cm/s: Re $\approx 10^4$) and the temperature differences considerable ($\approx$ 10 K: Gr $\approx 10^{10}$), so it can be expected that the transfer of energy and mass (condensation of water vapour) are due to natural convection. At the outside, forced convection can be expected.

In literature no data on convective heat and mass transfer to and from saw-tooth surfaces could be found. It is not known a-priori if literature data on the transfer to and from flat plates can be applied in this case while the flow field over the surface will be transformed by the saw-tooth surface. Therefore from expriments the convective heat transfer to and from the cover was measured and the flow field over the cover was sampled. This yielded natural convection relations for the heat transfer at the inside and also at the outside for low wind speeds up to about 3 m/s. At higher wind speeds forced convection was found at the outside.

In general the energy transfer Q_{conv} (Wm^{-2}) between the flowing air and the surface can be epressed as:

$$Q_{conv} = \alpha \; (T_{air} - T_{surf}) \qquad (4)$$

with T_{air} and T_{surf} the air and surface temperature respectively (K) and α the heat transfer coefficient ($Wm^{-2}K^{-1}$). For natural convection α depends on the temperature difference itself and for forced convection α depends on the wind velocity. So again the non-linearities are in the transfer coefficients.

Exchange with the heating system

Most commonly a greenhouse is heated by a heating pipe system, distributing hot water from a central boiler. Various pipe arragements are in use, the most common arrangement is that of four 2" pipes per span at a small distance (5-10 cm) to the soil surface. The four pipes are combined in two pairs, with a distance in the pair of about 20 cm and the pairs at a central distance of 1.6 m. In this way the heating pipes can be used in practice as a rail system for the internal transport. With local air velocities of 10 cm/s, Re is low (≈ 250), but Gr will be $\approx 10^6$ so the value of Gr/Re^2 is ≈ 10 which is of the same order of magnitude as the criterion for pure natural convection $Gr/Re^2 \geq 14$ (Morgan, 1975). However, literature data on the pure natural convective exchange from horizontal cylinders cannot be used for the convective exchange from the horizontal heating pipes while the above given arrangement differs from the experimental conditions from literature on pure natural convection.
Experiments under greenhouse conditions yielded the heat transfer coefficient, indeed being due to natural convection and difffering from literature on pure natural convection under ideal conditions. With the heat transfer between pipes and air described in the same way as rel. 4 the coefficient again introduces non-linearities.

Exchange with and transport in the soil

In the energy budget of the greenhouse, the exchange with the soil is of minor importance. However, the soil surface exchanges thermal radiation with the other greenhouse components and the energy storage in the soil determines the dynamics of the greenhouse system on a daily base (Bot, 1989b). So the exchange to and the transport in the soil have to be represented in a proper way for the description of the greenhouse climate especially for the long term dynamics. The calculation via the natural convection exchange at the surface and the mechanism of conduction for the transport in the soil has proved to be of sufficient accuracy.

DYNAMIC MODEL

With the quantitative description of the main exchange processes, the energy and mass balances can be set up over representative parts of the greenhouse, resulting in a set of first order differential equations. The parts are the cover, the greenhouse air, the crop and the layered soil, exchanging energy and mass with each other and with the environment. The response of the greenhouse system to the time varying environmental conditons is then represented by the solution of the set of differential equations. This solution can be found by e.g. forward integration as is done by various simulation languages. In our case we applied CSMP (IBM,1966) or TUTSIM (Kraan, 1974) in combination with bond-graph representation. Presently ACSL is a promising language for dynamic simulation. The simulation results are realistic under various out-

door conditions (Bot, 1989a) both in the short and long term response. This is demonstrated in fig. 3 and 4 for a 24 hour period from noon to noon with strong fluctuating shortwave radiation during the first day and a bright second day. The model is succesfully applied in optimization studies (Van Henten and Bontsema, 1991) and is used to get insight in the complex interactions in the greenhouse system. Not only the state variables of the system, like temperatures and concentrations, are calculated but of course also the flow variables, like energy and mass flows. Implementation of the behaviour of the mixing valve of the heating system enables the incorporation of control algorithms for heating. Then the heating pipe temperature as boundary condition can be replaced by a set point for e.g. air temperature or a set point calculating scheme. In this way, control algorithms can be checked with the advantage over simple control models that the response of the various parts of the greenhouse system, e.g. the crop, can be calculated.

CONCLUSIONS

The greenhouse climate is a complex interaction of various physical processes. Quantification of this climate as affected by the dynamic outdoor weather conditions demands quantification of the various exchange processes. Then combination in a physical model yields a reliable description of the greenhouse climate. This can be applied in practical and theoretical studies on the optimization of greenhouse production.

REFERENCES

Bot, G.P.A. (1983). Greenhouse climate: from physical processes to a dynamic model. Ph.D. thesis, Agricicultural University Wageningen. 240 pp.
Bot, G.P.A. (1989a). A validated physical model of greenhouse climate. Acta Horticulturae,245,389-396.
Bot, G.P.A. (1989b). Greenhouse simulation models. Acta Horticulturae,245, 315-325.
Bot, G.P.A. and H. Challa (1991). Crop growth and climate control of greenhouses. In Y.P.S. Bajaj (Ed.), Biotechnology in Agriculture and Forestry,Vol. 17, High-Tech and Micropropagation I. Springer Verlag, Heidelberg (in press).
Challa, H., G.P.A. Bot, E.M. Nederhoff and N.J. van de Braak (1988). Greenhouse climate control in the nineties.
Acta Horticulturae,230,459-470.
De Jong, T. (1989). Natural ventilation of large multispan greenhouses. Ph.D. thesis, Agricultural University Wageningen. 116 pp.
Gijzen, H.and J.A. ten Cate (1988). Prediction of the response of greenhouse crop photosynthesis to environmental factors by integration of physical and biochemical models. Acta Horticulturae,,229,251-258.

Gijzen, H., J.G. Vegter and E.M. Neder-
    hoff (1990). Simulation of greenhou-
    se crop photosynthesis: validation
    with cucumber, sweet pepper and to-
    mato. Acta Horticulturae,268,71-80.
Houter, G., H. Gijzen, E.M. Nederhoff and
    P.C.M. Vermeulen (1989). Simulation
    of CO_2 consumption in greenhouses.
    Acta Horticulturae,248,315-320.
Houter, G. (1990). Simulation of CO_2 con-
    sumption, heat demand and crop pro-
    duction of greenhouse tomato at dif-
    ferent CO_2 strategies. Acta Horti-
    culturae,268,157-164.
IBM (1966). IBM 1130 Continuous System
    Modelling Programme. IBM Manuals, SH
    20-0905 and H 20-0209.
Kondratyev, K.Ya. (1972). Radiation pro-
    cesses in the atmosphere.
    WMO report 309, WMO Geneva. 214 pp.
Kraan, R.A. (1974) THT SIM: A conversati-
    onal simulation programme on a small
    digital computer.
    Journal A,15,186-190.

Morgan, V.T. (1975). The overall convec-
    tive heat transfer from smooth cir-
    cular cylinders. Advances in heat
    transfer, Vol. 11,199-264. Academic
    Press, N.Y.
Stanghellini, C. (1987). Transpiration of
    greenhouse crops, an aid to climate
    management. Ph.D. Thesis Agricultu-
    ral University Wageningen. 150 pp.
Udink ten Cate, A.J., G.P.A. Bot and J.
    van Dixhoorn (1978). Computer con-
    trol of greenhouse climates.
    Acta Horticulturae,87,265-272.
Van Henten, E.J., J. Bontsema (1991).
    Optimal control of greenhouse clima-
    te. IFAC/ISHS 1st workshop on Mathe-
    matical and Control Applications in
    Agriculture and Horticulture. Mat-
    suyama-City, Japan.
Wartena, L., C.L. Palland and G.H.L. Vos-
    sen, (1973). Checking of some formu-
    lae for the calculation of long wave
    radiation from clear skies. Arch.
    Met. Geoph. Biokl. B,21,335-348.

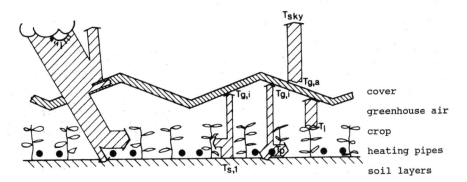

Fig. 1. Radiative processes: interception of direct and diffuse solar
    radiation by the greenhouse cover, the crop and the soil surface
    (left) and thermal radiative exchange between the various green-
    house components and with the sky (right).

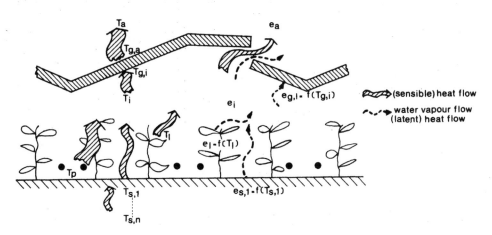

Fig. 2. Energy exchange (sensible and latent heat) between the various
    greenhouse components and to the outside air.

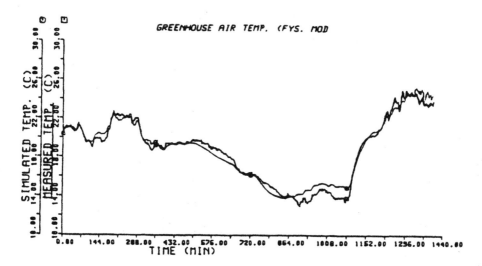

Fig. 3. Simulated and measured greenhouse air temperature during a 24 hr period with varying radiative conditions.

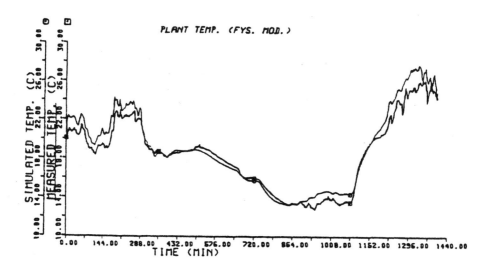

Fig. 4. Measured and simulated crop temperature during a 24 hr. period (same period as fig. 3.).

Copyright © IFAC Mathematical and Control Applications
in Agriculture and Horticulture, Matsuyama, Japan 1991

REFLECTIONS ABOUT OPTIMAL CLIMATE
CONTROL IN GREENHOUSE CULTIVATION

H. Challa* and G. van Straten**

*Dept. of Horticulture, Wageningen Agricultural University, P.O. Box 30, 6700 AA Wageningen,
The Netherlands
**Dept. of Agricultural Engineering and Physics, Wageningen Agricultural University,
Duivendaal 1, 6701 AP Wageningen, The Netherlands

Abstract. The cultivation in greenhouses is characterized by a strong influence of the
grower on the production process, made possible by the controlled crop environment. Up
till now the potentials for optimal control of the greenhouse/crop system, however,
have not been fully exploited. Given an optimal cultivation plan (sowing/planting date,
schedule of horticultural operations, planned temperature regime, harvest, etc.), the
diurnal control could be further optimized than is possible with present climate control systems. Improvements could be attained by incorporating present knowledge on
diurnal growth processes and detailed knowledge on physical processes in relation to
greenhouse climate in more intelligent control systems.
Diurnal climate control is one of the tools of the grower to control the production
process at the operational management level. The criteria for optimal climate control
should be derived from the goals of the grower and have to be formalized explicitly for
automatic control. To this purpose it is important to classify the major growth processes concerned according to response time and response type. In addition, interaction
with the grower, with his specific know-how and responsibility, needs to be settled for
proper integration with the management process and to provide information that cannot
be collected automatically. A possible architecture of an optimal control system based
on these principles is proposed.

Keywords. Crop growth; Dynamic optimization; Greenhouse climate; Modeling; Operational
management; Optimal control; Response time; User interface.

INTRODUCTION

In contrast to outdoor production, the growing conditions in greenhouse cultivation can be controlled
to a large extent. By means of ventilators and a
heating system, temperature and air humidity can be
regulated, and the CO_2 concentration of the air can
be increased by means of flue gasses or pure CO_2.
In the most sophisticated greenhouses, in addition,
there are possibilities to influence the level and
the duration of radiation by means of screens
and/or supplementary lighting. Also, for many crops
it is nowadays feasible to control the root environment with respect to temperature and availability of water and nutrients.

The grower needs these wonderful and sophisticated
technical achievements to control the production
process. Present control systems, however, do not
fully exploit the great potentials of modern greenhouse technology (Challa and others, 1988). Further
improvements are principally feasible in the diurnal climate control, if more knowledge on plant
physiological and physical processes could be
incorporated in new, more intelligent control systems (Challa, 1990). The solution of this problem
is complex for a number of reasons:
- the production process, as will be pointed out
  later, is a complex of a multitude of processes,
  taking place simultaneously, reacting with different response times and different response patterns to the environmental factors, and characterized by many interactions
- the grower has several objectives that often
  require, at least partly, conflicting actions
- climate control is a process that takes place
  without interference of the grower over appreciable periods of time where complex information has
  to be dealt with in real time under environmental

conditions that may deviate considerably from the
predictions
- there is a general lack of quantitative knowledge
  on the greenhouse/crop system required to translate the objectives of the grower into the proper
  actions

The problem of how to control greenhouse climate is
obviously an optimization problem, because of the
conflicting objectives, the fact that each climate
factor is affecting several processes simultaneously (Challa, 1990) and because the balance of
costs and economic returns has to be taken into
consideration. So far this problem has been dealt
with separately from different disciplines, such as
engineering, management, or horticulture. The purpose of the present study is to investigate optimal
climate control in greenhouses in a wider context
linking these disciplines together.

OPTIMAL CONTROL, AN ANALYSIS

What is Optimal Control?

Optimal control may be defined as the control that
satisfies most the goals set by the grower, which
is obviously a definition that in its implicit
truth does not contribute to any practical solution. For a theoretical analysis of the problem it
is thus necessary to elaborate further on these
goals and how they are achieved in the management
of the nursery. Furthermore the criteria for control, and the relation with processes to be controlled have to be considered more closely, before
investigating optimization of climate control.

Management

Management may be defined as the collection of activities directed to reach certain goals. One of the goals of a grower, as an entrepreneur, in general is to maximize his profit. Climate factors strongly affect the production process as well as the cost of operation. Climate control in greenhouse culture is then one of the tools the grower has to manage his nursery, and thus should be considered as a part of the overall management rather than as an isolated activity.

In management theory different management levels are usually distinguished (Davis and Olsen, 1985): strategical, tactical and operational level, depending on the planning horizon (Fig. 1). The argument for this distinction is that, in spite of interactions between different levels, basically the decisions at each level are made independently. E.g. there is no point in considering the question of building a new greenhouse (a strategical decision), or changing to another cultivar (a tactical decision) once the crop is established, and likewise it is not feasible to deal with all possible situations that might occur when making a strategical or tactical plan. On the other hand it is clear that the usual management at the operational level does affect the planning at e.g. tactical level.

```
                                        ───────────> planning horizon
24h    day-weeks   year   many years
└────┴─────────┘   │       │
      │            │       │
 operational   tactical  strategical management
```

Fig. 1. Management levels in relation to the planning horizon; the time scale is indicative rather than absolute.

With respect to climate control, the management levels distinguished have different implications. At the strategical level decisions on capital investments for equipment determine the technical possibilities for climate control. In addition decisions may be made concerning the long-term policy of the nursery, for example with respect to product quality.

Before the start of a new cultivation cycle a tactical plan has to be formulated, where the grower decides what crop and cultivar to cultivate, when to plant or sow the crop, as well as the temperature regime he plans to follow. This plan is, of course, made within the framework of the strategical plan. Connected with the tactical plan is an expectation of average climatic conditions, prices that the grower will receive for his product, as well as a "blueprint" of how the crop will grow, develop and produce as a function of time.

During implementation of the tactical plan considerable deviations may occur, which have to be dealt with at the operational level. Examples are the actual weather conditions, the behaviour of the crop, or unexpected developments in the market. Although deviations of the average do not necessarily have a negative effect, they were not planned and therefore the consequences have to be evaluated and some adaptations may be required. Climate control, because of its close relation with weather conditions and crop response, then is a tool of the operational management.

Operational management is often described as a cycle, progressing in time, of planning-implementation-control (Boehlje and Eidman, 1984). In this cycle "control" refers to the management process and should not be confounded with climate control.

It is clear that control of the management process requires criteria. These criteria are derived from the goals set at the tactical and strategical levels and will be further discussed within the scope of the problem of optimal climate control.

Criteria for Operational Management in Relation to Climate Control

With respect to climate control the following criteria in operational management can be formulated: yield (in economic terms), crop quality, product quality, timing of the production process, production costs and production risks. These criteria will often give rise to conflicting climate requirements (e.g. yield versus quality, yield versus costs). These conflicts have to be solved, explicitly or implicitly, at the tactical level, but they play also a role at the operational level. The criteria will now be briefly reviewed.

Economic yield. Yield is strongly affected by the climate conditions and as such it is a major criterion for climate control. At the operational level it is important, however, to realize that short-term yield increase may have negative long-term implications (see crop quality), an aspect considered at the tactical level. When yield increase requires extra economic inputs, as may be the case with e.g. CO_2 enrichment, additional yield and associated extra costs have to be compared (Challa and Schapendonk, 1986). Photosynthesis is a major yield determining process, that is primarily sensitive to radiation and CO_2 concentration, but to a lesser extent also influenced by air humidity (Bakker, 1991) and temperature (Challa, 1989).

Crop quality. Especially with long season crops the grower is anxious to keep his crop in a good condition for production. To maintain the productivity of the crop the internal balance between vegetative and generative growth is an important criterion with many crops (Ho and Hewitt, 1986). Temperature is the major climatic factor controlling this balance.

Product quality. Quality is a concept with a wide scope. In relation to climate control the external quality (e.g. size, weight, shape, colour) and the absence of visible injury are particularly relevant for the selling price. With respect to internal quality keeping-quality and taste are probably influenced by the climatic conditions during cultivation. The relation between climate factors and product quality are highly crop specific and besides are often poorly understood and documented.

Timing. The market may show predictable patterns with some crops. Known examples are Christmass (Poinsettia), Easter, Mother's Day, etc. In these cases timing is extremely important. Timing, besides, is also crucial for the cost of production in relation to labour requirements and space utilization in greenhouses (potplants). After establishment of the culture the production process can be advanced or delayed to a certain extent through temperature. With daylength sensitive crops flowering can be controlled by the duration of the light period. In practice days are shortened by means of darkening screens, whereas longer days can be provided through low level supplementary radiation.

Production costs. Part of the production costs can be directly attributed to climate control, e.g. heating, CO_2-enrichment, electricity consumption for supplementary lighting. In addition there are indirect effects of climate control due to e.g. cost of labour, or length of the production cycle.

Risk prevention. During cultivation there is a continuous risk of damage to the crop and the product, due to pests, diseases, physiological disorders and environmental stress. Climate control to a certain extent contributes to prevention and control of these risks. Humidity (Hand, 1988) and temperature, but sometimes also radiation have to be kept within certain limits in order to prevent acute problems. Beside these instantaneous reactions there are long-term adaptations of the crop to the climatic conditions that determine its sensitivity to pests, diseases and environmental stress (Levitt, 1980).

A characteristic of many of the criteria mentioned is the absence of an exact standard and the difficulty of quantification in economic terms. There is a notion of an "ideal" and of unacceptable situations, but in between there is often a gradual range. We believe that this is one important reason why the grower has to be closely involved in the determination of control procedures.

Greenhouse Climate and Growth Processes

The production process. Crop production is a process where, as a first step, radiative energy of the sun is trapped and transformed into chemical energy. This chemical energy subsequently is used to reduce CO_2 molecules and to form the essential building blocks: sugars, amino acids and organic acids (Penning de Vries, 1975). Sugars in addition provide energy for growth and maintenance of the crop. The building blocks are transported to the growing centres in the plant and together with water and nutrients they are used for growth of cells and organs. While growing the crop is in continuous development: leaves are formed, maturing and ageing, new shoots are developing, the crop may pass from the vegetative to the generative phase and it may form storage organs. This complex of processes is what can be summarized as the production process. In relation to climate control it is worthwhile to classify this complex process with respect to response type and time.

Response types. Some crop reactions show a smooth, continuous response to one or more climatic factors within the range normally encountered in greenhouses. Other reactions, usually unwanted reactions, only become manifest if some boundary condition is exceeded. The boundary conditions, beside showing an interaction with other climatic factors, are often affected by pre-conditioning.

This distinction in smooth continuous and strongly non-linear threshold reactions is important, because smooth continuous reactions have always to be taken into account, whereas threshold reactions have to be considered only when the boundary conditions are approached.

Response time. The processes involved in production have largely deviating response times. Although this fact complicates the problem to a certain extent, at the same time it offers a clue to handle this complexity with respect to climate control: for processes with a response time > 24 h it is usually acceptable that, within the response time, in stead of a constant level of the factor in consideration, certain deviations do occur that can be compensated for later on (Cockshull, 1988). Requirements with respect to processes with a slow response time then can be formalized as a set of rules, rather than fixed set-points, and in this way room is left for requirements of processes with a short response time.

Examples of processes with a response time > 24 h are development of the crop, the distribution of assimilates over various plant parts, formation of

leaf area, flowers, fruits and tubers, the development of Ca-deficiency, weak/soft plants, vegetative/generative plants, etc. An example of rule based criteria is the prevention of Ca-deficiency, where a suitable model is still lacking (Aikman and Houter, 1990), but where practical experience could be summarized in a set of rules concerning transpiration requirements.

When dealing with processes with a response time < 24 h there is clearly no or only very limited interaction with the grower possible, and compensation in time is not possible either.

Typical processes with a short response time are photosynthesis and transpiration. In addition acute stress, due to temperature (chilling, overheating), radiation (e.g. photo-bleaching), and extreme water conditions (water stress, tissue damage due to too high water potential) fall in this category. Also the occurrence of condensation on plant parts or high humidity (e.g. poor release of pollen in tomato; Bakker, 1991) may require action within 24 h.

OPTIMAL CONTROL, A SYNTHESIS

Because climate control is so intimately linked to management, optimal control should be considered as one aspect of optimization of the management. At the operational level, the grower uses climate control as a tool to optimize the production process. Knowing the status of the crop, developments of the market, weather predictions and other relevant information, he decides about required actions during the next planning interval, including the desired climate conditions. Optimization of this process may be enhanced by decision support systems, a research topic of management and information science.

During implementation of the operational plan the climate control system has the task to control the greenhouse climate. A new element in the management process described so far is the absence of interaction with the grower for most of the time. Because the actual weather conditions may deviate considerably from expected, climate control should be optimized in real time, that means automatically. The control system, in order to accomplish this task, has to be informed explicitly about the criteria for optimization and requires, in addition, sufficient information about the processes to be controlled.

The criteria for climate control have been mentioned before, but cannot be used as such for automatic decisions. In interaction with the grower they should be translated in more specific criteria. In this translation step, the grower is forced to attribute values to these criteria and, while doing so, he implicitly or explicitly makes decisions about the relative importance of different goals. He could (and probably should) be supported in this process by the system by predictions of the consequences of his suggestions.

The analysis of characteristic response types and response times suggests that a suitable basis for establishing optimal control is one full 24 h day/night cycle. Based on the analysis presented before we believe that the criteria for optimal control over 24 h in general could be formulated as follows (a full account of the theoretical background of these criteria cannot been given within the scope of this paper):
1. all climatic factors: maximization of (crop photosynthesis - associated costs) (Challa, 1990)

2. CO_2: acceptable maximum concentration to avoid risks (Hand, 1990); technical boundary conditions: availability, both instantaneously and over a whole day (Nederhoff, 1990)

3. temperature: maximum and minimum value as related to other conditions; average daily temperature and average day and night temperature (Cockshull, 1988)

4. water vapour pressure of the air: prevention of condensation on plant parts during a given maximum period of time; minimum and maximum turgor (the water pressure within the cells); minimum transpiration rate; minimum and maximum relative humidity (pollination); transpiration integral by day, by night and over 24 h (Hand, 1988)

It should be noticed that for these criteria often no direct observation is available, which complicates the determination of boundary values as well as the control. Nevertheless we believe that the introduction of criteria based on processes will lead to a greater universality. Models will be required to relate measurable parameters to internal plant phenomena.

The assessment of an economic value to photosynthesis is a step requiring much attention, because it forms one of the bases for optimization. This problem has been discussed in specific cases by Challa and Schapendonk (1986) and Heuvelink and Challa (1989), but has to be worked out within a wider scope. Especially in single harvest crops there is a relatively long period where photosynthesis is not contributing directly to growth of the harvestable product. In our view the tactical plan could provide a basis to assess a value to photosynthesis in relation to crop stage in cases where a direct assessment is not possible.

A second remark concerns the character of the criteria: with the exception of (rate of photosynthesis minus associated costs), all other criteria have the form of constraints on climate factors, or processes, either instantaneously, or integrated over a period of time.

Finally it should be noticed that the boundary conditions formulated might give rise to conflicting solutions. The system therefore should in some way be informed about priorities.

Framework for Optimal Climate Control

We can now try to formulate a framework for an integrated climate control system (Fig. 2). The key process is the rate of photosynthesis p, an almost instantaneously responding process, fed by the external inputs (w, weather factors), and the climate factors (x) realized in the greenhouse through manipulation of the inputs (u). In addition to photosynthesis, there are the fast crop responses, such as transpiration, turgor, temperature of the crop, etc., indicated by the crop state z. Plant transpiration is of special interest because of its effect on air humidity.

The control system must be able to cope with these short term dynamics. For diurnal control, therefore, the central issue is the greenhouse dynamics and the fast plant response dynamics, not the slow production dynamics. Requirements from the slow dynamics can be implemented in terms of a set of rules on the climate factors, including band with requirements for integrated climate factors, and others as outlined before.

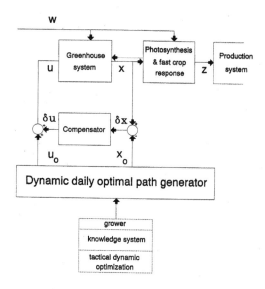

Fig. 2. Proposed framework for operational optimal greenhouse control. w = external inputs (weather); u = control inputs; x = climate factors; z = crop fast response state.

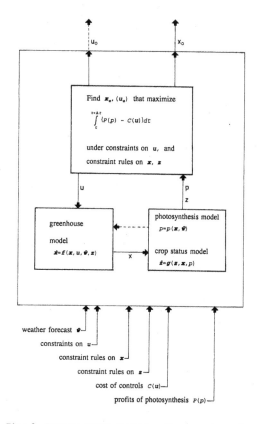

Fig. 3. Dynamic daily optimal path generator, also showing external information needs.

Now, given the greenhouse model [$\dot{x} = f(x,u,w)$], the crop photosynthesis model [$p = p(x,w)$], and the crop state model [$\dot{z} = g(p,z,x)$], the task is to construct an optimal path generator that calculates the optimal path of both the climate factors x_o and the steerings u_o over the optimization horizon (Fig. 3). This can be done by maximizing the profits associated with the rate of photosynthesis P(p) minus the costs of controls C(u), under constraints on the controls u, constraint rules on the calculated crop states z, and constraint rules, including integrated requirements, on the climate factors x. Several calculation methods are available to perform this task (e.g. Bryson and Ho, 1975) although the incorporation of integral constraints may require special attention.

A predictive element can be introduced in the path generator by using forecast of the weather $\hat{w}$ over the prediction horizon. In view of the break down of the crop response in < 24 h and > 24 h responses, 24 h seems to be a fair choice for the optimization horizon. Because of the importance of the first hour of the prediction horizon, it is probably advisable to think about a way of exploiting the relative accuracy of this hourly forecast. Of course, the optimal path can be updated at a rate dictated by the speed of the computing system, perhaps every five minutes or so. Optimizing over much longer periods than 24 h is not meaningful because of the implicit uncertainty in the external growth factors, and because of interference of the grower in the mean time.

Next the question should be asked how the calculated optimal pathways should actually be implemented. Three methods will be discussed.

One obvious way would be to use the classical single loop controllers, i.e. using the optimal path as a set-point path. This has all disadvantages of the usual control system of today, such as strong interaction between loops and a response to disturbances that is always late.

The disadvantage of the loop interaction can be removed by using a multivariable controller, for instance a linear quadratic (LQ) optimal controller (e.g. Kwakernaak and Sivan, 1972; van Henten, 1988). In this case there is some freedom for tuning the performance of the controller by manipulating the weight matrix of control and state deviations, but the cost/profit effects should not be too different from the assumptions made while calculating the optimal set-point path.

Finally, since the optimal path generator actually does yield the optimal steerings, another quite natural idea is to use these straight away. In this approach the feed-forward effect from the weather prediction is fully exploited. As such, this approach is not robust because it is open loop. The open loop character is mitigated if the optimal path generator computes the path with a receding horizon, using updates (Bitmead, Gevers and Wertz, 1990). In addition, robustness can be enhanced by implementing a compensator that compensates for model errors and deviations from the expected weather. A good option is a state feed-back perturbation regulator (e.g. Athans, 1971).

In the case of greenhouse control all relevant states (climate factors) and weather inputs can be observed with fair accuracy, so that perhaps no state estimator will be needed, the more so since the crop states are only needed to check on violations of the threshold values. This issue will require further research.

Information Resources

In this complex optimization problem the information required comes from different resources: actual climatic conditions inside and outside the greenhouse are provided by sensors, detailed daily weather forecasts can be obtained nowadays commercially in electronic form, and qualitative and quantitative models describe the relations between environmental conditions, actuators and greenhouse and crop reactions. The grower should provide information on the crop, price forecasts and should be able to inform the system about his policy and personal assessment of the criteria used for control. One way of doing this, suggested by the framework above, is in the form of constraint rules.

The structure of the knowledge system that should handle the complex information requirements deserves much attention. In particular the interaction with the grower is an essential aspect, because he bears the final responsibility for the management of the production process, and he needs to feel confident with the solutions provided by the system. The system then should be able to inform the grower what the consequences of his decisions will be, as well as an explain facility to provide the grower insight in the control strategy.

CONCLUSIONS

Climate control should be considered within the framework of the management of the nursery, where it is primarily a tool in operational management. Optimization of climate control requires criteria that are formulated explicitly, because this process has to proceed automatically. The criteria, derived from the growers goals, have to be tuned in interaction with the grower. The main criterion for optimality is the rate of photosynthesis, expressed in economic terms, minus associated costs. Other criteria for control can probably be formulated as constraints.

In order to formulate these constraints some characteristics of crop responses have to be taken into account: a distinction can be made with respect to response time (<24 h and > 24 h) and response type (smooth, continuous versus strongly non-linear threshold reactions). For crop reactions with a response time > 24 h it is usually acceptable that climatic conditions vary within a 24 h period. This characteristic will then give rise to constraints on diurnal integrals, more than on momentaneous values.

Due to the integral constraints an optimization horizon of 24 h is required. Optimization has to be implemented at two levels: an optimal path over the whole optimization period has to be generated, while short term control has to deal with the actual conditions, that may deviate from the prognoses, due to the weather conditions and model errors.

A sophisticated knowledge system based on quantitative and qualitative models and a high level user interface is required to deal adequately with the complex information requirements of this proposed climate control system.

REFERENCES

Aikman, D.P. and G. Houter (1990). Influence of radiation and humidity on transpiration: implications for calcium levels in tomato leaves. *J. Hort. Sci.* 65(3), 245-253.

Athans, M. (1971). The role and use of the stochastic linear-quadratic-gaussian problem in control system design. *IEEE Transactions on Automatic Control*, AC-16 (6), 529-552.

Bitmead, R., M. Gevers and V. Wertz (1990). *Adaptive Optimal Control*. Prentice Hall International Series in Systems and Control Engineering, London.

Bakker, J.C. (1991). *Analysis of humidity effects on growth and production of glasshouse fruit vegetables*. Diss. Wageningen Agric. Univ., Wageningen.

Boehlje, M.D. and V.R. Eidman (1984). *Farm Management*. John Wiley & Sons, New York.

Bryson, A.E. and Y.C. Ho (1975). *Applied Optimal Control - Optimization, Estimation and Control*; John Wiley & Sons, New York.

Challa, H. (1989). Modelling for crop growth control. *Acta Hort.* 248, 209-216.

Challa, H. (1990). *Crop growth models for greenhouse climate control*. In R. Rabbinge, J, Goudriaan, H. van Keulen, F.W.T. Penning de Vries and H.H. van Laar (Eds.). Theoretical production ecology: reflections and prospects. Simulation monographs, 34, PUDOC, Wageningen. pp. 125-145

Challa, H., G.P.A. Bot, E.M. Nederhoff and N.J. van de Braak (1988). Greenhouse climate control in the nineties. *Acta Hort.* 230, 459-470.

Challa, H. and A.H.C.M. Schapendonk (1986). *Dynamic optimalization of CO2 concentration in relation to climate control in greenhouses*. In H.Z. Enoch en B.A. Kimball (Eds.). Carbon dioxide enrichment of greenhouse crops. CRC press, Inc., Boca Raton. pp. 147-160.

Cockshull, K.E. (1988). The integration of plant physiology with physical changes in the greenhouse climate. *Acta Hort.*, 229, 113-123.

Davis, G.B. and M.H. Olsen (1985). *Management information systems*. McGraw-Hill Book Company, New York.

Hand, D.W. (1988). Effects of atmospheric humidity on greenhouse crops. *Acta Hort.*, 229, 143-158.

Hand, D.W. (1990). CO_2 enrichment in greenhouses: problems of CO_2 acclimation and gaseous air pollutants. *Acta Hort.*, 268, 81-101.

Henten, E.J. van (1989). Model based design of optimal multivariable climate control systems. *Acta Hort.*, 248, 301-306.

Heuvelink, E. and H. Challa (1989). Dynamic optimization of artificial lighting in greenhouses. *Acta Hort.*, 260, 401-412.

Ho, L.C. and J.D. Hewitt (1986). *Fruit development*. In J.G. Atherton and J. Rudich (Eds.). The tomato crop. Chapman and Hall, london. pp. 201-239.

Kwakernaak, H. and R. Sivan (1972). *Linear Optimal Control Systems*. Wiley-Interscience, New York.

Levitt, J. (1980). *Responses of plants to environmental stresses*. Academic Press, New York.

Nederhoff, E.M. (1990). Technical aspects, management and control of CO_2 enrichment in greenhouses. *Acta Hort.*, 268, 127-138.

Penning de Vries, F.T.W. (1975). *Use of assimilates in higher plants*. In J.B. Cooper (Ed.). Photosynthesis and productivity in different environments, Cambridge University Press. pp. 459-480.

18

GREENHOUSE PRODUCTION IN ENGINEERING ASPECTS

T. Takakura

*Dept. of Agricultural Engineering, University of Tokyo,
Bunkyo-ku, Yayoi 1-1-1 Tokyo, Japan*

Abstract.
Greenhouses are the basic configurations for food production in the future. In engineering aspects, recent developments of computer control, new sensors, automation and robotics and information networks support this movement well. New aspects of greenhouse productions are plant factories, Biosphere II project and space farming.

Keywords. Computer control; Hydroponics; Robot; Non-destructive; Reading plants' face; Plant factory; Space farming.

INTRODUCTION

Environmental control is essential to greenhouse production in engineering aspects. This technique is not only for greenhouses but also applicable to tissue culture and space farming as well.

Recent environmental control has been fully supported by computer and sensor developments. Greenhouse production including plant factory and space farming can not be described without it.

COMPUTER CONTROL

Greenhouse climate control has usually been accomplished by commercial analog controllers which mainly provide on/off control for heating units. Other processes, *e.g.* ventilators, movable thermal screens and carbon dioxide enrichment have in the past needed their own controllers, which may duplicate some functions. Because of piecemeal development, controllers were not centralized, and provision was rarely made for regular monitoring and recording of the accuracy of the control achieved.

The early stage of computer use in this area was characterized by the optimization of plant growth using a minicomputer (the so-called "speaking plant" techniques defined by Udink ten Cate *et al.*,1978), and the on-line process control of plant growth have been developed (Takakura, 1973; Takakura *et al.*, 1974; Takakura, 1975; Takakura *et al.*, 1975; Takakura and Ohara, 1976; Takakura *et al.*, 1978). this series of studies showed the potential for adaptation of digital computers to greenhouse climate control, and it is being continued by some other researchers (*e.g.*, Hashimoto *et al.*, 1983; Harazono *et al.*, 1984). However, it was not extended directly to commercial applications because of the cost of minicomputer hardware and the software logic which involves plant reaction in the feedback loop.

Modern cultivation methods have increasingly demanded more precise control to achieve maximum production of marketable products at minimum cost, particularly important since the energy crisis. Microcomputer for controlling greenhouse climate was developed in the Netherlands first. The first was installed in the practical greenhouse in Japan by the author's group (Takakura *et al.*, 1979). At the present time, there are more than 1,000 commercial microcomputer controllers installed in Japanese commercial greenhouses. Most of them are stand-alone types which handle only one compartment. Controllers for hydroponic systems have also been commercially available and been installed in some greenhouses (Horti-Mation, 1989).

AUTOMATION & ROBOTICS

Transportation systems of potted and bedding plants are commercially available (NTV, 1991; CBT, 1987) and becoming common in large scale greenhouses in the Netherlands and other countries. Several kinds of so called "agri-robot" transportation systems are operated by computers. Hanging cranes and battery cars are typical ones.
Harvesting robots are being investigated for greenhouse tomatoes and cucumbers in Japan (Kawamura *et al.*, 1984 and 1985; Amaha *et al.*, 1989). Transplanting robots are also being developed in U.S.A. (Ting *et al.*, 1990 a, b; Yang *et al.*, 1989) and Japan (Toshiba, 1990).

SENSOR DEVELOPMENT

Sensors are essential to monitor and control non-destructively the changes in plants and their environments that affect them significantly. Direct measurement by sensors of plant growth and development is not easy and is still one of the most important tasks to be solved in research as well as in practical fields. Plant growth and development is the integration of all physiological reactions of plants.

Another difficulty is measuring more than one plant, for example a plant canopy.

Measurement of one plant is valid for research purposes, but it is not enough in the practical sense. Monitoring several plants from a canopy in a greenhouse, for example, as indicators of an average could be a practical solution, but the problem would be how to select such indicators.

Image analysis is one of the promising methods in the future but has a very broad range. Computer vision is now commonly investigated in mechanization and automatization of plant management (e.g., Toshiba, 1990; Amaha et al., 1989). Fruits can be positioned for machine harvesting. Growth can also be detected by CCD camera imaging (e.g., Eguchi et al., 1979).

Over-all detection for whole plants in an enclosure can be done with gas balance techniques, but it is not possible to detect change in individual plants in a canopy. Another point to be noted is that some kind of plant growth models are needed because gas balance is not directly related to plant growth.

Monitoring and management of individual plants can only be done by image processing with either computer vision or infrared thermometry, just as expert growers are doing every day in their greenhouses. This technique can be called a "reading plants' faces" instead of a "speaking plant approach" (Takakura, 1991).

To cover all plants in an enclosure, the problems to be solved involve the resolution of sensors such as CCD cameras and infrared thermometers, the speed of processing computers and the size of their memories. Costs are still expensive, too. Such non-contact methods as gas balance and image processing will be the promising methods supported by the recent development of electronic devices.

Software developments that can be used to analyze images are another research target. Overlapping of plant leaves and hidden fruits behind leaves are examples to be analyzed in the future. An air blow method to move leaves blocking cucumber fruit from the view of a CCD camera is one of the attempts (Amaha et al., 1990). A combination of this kind could overcome the shortcomings of vision analysis.

PLANT FACTORY
Plant production systems in greenhouses whose operations are mostly automated or in darkrooms with artificial illuminations are called plant factories. Most of them are in hydroponics. Several factory style vegetable production systems are operated in several countries, such as U.S.A., The Netherlands, Sweden, Denmark and Japan.

Japanese radish sprout production is completed in 5-6 days in greenhouses. Large-scale production is mostly automated. Seeding is done by machine in styrofoam trays. These trays are stored vertically in stacks of 10 in the room with less light and then are carried automatically on a chain conveyer to the conventional greenhouse and are placed horizontally to be exposed to natural light for 2 days before harvesting (Takakura, 1985).

Watering is done automatically in two different processes. In the first stage where trays are piled vertically, spray nozzles move automatically to the approximate location after trays are spaced for these nozzles by a synchronized are operation. In the second process, spray nozzles mounted on gantry move over trays which are placed horizontally.

Fully insulated darkrooms with artificial light sources and air-conditioning units are used to produce lettuce in U.S.A. and Japan. Some other crops are also cultivated, but lettuce are mostly common. One of the serious problems to be solved is the management of nutrient solution. To maintain pH and EC at a certain level is not enough for better growth and to prevent rapid accumulation of undesirable salts in nutrient solution are the problems. Tipburn, physiological disorder, is another problem. All factors favorable to rapid growth are also the causes of tipburn. However, prevention of tipburn has been successfully conducted (Goto and Takakura, 1990; U.S., U.K., The Netherlands and Japan Patents are pending).

INFORMATION NETWORK
The recent expansion of microcomputer use has affected not only protected cultivation but also the other agricultural fields. Animal production and post-harvest operations such as sorting and packing are typical areas. Therefore, the future use of digital computers will be in multi-tasking and multi-user systems, coupling a greenhouse with a storehouse by a network system, for example.

A software such as new MS-Windows will open the road. In NTV (Dutch Horticultural Trade Fair) in 1991, some companies demonstrated the use of personal computers for not only control but also other jobs. The control programs for greenhouses are, therefore, operated under MS-Windows (Microsoft, 1990). Then, switching from control to other programs is very easy.

Public network systems by which growers can accumulate useful information are similar to the Prestel system in England and Teridon in Canada. The Japanese CAPTAIN system is another version of these systems, and already existing CATV systems in the Japanese countryside are being used for one of the cores of further development of an information network of this kind. In Japan, we have an example that greenhouses apart from the owner's house are controlled by a personal computer in the house through a public telephone line. Personal BBS network systems in nationwide is becoming popular and more than 30 local networks are particularly for agriculture.

BIOSPHERE II & SPACE FARMING
Greenhouse production is the basic style to produce fresh vegetables by controlling environment factors. Plant factory is one expansion of greenhouse productions and Biosphere II project in Arizona, U.S.A. and space farming is the present objective of this kind of production systems.

Main configuration of Biosphere II is a huge greenhouse with air-conditioning units. It is not only a step to space farming but also the first attempt to investigate ecological interactions of organic and non-organic materials in an enclosure. Air tightness is a common factor to be investigated in these studies. The effect of total pressure on plant growth is also to be studied in addition to the gas component of the air (Goto et al., 1991). Artificial ecosystem in an enclosure and gravity are the next topics to be studied.

REFERENCES

Amaha, K., Shono, H. and Takakura, T. (1989). A harvesting robot of cucumber fruits. *ASAE Paper*, **897053**, 19pp.

CBT (1987). *Glasshouse vegetable growing in the Netherlands.* 15pp.

Eguchi, H., Hamakoga, M. and Matsui, K. (1979). Computer control of plant growth by image processing. IV. Digital image processing of reflectance in different wave length regions of light for evaluating vigor of plants. *Environ. Control in Biol.*, **17**, 67-77.

Goto, E. and Takakura, T. (1990). Prevention of lettuce tipburn by environmental control. *ASAE Paper*, **904033**, 14pp.

Goto, E., Iwabuchi, K. and Takakura, T. (1991). An environmental control system for growing plants under total pressures, *in this book*.

Horti-Mation (1989). *Japanese Horticulture Trade Fair in Tokyo.*

Kawamura, N., Namikawa, K., Fujiura, T. and Ura, M. (1984). Study on agricultural robot. (Part 1)-Microcomputer-controlled manipulator system for fruit harvesting- *J. of Soc. Agr. Mach., Japan*, **46**, 353-358.

Kawamura, N., Namikawa, K., Fujiura, T. and Ura, M. (1985). (Part 2)-Measurement of fruit location by MOS type color TV camera and fundamental experiment of fruit harvesting. *Ibid.*, **47**, 177-182.

Microsoft, Co. (1990). *Microsoft Windows* (v. 3.0).

NTV (1991). *The European Horticultural Trade Fair, Catalogue, Holland*, 62pp.

Takakura, T. (1973). Plant growth optimization. *Proc. VII Inter. Cong. of Cybernetics*, 501-505.

Takakura, T., Kozai, T., Tachibana, K. and Jordan, K.A. (1974). Direct digital control of plant growth. I. Design and operation of the system. *Trans. ASAE*, **17**, 1150-1154.

Takakura, T. (1975). Plant growth optimization using a small computer. *Acta Hort.*, **46**, 147-156.

Takakura, T., Ohara, G., Kurata, K. and Tachibana, K. (1975). Plant growth optimization under natural light condition. *Phytotronics III Phytotronics in agricultural and horticultural research*, (ed. by P. Chouard and N. de Bilderling), Gauthier-Villars, Paris, 80-88.

Takakura, T. and Ohara, G. (1976). Direct digital control of plant growth II. Physiological analysis of cucumber plants. *J. Agr. Met.*, **32**, 107-115.

Takakura, T., Ohara, G. and Nakamura, Y. (1978). Direct digital control of plant growth. III. Analysis of the growth and development of tomato plants. *Acta Hort.*, **87**, 257-264.

Takakura, T., Okada, M., Shimaji, H. and Nara, M. (1979). Development of a microcomputer-based multi-variable control system for greenhouses. *J. Agr. Met.*, **35**, 97-102.

Takakura, T. (1985). Japan: Current research and development, in *Hydroponics Worldwide: State of the art in soilless crop production* (Savage, A.J. ed.), 137-143.

Takakura, T. (1989). Microcomputer use for energy and labor savings in Japanese protected cultivation. *Acta Hort.*, **257**, 79-85.

Takakura, T. (1991). Sensors in controlled environment agriculture: Measuring growth and development. *Ibid.*,(in press).

Ting, K.C., Giacomelli, G.A. and Shen, S.I. (1990a). Robot workcell for transplanting of seedlings. Part I-Layout and materials flow. *Trans. ASAE*, **33**, 1005-1010.

Ting, K.C., Giacomelli, G.A., Shen, S.I. and Kabala, W.P. (1990b). Part II-End-effector development. *Ibid*, **33**, 1013-1017.

Toshiba Co. (1990). *Demonstration at International Fair of Flowers and Greenaries in Osaka, Japan.*

Udink Ten Cate, A.J., Bot, G.P.A. and van Dixhoorn, J.J. (1978). Computer control of greenhouse climates. *Acta Hort.*, **87**, 265-272.

Yang, Y., Ting, K.C. and Giacomelli, G.A. (1989). Factors affecting a robot workcell for flexibly automated seedling transplanting. *ASAE Paper*, **897055**, 9pp.

THE PATH TO DYNAMIC OPTIMIZATION OF CARBON DIOXIDE IN THE GREENHOUSE

W. Day*, Z. S. Chalabi*, B. J. Bailey* and D. Aikman**

*AFRC Institute of Engineering Research, Wrest Park, Silsoe,
Bedford, MK45 4HS, UK
**Horticulture Research International, Worthing Road, Littlehampton, West Sussex, BN17 6LP, UK

Introduction

Horticultural production seeks to supply a market with quality produce using the necessary inputs such that the profit margin is maximised. This maximisation of return is subject to some assessment of risk and forecasting of environmental and market conditions, but in principle can be achieved by choosing an appropriate combination of inputs. In protected crop production, many of the input variables are under the grower's direct control. The grower must choose a temperature regime, can instal and use thermal screens to save energy, and can even select the greenhouse structure in relation to its ability to retain heat and transmit radiation. All these decisions reflect an attempt to optimise the crop environment for profitable production. Such decisions can be made on the basis of standard experimentation on production systems.

For some variables, however, the optimal level of the set point value may change on quite short timescales. Studies by Hurd and Graves (1984), for example showed that the growing temperature did not need to be held constant, but only to achieve a constant average value over some integration period. This provides the opportunity to choose a dynamically varying temperature regime such that energy inputs are minimised (Bailey and Seginer, 1989, Aikman and Picken, 1989).

Similar concepts of dynamically varying set points can be applied to the carbon dioxide concentration in the greenhouse. Unlike the temperature set point, however, the carbon dioxide set point for maximum production is not well defined, as crop photosynthesis and hence potentially growth and yield are an asymptotically increasing function of carbon dioxide concentration. The set point for maximum return however is strongly dependent on the cost of achieving high concentrations. The task of dynamic optimisation is therefore to choose the concentration at which the financial benefits of extra yield are balanced by the extra costs of increasing the concentration further.

The principal factors that influence this optimisation calculation can change on timescales from minutes (radiation, ventilation) to weeks (market price). This paper outlines the essential strategy that is required to achieve dynamic optimisation. It is strongly dependent on the use of mathematical models to define both the greenhouse environment and the crop response. Its success therefore, depends on the continuing progress of mathematical modelling of

horticultural systems, and the development of grower confidence in control based on suchmodels. Research at Silsoe and Littlehampton is developing major elements in this strategy.

The path and its steps

The goal of dynamic optimization of carbon dioxide concentration can only be achieved if all the steps along the way are identified and tackled. As well as defining the key system components of cost and benefit in quantitative terms, the optimisation itself must be developed in a form that is capable of real time control, and the end result must be formulated in such a way that the industry can be convinced of the benefit.

Models of component processes: crop response.
It is not straightforward to measure directly the crop response to particular patterns of extra CO_2 in a commercial production greenhouse. It is therefore necessary to produce a robust mathematical model of the crop response that can form the basis of the estimation of the benefit from additional CO_2. Dynamic models of crop photosynthesis have been developed for many crops, including models specifically tailored for CO_2 optimisation (Nederhoff et al., 1989). The particular requirement on these models is that they accurately reflect the change in photosynthesis with increased CO_2 concentration. As part of the dynamics will be driven by changing light levels, the model should reflect the changes in CO_2 responsiveness as radiation intensity changes. A major constraint on this modelling is therefore likely to be the availability of experimental data that discriminates these responses and interactions. The problem is complicated by the heterogeneous radiation environment in the greenhouse, with some leaves in direct sun and a wide variation in diffuse radiation between the top and bottom of the leaf canopy. For example, Critten (1990) has developed a detailed numerical model to describe light penetration into the canopy. The model takes into account the geometry of the canopy and greenhouse, the light losses due to transmission into the greenhouse and scattering within the canopy. To be usable, it has been necessary to develop a semi-analytical form from this model, giving crop photosynthesis at different CO_2 levels.

The benefit of this extra photosynthesis will be turned into a real return in the form of extra yield over period of a few weeks. The delay between the start of a treatment affecting photosynthesis and its effect being evident in the yield appears comparable to the temperature-dependent time interval between maximum growth rate and ripening of the tomato fruit. In principle, the detailed physiological response can provide important secondary information on the potential benefit. Thus, increased leaf area for a young crop or increased fruit set for a more mature crop may have an effect on the magnitude of the yield return from a given change in crop photosynthesis.

A mechanistically based model has been developed and is currently being validated. It predicts canopy photosynthesis and the distribution, over subsequent time, of increments of yield. Coupled with a function estimating future market values, these give the anticipated financial benefit.

Models of component processes: input costs.
The simplest system in which to determine input costs is where the extra CO_2 comes from bought-in liquid CO_2. The rate of supply of CO_2 will depend upon the CO_2 concentration enhancement, the rate of air exchange between the inside and outside of the greenhouse and net assimilation by the crop. The concentration is directly measured, and in principle the exchange could be determined from rate of supply. However, an alternative approach is required, both to avoid compounding errors in the estimation of photosynthetic rate and also to provide a method for periods when there is no CO_2 enrichment.

Fernandez and Bailey (1991) have analysed two approaches to such prediction of ventilation rate, and compared their models to data from tracer gas experiments. An empirical equation relating air exchange to wind speed for a range of ventilator openings gave good results independent of wind direction, and the form of the equation was close to that derived in similar experiments by de Jong (1990). A more mechanistic approach was also used, estimating ventilation rate from a steady-state energy balance model of the greenhouse. The energy balance method was reliable when ventilator openings were greater than 15° but for smaller openings the empirical equations based on degree of ventilator opening were more accurate.

Liquid CO_2 is used as the source of CO_2 enrichment in many commercial greenhouses in the UK. However, CO_2 from burning natural gas is also widely used. For liquid CO_2 as source, every unit of added CO_2 can be ascribed a (constant) cost, but this is not the case for CO_2 from gas combustion. In the winter and much of the spring, the heat from the gas is necessary and the CO_2 can be deemed to be free given a fixed temperature set-point. There may be some additional capital cost if gas is not the prime power source. When temperatures rise the heat is not required so the CO_2 production would have a cost, and lead to enhanced ventilation to keep temperature under control. A final element in the complexity with this CO_2 source is the availability of heat storage methods which can allow the CO_2 to be used in the day, and much of the heat to be available in the night when it is needed. At present our research is not tackling this component in the optimization problem.

Optimization procedure. A number of CO_2 optimization strategies have been developed recently (Challa and Schapendonk, 1984; Gal et al., 1984; Seginer et al., 1986; Nederhoff, 1988; Reinisch et al., 1989). Critten (1991) developed an optimization strategy for CO_2 enrichment of a greenhouse lettuce crop that reduced the system description to an analytical form. Mathematical models of crop growth and CO_2 loss were combined in an equation for grower income. Analytical relationships were then derived for optimal CO_2 concentration as a function of incident radiation, ventilation rate, cost of CO_2 and crop value. Though analytical formulations have considerable advantages, it is also important to have flexible methods as process modelling concepts change. Chalabi (1991) has, to this end, developed a generalized strategy for optimal CO_2 enrichment. It is constructed in terms of general formulations of crop photosynthesis, crop growth and ventilation rates. The strategy is derived using optimization theory, and the optimal CO_2 trajectory is given by the solution of a functional differential equation.

The approaches to optimization referred to above have predominantly been paper exercises to establish that the method will work in principle. As practical implementation is considered, other factors must be taken into account. Chalabi and Fernandez (1991) have investigated the spatio-temporal responses of a Venlo glasshouse to gaseous (nitrous oxide) enrichment, using linear system identification techniques. The first objective of the analysis was to determine the upper limit of the frequency at which the CO_2 set point trajectory can be updated for the proper execution of on-line CO_2 optimization. It was shown that the glasshouse response times to gaseous enrichment were of the order of 20-25 minutes when the ventilators were closed and 10-15 minutes when the ventilators were 10% open. The second objective was to confirm that the standard method used in the UK for enriching greenhouses with pure CO_2 gives adequate gaseous mixing. It was also shown that the method for distributing CO_2 in the glasshouse by using a low level ring main with a distribution pipe along the crop rows is adequate for uniform mixing.

The time response of the system also has implications for the acquisition (or prediction) of the microclimate variables that are used in the optimization algorithm. Chalabi and Critten (1990) evaluated the influence of real time filtering strategies for microclimate variables, as used in Critten's (1991) optimization algorithm, on the computed optimal CO_2 levels. They showed that the width of the time window of the digital low pass filter acting on these variables should be of the same order of magnitude as the response time of the greenhouse to CO_2 enrichment.

Implementation of dynamic control. The computing and control environment that is being used in the development of dynamic control algorithms at Silsoe and Littlehampton is inevitably complex, involving separate but intercommunicating climate control and data logging systems. When control algorithms are implemented in commercial practice, it will be possible to operate in a considerably simpler environment. The key feature will be the optimal control computer, able to take data on current microclimate conditions from the data logging unit and communicate optimal set points to the climate control unit in the greenhouse control computer system. In the current commercial units, data logging and climate control are in a single system and their separation in our experimental system reflects the higher demands for detailed monitoring and analysis during development.

A specific demand for any implementation of dynamic control will be sufficient speed of calculation to produce set point values in times to match the response time of the greenhouse environment to CO_2 enrichment. The complexity of some of the component models that have already been referred to will put considerable demands on processing speed. In particular, this complexity will require that computational techniques ensuring fast convergence of the optimization algorithm are developed. The availability of suitable hardware is less likely to be a problem, as processing speeds are estimated to have doubled every three years while memory density has doubled every two years in recent times (Oakley, 1990).

The transfer of set point definition from the grower to a computer operating a complex suite of programmes is a major step. There is a range of issues arising from the concept of such a transfer that requires careful assessment.

First the principle of transferred control probably needs independent demonstration in relation to a set point optimization that is more readily interpreted by the grower. The example of dynamic control of temperature, using the temperature integration concept, in response to wind speed to achieve optimization of energy use could be used for such a demonstration. The criteria for set point variation, in terms of temperature range and integration time, can be stated simply and the potential benefit in energy use can be calculated straightforwardly.

Secondly, the grower's information input is a crucial component in some aspects of the optimization. At a simple level, the cost of his CO_2 supply is required. Potentially more complex is his expectation of both yield and crop value. The implementation of a model of crop growth and CO_2 response does not remove the differences that exist between growers in potential yield and quality. Even if the cause of differences cannot be specified, the models used must allow for the grower's expectations to be taken into account.

Thirdly, and most importantly, the implementation of such control methods will require the grower to have confidence in the decisions the algorithm is making. The interface between the grower and the intelligent machine will need to be carefully designed. This is likely to involve the development of an expert system that can both interrogate aspects of the optimization routines to establish the factors leading to particular set point decisions, and also allow the grower to evaluate these factors in relation to his experience and crop management expertise. Jones et al. (1989) have developed experimental expert systems that can translate decisions made by the computer into a language that the grower can understand and react to.

The future

The translation of dynamic CO_2 optimization from a mathematical concept to a proven and commercially viable practice has still got some way to go. Our current research programme is concentrating on the definition of the component models and establishing a realistic optimization strategy that can be used in real-time. The methods are being developed in as general a format as possible to permit adaptation to other definitions of the component processes. Some aspect of system optimization will always need to be taken outside the context of dynamic control. Thus the choice between pure CO_2 enrichment or the use of combustion gases must be made separately, though in the light of the potential benefit from dynamic control.

The concept of CO_2 optimization being developed here is in terms of single variable control in parallel with separate control of other environmental variables. The influence of a control action, such as ventilation, to decrease temperature or humidity, on CO_2 concentration or the use of gas burners, to generate CO_2, on the temperature and humidity, suggest that interactive and multivariable control will need careful appraisal. Thus, it may be possible to decrease humidity to ensure adequate transpiration by ventilating at times when the benefits of extra CO_2 are small, thus allowing the benefit from extra CO_2 to be maximised. However, such complex control options will require a detailed experimental analysis of plant and crop responses. The biology may not be as simple as the engineering and mathematics has required it to seem to meet present challenges.

References

Aikman, D.P. & Picken, A.J.F. (1989). Wind-related temperature setting in glasshouses. Journal of Horticultural Science 64, 6, 649-654.

Bailey, B.J. & Seginer, I. (1989). Optimum control of greenhouse heating. Acta Horticulturae 245, 512-518.

Chalabi, Z.S. & Critten, D.L. (1990). The influence of real time filtering strategies on optimal CO_2 concentration for a greenhouse lettuce crop. Acta Horticulturae 268, 139-148.

Chalabi, Z.S. (1991). A generalized optimization strategy for dynamic CO_2 enrichment in a greenhouse. European Journal of Operational Research (in press).

Chalabi, Z.S. & Fernandez, J.E. (1991). Spatio-temporal responses of a glasshouse to gaseous enrichment. Submitted to Journal of Agricultural Engineering Research.

Challa, H. & Schapendonk, A.H.C.M. (1984). Dynamic optimization of CO_2 concentration in relation to climate control in greenhouses. In H.Z. Enoch and B.A. Kimball (eds), Carbon Dioxide Enrichment for Greenhouse Crops, CRC Press Inc., Boca Raton, Florida, pp 147-160.

Critten, D.L. (1990). Light penetration and photosynthesis in tall greenhouse crops. Divisional Note, DN.1558, AFRC Institute of Engineering Research, Silsoe, Bedford, United Kingdom, 54 pp.

Critten, D.L. (1991). Optimizing CO_2 concentration in greenhouses: a modelling analysis for the lettuce crop. Journal of Agricultural Engineering Research (in press).

Fernandez, J.E. & Bailey, B.J. (1991). Measurement and prediction of greenhouse ventilation rates. Submitted to Agricultural and Forest Meteorology.

Gal, S.; Angel, A. & Seginer, I. (1984). Optimal control strategies of greenhouse climate: methodology. European Journal of Operational Research 17, 265-281.

Hurd, R.G. & Graves, C.J. (1984). The influence of different temperature patterns having the same integral on the earliness and yield of tomatoes. Acta Horticulturae 148, 547-554.

Jones, P.; Roy, B.L. & Jones, J.W. (1989). Coupling expert systems and models for the real-time control of plant environments. Acta Horticulturae 248, 445-452.

de Jong, T. (1990). Natural ventilation of long multi-span greenhouses. Doctoral thesis. Agricultural University, Wageningen, 116 pp.

Nederhoff, E.M. (1988). Dynamic optimization of the CO_2 concentration in greenhouses: an experiment with cucumber (cucumis sativus L.). Acta Horticulturae 229, 341-348.

Nederhoff, E.M.; Gijzen, H.; Vegter, J.G. & Rijsdijk, A.A. (1989). Dynamic model for greenhouse crop photosynthesis: validation by measurements and application for CO_2 optimization. Acta Horticulturae 260, 137-147.

Oakley, B. (1990). The limits of growth in IT. Computing and Control Engineering Journal, 1, 7-14.

Reinisch, K.; Arnold, E.; Markert, A. & Puta, H. (1989). Development of strategies for temperature and CO_2 control in the greenhouse production of cucumbers and tomatoes based on model building and optimization. Acta Horticulturae 260, 67-75.

Seginer, I.; Angel, A.A. & Katz, D. (1986). Optimal CO_2 enrichment strategy for greenhouses: a simulation study. Journal of Agricultural Engineering Research 34, 285-304.

Copyright © IFAC Mathematical and Control Applications
in Agriculture and Horticulture, Matsuyama, Japan 1991

OPTIMAL CONTROL OF GREENHOUSE CLIMATE

E. J. van Henten* and J. Bontsema**

**Agricultural University, Dept. of Agricultural Engineering and Physics, c/o IMAG, P.O. Box 43,
NL-6700 AA Wageningen, The Netherlands*
***Agricultural University, Dept. of Agricultural Engineering and Physics, Duivendaal,
NL-6701 AP Wageningen, The Netherlands*

Abstract. Using mathematical process models and dynamic optimization
theory optimal temperature and CO2 strategies for the cultivation of a
lettuce crop have been calculated. The application of long term mean
weather data as weather forecast results in a lower energy and CO2
consumption compared to static strategies containing no information
about the weather. The applicability of the optimal strategies as open-
loop control of a real process depends on the differences between the
expected and the actual outdoor climate, modelling errors and the
accuracy of the climate control. The effect of an inaccurate long term
weather forecast is analyzed. Repeated optimization, in literature
reffered to as a way to cope with differences between the expected and
the actual weather, does not lead to a lower energy and CO2 consumption.

Keywords. Optimal control; Optimization; Nonlinear programming;
Greenhouse climate; Vegetable production

INTRODUCTION

Dutch horticultural crop production has
been faced with saturating market
demands, increasing competition within
Europe, increasing production cost and
stringent governmental demands with
respect to environmental pollution,
through-out the last decade. These are
major reasons why more efficient ways of
crop production have been looked for by
the production sector itself and agricultural research as well. The research
presented in this paper aims at the
improvement of production efficiency by
the optimization of greenhouse climate
control.

The improvement of greenhouse climate
control may be accomplished by means of
the application of more advanced
algorithms for the control of the
greenhouse climate (Van Henten, 1989) and
by the calculation of optimal climate
setpoint strategies using mathematical
models of crop growth and greenhouse
climate, hereafter referred to as

greenhouse climate optimization (Gal,
Angel and Seginer, 1984, Schmidt and
others, 1987). This paper is devoted to
the latter approach.

The objective of greenhouse climate
optimization is to maximize the net
profit. The net profit is defined as the
difference between the income from the
crop and the operating cost of the
climate conditioning equipment (Gal,
Angel and Seginer, 1984, Schmidt and
others, 1987). Because of the dynamic
character of the crop growth process,
greenhouse climate optimization
constitutes an optimal control problem.

Greenhouse climate optimization was
treated by Gal, Angel and Seginer (1984),
Schmidt and others (1987) and Berg and
Lentz (1989) as an open-loop optimal
control problem in which the optimization
results in time trajectories of the
controllable input variables of the
process, like the air temperature, for
the time interval considered. It was
assumed that the process was perfectly

described by the models used and that trajectories of the uncontrollable input variables of the process, like solar radiation, are known in advance. The performance of these climate strategies applied as open-loop control to a real process may be deteriorated by modelling errors, deviations of the estimated outside climate conditions from the real climate and improper climate control. Schmidt and others (1987) proposed repeated optimization as a way to circumvent this problem.

In this paper the climate optimization problem with respect to the cultivation of a lettuce crop is studied. First the general optimal control problem and it's solution will be treated. Secondly the lettuce growth model and the greenhouse climate model will be briefly described. Thirdly the optimal control problem with respect to the lettuce cultivation will be solved and the applicability of the resulting climate strategies as open-loop control of a real process will be examined using simulations. The effectiveness of repeated optimization to deal with differences between the expected and the actual outdoor climate will be analyzed.

THE OPTIMIZATION PROBLEM

The general constrained optimization problem which is considered in this research includes an objective functional of the form:

$$J(u(t)) = g(x(t_f), t_f) - \int_{t_b}^{t_f} h(x(t), u(t)) dt \quad \textbf{(1)}$$

where $x \epsilon \mathbf{R}^n$: the state vector, $u \epsilon \mathbf{R}^m$: the control vector, $t \epsilon \mathbf{R}^1$: time, $g: \mathbf{R}^{n+1} \to \mathbf{R}^1$, $h: \mathbf{R}^{n+m} \to \mathbf{R}^1$ and t_b, t_f are the begin time and the finish time of the optimization interval respectively.

The optimization problem is defined as to find

$$u^*(t) = \max_{u(t)} J(u(t)) \quad \textbf{(2)}$$

such that

$$f(\frac{dx(t)}{dt}, x(t), u(t), v(t)) = 0 \quad \textbf{(3)}$$

and

$$u_{min} \le u(t) \le u_{max} \quad \textbf{(4)}$$

where $v \epsilon \mathbf{R}^o$: the uncontrollable inputs of the process, $f: \mathbf{R}^{n+m+o} \to \mathbf{R}^n$: the nonlinear process model and u_{min}, u_{max} are the lower

and upper bounds of the controllable process inputs respectively.

Constraint (3) encompasses the type of models used to describe crop growth and production as well as greenhouse climate. Constraint (4) emphasizes the physical bounds usually encountered in practical control systems.

The solution of the constrained variational problem as defined by equations (1), (2), (3) and (4) is quite cumbersome (Renfro and others, 1987). In this research the variational problem is converted to an approximating nonlinear programming (NLP) problem which is solved using NLP methods (Renfro and others, 1987). The conversion is accomplished using a piecewise constant function form to approximate the continuous independent variable u(t). The independent variable changes value at N equally spaced instances $t_0 + kT$, $k=0,1,...,N-1$, and is constant in between, $u(t) = u_k$, $t_0 + kT \le t \le t_0 + (k+1)T$, $k=0,1,...,N-1$, and T the sample interval. So the input function u(t) is completely specified by N values $u_0, u_1, ..., u_{N-1}$.

The optimization problem is solved by sequentially solving the system equations (3) and searching for the N values of u within the admissible region defined by (4), that maximize the objective function (1). In this research a search routine based on the conjugate gradient algorithm of Fletcher-Reeves is employed, with an active set strategy to cope with the linear constraints with respect to u as defined in (4) (Gill, Murray and Wright, 1981).

THE LETTUCE CROP GROWTH MODEL

The lettuce crop is described by a 4-dimensional growth model including the state variables non-structural dry weight, structural dry weight of the roots, structural dry weight of the shoot and the leaf area index. Crop growth is described by non-linear differential equations of the form:

$$\frac{dx(t)}{dt} = f(x(t), u(t), v(t)) \quad \textbf{(5)}$$

where x(t) are the four state variables mentioned before, u(t) are the controllable inputs of the crop growth process like the air temperature and the CO_2 concentration, v(t) is the uncontrollable input of the crop growth process i.e. the solar radiation and t denotes time.

The crop growth model is based on the work of Sweeny and others (1981). It is extended with a description of canopy photosynthesis, maintenance respiration and leaf area expansion according to Goudriaan and others (1985), Spitters and others (1989) and Seginer and others (1988).

THE GREENHOUSE CLIMATE MODEL

For the description of the greenhouse climate a static model is used. It consists of energy and mass balances for the greenhouse air. The following energy balance is applied:

$$Q_{heat}+Q_{sun}-Q_{vent}-Q_{trans}=0 \qquad (6)$$

with Q_{heat}: the energy input by the heating system [W.m-2], Q_{sun}: the part of the solar radiation which is converted to sensible heat [W.m-2], Q_{vent}: the energy transport caused by natural ventilation [W.m-2], Q_{trans}: the energy transport caused by transmission through the greenhouse cover [W.m-2].

The CO2 mass balance reads:

$$\varphi_{CO2,supl}-\varphi_{CO2,vent}-\varphi_{CO2,photo}=0 \qquad (7)$$

with $\varphi_{CO2,supl}$: the CO2 supply to the greenhouse [kg.m-2.s-1], $\varphi_{CO2,vent}$: the CO2 transport caused by natural ventilation [kg.m-2.s-1], $\varphi_{CO2,photo}$: the CO2 transport as a result of net crop photosynthesis [kg.m-2.s-1].

Energy and mass transport are described according to Bot (1983) and De Jong (1990). The model of Goudriaan and others (1985) is used to calculate crop photosynthesis.

DETERMINISTIC OPTIMAL CONTROL

The lettuce crop considered in the simulations is planted at 1 January and grown for a period of 80 days. The desired harvest weight is 250 g per head. The following objective functional is used:

$$J(T,CO_2)=\alpha*(fw_{t_f}-fw(t_f))^2 \qquad (8)$$
$$+\int_{t_b}^{t_f}\beta*Q_{heat}+\gamma*\varphi_{CO2,supl}dt$$

in which T and CO_2 are the setpoint trajectories of the air temperature and CO2 concentration, t_b is the starting time of the cultivation period and t_f is the harvest time [day], fw_{tf} and $fw(t_f)$ are the desired and the realized fresh weight at harvest time t_f respectively

[kg.m-2], Q_{heat} is the heating energy input to the greenhouse [W.m-2], $\varphi_{CO2,supl}$ is the CO2-supply to the greenhouse [kg.m-2.s-1], α determines the necessity to reach the desired fresh weight and β [dfl.W-1.m2] and γ [dfl.kg-1.m2.s] are the unit prices of the heating energy and the CO2 supply respectively.

The optimal control problem is defined as to find daily temperature and CO2 setpoint trajectories that minimize the objective function (8).

The optimal control problem is solved using smoothed mean weather data of the period 1975-1989. The results are compared with practical strategies. In fig. 1 both the optimal and practical strategies of the day and night temperature are shown.

In the practical strategies a day temperature of 10 °C is applied the first 14 days of the cultivation period after which it is increased to 12 °C. The night temperature is kept at 10 °C the first 14 days and 7 °C during the rest of the cultivation period. The optimal day temperature increases with time. The optimal night temperature is kept at a constant level of 5 °C. Increasing the night temperature apparently has no positive effect on crop growth but has a considerable effect on the energy consumption.

In fig. 2 the practical and optimal strategies of the CO2 concentration are shown. A constant level of 700 ppm is used as a practical strategy for the CO2 concentration. The optimal CO2 concentration is increasing with time.

The higher day temperature and CO2 levels applied in the practical strategies during the first half of the cultivation period result in a higher crop growth rate compared with the growth rate as a result of the optimal climate strategies while during the second half the growth rate is lower. The practical and optimal strategies lead within 5% to the desired harvest weight of 250 g per head.

Evaluated on a relative basis the optimal climate strategies lead to a 35 % lower production cost as a result of a more efficient application of energy and CO2.

From these simulations it may be concluded that the dynamic optimal climate strategies result in a better performance with respect to energy and CO2 consumption than the static practical strategies.

OPEN-LOOP CONTROL

The applicability of the optimal climate strategies shown in fig. 1 and 2 as an open-loop control of a real process depends on the effect of differences between the expected and the real weather, modelling errors and the accuracy of the climate control. The effect of incomplete knowledge of the outside weather is analyzed in the following simulation.

The before mentioned strategies were used in a simulation in which weather data of 1976 were applied. The performance of the practical and the optimal strategies, the latter hereafter referred to as the open-loop strategies, were evaluated and compared with optimal climate strategies which were calculated for the weather data of 1976 specifically, representing the best possible performance with respect to energy and CO_2 consumption. In fig. 3 the solar radiation sum, I, of the first 80 days of 1976 is shown.

In fig. 4 and 5 the optimal temperature and CO_2 strategies calculated on the basis of the weather data of 1976 are presented. There is a strong correlation between the solar radiation, the day temperature and the CO_2 concentration.

The effect of the practical climate strategies, the open-loop strategies and the optimal climate strategies calculated on the basis of the weather data of 1976 on the crop growth, is shown in fig. 6. The desired harvest weight is not reached when applying the practical and the open-loop strategies although the difference is less then 10%.

A significant difference is found in the overall energy and CO_2 consumption. Evaluated on a relative basis and compared with the performance of the climate strategies calculated for 1976, the practical and open-loop strategies result in a 42% and 37% higher production cost respectively.

On the basis of these simulations it may be concluded that the application of some information about the future trend in the outside weather as was done in the open-loop strategies results in a better performance than the static practical strategies containing no information about the future weather. But compared to the performance of the optimal control calculated with complete foreknowledge of the outside weather, an unrealistic situation for sure, there is still room for improvement.

OPEN-LOOP FEEDBACK CONTROL

Schmidt and others (1987) reported favourable results of repeated optimization to cope with the differences between the expected and the real outside climate. In that approach first an open-loop control is calculated for the whole growing period $[t_b, t_f]$ on the basis of a long term weather forecast as described above. This open-loop control is applied to a process which is also influenced by the actual outside climate. After a fixed period of time Δt, the state of the crop $x(t_b + \Delta t)$ is evaluated and used as an initial state for a repeated optimization with a shortened horizon $[t_b + \Delta t, t_f]$. The resulting open-loop control is applied to the process for the next time interval Δt. This sequence is repeated until the harvest time t_f is reached. This approach may be referred to as open-loop feedback control because every time $t = t_b + k\Delta t$ the systems state is used in a feedback way to calculate a new open-loop control for the remaining time $t \epsilon [t_b + k\Delta t, t_f]$.

Open-loop feedback strategies have been calculated using the smoothed mean weather data of the period 1976-1989 as weather forecast, the data of 1976 as actual climate data and a repetition interval $\Delta t = 10$ days. Figure 7 shows the resulting CO_2 strategy.

The open-loop feedback strategies have as a consequence a 45% higher production cost compared to the optimal control calculated for 1976, which is even higher than the production cost resulting from the application of the practical strategies and the open-loop strategies considered before.

In contradiction with the results reported by Schmidt and others (1987) the concept of repeated optimization applied in a simulation using the weather data of 1976 and a repetition interval of 10 days does not result in a better performance than the open-loop strategies.

CONCLUSIONS

In this paper it was shown that optimal strategies for the night temperature, the day temperature and the CO_2 concentration can be calculated using a mathematical process model. The calculated dynamic strategies perform better than the static practical strategies. Although the application of long term mean weather data as weather forecast result in a better performance with respect to energy and CO_2 consumption compared with strategies containing no information about the outdoor climate, there is still room for improvement. The concept of repeated optimization resulting in open-loop feedback strategies does not seem the right way to cope with the effect of differences between the expected and the actual weather.

REFERENCES

Berg, E., and W. Lentz (1989). Dynamic optimal control of plant production. Acta Horticulturae,248,223-241.

Bot, G.P.A. (1983). Greenhouse climate: from physical processes to a dynamic model. Ph.D. Thesis, Agricultural University, Wageningen.

Gal, S., A. Angel, and I. Seginer (1984). Optimal control of greenhouse climate: methodology. European Journal of Operational Research, Vol. 17,45-56.

Gill, P.E., W. Murray and M.H. Wright (1981). Practical optimization. Academic Press, London.

Goudriaan, J., H.H. van Laar, H. van Keulen and W. Louwerse (1985). Photosynthesis, CO2 and plant production. In W. Day and R.K. Watkin (eds.), Wheat growth and modelling, NATO Asi Series, Series A: Life Sciences, Vol.86. Plenum Press, New York.

Henten, E.J. van (1989). Model based design of optimal multivariable climate control systems. Acta Horticulturae,248,301-306.

Jong, T. de (1990). Natural ventilation of large multispan greenhouses. Ph.D. Thesis, Agricultural University, Wageningen.

Renfro, J.G., A.M. Morshedi and O.A. Asbjornsen (1987). Simultaneous optimization and solution of systems described by differential/algebraic\ equations. Comput. chem. Engng., Vol.11,No.5,503-517.

Schmidt, M., K. Reinisch, H. Puta, and A. Markert (1987). Determining climate strategies for greenhouse cucumber production by means of optimization. Preprints of the 10th World Congress on Automatic Control, IFAC, Munich, Vol.2,350-355.

Seginer, I., G. Shina, L.D. Albright and L.S. Marsh (1988). Optimal temperature setpoints for greenhouse lettuce. ASAE paper nr. 88-4525.

Spitters, C.J.T., H. van Keulen and D.W.G. Kraalingen (1989). A simple and universal crop growth simulator: SUCROS87. In R. Rabbinge, S.A. Ward, H.H. van Laar (eds.), Simulation and system management in crop production. Pudoc, Wageningen.

Sweeny, D.G., D.W. Hand, G. Slack and J.H.M. Thornley (1981). Modelling the growth of winter lettuce. In D.A. Rose and D.A. Charles-Edwards. Mathematics and plant physiology. Academic Press, New York.

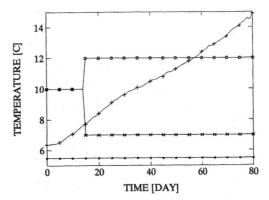

Fig. 1. The optimal day (-+-) and night (-*-) temperature strategy and the practical day (-o-) and night (-x-) temperature strategy.

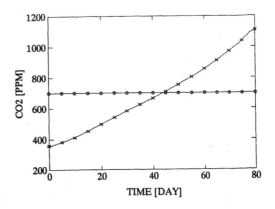

Fig. 2. The optimal (-x-) and practical (-o-) CO2 strategy.

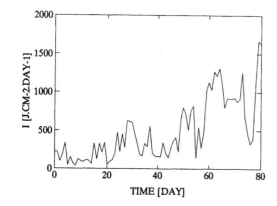

Fig. 3. The solar radiation sum, I,
during the first 80 days of 1976.

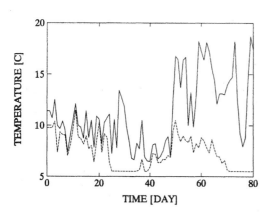

Fig. 4. The optimal day (-) and night
(--) temperature strategies
calculated on the basis of the
weather data of 1976.

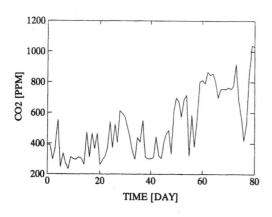

Fig. 5. The optimal CO2 concentration
calculated on the basis of the
weather data of 1976.

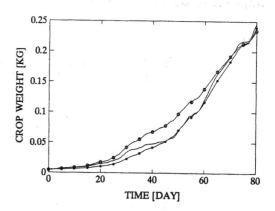

Fig. 6. Crop growth as a result of the
outdoor weather data of 1976, the
practical strategies (-o-), the
open-loop strategies (-*-) and
the climate strategies calculated
specifically for 1976 (-).

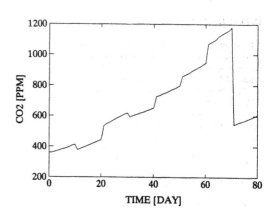

Fig. 7. The optimal CO2 strategy as a
result of open-loop feedback
control.

SELF-ADAPTIVE AND SELF-TUNING CONTROL OF A NUTRIENT FILM TECHNIQUE (NFT) SYSTEM

A. Chotai and P. C. Young

Centre for Research on Environmental Systems, University of Lancaster, Lancaster, LA1 4YQ, UK

Abstract There has been considerable interest during the past few years in the development of self-adaptive and self-tuning control methods for systems whose dynamic characteristics change over time. In this paper, we report the develoment of new approaches to adaptive control based on the Proportional-Integral-Plus (PIP) controller applied to Nutrient Film Technique (NFT) Systems used in glasshouse horticulture.

Keywords Instrumental variable method, Nutrient Film Technique (NFT) System, Proportional-Integral-Plus (PIP) control, recursive identification and estimation, Self-adaptive and Self-tuning Control.

INTRODUCTION

The advent of digital computers with considerable processing power has provided control engineers with the opportunity to consider sophisticated control system concepts, such as self-tuning (STC) and self-adaptive control (SAC), which could not have been implemented in practice only a few years ago. STC and SAC are attractive to the control systems designer because they offer the possibility of improved control for systems whose dynamics are poorly defined or change radically over time. In this manner, the performance of the system should be maintained at a high level, despite the changing nature of the controlled system, and this should lead to savings in commissioning and running costs. However, these potential advantages are obtained with some penalty: the STC and SAC systems are naturally more complicated than conventional, fixed gain control systems, since they normally require some explicit or implicit procedure for continually updating estimates of the controlled system model parameters. As a result, the effects of this increased complexity must be fully evaluated in the design studies to ensure that the system will not suffer from problems of stability and reliability .

The Nutrient Film Technique (NFT) system used for the cultivation of protected crops in glasshouses is an obvious candidate for SAC: its dynamic behaviour changes considerably over time as the result of various factors, such as the variations in nutrient uptake by the plants, and the seasonal changes in the transportation time delays and time constants caused by plant root growth. In this paper, we outline the major steps in the design of a computer-based, SAC system for an NFT process. This has been one of the main examples in a research program on self-adaptive and self-tuning control carried out over the past few years in the Centre for Research on Environmental Systems at Lancaster.

The approach to SAC developed in this research program is based on the concept of "True Digital Control" (TDC), as described in our previous publications on this topic (Young et al., 1987b, 1988; Young, 1989). Here, all aspects of the design process, from off-line modelling, through control system design and evaluation, to practical implementation, are considered *overtly* in discrete-time terms, with little reference to continuous-time concepts, except where these aid in the physical interpretation of the models and methods. The TDC approach to NFT control developed at Lancaster (Behzadi, 1989; Young et al., 1987a; Behzadi et al., 1989) exploits new methods of control systems analysis and design which are based on the definition of a novel Non Minimum State Space (NMSS) representation of the NFT system. Here, the state variables are defined as the present and past sampled values of the system input and output signals, all of which are available for direct measurement and can be used to implement State Variable Feedback (SVF) control laws. Although the resulting "Proportional-Integral-Plus" (PIP) controller has a superficial resemblance to conventional PI and PID controllers, it is inherently much more flexible and sophisticated in its ability to exploit the power of SVF for closed pole assignment or optimal LQ control.

These PIP self-tuning and adaptive control studies have been carried out both by digital computer simulation of the NFT system, and on a small scale physical model of the NFT flow system constructed at Lancaster. This pilot plant is *dynamically similar* to the full system in flow terms but has much shorter dominant time constants and so is much easier to use for control experiments. The control systems have not, as yet been implemented on the full NFT system, but similar PIP controllers have been applied successfully to a number of practical systems (Young et al., 1988; Chotai, 1991a; Billington et al., 1991; Ng et al., 1991; French et al., 1991).

THE NFT SYSTEM

The Nutrient Film Technique (NFT) is a water culture system, normally used in protected environments like glasshouses, where a shallow stream of nutrient solution is recirculated over the bare roots of the growing plants in order to provide adequate water, nutrient and aeration. The plants are sown in small blocks of a substrate called "rockwool" and are placed in gently sloping channels. The nutrient solution is collected in a tank, or settling trench, and then recirculated by an electrically driven pump back into the channels, so forming a positive feedback loop. An extensive exposition of the NFT method is given in Graves (1983).

In the NFT, temperature, ion concentration and acidity of the solution are monitored and controlled in some manner. The NFT system requires the addition of up to twelve nutrients, although most of them are needed in only small quantities. On the other hand, nutrients like potassium and nitrogen are taken up by the plants in comparatively large quantities.

NFT culture is particularly successful for tomato plants (Cooper, 1967). Such plants can tolerate quite a range of nutrient concentrations but fruit quality can suffer at low concentrations (Winsor and Massey 1978). Fig. 1 shows the simplified block diagram of NFT sysytem, which is dominated, in a dynamic sense, by the nutrient delivery flow system, in which the nutrient mixture is circulated continuously through the parallel growth channels and the catchment trench.

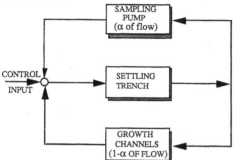

Fig. 1 Simplified block diagram of the nutrient film system.

The NFT system with its positive feedback mechanism can, of course, be modelled with various levels of complexity. An important first stage in modelling is often the development of an adequate computer simulation model. In the present study a relatively high order, simulation model of NFT system was constructed using a combination of experimental investigation (based on tracer experiments to model the flow dynamics) and normal, physically-based, mathematical modelling (Young et al 1987a; Behzadi 1989).
Computer simulation models are often too complex to provide a direct basis for control system design. In the present study, a variety of simpler control models were considered. In order to obtain these control models, the channels and the trench were initially represented in overall, aggregative terms as first order, discrete-time systems with associated time constants and pure time delays, as shown in figure 2, where the sampling interval is 15 minutes.

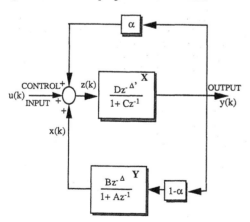

Fig.2 Block diagram of simple, highly aggregated TF model for NFT flow system.

The time-constants and time delays characterising these models are then selected in order to match the dynamic characteristics of the high order simulation model.

In figure 2, the forward block **X** and the feedback block **Y** represent the settling trench and the channel system, respectively. The trench characteristics are defined by the aggregated time delay Δ' and the parameters C and D: these parameters are mainly functions of flow, which will normally remain fairly constant . The aggregated time delay Δ and the parameters A and B of the channel system, on the other hand, will change considerably as the result of plant growth over the growing season.

In Fig.2, x(k) is the nutrient concentration entering the trench; y(k) is the measured concentration leaving the trench; u(k) is the controlled input signal (i.e. the measured inputs of salts, acid and other control chemicals that directly affect the controlled nutrient); and z(k) represents the sum of u(k). and x(k). Note that, here, we consider only a single input variable whereas, in practice, it is necessary to control the levels of several variables. This has proven necessary both because of our inability to conduct experiments on the full NFT system (see below) and because parallel research on ion sensitive electrode measurement of the various nutrient concentrations had not reached a stage where reliable measurement of the variables was possible. This implies, of course, that the controllers discussed here could, in the first instance, only be applied in single input, single output (SISO) terms, each operating on a single controlled variable (see Conclusions).

The system in Fig.2 can be reduced by normal block diagram manipulation, to a single backward shift (z^{-1}) operator transfer function (TF) equation of the following form,

$$y(k)=\frac{Dz^{-\Delta'} - ADz^{-\Delta'-1}}{1-(A+C)z^{-1}+ACz^{-2}-\alpha Dz^{-\Delta'}+\alpha ADz^{-\Delta'-1}-(1-\alpha)BDz^{-\Delta-\Delta'}}u(k)$$
$$(1)$$

where z^{-i} is the backward shift operator (i.e. $z^{-i}y(k) = y(k-i)$). The model is clearly a highly simplfied representation, since it only assumes a lumped outflow from the channel system into the catchment trench. In effect, we are replacing the multi-order, time delay elements of the system by a single first order delay system whose residence time and delay characteristics are equivalent, in an aggregate sense, to those of the original model. For a 15 min. sampling interval, the parameters in equation (1) were estimated as follows,

A = -0.794; B = 0.197; Δ = 9; C = -0.727; D = 0.273; Δ' = 3; α = 0.45

These parameter values correspond to the following aggregate steady state gains and time constants,

CHANNEL: SSG = 0.956 ; Time const. = 65 min. ; Time delay = 45 min

TRENCH : SSG = 1.00 ; Time const. = 47 min. ; Time delay = 135 min.

The TF model for the system is obtained from equation (1). With the above parameter values and it takes the form,

$$y(k)=\frac{0.273(1-0.794z^{-1}) z^{-3}}{1-1.521z^{-1}+0.577z^{-2}-0.123z^{-3}+0.098z^{-4}- 0.0296z^{-12}} u(k)$$
$$(2)$$

We find that the "open loop" steady state gain (SSG) gain of the system is high because the plants remove only a small proportion of the nutrients which pass over the roots. Also, this high gain, together with the inherent positive feedback introduced by the pumped nutrient recirculation, have produced a system which is dominated by a combination of two primary modes of behaviour, as illustrated by the impulse response shown in Fig. 3: a short term oscillatory mode associated with the recirculation; and a very long period mode which can appear, in the short term, to be functioning as a pure integrator (as, indeed, it is if there are no nutrient losses due to plant uptake or other causes).

A Simple First Order Model

If we consider only the long term dynamics of the model (2), then we find that the system can be represented by the following first order model with three period delay,

$$y(k) = \frac{0.0987 z^{-3}}{1 - 0.9976 z^{-1}} u(k) \qquad (3)$$

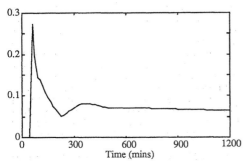

Fig.3 Impulse response of NFT system (2)

This model can be obtained from the 12th order model (2) by simply matching the impulse response decay and steady state gain of the two models; or, more systematically, by using the model order reduction procedure. In this latter case, as used here, the simplified refined instrumental variable (SRIV) method of identification and estimation (Young 1985, 1989, 1991; Tych and Young, 1990) was used to find the lower order model based on the impulse response of the higher order model.

The main limitation of the first order model with a pure time delay of three sampling intervals, as given by equation (3), is its inability to represent the short term oscillatory behaviour introduced by the re-circulation. Nevertheless, it has the same SSG of 41.13 as the model (2) and its time constant of 104.04 hrs. characterises the long period mode very well. In this sense, it provides a satisfactory vehicle for the initial design of the control system. Such a system would clearly be considered rather crude in conventional control system terms. However, as we shall see, if the control system is properly designed, the short period modes will be "well behaved" and not excited by the feedback. The basic control requirements will then be satisfied, with the system able to maintain nutrient and acidity levels in the region of the desired values, despite the continually changing dynamic characteristics of the NFT process.

PIP Control System Design

In order to design a control system for the NFT process, we have used the proportional-integral-plus (PIP) design method of Young and et al. (1987b) as shown in Fig.4. Here, $y_d(k)$ is the desired nutrient concentration set point, while the gains f_0, g_1, g_2 and k_I are chosen either to ensure specified closed loop pole assignment for model (3) or to satisfy a Linear Quadratic (LQ) optimality criterion (Tych et al., 1991).

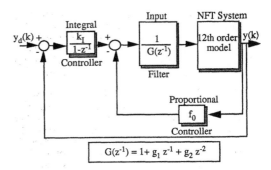

Fig.4 Block diagram of fixed gain NFT control system

Simulation results

The model reduction and PIP control system design for the system shown in Fig.4 was carried out using the TDC Computer Aided Control System Design (CACSD) package developed at Lancaster (Tych et al., 1991). For the PIP design, the desired closed loop characteristic polynomial was chosen as $(z-0.7)^4$ (i.e. all four poles assigned to 0.7 in the complex z domain), in order to yield well damped rapid response to command inputs and reasonable rejection of disturbance effects. The design was also evaluated by simulation in the TDC package using the full 12th order simulation model, as shown in Fig.4. Here, the closed-loop step response is shown in Fig.5, where we see that the design yields an acceptable response time but that, not surprisingly, the higher order mode associated with the recirculation is not being fully regulated because it was not incorporated in the control model (3).

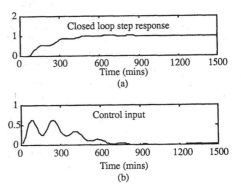

Fig.5 Control system simulation results for full NFT model (a) Closed loop step response (b) Control input

In the light of the these results, three possibilities exist: (a) as discussed previously (Behzadi, 1989; Young et al., 1987a; Behzadi et al., 1989), the sampling interval could be increased substantially, so that the higher frequency oscillations are not "observed"; (b) the 15 min. sampling interval could be retained, but the design based on a higher order model, e.g the model (2) or preferably an intermediate reduced order model; or, finally, (c) the design could be accepted as "satisfactory" provided further evaluation could establish that the oscillatory characteristics are likely to remain well behaved. Option (b) is preferable and we consider this further in the next Section. But the latter approach (c) has some merit since it would allow us to retain the simple PI-type control structure of Fig. 4.

In the TDC package, the method used for evaluating the sensitivity of the closed loop system to uncertainty is based on stochastic or "Monte Carlo" simulation analysis (seeTych et al., 1991), with the uncertainty quantified in terms of the probability distribution associated with the estimates of the model parameters (i.e. in the Gaussian case, the first two moments in the form of the estimates and their associated covariance matrix). The sensitivity to this uncertainty is then assessed by reference to the ensemble of closed loop characteristics obtained from multiple stochastic realisations generated in this manner. The results obtained for the present example are given in Fig. 6: the ensemble of closed loop step responses are shown in (a); the ensemble of control inputs in (b); and the stochastic root loci (closed loop pole locations) in (c).

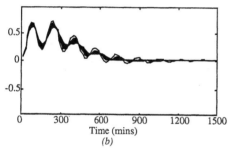

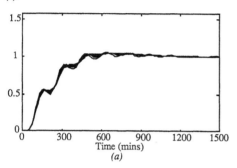

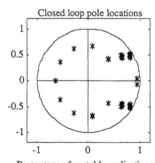

Percentage of unstable realisations: 0

(c)

Fig. 6 The closed-loop (a) step responses (b) control values (c) stochastic root loci.

From these results, we see that, in this case, the system seems to be very robust in the presence of uncertainty, with the ensemble showing only minor perturbation about the nominal responses and the stochastic root loci indicating a large stability margin. Thus, if the response time and the presence of the oscillations in the response is considered acceptable, the simple design in Fig. 4 could provide a basis for further evaluation. However, the results described in the next Section, based on more extensive experimental studies, suggest that a better practical solution is obtained using a higher order model coupled with self-adaptive adjustment of the PIP control gains.

MODELLING AND CONTROL OF A PILOT SCALE FLOW MODEL OF THE NFT SYSTEM

Most practical evaluation of the PIP designs was carried out on a small scale physical model of the NFT flow system which was constructed at Lancaster following difficulties experienced during initial experiments on the full scale NFT system. In particular, the extremely slow dynamics of the full scale system necessitated unacceptably long experimental times, which interfered with the normal glasshouse operation and seriously limited the nature of the control studies.

The NFT flow dynamics are dictated by a closed loop combination of time constants and time delays. It was found that the simplest mathematical model which replicates the physical characteristics (pure advective time delay and dispersive time constant) of the NFT system flow system is the aggregated dead zone (ADZ) model developed at Lancaster to model the transportation of solutes in river channels (Beer and Young, 1983; Wallis et al., 1989; Young and Wallis, 1986). In the laboratory, a physical version of the ADZ model can be constructed as the combination of a continuous stirred tank reactor (CSTR), to simulate the dispersive time constant, and an associated pipeline, to simulate the advective time delay. The time constant (or residence time) of the CSTR is then obtained from the relationship

$$T = \frac{V}{Q} \qquad (4)$$

where, T is the time constant (sec), Q is the flow rate (ml /sec) through the CSTR, and V is volume of the solution in the tank. The time delay is defined by Q and the pipe length.

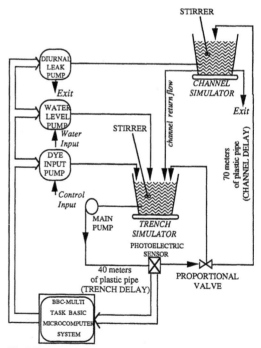

Fig.7 The schematic diagram of the pilot scale plant of the NFT system

A diagram of the complete model flow system is shown in Fig.7. The long time delays required to simulate the advective transport processes associated with the channels and the trench of the full scale NFT system are constructed from considerable (70 metre and 40 metre) lengths of plastic pipe. These connect the two CSTR's which simulate the dispersive characterists of the channels and trench, respectively. The pump rate can be varied to simulate major variation in the pure time delays; and a sinusoidal leak of liquid from the system is incorporated to simulate the diurnal losses of nutrient that would occur on the full size NFT plant, due to uptake by the growing plants. The nutrient itself is simulated by a black dye: the dye delivery system, which constitutes the main control signal to the system, is in the form of an electrically driven peristaltic pump; while the concentration of dye, which represents the system output, is measured by a photoelectric sensor positioned as shown in Fig.7.

The pilot plant is "*dynamically*" similar to the full system in flow terms, but has much shorter dominant time constants and time delays (in the ratio 1:36, with a plant sampling interval of 25 seconds being equivalent to 15 minutes in the full system). The system dynamics may be easily altered by changing the flow rates, the volume of solution in any of the tanks, the system leak (simulating the rate of uptake), the return flow ratio valve, or any combination of these factors. The system has several advantages over the numerical computer simulation models: for example, it represents a full analog simulation of the flow system and it is subject to realistic inter-sample stochastic disturbances and non-linearities.

Both open and closed loop experiments have been conducted on the pilot plant, with data acquisition and subsequent digital control based on a BBC microcomputer, using a multi-tasking operating system. The system input (the microcomputer generated signal to the peristaltic pump), and output (the dye concentration measured by the photoelectric system) provide the main data used in subsequent analysis and modelling. Various other data, such as the sinusoidal leak control signal and the flow rate of water through the system, are also monitored throughout the experiments.

The data from the experiments are processed in two ways. First, they are stored on disk in the BBC micro and presented graphically on the visual display, both for on-line monitoring of the experiment and, following a "screen dump" to the printer, for later reference. Secondly, they are transferred to an IBM-PC AT, by Kermit, for later analysis and modelling studies using the microCAPTAIN (Young & Benner 1986, 1989) and TDC (Tych et al., 1991) program packages . Since on-line recursive estimation algorithms are programmed in the BBC for self-tuning/adaptive control studies, it is also possible to monitor the parameter estimation results during the open or closed loop experiments.

Data-Based Modelling

All the modelling analysis was carried out using the microCAPTAIN computer program based on input-output data monitored at a sampling interval of 1.25 mins. Full details of the identification and estimation results are given in Young et al.(1988) and Behzadi (1989). The most effective control model was identified as the following 5th order TF,

$$y(k) = \frac{G (b_1 z^{-1} + b_2 z^{-2} + b_3 z^{-3})}{1 + a_1 z^{-1} + a_2 z^{-2} + a_3 z^{-3} + a_4 z^{-4} + a_5 z^{-5}} u(k) \qquad (5)$$

where:

$a_1 = -1.092$; $a_2 = 0.329$; $a_3 = -0.09$; $a_4 = -0.032$; $a_5 = -0.1$;

$b_1 = 0.009$; $b_2 = 0.07$; $b_3 = -0.032$ and $G = 3.44$.

Here G is the "*system gain*", which is defined as the overall scaling effect between the actual concentration levels in the system, and their numerical representations in the digital computer program. In this case, the black dye used to represent the nutrient is actually a suspension which leads to the thickening of the input solution towards the bottom of the supply container. In this sense, it is very difficult to exactly match the concentration of the input solution on replacement of the dye supply; or even maintain a uniform control input concentration during an experiment unless continuous, vigorous stirring is used. The system gain, G of the pilot plant is, therefore, generally unknown and time variable and has to be estimated by manual tuning, in a fixed gain controller, or by recursive estimation in a self tuning or self-adaptive control system.

As might be expected, the impulse response of the model (5), shown in Fig.8, is similar to that of the high order computer simulation model and exhibits the major dynamic characteristics of the NFT flow system. The long term response is dominated by the first order mode with a very large time constant. This arises from the low losses in the system: except for the sinusoidal leak, all dye entering the system is retained, so that the system is always quite close to being an integrator. The shorter term behaviour is more interesting and is dominated by the circulatory flow system, with the three decaying peaks on the impulse response indicating that the initial impulsive input is being transported and progressively dispersed around the system.

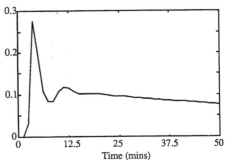

Fig.8 Impulse response of model (5).

Fixed Gain Control System Design

Typical results obtained using a fixed gain PIP pole assignment design (Young and et al., 1987b, 1988) based on the model (5) are shown in Fig. 9. The desired closed loop poles were set to provide a damped second order response with an overall time constant of 5 minutes (equivalent to 3 hrs. in the full system). A sinusoidal leak (with its maximum and minimum levels corresponding to the theoretical channel simulator steady state gains of approximately 0.94 and 0.98, respectively), is also included in order to simulate the diurnal uptake variations. From Fig. 9, it is evident that the closed loop response is satisfactory, with the small initial overshoot due to the sub-optimal tuning of the system gain. The control system also shows some sensitivity to the modelling errors in the longer term.

37

Self-Tuning and Self-Adaptive Control

The NFT system is known to undergo dynamic changes, either because of the plant growth in the channels and the settling tank, or the changing behaviour associated with plant nutrient uptake. Our experimentation with NFT systems so far has indicated that such changes can result in radical model structure alterations, ruling out any assumption of the *global* reliability of any fixed gain controller. In other words, a *"general* "solution to the NFT system problem is unlikely from off-line modelling and control system design exercises, particularly if tight control is required.

In practice, therefore, a fixed gain approach may well result in the long term deterioration of the control system performance, to a degree which will depend on the complexity of the particular control system design strategy adopted. In such cases, unless the poorer performance can be tolerated in practice (as may be the case for the growth of plants), a more sophisticated self-tuning or self-adaptive control system design must be adopted. Such a design can be achieved in two ways; either the system may be regularly remodelled off-line and the computer-based part of the control system re-designed if significant changes are encountered; or a complete on-line model parameter estimation procedure can be used to continually update the PIP design. In this latter case, the implementation can involve self-tuning control (STC) which is reactivated at regular intervals; or full self-adaptive control (SAC), where the model parameters are all recursively updated, and new control gains computed, at regualr intervals (e.g. each sampling instant).

Self-tuning and self-adaptive PIP controllers rely on the specification of a system model *structure* within the computer program. The parameters of this model structure, which is initially defined during the off-line modelling experiments that are an essential part of TDC design, are then estimated recursively based on present and past values of the measured system input and output signals. This *time series analysis* of the input-output data is normally carried by the recursive least squares (RLS) method; or alternatively, in the presence of heavy noise by the recursive instrumental variable (IV) approach (see e.g Young, 1984). Using these recursive parameter estimates, a new PIP feedback gain vector is then computed, and the control system is suitably updated to reflect the changes in the gain vector.

The fifth order model (5), was found to be the best control model representation of the pilot scale NFT flow system. Based on this knowledge, therefore, we specify a similar fifth order estimation model structure for the STC system design, where the estimated parameter vector $\hat{a}(k)$ is composed of the 8 unknown model parameters, i.e.,

$$\hat{a}(k) = [\, a_1 \quad a_2 \quad a_3 \quad a_4 \quad a_5 \quad b_1 \quad b_2 \quad b_3 \,]^T$$

and the NMSS state vector $x(k)$, is of the form (Chotai et al., 1991a; Young et al., 1987b, 1988,),

$$x(k)=[y(k) \; y(k-1) \; y(k-2) \; y(k-3) \; y(k-4) \; u(k-1) \; u(k-2) \; z(k)]^T$$

where $z(k)$ is the *integral-of-error* state defined by the following equation,

$$z(k) = z(k-1) + \{y_d(k) - y(k)\}$$

while $y_d(k)$ is the reference or command input at the kth sampling instant. With this model structure, we obtain a feedback control system with a 8-dimensional control gain vector, which is continuously updated as the recursive

parameter updates are obtained from the RLS estimation algorithm. In this NMSS context, the control system design can be based on any SVF approach. These include pole assignment, as discussed above; or optimal LQ or LQG control, as described in other recent papers (Young et al., 1991; Tych et al., 1991; Billington et al., 1991). The latter would provide a more natural setting for control system design if energy-efficient solutions were being sought.

Practical Results

Figs. 10 and 11 show two typical examples of the self-tuning PIP controller applied to the NFT pilot scale plant: in Fig. 10 the tuning is applied to all 8 parameters of the fifth order model. The initial fluctuations in the input signal are deliberately created to *"excite"* the system during the initial start-up and provide richer information to the recursive estimation algorithm. The plot of the on-line parameter estimates, as shown in Fig.10, clearly indicates the good initial convergence of the parameter estimates.

The results in Fig. 11 are obtained with a much simpler adaptive gain control system, in which only the gain G in (5) is recursively updated, with the other parameters set at their *a priori*, off-line estimaed values. Clearly, both controllers maintain tight maintenance of the set point despite the sinusoidal leak (which is introduced to model the effects of nutrient uptake by plants which would be growing in the full scale system). Full self-adaptive control is similarly effective but is difficult to illustrate because of the slow rates of change associated with the model parameters of the NFT system.

CONCLUSIONS

The NFT process presents an challenging modelling and control problem. But the TDC approach to solving this problem, as outlined in the present paper, is not restricted to the NFT process, or even to other glasshouse systems. It provides an attractive, unified approach to control system design in many areas of agriculture and horticulture; an approach which is sophisticated in theoretical terms but which yields control system designs that are both flexible and easy to implement in practice. Indeed, we believe that the PIP family of TDC systems introduced in this paper provides a logical successor to the ubiquitous "three term" PID controller which still dominates most commercial control systems.

It should be emphasised that, because of its state-space formulation, PIP control system design is not limited to the kind of single variable (SISO) control discussed in the present paper. The NFT system, for example, should really be treated as a multivariable (MIMO) problem and we have considered it in SISO terms only because of the restrictions imposed by the measurement system. It would not be difficult, however, to design a multivariable PIP control system (Chotai et al., 1990, Wang, 1988) if a multivariable measurement system was available, and coupling between the variables in the system justified such multivariable control action. And such a MIMO system could be based on any SVF design principle, such as pole assignment or optimal LQG (Chotai et al., 1990).

Finally, rapidly sampled, near continuous-time operation is sometimes demanded by users who may be sceptical of control systems operating at a fairly coarse sampling interval. If higher sampling frequencies are utilised, however, they can lead to sensitivity problems with the kind of z^{-1} operator PIP designs discussed in the present paper. A more appropriate solution in these circumstances is the equivalent discrete differential or delta (δ) operator PIP control system design, which we discuss in a number of other recent papers (Chotai et al., 1991b; Young et al., 1991).

REFERENCES

Beer, T. and P. C. Young (1983) Longitudinal dispersion in natural streams, *American Society of Civil Engineers Journal of Environmental Engineering*, **109**, 1049-1067.

Behzadi, M. A. (1989) True digital control of greenhouse systems, PhD Thesis, Centre for Research on Environmental Systems, Lancaster University.

Behzadi, M. A., P. C., Young, and A. Chotai. (1989) Modelling and coarse sampling control of a nutrient film system, *Systems Science*, Vol 15, No. 2.

Billington, A. J., A. R. Boucher and C. S. Cox. (1991) Optimal PIP control of scalar and multivariable processes, *I.E.E. Conference Publication No. 332: 3rd Int. Conf.'Control 91'*, pp 574-580.

Chotai, A., P.C.Young, and W.Tych. (1990) True digital control of multivariable systems by input/output state variable feedback", Report No.TR86/(1990), CRES, University of Lancaster.

Chotai, A., P. C. Young and M. A. Behzadi. (1991a). The self-adaptive design of a nonlinear temperature control system, in the Special Issue on *Self Tuning Control, Proc. I.E.E.*, Pt.D., vol. 138, 41-49.

Chotai, A., P. C. Young and W. Tych. (1991b) A non-minimum state space approach to true digital control based on δ operator models, *I.E.E. Conference Publication No. 332: 3rd Int. Conf. ' Control 91'*, pp 567-573.

Cooper, A. J. (1967) Tiered through and nutrient film methods of tomato growing, *Annual Report of the Glasshouse Crops Research Institute*, 59.

French, I. G., I. Fletcher and K. C. S. Ng. (1991) Control of an industrial waste gas scrubber process, *I.E.E. Conference Publication No. 332: 3rd Int. Conf.' Control 91'*, pp 596-601.

Graves, C.J.(1983) The nutrient film technique, *Horticultural Reviews 5* 1-10.

Ng, K. C. S., W. J. B. Arden and I. G. French (1991) Pressure control within an industrial plasticating extruder, *I.E.E. Conference Publication No. 332: 3rd Int. Conf.' Control 91'*, pp 612-616.

Tych, W. and P. C. Young (1990) A refined instrumantal variable approach to model reduction for control systems design, Report No.TR81/(1990), CRES, University of Lancaster.

Tych, W., A. Chotai, and P. C. Young (1991) Computer aided design package for true digital control (TDC) systems, *I.E.E. Conference Publication No. 332: 3rd Int. Conf.' Control 91'*, pp 288-293.

Wallis, S. G., P. C. Young and K. J. Beven (1989) Experimental investigation of the Aggregated Dead Zone model for longitudinal solute transport in stream channels, Proc. Inst. Civ. Engrs. (U.K.), vol. 87, pp 1-22.

Wang, C. L. (1988) New methods for the direct digital control of discrete-time systems, PhD Thesis, Univ. of Lancaster.

Winsor, G. W. and D. M. Massey. (1978) Some aspects of the nutrition of tomatoes grown in recirculating solution, *Acta Horticultural, 82*, 121-132.

Young, P. C. (1984). *Recursive Estimation and Time Series Analysis*. Communication and Control Engineering. Springer-Verlag: Berlin.

Young, P. C. (1985). The instrumental variable method: a practical approach to identification and system parameter estimation; appears in H.A.Barker and P.C.Young (ed.). *Identification and System Parameter Estimation 1985, Vols 1 and 2*. Pergamon: Oxford, 1-16

Young, P. C. (1989). Recursive estimation, forecasting and adaptive control; appears in C.T.Leondes (ed.). *Control and Dynamic Systems*. Academic Press: San Diego, 119-166.

Young, P. C. (1991) Identification, estimation and adaptive control of glasshouse systems, to appear in IFAC Workshop on Mathematical and Control Applications in Agriculture and Horticulture Matsuyama, Japan.

Young, P. C. and S. Benner. (1986) microCAPTAIN Handbook: Version 1.0, Centre for Research on Environmental Systems, Lancaster University.

Young, P. C. and S. G. Wallis. (1986) The Aggregated Dead Zone (ADZ) model for dispersion in rivers, Proc. BHRA Int. Conf. on *Water Quality Modelling in the Inland Natural Environment*, Bournemouth, England.

Young, P. C. and S. Benner. (1989) microCAPTAIN Handbook: Version 2.0, Centre for Research on Environmental Systems, Lancaster University.

Young, P. C., M. A. Behzadi, A. Chotai and P. Davis. (1987a). The modelling and control of nutrient film systems; appears in J. A. Clark, K. Gregson and R. A. Scafell (ed.). *Computer Applications in Agricultural Environments*. Butterworth: London, pp.21-43.

Young, P. C., M. A. Behzadi, C. L. Wang and A. Chotai. (1987b). Direct digital control by input-output, state variable feedback pole assignment. *Int. Jnl. Control*. **46**, 1867-1881.

Young, P. C., M. A. Behzadi, and A. Chotai. (1988) Self tuning and self adaptive PIP control systems; appears in K. Warwick (ed.). *Implementation of Self-Tuning Controllers*. Peter Perigrinus: London, 220-259

Young, P. C., Chotai, A., and Tych, W. (1991) True Digital Control: A Unified Design Procedure for Linear Sampled Data Systems; to appear as a chapter in K.Warwick (ed.), *Advanced methods in adaptive control for industrial applications*, Springer-Verlag: Berlin.

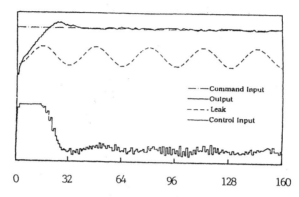

Fig. 9 Performance of fixed gain PIP design of the NFT pilot scale plant .

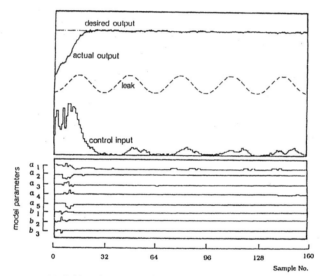

Fig. 10 Self-tuning control of solute concentration in the NFT pilot plant.

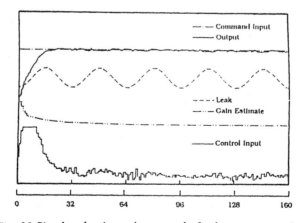

Fig. 11 Simple adaptive gain control of solute concentration in the NFT pilot plant.

TRUE DIGITAL CONTROL OF GLASSHOUSE SYSTEMS

A. Chotai*, P. C. Young*, P. Davis** and Z. S. Chalabi**

*Centre for Research on Environmental Systems, Lancaster University, Lancaster, LA1 4YQ, UK
**AFRC, Institute of Engineering Research, Wrest Park, Silsoe, Bedford, MK45 4HS, UK

Abstract The paper first outlines the Non-Minimum State-Space (NMSS) approach to Proportional-Integral-Plus (PIP) control system design, which allows for state variable feedback (SVF) control involving only the measured input and output variables of the system. It then describes two practical examples; the first example is concerned with controlling the environmental temperature in a glasshouse; and the second involves the design and implementation of a self-adaptive PIP temperature control system for a nonlinear heated bar system.

Keywords Nonlinear systems, Proportional-Integral-Plus (PIP) control, Pole assignment, Self-adaptive control, State variable feedback, Temperature control, True Digital Control (TDC).

INTRODUCTION

In this paper, we describe the True Digital Control (TDC) design philosophy (Young et al., 1987b, 1988, 1991; Chotai et al., 1991) for linear, single input, single output (SISO) systems described by discrete-time equations or the equivalent backward shift (z^{-1}) operator transfer function models. The control system design method can be considered as a direct development of the multivariable continuous-time servomechanism design procedures for continuous-time systems suggested by Young and Willems (1972). The simplicity of the implemantation is achieved by the definition of a suitable Non-Minimum State Space (NMSS) form for z^{-1} model. This NMSS model representation is quite unusual in control system terms and has certain advantages when the model is used as the basis for state variable feedback (SVF) control system design. In particular, the definition of the state vector means that the resulting control law involves only the present and past sampled values of the input and output signals, together with an *"integral of error"* state which allows for the required type one servomechanism performance.

The NMSS synthesis procedure is attractive in practical terms not only because it is easy to implement, but also because it provides a logical extension to conventional direct digital control (DDC) algorithms. In particular, the simplest mechanisation of the proposed design procedure, for a first order model with a single sample pure time delay, has the conventional proportional-Integral (PI) form; while higher order implementations retain these PI elements but introduce additional feedback terms, some of which can be interpreted as providing first and higher order derivative action. For this reason, we refer to the general design as a *Proportional-Integral-Plus* (PIP) control system. This PIP design procedure is generally applicable and is currently being applied to a wide variety of different processes. In this paper, however, we consider its potential for application to horticultural

systems (Young et al., 1987a) and, in particular, to the control of the environmental temperature in glasshouses.

Many researchers (Saffell and Marshall, 1983; Hurd and Graves, 1984; Bailey, 1985; Miller et al., 1985) have suggested control methods which can save greenhouse heating energy by reducing the temperature setpoint when heat losses are high (windy, no thermal screen), and increasing it when the losses are lower. In order to obtain appreciable energy conservation with such methods, the accuracy of temperature control must be good, so that the setpoint can be reduced without any risk that the temperature will fall below an allowed limit anywhere within the crop. The PIP design is perfectly suited to these requirements: indeed, its inherent use of state variable feedback means that the controller can satisfy the kind of optimal Linear-Quadratic-Gaussian (LQG) criteria that can be related directly to factors such as energy utilisation and conservation. Also, it is straightforward to introduce self-tuning or self-adaption into the PIP design, so ensuring that it retains this optimal operation despite any changes that may occur to the controlled system over the passage of time.

In the present paper, we discuss two practical examples. The first example is quite straightforward and concerns the well known problem of controlling the environmental temperature in a glasshouse. The second example is more unusual: it involves the design and implementation of a self-adaptive PIP temperature control system for a nonlinear heated bar system. Although nominally not related to glasshouse environmental systems, there are close similarites between the dynamic characteristics of this system and heating/ventilation systems in glasshouses. In particular, the control system involves temperature regulation via electrical heating and forced ventilation; it is a "bang-bang" system in both of these control modes; and it can subjected to severe disturbances caused by changing environmental conditions. As a result, the design of an adaptive temperature control system for the heated bar represents an admirable surrogate for the glasshouse system; one which poses the same kind of control problems that are encountered in the glasshouse context and yet allows much more freedom in experimental research and development. Techniques perfected in this low cost, experimental environment can then be evaluated and perfected in a more time and cost efficient manner on the glasshouse system itself.

THE NON-MINIMUM STATE SPACE (NMSS) MODEL

Consider the following, general, discrete-time transfer function (TF) representation of a pth order single input, single output (SISO) system :

$$y(k) = \frac{b_1 z^{P-1} + \ldots + b_P}{z^P + a_1 z^{P-1} + \ldots + a_P} u(k)$$

Note that above TF can also be written in the backward shift operator form, i.e.

$$y(k) = \frac{B(z^{-1})}{A(z^{-1})} u(k) \qquad (1)$$

Here, $u(k)$ and $y(k)$ are, respectively, the input and output of the system measured at the kth sampling instant; while $A(z^{-1})$ and $B(z^{-1})$ are polynomials in the backward shift operator z^{-1} (i.e. $z^{-i}x(k) = x(k-i)$) of the following form,

$$A(z^{-1}) = 1 + a_1 z^{-1} + \ldots + a_P z^{-P}$$
$$B(z^{-1}) = b_1 z^{-1} + \ldots + b_P z^{-P}$$

No prior assumptions are made about the nature of the transfer function $B(z^{-1})/A(z^{-1})$, which may be marginally stable, unstable, or possess non-minimum phase characteristics. However, if the input-ouput behaviour of the system is characterised by any pure time (transport) delay of τ sampling intervals, then these are accommodated by assuming that the first τ coefficients of the $B(z^{-1})$ polynomial, i.e. $b_1, b_2 \ldots b_\tau$ are all zero.

It is well known that a linear state variable feedback (SVF) control law can, in theory, arbitrarily assign the closed loop eigenvalues, provided the dynamic system is controllable (Popov, 1964, Wonham, 1967, Young and Willems, 1972). One obvious drawback of SVF system design, which has almost certainly restricted its practical applications, is the apparent need to measure all of the state variables which characterise the state vector. This requirement can be obviated by the use of a state-reconstructor (observer); or its stochastic equivalent, an optimal state estimator (Kalman filter). In the discrete-time system considered here, however, it is possible to avoid the complexity of such state reconstruction by defining a non-minimal state-space (NMSS) form, whose variables consist of the present and past values of the system input and output signals (Young et al., 1987b, 1988). The NMSS model of (1) is given by the following;

$$x(k) = Fx(k-1) + gu(k-1) + dy_d(k) \qquad (2)$$

where,

$$F = \begin{bmatrix} -a_1 & -a_2 & \ldots & -a_{P-1} & -a_P & b_2 & b_3 & \ldots & b_{P-1} & b_p & 0 \\ 1 & 0 & \ldots & 0 & 0 & 0 & 0 & \ldots & 0 & 0 & 0 \\ 0 & 1 & \ldots & 0 & 0 & 0 & 0 & \ldots & 0 & 0 & 0 \\ & & & & \cdot & & & & & & \\ 0 & 0 & \ldots & 1 & 0 & 0 & 0 & \ldots & 0 & 0 & 0 \\ 0 & 0 & \ldots & 0 & 0 & 0 & 0 & \ldots & 0 & 0 & 0 \\ 0 & 0 & \ldots & 0 & 0 & 1 & 0 & \ldots & 0 & 0 & 0 \\ 0 & 0 & \ldots & 0 & 0 & 0 & 1 & \ldots & 0 & 0 & 0 \\ \dot{0} & \dot{0} & \ldots & \dot{0} & \dot{0} & \dot{0} & \dot{0} & \ldots & \dot{1} & \dot{0} & \dot{0} \\ a_1 & a_2 & \ldots & a_{P-1} & a_P & -b_2 & -b_3 & \ldots & -b_{P-1} & -b_P & 0 \end{bmatrix}$$

$$g = [\, b_1 \ 0 \ \ldots \ 0 \ 1 \ 0 \ 0 \ \ldots \ 0 \ 0 \ -b_1 \,]^T$$

$$d = [\, 0 \ 0 \ \ldots \ 0 \ 0 \ 0 \ 0 \ \ldots \ 0 \ 0 \ 1 \,]^T$$

with the non-minimal state vector $x(k)$ defined as

$$x(k) = [y(k), y(k-1), \ldots, y(k-p+1), u(k-1), \ldots, u(k-p+1), z(k)]^T$$

where $z(k)$ is the "integral-of-error" state defined by the following,

$$z(k) = z(k-1) + \{y_d(k) - y(k)\} \qquad (3)$$

and $y_d(k)$ is the reference or command input at the kth sampling instant.

The PIP Control Algorithm

It can be shown (Young et al., 1987b, 1988; Wang, 1988) that the conditions for controllability of the NMSS model (2) are similar to those of minimal state realisations. The main consequences of defining the state in this manner are the increased order of the system and the definition of a state vector in which all of the state variables are directly available for measurement at each sampling instant. As a result, the power of state variable feedback control can be exploited to achieve the desired control objectives. The SVF control law is defined in the normal manner, i.e.

$$u(k) = - k^T x(k)$$

where, the state feedback control gain vector k is defined in the form,

$$k = [f_0 \ f_1 \ \ldots f_{P-1} \ g_1 \ \ldots g_{P-1} \ k_I]^T$$

and the closed-loop system, in the block diagram terms, is shown in Fig. 1. This reveals that the NMSS control strategy has resulted in a control system which can be related structurally to more conventional designs, such as PI and PID controllers. In this regard, note that the PI pathways defined by the SVF law and shown in Fig. 1 have certain advantages over the more conventional PI designs, where the P and I elements are normally applied in parallel to the error signal. In particular, the present design inherently avoids problems such as "integral wind-up".

From Fig. 1, the control input $u(k)$ can be written as

$$u(k) = \frac{I}{G(z^{-1})} [\, y_d(k) - y(k) \,] \ \frac{F(z^{-1})}{G(z^{-1})} y(k) \qquad (4)$$

so that equations (1) and (4) can be written as

$$\begin{bmatrix} A(z^{-1}) & -B(z^{-1}) \\ F(z^{-1}) + I(z^{-1}) & G(z^{-1}) \end{bmatrix} \begin{bmatrix} y(k) \\ u(k) \end{bmatrix} = \begin{bmatrix} 0 \\ I(z^{-1}) \end{bmatrix} y_d(k) \qquad (5)$$

The characteristic polynomial of the closed-loop system (CLCP) is then obtained by setting the determinant of the matrix on the left hand side of (5) equal to zero.

Any SVF control system design procedure can now be utilised for control system synthesis. For example, in the simplest case of pole assignment design, the desired CLCP $d(z^{-1})$ is first defined by reference to the chosed closed loop pole positions, i.e.,

$$d(z^{-1}) = 1 + d_1 z^{-1} + \ldots\ldots d_{2P} z^{-2P}$$

and the control gains are then obtained by setting the CLCP equal to $d(z^{-1})$, i.e.,

$$A(z^{-1}) G(z^{-1}) + B(z^{-1})F(z^{-1}) + B(z^{-1})I(z^{-1}) = d(z^{-1}) \quad (6)$$

Equating the coefficients for like powers of z^{-i} in (6), results in the solution of a particularly simple set of linear simltaneous equations (Young et al., 1987b, Chotai et al., 1991). Alternatively, the SVF control law can be used as a basis for the design of alternative Linear Quadratic (LQ) optimal PIP controllers (Young et al., 1991; Tych et al., 1991) which, in the infinite time case, involves the solution of the appropriate matrix Riccati equation

RELATIONSHIP BETWEEN PIP AND CONVENTIONAL STATE VARIABLE FEEDBACK SYSTEMS

The most obvious difference between the PIP system described above and conventional 'minimal' state variable feedback system design is the explicit introduction of the input and its past values into the definition of the enlarged state vector. While this is unusual, it is easy to see how the NMSS control law can be related directly to the more conventional SVF law. For example, suppose that the system is modelled in the following, more conventional canonical form,

$$z\chi(k-1) = \chi(k) = A\chi(k-1) + bu(k-1) + hy_d(k) \quad (7)$$

in which,

$$A = \begin{bmatrix} -a_1 & 1 & 0 & \ldots & 0 & 0 \\ -a_2 & 0 & 1 & \ldots & 0 & 0 \\ \vdots & \vdots & \vdots & & \vdots & \vdots \\ -a_{P-1} & 0 & 0 & \ldots & 1 & 0 \\ -a_P & 0 & 0 & \ldots & 0 & 0 \\ a_1 & -1 & 0 & \ldots & 0 & 1 \end{bmatrix}$$

$$b = [b_1 \quad b_2 \quad b_3 \ldots\ldots b_P \quad -b_1]^T$$
$$h = [0 \quad 0 \quad 0 \ldots\ldots 0 \quad 1]^T$$

This p+1 dimensional state vector $\chi(k)$ is then related to the 2p dimensional NMSS vector x by the equation,

$$\chi = [P \quad Q] x$$

with

$$P = \begin{bmatrix} 1 & 0 & \ldots & 0 & 0 \\ 0 & -a_2 & \ldots & -a_{P-1} & -a_P \\ \cdot & \cdot & \cdot & \cdot & \cdot \\ 0 & -a_P & \ldots & 0 & 0 \\ 0 & 0 & \ldots & 0 & 0 \end{bmatrix} \quad Q = \begin{bmatrix} 0 & 0 & \ldots & 0 & 0 \\ 0 & b_2 & \ldots & b_P & 0 \\ \cdot & \cdot & \cdot & \cdot & \cdot \\ 0 & b_P & \ldots & 0 & 0 \\ 0 & 0 & \ldots & 0 & 1 \end{bmatrix}$$

It is clear, therefore, that the SVF control law in this case involves the output and input and their past values up to the appropriate order, as in NMSS PIP control system; but that the NMSS design provides both more flexibility in design and avoids the need for state reconstruction system (observer) design.

EXAMPLE 1: TEMPERATURE CONTROL IN A GLASSHOUSE SYSTEM

In this example, we consider a simple implementation of the PIP system for a glasshouse temperature control system. The PIP design method is used because of its ability to handle a wide range of processes, including those with large time delays. Fig. 2 shows some open loop temperature responses to valve adjustments in a single glass greenhouse. The details of the modelling analysis for this system is given in Davis and Hooper (1991). For a sampling interval of 10 minutes, the overall discrete-time model between valve aperture and internal air temperature is first order, with a one sample period pure time delay, i.e.,

$$T(k) = \frac{1.22z^{-1}}{1 - 0.922z^{-1}} u(k) + \xi(k) \quad (8)$$

where $T(k)$ is the internal air temperature in °C, $u(k)$ is the control input (the fractional valve aperture of the heating system) and $\xi(k)$ is a term to account for the noise in the system. From equation (8), we see that the natural (open loop) system is first order, with a time constant of 123 minutes, steady state gain of 15.6, and pure time delay of 10 minutes.

The sampling interval of 10 minutes was selected following model identification studies carried out both at Lancaster and at the AFRC Institute of Engineering Research, Silsoe, England (Davis, 1986, Davis and Hooper, 1991). In contrast to the situation at shorter sampling intervals, most of the changes in measured internal air temperature between samples were statistically significant. Moreover, this sample rate yields data-based models with the best defined parameter estimates and allows, most conveniently, for the dead-time characteristics of the process.

For the system in equation (8), the PIP design reduces to a PI design. In contrast to the conventional PI controller used previously in the greenhouse, however, the PIP design has numerous advantages: its PI structure exploits the power of SVF; it is based directly on the discrete-time model of the heating system and not on the digitisation of a continuous-time design; and the vagueries of manual tuning are replaced by objective, model-based synthesis.

The NMSS model of (8) in noise free case is given by

$$\begin{bmatrix} T(k) \\ z(k) \end{bmatrix} = \begin{bmatrix} -0.926 & 0 \\ 0.926 & 1 \end{bmatrix} \begin{bmatrix} T(k-1) \\ z(k-1) \end{bmatrix} + \begin{bmatrix} 1.22 \\ -1.22 \end{bmatrix} u(k-1) + \begin{bmatrix} 0 \\ 1 \end{bmatrix} y_d(k)$$

and the SVF control law takes the form,

$$u(k) = -f_0 T(k) - k_I z(k)$$

where is f_0 the proportional gain and k_I is the integral gain.

As a result, the closed loop TF in the noise free case is

$$T(k) = \frac{1.22 k_I z^{-1}}{1 + (1.22 f_0 + 1.22 k_I - 1.926)z^{-1} + (0.926 - 1.22 f_0)z^{-2}} y_d(k)$$

43

If the desired closed-loop polynomial is defined as,

$$d(z^{-1}) = 1 - 0.5z^{-1}$$

then the system closed loop response will be dominated by a first order mode with unity steady state gain to input commands $y_d(k)$, and a time constant of 14.4 minutes, which can be contrasted with the open loop dynamics. The resultant control gains are computed straightforwardly as $f_0 = 0.759$ and $k_I = 0.41$.

Unlike the PIP system, the conventional controller used in the glasshouse is a digitised continuous-time PI control system operating on the much smaller sampling interval of one minute. It is typical of commercially available controllers (although some controllers make adjustments even more frequently; see Sanger, 1985) and, while it was probably not tuned optimally, the control parameters were set at what were considered to be reasonable values. The performance of the two controllers is compared in Fig. 3, which shows them responding to similar, but not identical, set point changes over a period of some 55 hours. The failure of both systems to maintain the set points during parts of the day-time is due to the nature of the system: only heating action is available so, during the day when natural temperatures rise because of solar irradiance, the controllers provide their minimum zero input most of the time. However, at other times, and particularly during the cold night hours, the controllers are fully operative. It is here that we see the superiority of the PIP control system, which quickly compensates for the disturbances and maintains the desired temperature level throughout the night.

In contrast, the conventional PI system, despite its more rapid sampling frequency, is quite sluggish and oscillatory in its response to both set point changes and disturbances. This apparently paradoxical behaviour is quite easy to understand. First, the rapid sampling rate makes the system more sensitive to the time-delays in the system and it is well known that conventional, continuous-time PI controller leads to oscillatory behaviour when significant time delays are present. Second, the PIP pole assignment design in this case is intended to provide rapid and critically damped response and achieves this because of its model-based design and SVF control law.

For some insulated greenhouses, it has been found (Davis and Hooper, 1991) that still better and more robust PIP control is obtained, if heating pipe temperatures are included in the PIP feedback design. It is straightforward to extend the PIP design procedure to accomodate this additional measured variable: the overall system is then of higher order and control law simply incorporates present and past values of the heating pipe temperature, in addition to the air temperature measurements. Fig. 4 shows the response of such a system and more details are given in Davis and Hooper (1991).

EXAMPLE 2: SELF-ADAPTIVE PIP CONTROL OF A NON LINEAR HEATED BAR SYSTEM

The PIP approach described above can also form the basis for the design of controllers for certain nonlinear systems characterised by piecewise linear or bilinear behaviour. This is demonstrated by recent research at Lancaster on the adaptive temperature control of a nonlinear heated bar system in which the bar is heated by an electrical element and cooled by a small fan. Although apparently disimilar, this system does, in fact, provide a reasonable test-bed for greenhouse temperature control system designs. In particular, the non-linear heater-fan system can be considered analogous to a small scale heating-ventilation system, with behavioral

similarities to greenhouse environmental control systems but operating on a much faster and experimentally convenient time-scale. This similarity means that the self-adaptive PIP control approach proposed below could form the basis for greenhouse temperature control system design, particularly where only "bang-bang" type of control operation is possible.

In this self-adaptive control (SAC) system, an online PIP pole assignment control algorithm is utilised to implement pulse-width modulation (PWM) of the "bang-bang" control signals to either the electric heater or cooling fan, which provide the mechanism for controlling the bar temperature. Adaption is based on a novel recursive time variable parameter (TVP) estimation algorithm capable of tracking the extremely rapid variation of model parameters that occur when the control signal changes sign (see, Chotai et al, 1991 and Behzadi, 1989).

This rather novel SAC system is based on a earlier adaptive system for an airborne vehicle (Young, 1980). It is a computationally efficient solution, in the sense that a single recursive estimator is able to track the changes in the parameters of what are, in effect, the two, quite different dynamic systems: namely, the bar temperature when controlled rather sluggishly by the heater, and the same temperature controlled much more rapidly by the fan. Clearly a SAC system based on two separate recursive estimation algorithms would provide an alternative but computationally somewhat slower and more complex solution.

A block diagram of the overall SAC heater-fan system is shown in Fig. 5. Certain details of the design that are specific to the heated bar example should first be noted. Most importantly, the first order models for the heater and fan sub-systems are different in structural terms: they are defined at different sampling rates; and they have different order numerators and time delay characteristics. In particular, the heater sub-system model is defined for a sampling interval of 10 secs., with a pure time delay of two sampling intervals (20 secs), and two numerator coefficients b_2 and b_3 ; while, for the fan sub-system model, the sampling interval is 3 secs., the time delay is only one sampling interval (3 secs.), and there is only one numerator polynomial coefficient b'_1 .

Since the time constants of both system models remain reasonably constant, they have been assumed stationary for the purposes of the present design (although this is not essential and could be assumed otherwise in a more general situation; see Behzadi, 1989, Young et al., 1988). In other words, the denominator polynomial coefficients a_1 and a'_1 in each of the sub-system models are frozen at their initial, off-line, estimated values and only the rapidly time variable numerator polynomial terms (i.e. the b_2 and b_3 parameters for the heater system and b'_1 for the fan system) are recursively updated in real-time.

Fig. 6 shows a typical run using the SAC system based on the single recursive estimator: here, the system is responding to a "staircase" type input, chosen to move the bar temperature over a fairly wide range. In this manner, the adaptive action is required not only to allow for the rapidly changing dynamics associated with reversals in the control action, but also to compensate for the longer term, slower changes in the system dynamics which arise either because of the increased heat losses at higher bar temperatures, or the lower cooling effect of the fan at lower temperatures.

The plot of the control input signal switching above and below the zero level indicates which one of the SAC systems is operational at a particular instant. This can also be concluded from the movements of the recursive parameter estimates in part (a) of Figure 6, where the estimates of the sub-system parameters are shown. The denominator parameters for both heating and cooling models (i.e. a_1 and a'_1) are frozen at their off-line specified values, -0.933 and -0.913, as discussed previously. The on-line variations of the first numerator (i.e. the heater system (positive) b_2 parameter and the fan system (negative) b'_1 parameter, are shown together in part (b). While the rapid switching between these parameters indicates changes in the operational mode of the overall system from the heater to fan and vice-versa, the obvious longer term envelopes illustrate the variations of the individual parameters (i.e. b_2 and b'_1) with bar temperature. Note how the b'_1 parameter estimates start with small values at low temperatures, but are significantly larger (i.e. more negative) at higher temperatures as the heat losses increase.

Since the fan SAC model only includes the single numerator parameter, only the variations of the second heater system numerator parameter (i.e. b_3) are plotted in part (c) of Figure 6. A similar pattern of long-term variation to those observed for b'_1 occurs again, with the initially larger estimates of this parameter at lower temperatures gradually decreasing in value as the bar temperature increases (although some transient increases can also be observed).

These results do not demonstrate the performance of the control system in the face load disturbances or input signal saturation. In both cases, however, the system performs well: the inherent integral action ensures that the effects of constant load disturbances are rejected; and the PIP control law (Young et al., 1987, Behzadi, 1989) avoids integral wind-up when large command inputs lead to input signal saturation. Further details of this heated bar control system are given in Behzadi (1989) and Chotai et al (1991), including those obtained with the more complex SAC implementation involving two separate recursive algorithms, one for each sub-system. However, this involved greater complexity and yielded no improvement in system performance.

CONCLUSIONS

In this paper, we have introduced the concept of True Digital Control (TDC) and outlined the main aspects of a TDC design procedure based on the definition of a novel, Non-Minimum State Space (NMSS) model. The practical potential of the resulting Proportional-Integral-Plus (PIP) control system for application in glasshouse horticulture has been illustrated by two temperature control examples which demonstrate that, despite their simplicity, the PIP controllers have many advantages over conventional analog or digital PI and PID controllers.

The NMSS design procedure and the PIP control system are not restricted to the simple systems described in the present paper. Their state-space formulation means that they can take advantage of all the theory developed previously for minimal state space models and develop this to advantage within the more flexible non-minimal context. Thus, in addition to the simple, discrete-time, SISO system design procedures discussed in the present paper, it is straightforward to develop non-minimal PIP designs for rapidly sampled, near continuous-time systems described by delta (δ) operator models (Young et al, 1991), as well as multivariable (MIMO) and stochastic systems (Chotai et al, 1990). Research on all these topics is continuing at Lancaster and some of this is reported in other papers at this Conference (Young, 1991; Chotai and Young, 1991).

REFERENCES

Bailey, B. J. (1985) Wind dependent control of greenhouse temperature, *Acta Horticulturae*, 174, 381-386.

Behzadi, M.A., (1989) True digital control of greenhouse systems, PhD Thesis, Univ. of Lancaster.

Chotai, A., P.C.Young, and W.Tych. (1990) True digital control of multivariable systems by input/output state variable feedback", Report No.TR86/(1990), CRES, University of Lancaster

Chotai, A., P. C. Young and M. A. Behzadi. (1991). The self-adaptive design of a nonlinear temperature control system, in the Special Issue on *Self Tuning Control*, *Proc. I.E.E.*, Pt.D., vol. 138,41-49.

Chotai, A., and P.C. Young. (1991) Self-adaptive and Self-tuning Control of a Nutrient Film Technique (NFT) System, to appear in IFAC Workshop on Mathematical and Control Applications in Agriculture and Horticulture Matsuyama, Japan.

Davis, P. F. (1986) Pers. Comm. relating to confid. paper by Davis and Hooper, Bedfordshire, England.

Davis, P. F. and A.W. Hooper (1991) Improvement of greenhouse heating control, to appear in *Proc. I.E.E.*, Pt.D.

Hurd, R. D. and C. J. Graves (1984) The influence of different temperature patterns having the same integral on the earliness and yield of tomatoes, *Acta Horticulturae*, 148, 547-554.

Miller, W. B., R. W. Langhans, and L. D. Albright (1985) Plant growth under averaged day/night temperatures, *Acta Horticulturae*, 174, 313-320.

Popov, V.M., Hyperstability and optimality of automatic systems with several control functions, *Rev. Roum. Sci. Tech., Ser. Electrotech. Energ.* 1964, **9**, (4), pp. 629-690.

Saffell, R. A., and B. Marshall (1983) Computer control of air temperature in a glasshouse, *Jnl. of Agricultural Engineering Research*, 28, 469-478.

Tych, W., A. Chotai, and P. C. Young (1991) Computer aided design package for true digital control (TDC) systems, *I.E.E. Conference Publication No. 332: 3rd Int. Conf.'Control 91'*, pp 288-293.

Wang, C. L. (1988) New methods for the direct digital control of discrete-time systems, PhD Thesis, Univ. of Lancaster.

Wonham, W. M., On pole-assignment of multi-input controllable linear systems, *IEEE Trans.*, 1967 , **AC-12**, pp. 660-665.

Young, P. C. (1991) Identification, estimation and adaptive control of glasshouse systems, to appear in IFAC Workshop on Mathematical and Control Applications in Agriculture and Horticulture Matsuyama, Japan.

Young, P. C., M. A. Behzadi, A. Chotai and P. Davis. (1987a). The modelling and control of nutrient film systems; appears in J. A. Clark, K. Gregson and R. A. Scafell (ed.). *Computer Applications in Agricultural Environments*. Butterworth: London, pp.21-43.

Young, P. C., M. A. Behzadi, C. L. Wang and A. Chotai. (1987b). Direct digital control by input-output, state variable feedback pole assignment. *Int. Jnl. Control*. 46, 1867-1881.

Young, P. C. (1980). A second generation adaptive autostabilisation system for airborne vehicles. *Automatica*. **17**: 459-469.

Young, P. C., Behzadi, M.A., and Chotai, A. (1988) Self tuning and self adaptive PIP control systems; appears in K. Warwick (ed.). *Implementation of Self-Tuning Controllers*. Peter Perigrinus: London, 220-259

Young, P. C., Chotai, A., and Tych, W. (1991) True Digital Control: A Unified Design Procedure for Linear Sampled Data Systems; to appear as a chapter in K.Warwick (ed.), *Advanced methods in adaptive control for industrial applications*, Springer-Verlag: Berlin.

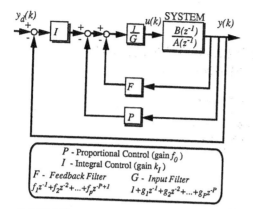

$$y_d(k)$$

SYSTEM

$$u(k) \quad \frac{B(z^{-1})}{A(z^{-1})} \quad y(k)$$

P - Proportional Control (gain f_0)
I - Integral Control (gain k_I)
F - Feedback Filter $\qquad G$ - Input Filter
$f_1z^{-1}+f_2z^{-2}+...+f_pz^{-P+1} \qquad 1+g_1z^{-1}+g_2z^{-2}+...+g_pz^{-P}$

Fig.1 Block Diagram Representation of PIP Servomechanism Control System

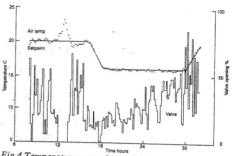

Fig.4 Temperature control in a Melinex lined glasshouse using air and pipe temperatures in the feedback.

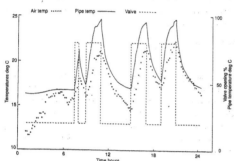

Fig.2 Open loop temperature responses of a single glass greenhouse

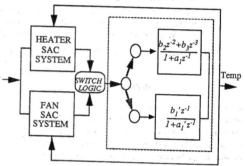

$$\frac{b_2z^{-2}+b_3z^{-3}}{1+a_1z^{-1}}$$

$$\frac{b_1'z^{-1}}{1+a_1'z^{-1}}$$

Temp

Fig.5 Block diagram of the Heater/Fan self-adaptive control (SAC) system

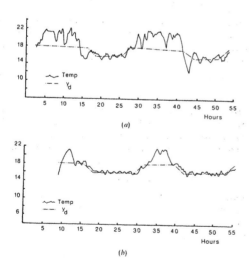

(a)

(b)

Fig.3 Temperature control in a glasshouse: comparison of PIP and conventional PI control systems: (a) conventional PI control; (b) PIP control.

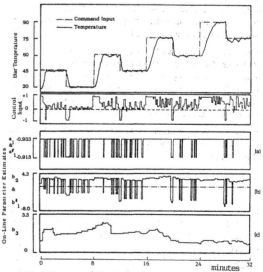

Fig.6 Heater and Fan self-adaptive control (SAC) system design performance.

SIMULATION MODEL TO ESTIMATE FRESH
MASS OF A PLANT

H. Shimaji* and A. Kano**

*Dept. of Protected Cultivation, National Research Institute of Vegetables,
Ornamental Plants and Tea, Aichi, 470-23, Japan
**Faculty of Agriculture, Shizuoka University, Shizuoka, 422, Japan*

Abstract. A simulation model of plant growth, including changes in fresh mass, leaf area, and shoot elongation, was developed. The model is based on the tussusception growth theory of plant cells. To express realistic behavior of general plant growth, submodels of transpiration, photosynthesis, and carbohydrate translocation were added to the basic model. To verify behavior of the model, growth rate of rose shoots was measured and compared to the simulation outputs. The model reasonably estimated the shoot elongation. The model can be used for studies in plant science as well as greenhouse control systems to maximize yield and quality of the products.

Keywords. Simulation; Models; Plant growth; Phsiological models; Agriculture; Biology.

INTRODUCTION

A number of simulation models as well as statistical models to estimate crop yield or to analyze the growth of plants have been developed in the last thirty years (Thornley, 1976). Many of them are photosynthesis and dry mass production models. They are suitable for grain crop growth simulation but not for vegetable or flower crop production, in which fresh mass, rather than dry mass, is important to the yield and quality of the crop.

Because of recent greenhouse control technology, most environmental factors can be controlled. For efficient environmental control of greenhouses, where vegetables and flowers are major crops to be cultivated, it is required to know the effects of environment on plant growth, especially fresh mass change. Therefore, we developed a model for simulating change in plant fresh mass, using environmental conditions as inputs.

This growth model is based on the tussusception growth theory for plant cells (Ray, 1962). The theory was assumed to be applicable to larger tissues and organs to build a whole plant growth model and was modified to express realistic growth behavior. Submodels of transpiration, photosynthesis, and carbohydrate allocation (translocation) were added to the basic model.

To test the model's behavior against actual plant growth, the simulated results were compared with growth of a young shoot of a rose plant in a greenhouse.

STRUCTURE OF THE MODEL

It is known that elongation of plant cells is induced by the irreversible, plastic extension of the cell wall. Its driving force is positive internal pressure of the cells or turgor. Regarding cell expansion, Lockhart (1965) suggested Eq. (1), showing relative cell expansion rate was proportional to effective turgor (P-Y). However, this equation shows only instantaneous rate of expansion, not continuous cell expansion.

$$1/V \cdot dV/dt = m(P-Y) , \tag{1}$$

where V is volume of the cell, P is turgor potential and Y is a threshold value.

In this study, we used a theory which assumes that the following process occurs during cell growth (intussusception growth theory). First, the cell wall is extended by the effective turgor and then the cell wall is relaxed because of its plastic extension. Second, photosynthate fills the loosened cell wall, making further cell wall extension possible. These two steps are repeated continuously as long as the cell is extensive.

To express this process in a mathematical model, virtual thickness of the cell wall, h, was introduced. Thickness before the extension is h0, which reduces to h1 after the cell wall is extended due to the effective turgor, Pe (Eq. (2)).

$$h1 = h0(1-k \cdot Pe) , \tag{2}$$

where k is an extension coefficient that is inversely proportional to h. Cell wall synthesis defined by the above theory depends on the photosynthate pool and a coefficient. Considering cell wall decomposition, Eqs. (3) and (4) shows this relationship.

$$fg = kg \cdot C - h/Td , \tag{3}$$

$$\text{if } h>h0, \ h0 = h , \tag{4}$$

where fg is rate of change in cell wall thickness, C is the photosynthate pool concentration, kg is a function of temperature, and Td is the time constant for the decomposition. If we set Δh=h-h1, plastic extension of cell wall can occur when Δh>0. Therefore, rate of decrease in cell wall thickness can be expressed by Eq. (5).

$$fe = \Delta h/Te , \quad \text{and if } \Delta h<0, \ fe = 0 , \tag{5}$$

where Te is a time constant for cell expansion. From Eqs. (3) and (5), rate of change in cell wall thickness caused by the photosynthate packing into the cell wall and the plastic expansion, dh/dt, is given by Eq. (6).

$$dh/dt = fg - fe , \tag{6}$$

where t is time.

Generally cell elongation does not occur evenly in all directions. For example, in the case of a leaf, cell enlargement in the direction of expanding leaf area is much greater than that for increasing its thickness, and in stems and roots, axial elongation is dominant.

Since increase in fresh mass is considered to accompany cell expansion, relative rate of fresh mass change can be expressed as in Eq. (7).

$$1/W \cdot dW/dt = 1/h \cdot fe \ , \tag{7}$$

where W, fresh mass, is the capacity of the plant corrected to the value when there is no water stress due to transpiration.

To simplify the model, we assumed that growth (cell expansion) occurs two dimensionally in leaves and one dimensionally in stems and roots. Therefore, the thickness of the leaf was alternatively expressed as structural dry mass per unit leaf area, Dl. Consequently, the relative rate of increase of leaf area, A, is expressed in Eq. (8).

$$1/A \cdot dA/dt = 1/Dl \cdot fe \tag{8}$$

In the stem, h is defined as the structural dry mass of the stem, Ds, and, therefore, relative rate of increase in stem length, Ls, is expressed as follows.

$$1/Ls \cdot dLs/dt = 1/Ds \cdot fe \tag{9}$$

From these equations, this intussusception growth model uses effective turgor pressure, Pe, and photosynthate pool concentration (sucrose concentration), C, as boundary conditions and outputs state variables of the plant, such as leaf area, A, and stem length, Ls.

To construct a growth model of an entire plant, organs, such as leaves, stems, and roots, are connected through xylem and phloem systems as shown in Fig. 2.

Transpiration rate, Tr, is expressed by Eq. (10).

$$Tr = (Cs(Tl)-Ca)/(ra+rl) \ , \tag{10}$$

where Cs is vapor density in stomatal cavity, taken as saturation vapor density at leaf temperature (Tl), Ca is vapor density in the ambient air, and, ra and rl are aerodynamic resistance and leaf resistance to vapor defusion, respectively. The water potential of plant organs, ψ can be calculated from the relations between the transpiration rate and resistance to water flow at different parts of the plant. Effective turgor pressure of each part of the plant, Pe, was simply defined as the difference between ψ and a threshold value, Y.

An equation concerning photosynthate pool concentration, C, can be obtained from the balance of incoming synthates (photosynthesis, Pg) and outgoing synthates (organ synthesis, fg, and translocation,

ft) as shown in Eq. (11).

$$Vpool \cdot dC/dt = Pg - fg - ft \ , \tag{11}$$

where Vpool is the pool capacity. Photosynthesis rate, Pg, is expressed as a hyperbolic function of irradiance, leaf temperature, and CO_2 concentration in the stomatal cavity. In addition to dark respiration, carbohydrate consumption for organic synthesis (growth respiration) was considered to be included in fg.

Pressure flow theory was introduced into the model to explain translocation of photosynthetic products. In the model, it is assumed that an amount of photosynthate (sucrose) is loaded into the phloem system at a rate proportional to the pool concentration (with a rate constant, Kload), accompanied by respiration.

Sucrose is also carried into the phloem with the pressure flow or volume flow with water. If we assume a non-ideal semipermeable membrane between plant organs and the phloem, rate of sap flow into, out of, and through the phloem system, Jv, is expressed by Eq. (12).

$$Jv = Lp(\Delta P - s \Delta \pi) \tag{12}$$

where Lp is hydraulic conductivity, s is reflection coeffcient, and

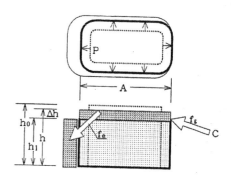

Fig. 1. Basic model of cell expansion.
See text for parameters.

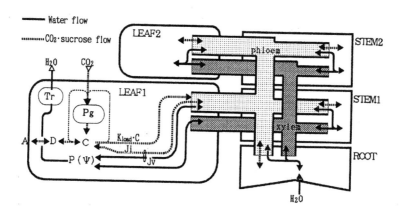

Fig. 2. Relations among submodels of transpiration (Tr), photosynthesis (Pg), and translocation. See text for parameters.

ΔP and $\Delta \pi$ are turgor and effective osmotic potential difference between sites, respectively.Rate of sucrose transport due to the volume flow, Ji, is calculated by Eq. (13).

$$Ji = (1-s)Cave \cdot Jv , \qquad (13)$$

where Cave is the average sucrose concentration between sites. Therefore, total sucrose into or out of the organ, ft, is expressed in Eq. (14).

$$ft = Kload \cdot C + Ji \qquad (14)$$

In non-photosynthetic organs, such as shoots and roots, the same structure of sucrose pools was applied in the model, except the photosynthesis rate, Pg, was set to zero. Pressure increase because of volume flow from the source organs to the phloem is transmitted through the network, and at non-photosynthetic organs (sink organs), according to Eq. (12), volume flow from phloem occurs because of the low pool concentration. In this model, it was assumed that only the pressure gradient within the phloem system causes the mass flow of sucrose; in other words, s=0 in Eq. (12) was assumed. Relative concentration among different organs determines whether the organs are sinks or sources.

In summary, this model was developed to analyze effects of environmental conditions, such as irradiance, temperature, humidity, CO_2 concentration, and soil moisture, on plant dry and fresh mass, systematically based on internal water status and photosynthate accumulation.

PHYSICAL EXPERIMENT

For the verification of the model, rose plants (cultivars Madam Violet and Golden Emblem) were grown in a greenhouse located in Shizuoka, Japan. Shoot elongation rate was measured on three to five rapidly growing shoots on July 3 and 16, 1990. Length of each shoot was measured by hand using a caliper on an hourly basis.

Environmental parameters, such as air temperature, relative humidity, solar irradiance, and soil water potential were measured and recorded every 10 minutes.

RESULTS AND DISCUSSION

Figure 3 shows basic behavior of the model under different water potentials, simulating shoot growth under solutions with different osmotic potentials. In this example, threshold value to obtain effective turgor pressure was set to -8 bar, and photosynthate pool was set to be constant. For the purpose of comparison, coleoptile elongation of rice plants under solutions of different osmotic potentials is also shown (Wada, 1969). The model reasonably simulated the coleoptile growth. When water potential decreased, the growth halted for several minutes. This is because under such conditions, h1 becomes larger and Δh remains negative during the period (see Eqs. (5) and (6)). Afterward, the elongation rate depends on the water potential, Ψ. If Ψ is less than the threshold, Y, the elongation ceases, but thickness of cell wall irreversibly increases (Eq. (4)). When Ψ recovers to zero, according to Eq. (2), h1 decreases, and Δh increases; therefore, the growth rate increases. However, the growth rate then gradually decreases until fg and fe balance out. In Lockhart's model, growth rate is solely dependent on water status of the plant, but our model showed that cell wall synthesis rate, fg, also had a large effect on cell expansion. Since fg is expressed as a function of temperature and photosynthate pool concentration, this model indicates that growth rate is influenced by plant temperature and photosynthesis, as well as water status.

Results of a simulation using the model described in Fig. 2 are shown in Fig. 4, in which temperature, irradiance, and soil water

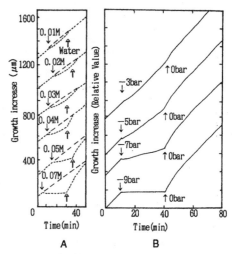

Fig. 3. Simulation outputs from the intussusception growth model (B) and results of rice coleoptile growth (A), after Wada(1969).

potential were changed during simulation. On Day 1 (0-24h), a clear day was assumed with soil water potential at 0 bar. During the daytime, because transpiration caused leaf water potential to drop below the threshold value, no increase in leaf area and stem length were seen. When irradiance became low in the afternoon, leaf area and stem length started increasing, but, later, their rates decreased due to consumption of photosynthates.

On Day 2 (24-48h), irradiance was set to simulate a cloudy day. Under such low light, water potential during the daytime did not decrease, so that growth started at the time when photosynthates accumulated up to a certain level of the pool concentration. Stem growth started several hours after leaf expansion. This lag occurred because in the stem it took longer for the pool concentration to reach a level sufficiently high enough to generate elongation. On the other hand, root length constantly increased because h1 was kept relatively low at the roots, where water potential did not decrease as much as in leaves and stems. The pool concentration in the root was also low, so that the roots were a sink for sucrose (see Eq. (12)).

On Day 3 (48-72h), temperature was lowered by 10°C. The cell wall synthesis rate, fg, was low, so that general growth was also low. However, the decrease in photosynthesis rate was not as obvious as the decrease in growth rate.

Soil water potential was lowered by 3 bars on Day 4 (72-96h). As a consequence, the leaf area did not increase as much as it had on Day 2 because of low leaf water potential in leaf.

We tested the model's validity against rose shoot growth, which has a very high elongation rate, typically 30 to 150 mm/day. Measured and simulated results of the experiment are shown in Fig. 5. On the rainy day, July 3, the maximum irradiance was lower than 200W/m² and relative humidity was near 100 % throughout the day. The degree of water stress on such a day is small and shoot elongation rate is relatively high. Shoot A showed its growth according to the cumulative irradiance and was similar to the simulated growth. Growth of shoot B was relatively constant, showing a possibility that photosynthate accumulation was high due to previous translocation.

On a clear day, July 16, there was a rise in air temperature and a drop in relative humidity, creating a condition under which water deficit in the plant occurred. Elongation of shoot A started increasing after 12:00; on the other hand, shoot B showed its growth prevented during the daytime and started elongation after sunset. It is not certain what caused this difference in growth pattern, but difference in turgor and sucrose concentration in the shoots are possibilities. In any event, the model simulated reasonable relative behavior of rose shoot under the given environment.

In terms of plant growth or fresh mass change, it is considered that the rate is regulated by the hydraulic condition in the expanding tissues, if there is enough carbohydrate. Similarly, the model suggested that growth would be prevented if photosynthate accumulation is below the level at which the cell wall could not be thick enough to extend. Although this was not shown by our experiment, as commonly seen in nature, plant growth would be restricted even under high water potential if the plant is starved for carbohydrates. The model can explain the so-call "spindling growth", which occurs under high nighttime temperature as well, because the intussusception of carbohydrate is expressed as a function of temperature (and other variables) in Eq. (3). However, other factors not considered in this model, such as cell division rate and hormone effects, can affect growth and need to be considered in the future studies.

CONCLUSIONS

Using the model developed by this study, plant fresh mass, leaf area, and shoot elongation can be simulated. The model also includes a submodel to estimate change in dry mass of the plant. Therefore, it can be used to analyze environmental effects (irradiance, air temperature, humidity, CO_2 concentration, and soil water potential) on plant growth by summarizing internal parameters, such as plant temperature, water potential, and carbohydrate accumulation. Also since the model directly outputs leaf area, it can be extended to a larger model for analyzing plant or tree shape. Environmental control systems can utilize the model to control greenhouse environment according to dry mass / fresh mass ratio, which is highly correlated with the quality of greenhouse products.

REFERENCES

Lockhart, J.A. (1965). An analysis of irreversible plant cell elongation. *J. Theor. Biol.* 8:264-275.

Ray, P.M. (1962). Cell wall synthesis and cell wall elongation in oat coleoptile tissue. *Am. J. Bot.* 49:928-939.

Thornley, J.H.M., (1976). *Mathematical models in plant physiology*, Academic Press, London, pp.318.

Wada, S. (1967) In M.Furuya(Ed.) *Shokubutsu seirigaku(Plant physiology)*, Vol.7.Asakura shoten, Tokyo, 214-219.

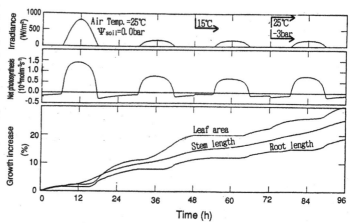

Fig. 4. Results of a simulation for four continuous days with different environmental conditions.

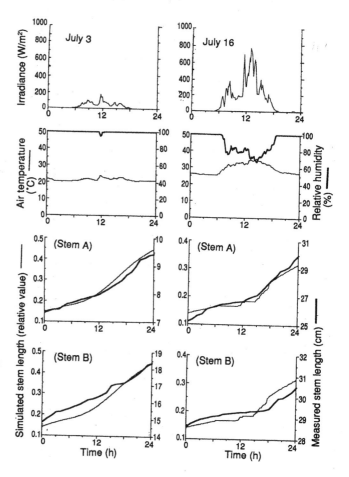

Fig.5. Measured and simulated rose shoot growth under two different environmental conditions on July 3 and July 16.

PERFORMANCE OF ZONE COOLING SYSTEM IN GREENHOUSE: AN EVALUATION OF EXPERIMENTAL FACTORS

K. Kojima, T. Matsuoka and H. Suhardiyanto*

Faculty of Agriculture, Kochi University, Nankoku, Kochi 783, Japan

Abstract. In the previous study, a zone cooling system had been tested to reduce air temperature inside a model-sized greenhouse during the daytime in summer. Cool air was provided by a spot cooler, and blown into the lower layer in the greenhouse. In the present study, the effect of experimental factors was evaluated quantitatively. Experimental data were collected during 24 days in summers of 1989 and 1990. Data on 8 days in summer of 1989 were taken from the published report of the previous study. Results showed that under cheesecloth shading in wet soil condition, the cooling system was capable of reducing the maximum air temperature at the lower layer to about $32 \sim 35$ °C compared to about $33 \sim 38$ °C at the upper layer in the greenhouse. The distribution and change of dimensionless temperature inside the greenhouse was presented for typical maximum values. Using the method of variance analysis, it was proved how the cooling performance of the system is affected by ventilation rate, soil wetness and shading application.

Keywords. Environment control in greenhouse; temperature control; cooling; ventilation; variance analysis.

INTRODUCTION

Many greenhouse managers left their greenhouses vacant in summer for some difficulties in reducing air temperature inside the greenhouse during the daytime. In summer nighttime, reducing inside air temperature had been succesfully done using heat pump system (Hayashi, 1983; Kozai and colleagues, 1985) whereas reducing temperature inside the greenhouse during a clear day in summer using reasonable energy remains unsolved. In addition, it is very impractical to cool the entire greenhouse air in summer daytime using refrigeration system as it will require too large amount of energy (Hanan, Holley and Goldsberry, 1978).

In the previous study (Kojima and Suhardiyanto, 1991), a zone cooling system rather than cooling the entire greenhouse air, using cool air blowing and opening the ridge window, had been tested in a model-sized greenhouse. In their experiment, the maximum air temperature at the lower layer in the greenhouse in wet soil condition could be maintained at about 35 °C. They used the steady state heat balance to describe heat fluxes in their greenhouse. However, the effect of experimental factors in their experiment had not been evaluated quantitatively. Evaluation of experimental factors is essential in further development of the cooling system.

The objective of the present study was to evaluate the effect of experimental factors on the performance of the zone cooling system quantitatively. Experimental factors to be evaluated were ventilation rate, soil wetness and shading application.

----
* Permanent address : Department of Agricultural Engineering, Bogor Agricultural University, Kampus Darmaga, PO Box 122, Bogor, Indonesia.

MATERIALS AND METHODS

Greenhouse and the Zone Cooling System

The experiments were carried out in a 7.2 m^2 model-sized greenhouse (Fig. 1), located at the Faculty of Agriculture, Kochi University, Japan. The greenhouse has standard-peak structure with metal frame work and vinyl chloride cover (thickness : 3 mm). The cover area of the experimental greenhouse was 14.3 m^2, and the air volume was 6.3 m^3. The height of the ridge was 110 cm. The soil in the experimental greenhouse was tilled to the depth of approximately 20 cm, and left uncultivated and bare.

Cool air was provided by a spot cooler (0.97 kW), and blown into the lower layer in the greenhouse from the centre of one wall end of the greenhouse at about 25 cm high. Air temperature blown into the greenhouse varied since temperature of outside air as input air into the cooler varied during the daytime. The warmer air was discharged from the greenhouse through ridge windows which were opened throughout the experiment.

Temperature and Heat Flux in Greenhouse

In this study, temperature distribution and characteristics of heat flux in the greenhouse were evaluated to describe the performance of the zone cooling system. The greenhouse air was considered to be divided into three horizontal layers, namely lower layer, middle layer, and upper layer. In this approach, the assumption was made that air temperature at horizontal layers inside the greenhouse and outside air temperature are uniform, and can be represented by the average value of the relevant features.

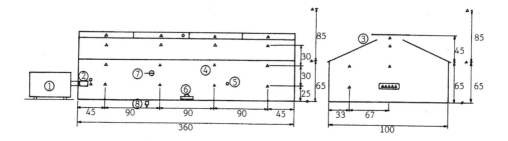

Fig. 1. Schematic diagram of the model-sized greenhouse
(dimensions are in cm.) and arrangement of sensors
during experiment in summer 1990.

1 : spot cooler
2 : inlet
3 : outlet (ridge window)
4 : thermocouple

5 : humidity meter
6 : pyranometer
7 : net radiometer
8 : heat flux meter

Heat balance of the experimental greenhouse was
analyzed in the form of heat flux per unit floor
area. The steady state heat balance selected
for this study was composed of the heat generated by solar radiation, the change in heat
associated with air blowing into and flowing out
of the greenhouse, the vertical heat flow in
surface layer of soil floor and the overall heat
transfer through the greenhouse cover with or
without shading material. The relationships
used in calculating these heat fluxes and the
overall coefficient of heat transfer through the
cover and method of calculating these parameters
are summarized in the Appendix.

Measurements

Measurements were carried out in summers of 1989
and 1990 to provide the data required in
describing the performance of the zone cooling
system. Air temperature inside the greenhouse
was measured by means of 30 and 20 copper-
constantan thermocouples in experiments during
summers of 1989 and 1990, respectively. Outside
air temperature was measured by 3 thermocouples. Air temperature at inlet and outlet
(ridge window opening) were each measured by
means of 4 thermocouples. A pyranometer was
placed horizontally inside the greenhouse to
measure the inside solar radiation. Net radiation inside the greenhouse (both short and long
wave radiations) was measured by means of a net
radiometer. In experiment of summer 1989, the
change in latent heat associated with incoming
and outgoing air was separated from that of sensible heat, and calculated from evaporation rate
at greenhouse soil surface. However, in summer
of 1990, the change in latent heat was calculated simultaneously with that of sensible heat
through enthalpy difference between incoming and
outgoing air (relative humidity was also
measured by means of digital humidity meters at
the two positions). In experiment of summer
1989, the vertical heat flow in soil floor was
obtained through temperature gradient method.
However, in summer of 1990, the heat flow was
measured by means of a heat flow meter. Temperature, humidity and heat flow data were recorded hourly using digital recorders, and data
on solar radiation and net radiation were recorded hourly by means of an analog recorder.
Hourly meteorological informations were obtained
from the meteorological station of the Faculty
of Agriculture, Kochi University.

Evaluation of Experimental Factors

The zone cooling system was tested for three experimental factors : (1). ventilation rate, (2).
soil wetness and (3). shading application. The
cooling system was operated at ventilation rate
of 35 and 70 times per hour during the summers
of 1989 and 1990, respectively. Moisture content of soil surface layer in the greenhouse
ranged 2 ~ 5 % (dry base) in dry soil condition.
In wet soil condition after irrigation in the
previous day, the moisture content ranged 8 ~ 15
% (dry base). In this study, haze vinyl film,
white cheesecloth, and black cheesecloth were
used as shading materials. The shading material
was placed adjacent to outer surface of the
cover. The overall transmissivity of the greenhouse cover (with or without shading material
depend on the experimental series) was obtained
from solar radiation data outside and inside the
greenhouse. The evaluation was conducted for 12
series. The levels of the experimental factors
are described in Table 1.

TABLE 1 The Experimental Factors to be Evaluated
and its Levels

Series No.	VR[*] (/hr.)	MC[**] (%)	t_{cs}[***] (%)	Month /Day
				(1989)
1[****]	35	2 ~ 5	80 ~ 85	8/21,8/22
2[****]	35	2 ~ 5	60 ~ 77	8/24,8/25
3[****]	35	2 ~ 5	50 ~ 53	8/23,8/26
4[****]	35	8 ~ 15	80 ~ 85	9/11,9/12
5[****]	35	8 ~ 15	60 ~ 77	9/8,9/9
6	35	8 ~ 15	50 ~ 53	8/28,8/29
				(1990)
7	70	2 ~ 5	80 ~ 85	7/18,7/19
8	70	2 ~ 5	60 ~ 77	7/29,7/30
9	70	2 ~ 5	50 ~ 53	8/5,8/6
10	70	8 ~ 15	80 ~ 85	8/16,8/18
11	70	8 ~ 15	60 ~ 77	8/20,8/21
12	70	8 ~ 15	50 ~ 53	8/26,8/27

[*] Ventilation rate through the greenhouse
(air flow rate was measured at inlet).
[**] Moisture content of soil surface layer in
the greenhouse (dry base).
[***] Overall transmissivity (single cover or
combination of cover and shading, depend on
experimental series) to solar radiation.
[****] Taken from Kojima and Suhardiyanto (1991).

The effect of experimental factors on air temperature at lower layer in the greenhouse, heat removed by cool air blowing, and overall coefficient of heat transfer through the greenhouse cover were analyzed using the method of variance analysis. Air temperature at lower layer in the greenhouse is expressed in dimensionless form based on outside air temperature. In the present study, the heat removed by cool air blowing during the daytime is expressed in dimensionless form based on the solar radiation in the relevant day. Kessey and Glockner (1983a, 1983b) described examples of dimensionless energy and temperature in their investigation regarding energy transfer and effect of environmental temperatures on the inflatable greenhouse.

RESULTS AND DISCUSSION

Temperature and Heat Flux in Greenhouse

Cool air reduced maximum air temperature at the lower layer in the greenhouse in dry soil condition without shading to about 44 °C. It was about 12 °C lower than that of without cool air blowing. In combination of wet soil condition and shading application, the daily maximum air temperature at the lower layer in the greenhouse could be maintained to 31.7 ~ 35.0 °C compared to 32.6 ~ 38.3 °C at upper layer. The maximum air temperature at the lower layer is at a level may tolerated by plants. Maximum air temperature tolerated by some species of Solaneous and Cucurbitaceous plants was noted to be 35 °C (Inden, 1977). The effect of such cool air blowing on plants is needed to be examined. Table 2 gives a list of air temperature at the lower layer in the greenhouse (T_1) and air temperature at the other layers when T_1 reaches its maximum. All temperatures are the average values. Even in wet soil series, the relative humidity of air in the greenhouse was low, as it ranged 25 ~ 50 % during the daytime.

As the path of air blowing from cool air inlet increased, T_1 tended to increase. In addition, an increasing temperature was also observed at point of measurement at the left and right hands of the inlet. An improvement to the type of

TABLE 2 Air Temperatures of Horizontal Layers[*] in the Greenhouse and the Outside Air Temperature (T_o) when Air Temperature at the Lower Layer Reachs its Maximum (°C)

Series No.	Month /Day	Time (JST)	T_1 max.	T_m	T_u	T_o
	(1989)					
1[**]	8/21	12:00	43.5	45.3	44.4	33.6
1[**]	8/22	13:00	44.2	46.2	45.5	33.2
2[**]	8/24	11:00	44.5	47.0	45.7	34.2
2[**]	8/25	13:00	40.9	43.9	43.0	32.2
3	8/23	14:00	43.0	44.1	44.7	33.5
3	8/26	12:00	35.1	36.0	36.1	29.8
4[**]	9/11	12:00	34.9	37.6	37.0	31.5
4[**]	9/12	13:00	35.9	39.0	38.7	32.3
5[**]	9/8	14:00	34.9	37.0	38.2	32.9
5[**]	9/9	13:00	34.8	37.6	37.9	32.4
6	8/28	14:00	35.0	37.3	38.3	34.0
6	8/29	15:00	33.5	35.8	36.7	31.2
	(1990)					
7	7/18	13:00	40.7	42.3	42.2	33.3
7	7/19	13:00	41.0	42.3	42.6	34.0
8	7/29	14:00	40.5	41.5	41.7	33.6
8	7/30	14:00	41.4	42.6	42.8	35.0
9	8/5	13:00	39.2	40.5	40.7	33.0
9	8/6	14:00	40.3	40.9	41.2	36.0
10	8/16	14:00	37.8	39.1	39.2	35.0
10	8/18	14:00	37.1	38.6	38.9	35.2
11	8/20	14:00	32.4	33.1	33.2	33.0
11	8/21	14:00	33.8	34.9	35.3	35.5
12	8/26	12:00	31.7	32.6	33.0	33.8
12	8/27	14:00	31.9	32.6	33.3	33.0

[*] T_1 : Air temperature at the lower layer.
T_m : Air temperature at the middle layer
T_u : Air temperature at the upper layer.
[**] Taken from Kojima and Suhardiyanto (1991).

inlet such as using distribution ducts is required in the development of the cooling system. Figure 2 shows the three dimensional distribution of air temperature at the lower layer in the greenhouse in dimensionless form for typical maximum values of wet soil condition and shading with black cheesecloth. The pattern of temperature distribution at a level of ventilation rate may be different among greenhouse types and dimensions.

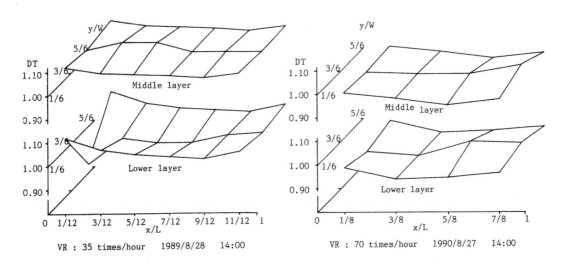

VR : 35 times/hour   1989/8/28   14:00

VR : 70 times/hour   1990/8/27   14:00

Fig. 2. Three dimensional distribution of DT.

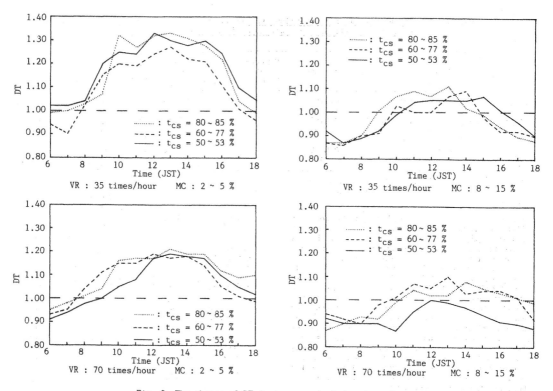

Fig. 3. The change of DT during a typical day in each series.

Air temperature at the lower layer in the green-house and the outside air used for expressing the dimensionless temperature (DT) is average value of the relevant temperature. The change of DT during a typical day in each series are presented in Fig. 3. As shown in Fig. 3, air temperature at the lower layer in the greenhouse in wet soil series was lower than the outside air temperature (DT < 1.00) only at about 06:00 to 10:00 JST (Japan Standard Time) and at about 17:00 to 18:00 JST. At about 10:00 to 17:00 JST air temperature at the lower layer was only slightly higher than outside air temperature. Examples of diurnal changes in temperature, humidity and heat fluxes in a model-sized greenhouse during zone cooling operation can be found in Kojima and Suhardiyanto (1991).

The heat fluxes and overall coefficient of heat transfer through the greenhouse cover with or without shading (U) during the daytime was calculated for 12 hours basis. The heat generated by solar radiation contributed the largest cooling load. On the other hand, the change of heat associated with air blowing was the largest portion of the removed heat. The heat removed from the greenhouse through air blowing during the daytime (06:00 to 18:00 JST) ranged 15 ~ 64 %, as compared to that of solar radiation.

The average U was 14.8 and 13.6 W/(m².K) for single cover and with shading application, respectively. Using U in describing the thermal performance of the greenhouse in case of windows were opened may be less merit than that for un-ventilated greenhouse in which the U value often used to describe coefficient of heating load or nighttime cooling load. The U value in this case does not express the coefficient of cooling load. The coefficient of cooling load for this type of operating condition should at least in-clude the coefficient of heat generation by solar radiation and the U value when inside air temperature is lower than outside air temperature. On the contrary, when inside air temperature is higher than outside temperature which may occurs in this type of cooling system during the daytime, the U value may express a reducing factor of heat load due to heat transfer through the greenhouse cover.

Effect of Experimental Factors

The effect of ventilation rate, soil wetness and shading application on dimensionless temperature of air at the lower layer in the greenhouse (DT) was evaluated for 288 data (hourly) collected from 24 days of experimental period. The effect of each experimental factor for DT was evaluated using the method of variance analysis and the result is shown in Table 3. DT was influenced significantly (1 %) by soil wetness and shading application, and at 5 % level of significance by ventilation rate. The interaction between ventilation rate and shading application and between ventilation rate and soil wetness influenced DT at 1 and 5 % level of significance, respectively.

Among the experimental factors, soil wetness contributed the strongest effect on DT. This shows that evaporative cooling reduced temperature of soil surface layer considerably. However, due to air flowing out from the greenhouse through ridge window, it was not accompanied by considerable increase in humidity of inside air. Ventilation rate influenced DT inferior to soil wetness or shading application. This suggests that the zone cooling system does not require too large ventilation rate except to facilitate replacing warmer air by cooler air.

TABLE 3 Variance Analysis of Dimensionless Temperature (Based on Outside Air Temperature) of Air at the Lower Layer in the Greenhouse

Source	Variance	Degrees of freedom	Unbiased estimator of population variance	Variance ratio
Ventilation rate (A)	0.0563	1	0.0563	7.52[*]
Soil wetness (B)	1.3930	1	1.3930	186.12[**]
Shading application (C)	0.2165	2	0.1083	14.46[**]
A x B	0.0495	1	0.0495	6.61[*]
A x C	0.2314	2	0.1157	15.46[**]
B x C	0.0019	2	0.0010	–
Error	2.1481	287	0.0075	

[*] 5 % level of significance
[**] 1 % level of significance

The effect of each experimental factor for dimensionless heat flux removed through air blowing (DH) is presented in Table 4. DH was introduced as the ratio of removed heat per unit floor area through air blowing in a day to daily solar radiation on horizontal surface outside the greenhouse (each heat flux is 12 hours basis from 06:00 to 18:00 JST). Soil wetness and shading application influenced DH significantly (5 %). The interaction between ventilation rate and soil wetness also influenced DH at 5 % level of significance. No significant effect was observed in ventilation rate. This suggests that in H_c value (heat removed through air blowing into and flowing out from the greenhouse), the latent heat portion as a result of evaporation at wet soil surface occupied considerable larger portion than the sensible heat.

No attempt was made to express the overall coefficient of heat transfer through the greenhouse cover with or without shading material (U) in dimensionless form. The variance analysis was conducted to 24 data (each U value is 12 hours basis from 06:00 to 18:00 JST). Table 5 shows the effect of each experimental factor on U value. Ventilation rate influenced U significantly (5 %). No significant effect was found in the soil wetness and shading application. It is well known that one of the factors affecting U value is the movement of air inside and outside the greenhouse. However, the collected data could not explain the effect of material properties (cover without or with shading application) on U value. This may caused by external noise which could not be monitored in this hourly data recording.

CONCLUSSION

Under cheesecloth shading in wet soil condition, the zone cooling system was capable of reducing maximum air temperature at the lower layer in the model-sized greenhouse to 32 ~ 35 °C compared to 33 ~ 38 °C at the upper layer.

TABLE 4 Variance Analysis of Dimensionless Heat Flux (Based on Outside Solar Radiation) Removed from the Greenhouse through Cool Air Blowing

Source	Variance	Degrees of freedom	Unbiased estimator of population variance	Variance ratio
Ventilation rate (A)	1.67×10^{-5}	1	1.67×10^{-5}	–
Soil wetness (B)	0.3037	1	0.3037	62.32[*]
Shading application (C)	0.0847	2	0.0424	8.68[*]
A x B	0.0140	1	0.0140	2.88[*]
A x C	0.0019	2	0.0010	–
B x C	0.0078	2	0.0039	–
Error	0.1121	23	0.0049	

[*] 5 % level of significance

TABLE 5 Variance Analysis of Overall Coefficient of Heat Transfer through the Greenhouse Cover with or without Shading Material

Source	Variance	Degrees of freedom	Unbiased estimator of population variance	Variance ratio
Ventilation rate (A)	78.1204	1	78.1204	3.81[*]
Soil wetness (B)	34.8004	1	34.8004	1.70
Shading application (C)	27.3633	2	13.6817	–
A x B	21.4704	1	21.4704	1.05
A x C	8.8033	2	4.4017	–
B x C	22.1433	2	11.0717	–
Error	472.0450	23	20.5237	

[*] 5 % level of significance

The experimental factors : (1) ventilation rate, (2) soil wetness and (3) shading application influenced DT (dimensionless temperature of air at the lower layer in the greenhouse based on outside air temperature) significantly. Among the experimental factors evaluated, soil wetness contributed the strongest effect on DT (1 % level of significance).

Soil wetness and shading application influenced DH (dimensionless removed heat flux per unit floor area based on solar radiation on horizontal surface outside the greenhouse) significantly (5 %). Ventilation rate influenced U (overall coefficient of heat transfer through the greenhouse cover with or without shading) significantly (5 %).

REFERENCE

Duncan, G.A., O.J. Loewer, Jr., and D.G. Colliver (1981). Simulation of energy flows in a greenhouse : Magnitude and conservation potential. Trans. ASAE, 24 : 197-201.
Hanan, J.J., W.D. Holley, and K.L. Goldsberry (1978). Cooling greenhouses. 'In' Greenhouse Management, Springer-Verlag, Berlin. pp. 175-196.
Hayashi, M., T. Kozai, and I. Watanabe (1983). Applications of heat pump to greenhouse environment control : (2) Greenhouse cooling in summer nighttime. J. Agric. Met., 39 : 181-189. (in Japanese with English summary)
Hillel, D. (1980). Water balance and energy balance in the field. 'In' Applications of Soil Physics, Academic Press, Inc. Ltd., London. pp. 197-214.
Inden, T. (1977). Shisetsu kasai no ondo kanri. 'In' Shisetsu Engei no Kankyou to Saibai, Seibundoshinkousha, Tokyo. pp. 85-108. (in Japanese).
Kasubuchi, T. (1973). Problems of the mechanism of heat conduction in soil. J. Agric. Met., 29 : 201-207.
Kessey, K.O. and P.G. Glockner (1983a). Experimental investigation of energy transfer in the inflatable greenhouse. J. Agric. Met., 38 : 395-402.
Kessey, K.O. and P.G. Glockner (1983b). Effect of environmental temperatures on the inflatable greenhouse. J. Agric. Met., 38 : 403-408.
Kojima, K. and H. Suhardiyanto (1991). Studies on the zone cooling system in greenhouse (1). Performance of the system in a model-sized greenhouse. Environ. Control in Biol. 29 : 1-10.
Kozai, T., M. Hayashi, T. Kodama, Z. Quan, I. Watanabe and H. Oosawa (1985). Applications of a heat pump system to greenhouse environmental control. (3) Operating characteristics of the cooling system in summer nighttime. J. Agric. Met., 41 : 231-240.

APPENDIX

Steady State Heat Balance

The steady state heat balance selected in calculating the heat fluxes per floor area in the greenhouse is

$$H_s + H_c + H_f + H_o = 0 \qquad (1)$$

where H_s is the heat generated by solar radiation, H_c is the change in sensible and latent heat associated with air blowing, H_f is the vertical heat flow to or from greenhouse floor, and H_o is the overall heat transfer through the greenhouse cover. All terms in the left hand of Eq. 1 are in W/m^2 (floor area).

Heat Generated by Solar Radiation

The heat generated by solar radiation was calculated through the following equation (Duncan, Loewer Jr. and Colliver, 1981).

$$H_s = t_{cs} \cdot a_f \cdot S_o \qquad (2)$$

where t_{cs} is the overall transmissivity of the greenhouse cover with or without shading material, a_f is the absorptivity of the greenhouse floor to solar radiation, S_o is the solar radiation on horizontal surface outside the greenhouse.

Heat Associated with Incoming and Outgoing Air

In experiment of summer 1989, the change in latent heat associated with air blowing into and out of the greenhouse was separated from that of sensible heat, and calculated from actual evaporation rate at greenhouse soil surface based on Hillel (1980). However, in summer of 1990, the change in heat associated with air blowing was calculated through enthalpy difference between incoming and outgoing air as follows

$$H_c = M_f (h_i - h_o) / A_f \qquad (3)$$

where M_f is the air mass flow through the greenhouse (kg/sec.), h_i is the enthalpy of incoming air (J/kg), h_o is the enthalpy of outgoing air (J/kg), A_f is the greenhouse floor area (m^2).

Vertical Heat Flow in Greenhouse Floor

The vertical heat flow in greenhouse floor (soil surface layer) was obtained through temperature gradient method (in summer 1989), and measured directly by means of a heat flow meter (in summer 1990). The relationship used in calculating the heat flow in soil surface layer through temperature gradient method is

$$H_f = k \cdot T_g \qquad (4)$$

where k is the soil heat conductivity $(W/(m \cdot {}^\circ C))$, T_g is the vertical temperature gradient in the soil layer $({}^\circ C/m)$. The soil heat conductivity was determined based on the moisture content of soil (Kasubuchi, 1973).

Overall Coefficient of Heat Transfer

After obtaining H_s, H_c, and H_f, H_o can be calculated from Eq. 1. The overall coefficient of heat transfer through the greenhouse cover (U) with or without shading was calculated from

$$U = (H_o \cdot A_f)/(A_c \cdot (T_o - T_i)) \qquad (5)$$

where A_c is the greenhouse cover area (m^2), T_o and T_i are the average temperature of outside and inside air $({}^\circ C)$, respectively.

INTRODUCING DUTCH SUBSTRATE SYSTEM TO JAPAN

T. Tachibana and T. Yamaguchi

Bio-System Division, ISEKI & CO., LTD., Yakura 1, Tobe-tyou, Iyo-gun, Ehime, 791-21 Japan

Abstract. ISEKI started the substrate business in Japan 6 years ago
with NFT system. Separately we tried the complete Dutch style rockwool substrate system and imported it in '87 with venlo-type glasshouse including Dutch growing techniques. We built two 60a houses
installed with rockwool system with many kinds of automatic devices.
The crop is tomato with Japanese variety named "momotaro" under
the govermental subsidiary in Kagawa prefecture. The results in the
-se three years, the average yield per 10a was 33-35 tons and annua
lsales amount to approx. 11-12 million yen per 10a. These figures
are fantastic for Japanese growers comparing with their ordinary
productivities of 15tons, futhermore nearly equal to the growers
index in 2000 announced in October,1990 by Japan Agriculture Administration. The big differences come from productivity and products
quality. There are many different points on growing control between
Dutch and Japanese. For example at the project we set the lowest
temperature in the night at 18°C and during daytime at 20°C in order to have a good growing and possibility to pollinate by vibrator
and also keep a high CO_2 level to stimulate the photosynthesis. To
do so the products are well solid and have long shelflife, get high
price in the market. ISEKI like to introduce the system to Japanese
growers as one of the solving way for their financial problem.

Keywords. Substrate system;automatic control;computer control;
CO_2 density;temperature control.

THE PRESENT CONDITION OF SOILLESS CULTURE IN JAPAN

The total area of greenhouse has been
increasing every year steadily.

TABLE 1 Changes of Greenhouse Area

[unit:100ha]

	1981	1983	1985	1987	1989
Glasshouse	17	18	19	20	21
Others	342	348	372	395	428
Total	359	366	391	415	449

The total area was approx. 45,000ha
at June,1989. However the majority of

it is simple plastic tunnels with
simple equipment. When we limit the
house with steel posts, we can only
count 10,000ha. The figure is nearly
equal to Dutch of it.

Soilless Culture in Japan

The house which is equipped with the
soilless culture facilities, covers
only 370ha, under 1% of the total
area of greenhouses.In detail, comparing with 1987,the total area of it
increased 25%. The hydroponic system
increased only 19%, on the other hand
rockwool system increased 60%.

TABLE 2   Soilless Culture in Japan

[unit:ha]

	Hydroponic		Rock-Wool System	Others	Total
	NFT	Others	System		
1987	50	181	41	27	299
1989	69	206	66	32	373

Reasons why the Soilless Culture does not spread in Japan

According to the questionaires about rockwool system which were carried out in 1987 by Agriculture Adminis- tration. Growers said;
1) Productivity is equal to or less than soil culture;
2) Needs big initial investment;
3) By above mentioned factors, dif- ficult to make profits;
4) Difficult to operate or control.

At the same time, they recognized;
5) Quality of products is better than the soil culture one;
6) Easy work and can save labour costs;
7) Less occur of insect diseases and illness than in case of soil cul- ture.

So Jpanese growers only have the im- pression about soilless culture that it is the best way to avoid diseases which occure in soil culture.

PROGRESS OF IMPORT A DUTCH SUBSTRATE SYSTEM

ISEKI started soilless culture busi- ness with NFT system from 1985. The fundamental studies have been in op- eration at our Tsukuba Laboratory. Especially in 1987 and 1988, we con- centrated on strawberry growing with NFT system. And we built 50 sites in Japan. But this business were not attractive for us, as these projects are small, average area is about 10a, and use existing plastic tunnels. That is we needed many labors to tea- ch growers how to operate and manage the system, futhermore usual plastic tunnel only have a heater. So we can not control the circumstances in it

well, in particular with humidity. At the results, we failed to get enough production and quality which bring profits to a grower.

Import of Dutch Rockwool System

On the other hand, we had been study- ing Dutch rockwool system separately from NFT system at that time. In 1987 we luckily took an opportunity to im- port from Netherlands for beeftomato cropping as a trial. There were many opposition like follows;
1) There are big differences in weather conditions between Nether lands and Japan, so it was not possible to grow sensitive Japa- nese variety;
2) We couldn't get products which vegetable markets accept, because a Dutch consultant didn't have an experience about Japanese variety and didn't know the Japanese mar- ket;
3) The Dutch system was too expen- sive to make profit for growers.

We already decided that we didn't im- port hardware only, but should also import software at the same time, in cooperation with a Dutch manufacturer. So we could take a trial overcoming many opposition.

RESULTS OF THE TRIAL INTRODUCE

The actual results of last two sea- sons were 35.0kg/m^2 and 33.3kg/m^2. The results are beyond of average one in Netherlands, but for Japanese ordinary tomato growers it is a fan- tastic figures. As "momotaro" is very difficult to grow and to keep a good balance between vegetative and gener- ative growth. The farmers union who manage the house gets total income at least 140 million yen every year. The amount comes from not only production volume but also quality that market evaluates superior as products have good overall coloring, fine luster, long shelflife and are well solid. About this season, the production will exceed the results of last two seasons, we can expect over 35kg/m². Now the union challenge 40kg/m² in production and 160 million yen in net income.

TABLE 3 Production Report of 1989

[unit:ton,%]

	First		Second		Third		Total	
	ton	%	ton	%	ton	%	ton	%
NOV	–	–	1.6	100	0	0	1.6	0
DEC	10.0	24	27.5	66	4.4	10	41.9	10
JAN	5.2	15	23.4	68	5.7	17	34.3	8
FEB	6.5	18	24.3	67	5.6	15	36.4	9
MAR	8.0	17	31.7	68	7.1	15	46.8	12
APR	13.9	21	42.7	65	9.3	14	65.9	16
MAY	10.5	15	49.7	71	9.4	14	69.6	17
JUN	10.4	16	44.8	69	9.7	15	64.9	16
JUL	4.6	10	35.7	76	6.9	14	47.2	12
TTL	69.1	17	281.4	69	58.1	14	408.6	100

area of house:12,288m²
crop type: one crop one year
variety: momotaro
sowing date: 13/09/1989
first harvest:25/11/1989
end of crop: 30/07/1990

THE OUTLINE OF INTRODUCED DUTCH SYSTEM

area of house:12,288m²
house type: venlo-type with trellis girder
eaves height: G.L.+3,500mm
crop type: one crop one year
variety: Japanese pink tomato "momotaro"
strength: wind velosity 40m/sec
               snow depth   22cm/m²
culture type: rockwool with drip irrigation

The Outline of main Facilities

Enviromental computer control.Due to the incresing cost of energy, it has become necessary for every grower to control more accurately the enviromental conditions in the greenhouse. The interaction between the control equipment and control devices,such as mixing valves,ventilation motors,shading screen,CO_2 dosing,lighting,irrigation and others, is becoming increasingly more intensive.The computer has climate control, irrigation control, CO_2 dosing and other programs and control with the proportion to the amount of light and also connected with evaporation meters.

Overview

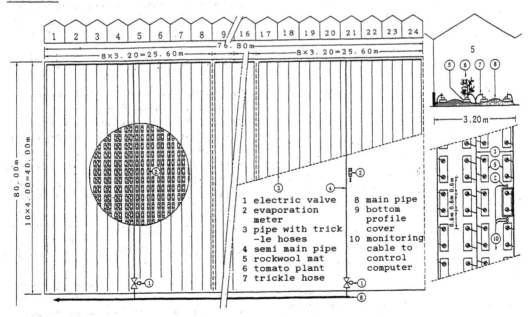

1 electric valve
2 evaporation meter
3 pipe with trickle hoses
4 semi main pipe
5 rockwool mat
6 tomato plant
7 trickle hose
8 main pipe
9 bottom profile cover
10 monitoring cable to control computer

Fig.1. Layout of the project.

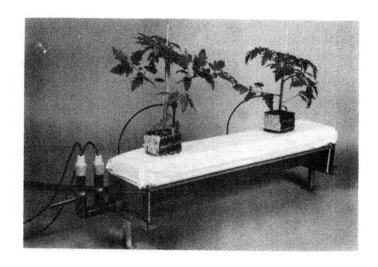

Fig.2.1.Evaporation meter.

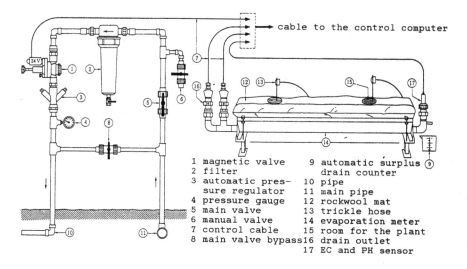

cable to the control computer

1 magnetic valve
2 filter
3 automatic pres-
   sure regulator
4 pressure gauge
5 main valve
6 manual valve
7 control cable
8 main valve bypass

9 automatic surplus
   drain counter
10 pipe
11 main pipe
12 rockwool mat
13 trickle hose
14 evaporation meter
15 room for the plant
16 drain outlet
17 EC and PH sensor

Fig.2.2.The structure of evaporation meter.

Evaporation meter.Depending on the data received from the evaporation meters, the computer regulates which group of plants get water and adjust fully automatically whether too much or not enough water has been supplied by connecting with dosing system and also has a drain counter.

CO_2 dosing.In the house we keep high CO_2 level at 400-600ppm to stimulate the photosynthesis and make leaves thicker to receive enough light.

Other features.We have been taking many special operarions in Kagawa project which are still rare cases in Japan. Such as;
  1) physical pollination by vibrator;
  2) using single fertilizer;
  3) high-wire cropping;
  4) hot water heating by pipes;
  5) working with electric driven pipe rail transporter;
  6) less pesticide.

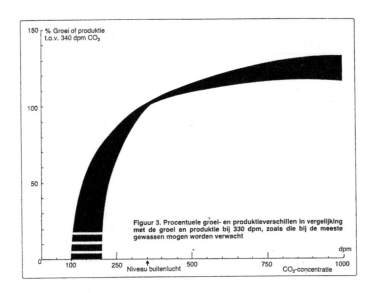

Figuur 3. Procentuele groei- en produktieverschillen in vergelijking met de groei en produktie bij 330 dpm, zoals die bij de meeste gewassen mogen worden verwacht

Fig.3.The relation between CO₂ density and tomato production.

Fig.4.Pollination by vibrator.

GROWERS INDEX IN 2000 IN JAPAN

Japan Agriculture Administration an-noced it in Oct.1990.

TABLE 4 The Prior Conditions

	Description in Index	Ocassion in Kagawa Proj.
District	some warm place in southern part of Japan	in it
Area	60a	60a x 2

TABLE 5 Growers Management Index

Cropping style	one year one crop (long term crop)	the same way
Labours	main-3persons parttimer- 5persons	main-4persons parttimer- 8persons
Necess- ary main tech- nique	*overall col- oring round shape varie- ty	*adopted
	*climate con- trol with light meter	*put into practice
	*also connet- ed with sub- strate and CO₂ dosing system	*put into practice
	*using solid bed like rockwool	*adopted
	*high produc- tivity with one crop one year	*put into practice
	*biological control and humid cont- rol to red- uce insecti- cide and pesticide *others	*under the considera- tion

	Index	Results in Kagawa proj
productivity	30kg/m²	33-35kg/m²
quality *sugar content *vitamin	8 % rich	6 % now testing

*long shelf- life	———	realized
production cost	¥135/kg (-30%)	unknown

CONCLUSION

The important point for tomato growing is to keep the circumstance which
tomato desire as much as we can, and
understand physiology and nutrition.
We think there are a few points depend on variety, but basical theory
is the same. So as to we need a house
well closed and high transmission of
light. At that time we can easily get
uniform circumstances. Otherwise we
can not do the uniform operation.

REFERENCES

Vogels,A.G.,and J.Mens (1989).
    Agrotherm horticultural engi-
    neering.Agrotherm B.V.,pp.58-83,
    111-120.

Harten,J.V.,and others (Jan.,1988)
    CO_2 indekas.Proefstation voor
    Tuinbouw onder Glas te Naaldwijk.
    pp.8-10.

Dept.Information and Statistics,Japan
    Agri., Admi. The present codi-
    tions of horticultural glass-
    houses and plastic tunnels,1990
    edition.Jpn.Hort.Assoc.,pp.2-5,
    43,65-76.

Hayashi,T.(Dec.,1990).Equipment and
    Horticulture.Jpn.Hort.Assoc.,Vol.
    71,pp.12-15.

TWO APPROACHES TO ENVIRONMENTAL CONTROL IN GREENHOUSES - HEAT BALANCE ANALYSIS AND SYSTEM IDENTIFICATION IN HEATING SYSTEM WITH HEAT PUMP

H. Nishina, I. H. Cho and Y. Hashimoto

Dept. of Biomechanical Systems, Ehime University, Matsuyama, Ehime, Japan

Abstract. In the greenhouse with heat pump, heat balance analysis, system identification of heating and performance of heat pump system of air-water type were examined. The experiment was carried out in a greenhouse with floor area of 114m² in the winter from 1990 to 1991.
The results of heat balance analysis and system identification show a good agreement, and it is considered that both heat balance analysis and system identification are effective in analysis of environment in heated greenhouse. Performance of heat pump of air-water type was also made clear.

Keywords. Agriculture; environmental control; system identification; greenhouse; heating; heat pump.

INTRODUCTION

In greenhouse engineering, many researches have been made for more effective control of environment in greenhouse. The greater part of the researches have been performed based on analysis of heat balance (Takakura et al., 1968; Nishina and Takakura, 1988). Analysis of heat balance in greenhouse is very effective in greenhouse engineering, especially in analysis and calculation concerning long time such as hours, days or months.

On the other hand, dynamical analysis of the environment in greenhouse have also been performed (Hashimoto and Morimoto, 1984; Uding ten Cate and van de Vooren, 1984). This approach is based on theory of system identification, and is effective in environmental control for short-time change.

Therefore, in order to achieve advanced control of environment in greenhouse, it is necessary to combine heat balance analysis with system identification.

While heat pump systems have been used for environmental control, performance of heat pump system of air(heat source)-water type has not been made clear. Heat pump system of this type has an advantage of wide applicability and is expected as air-conditioning system in advanced greenhouses and plant factories.

For the purpose mentioned above, the authors made experiment of heating in greenhouse in the winter from 1990 to 1991, and examined two approaches to environmental control, that is, heat balance analysis and system identification.

METHODS

Greenhouse

The greenhouse is a north-south and double-span glasshouse, which locates in College of Agriculture of Ehime University. The span and length of the greenhouse are 7.2m and 7.9m, respectively. The floor and surface areas are 114m² and 256m² respectively, and the volume is 364m³. The greenhouse has a movable covering system and one layer of thermal screen was furnished for heat insulation in the nighttime. The film used was PVC(polyvinylchloride) of thickness of 0.075mm. In the greenhouse, tomato plants were cultivated in hydroponic system.

Heating system with heat pump

The heating system consists of a heat pump, a water storage tank, heating pipes and three pumps. One pump is for circulating water between the heat pump and the water storage tank, and the other two pumps are for circulating water between the water storage tank and the heating pipes. The conceptional composition of the heating system is shown in Fig. 1.

The heat pump is air-water type, which absorbs heat from outside air and releases heat to water in case of heating. The water can be heated up to 50℃ in heating or cooled to 8℃ in cooling. Compressor and ventilating fan of the heat pump consume electricity of 7.5kW and 0.26kW, respectively. The heating capacity of the heat pump is 29000kcal/h, when temperature of water at inlet, water at outlet and outside air are 40℃, 45℃, and 7℃, respectively.

The water storage tank contains two hundred liters of water. The pump circulating water between the heat pump and the water storage tank consumes electricity of 0.5kW. Each of the two pumps circulating water between the water storage tank and heating pipes consumes 0.18kW, and provides the maximum water flow rate of 84 liters per minute. The total length of the heating finned pipes is 76.7m.

RESULTS

Heat balance analysis and performance of heat pump

Principle of heat balance analysis is expressed as follows.

$$Q_{in} - Q_{out} = 0$$

Qin is the sum of heat flowing into the system considered, and Qout is the sum of heat flowing out of the system.

Heat balance equation for this experiment is the following. The symbols are written in Fig. 1.

$$Q_p + Q_s - Q_v - Q_t = 0$$

Qp is heat released from the heating finned pipes, calculated as the product of temperature difference between water in pipe flowing into greenhouse and water in pipe flowing out of greenhouse and water flow rate. Qs is heat released from the soil, measured by heat flow plate. Qv is heat transfer due to infiltration, calculated as the product of enthalpy difference between inside and outside air and infiltration rate. Qt is overall heat transfer through covering, calculated from heat balance equation.

Table 1 shows the results of heat balance analysis. Table 1 shows the values on six hours from 22:00 each day to 4:00 next day. Heating degree hour is the sum of inside-outside temperature difference in six hours. Overall heat transfer coefficient is calculated as overall heat transfer divided by the product of heating degree hour and surface area of greenhouse. In this experiment, the values of overall heat transfer coefficient are lower than expected. This reason is considered that weather in this period was cloudy or rainy, resulting in low radiative heat loss from the surface of the greenhouse.

Table 2 shows the performance of the heating system with heat pump. Set point of water temperature is for the water storage tank. Consumed electricity was measured by reading watt-hour meter. Eh is electricity consumed by the heat pump only, and Es is electricity consumed by the heat pump and three water pumps. COP is coefficient of performance, and its definition is output energy divided by input energy. In this case, COPh is defined as Qp/Eh, and COPs is defined as Qp/Es.

In Fig. 2, relation between water-outside air temperature difference and inside-outside air temperature difference is shown. The regression equation calculated by method of least squares is as follows.

$$T_i - T_o = 0.38 \times (T_w - T_o) + 1.6$$

Ti is inside air temperature (℃), To is outside air temperature (℃), and Tw is temperature of water in pipe flowing into greenhouse (℃). Coefficient of correlation is 0.99. This means that if temperature of water in pipe flowing into greenhouse become 1℃ higher, inside air will become 0.38℃ higher.

Fig. 3 show relation between water-outside air temperature difference and COP. As water-outside air temperature difference increases, COPh decreases gradually. The regression equation calculated by method of least squares is as follows. Coefficient of correlation is -0.99.

$$COP_h = -0.035 \times (T_w - T_o) + 3.90$$

System identification

In this section, the authors discuss dynamic approach by system identification. We choose a system which have input of water-outside air temperature difference and output of inside-outside air temperature difference.

By expressing input signal as u(t), output signal as y(t) and noise as v(t), the following equation is obtained.

$$Y = U b + V$$

where

$$Y = [y(0), y(1), \cdots, y(N)]^T$$

$$U = \begin{bmatrix} u(0) & u(1) & \cdots & u(-n) \\ u(1) & u(0) & \cdots & u(1-n) \\ & & & \\ u(N) & u(N-1) & \cdots & u(N-n) \end{bmatrix}$$

$$b = [b_0, b_1, \cdots, b_n]^T$$

$$V = [v(0), v(1), \cdots, v(N)]^T$$

Using method of least square, impulse response can be obtained by the following equation.

$$\hat{b} = (U^T U)^{-1} U^T Y$$

The results of system identification are shown in Fig. 4. (a) is input, that is, water-outside air temperature difference. (b) is output, that is, inside-outside air temperature difference. (c) is impulse response with sixty order. (d) is step response calculated from impulse response, supposing that step input is 1℃. Output calculated from impulse response in (b), being compared with measured output.

Measured output and calculated output show a good agreement. Furthermore, values of latter half of step response are about 0.3℃, which is nearly equal to the value obtained by heat balance analysis (0.38℃). These mean that system identification is effective in calculation of environment in greenhouse.

CONCLUSION

The results show that both heat balance analysis

and system identification are effective in analysis of environment in heated greenhouse. Heat balance analysis is considered to be effective in dealing with long-time average values. On the other hand, system identification is good for dealing with short-time changing values. Therefore, combination of heat balance analysis and system identification is required and will make a great contribution to development of advanced greenhouse and plant factories.

ACKNOWLEDGMENTS

The authors wish to thank Mr. M. Miyoshi, Mr. H. Asaumi, Mr. M. Tanaka, Miss A. Matsumoto and Miss N. Taniguchi of our laboratory for their assistance.

REFERENCES

Hashimoto, Y., and T. Morimoto (1984). System Identification of Plant Response for Optimal Cultivation in Greenhouses. Proc. IFAC 9th Triennial Congress, 2039-2244, Pergamon Press, Oxford.

Nishina, H. and T. Takakura (1988). Solar Heating of a Commercial Greenhouse by Means of Latent Heat Storage. Acta Horticulturae, 230, 555-558.

Takakura, T., K. Tachibana, T. Kozai and K. Ikari (1968). Heat Balance of Glasshouses. J. of Agricultural Meteorology, 24, 115-118. (in Japanese with English summary).

Uding ten Cate, A.J., and J. van de Vooren (1984). Modelling of Greenhouse Temperatures Using Time-series Analysis Techniques. Proc. IFAC 9th Triennial Congress, 2033-2038, Pergamon Press, Oxford.

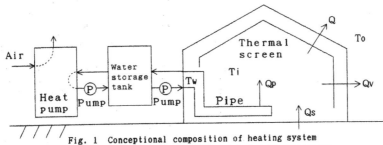

Fig. 1 Conceptional composition of heating system with heat pump and heat balance in greenhouse.

Table 1 Haet balance analysis of heating in greenhouse.

Date	Heat released from pipe Q_p (kWh)	Heat released from soil Q_s (kWh)	Heat transfer due to infiltration Q_v (kWh)	Overall heat transfer Q_t (kWh)	Heating degree hour DH (℃h)	Overall heat transfer coefficient K (Wm⁻²℃⁻¹)
Dec.31,1990	40.88	0.73	4.10	37.50	70.6	2.07
Jan. 1,1991	39.66	0.94	4.21	36.39	67.9	2.09
2	57.30	2.06	6.14	53.22	89.8	2.32
3	63.43	3.43	5.93	60.93	97.8	2.43
5	80.99	0.57	5.93	75.59	106.2	2.78
6	82.15	1.70	6.19	77.66	112.7	2.69

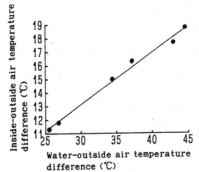

Fig. 2 Relation between water-outside air temperature difference and inside-outside air temperature difference.

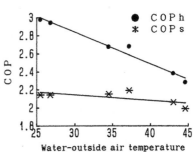

Fig. 3 Relation between water-outside air temperature difference and COP.

Table 2   Performance of heating system with heat pump.

Date	Set point of water temperature (℃)	Heat released from pipe Qp (kWh)	Electricity consumed by heat pump Eh (kWh)	Electricity consumed by heating system Es (kWh)	COP of heat pump COPh	COP of heatig system COPs	Water*-outside air temperature difference (℃)	Inside-outside air temperature difference (℃)
Dec.31,1990	35	40.9	13.9	19.1	2.94	2.14	26.8	11.8
Jan. 1,1991	35	39.7	13.3	18.5	2.98	2.14	25.5	11.3
2	40	57.3	21.4	26.6	2.68	2.15	34.5	15.0
3	40	63.4	23.7	28.9	2.68	2.19	37.2	16.3
5	45	81.0	34.1	39.3	2.38	2.06	43.0	17.7
6	45	82.1	36.1	41.3	2.28	1.99	44.6	18.8

*: Water in pipe flowing into greenhouse

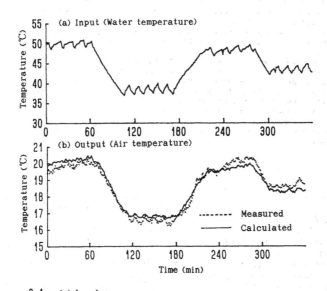

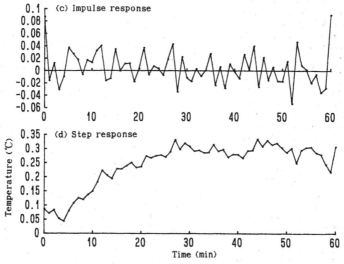

Fig. 4 Results of system identification of greenhouse heating.

LIGHT TRANSMISSIVE AND THERMALLY INSULATED WALLS FOR HORTICULTURE

K. Abe* and M. Nara**

*Mechanical Engineering Research Laboratory, Hitachi Ltd., Kandatsu, Tsuchiura, Ibaraki, Japan
**National Research Institute of Agricultural Engineering, Kan-nondai, Tsukuba, Ibaraki, Japan

Abstract. A new type of wall panel has been developed for horticultural structures with much improved characteristics over conventional greenhouse walls. Specifically, the wall consists of a light transmissive and thermally insulated airtight panel that incorporates a thermal reflecting film. The heat insulating properties of the wall are far better than conventional glass greenhouses with movable double curtains. Because the high thermal resistance of the panels doesn't vary, greenhouses constructed with the panels exhibit excellent solar heat collection properties even during wintertime. Moreover, since the walls effectively obstruct near-infrared rays, the thermal shading characteristics of the greenhouse during summer are also very good. In short, the wall exhibits excellent energy conservation all year-round without need of any adjustment.

In addition, the panels have a long service lifetime (and are thus resource conserving), and are largely maintenance-free because they tend to stay bright and clean. To evaluate the panels, they were used to construct a demonstration greenhouse in northern Japan, where winters are extremely cold.

Successful cultivation of three crops of melons a year in the greenhouse, demonstrated that the panels are highly effective for such applications.

Keywords. Agriculture; solar energy; environment control; greenhouse; thermal insulation.

1. Introduction

Japan had some 45 kha under protected cultivation in 1987, and it is estimated this figure increases about 1.7 times every decade. While this trend is likely to continue, there are a number of problems looming ahead that could check this progress; namely, declining natural resources and energy, and environmental deterioration.

As the source of energy for horticultural facilities requiring inputs of heat, we are overwhelmingly dependent on oil. The specific breakdown of energy sources is 95.7 % petroleum products, 2.3% oil-substitute fuels, 1.4% geothermal power, and a scant 0.06% solar power. Moreover, the lifetime of petroleum-based systems and equipment is short; such equipment is routinely scrapped after only one to two years of service. Of course once discarded, the material or equipment then becomes an environmental concern.

Given our current over-reliance on and great consumption of oil, coming up with viable countermeasures has emerged as an urgent concern. Again this year, the Ministry of Agriculture, Forestry and Fisheries has designated promotion of energy -efficient horticulture and exploitation of locally available natural sources of energy (solar, geothermal, etc.) as top-priority policies. It is hoped, that by focusing on these priority themes, genuine progress can be made toward resolving these difficult issues.

We believe that the present work-development of transmissive and insulting wall panels, a key component of horticultural structures-is a significant first step in the right direction. To satisfy the requirements of such a panel, we sought to develop a material (i.e., a film) that lets light through but is thermal reflecting. By combining this thermal reflecting film with another layer of light transmissive film, we succeeded in realizing airtight, light transmissive and thermally insulated wall panels that will here be designated LT^2I panels (Nara, M., and Abe, K., 1986).

What we sought to achieve was a wall panel that exhibits good insulating properties during winter so it retains heat and collects solar heat well, and yet also exhibits good thermal shading properties during summer (i.e., screens out near -infrared rays). In short, our intent was to develop a type of panel that was energy efficient all year-round (Abe, K., and Nara, M., 1987). The panels were also conceived to be efficient in other ways. They should hold up well over a long lifetime to conserve natural resources, and since they are fixed in place and don't need any adjustment, the panels are maintenance-free and thus labor-saving.

To evaluate the panels in situ, a solar greenhouse was initially constructed and used to cultivate a variety of crops including melons, cucumbers, and strawberries (Nara, M., Morinaga, M., Abe, K., and co-workers. 1988). The thermal properties of this demonstration greenhouse were also investigated (Abe, K., and Nara, M., 1989). More recently, the panels have been applied to a number of other agricultural structures (Abe, K., Kashimura, T., and Nara, M., 1991).

2. Insulating Properties of LT^2I Panels

2.1 Basic Structure

There are two fundamental requirements that walls for horticultural structures should satisfy: they should be transmissive to let light in for photosynthesis, and they should be insulating to conserve energy. While it is far from easy to satisfy both these requirements at once, the configuration shown in Fig. 1 was conceived to do so. As can be seen from the figure, a transmissive thermal reflecting film R is sandwiched between two transmissive films S with a separation of D.

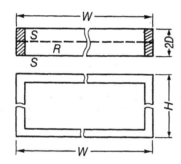

Fig.1 Basic structure of the light transmissive and thermally insulated panel

Heat transfer through the panel mainly consists of two components: a component due to convection and conduction air currents (gas heat transfer), and a component due to radiating heat (radiative heat transfer). Gas heat transfer can be reduced by implementing a sealed airtight layer between the constituent films and by increasing the thickness of that layer. Gas heat transfer can also be reduced by employing a gas with a large thermal resistance. On the other hand, radiative heat transfer can be reduced by endowing the film surface with enhanced thermal reflecting capability.

Typical properties of the thermal reflecting film are shown in Fig. 2. It is clear from the figure that over the visible range $0.4~\mu m \leqq L \leqq 0.7~\mu m$, transmissivity τ is high, and over the infrared range $L > 0.7~\mu m$, reflectivity ρ is high. In other words, the film doesn't stop many visible rays, but is highly resistant to the infrared region and thus

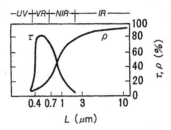

Fig.2 Optical properties of the thermal reflecting film

acts to suppress radiative heat transfer.

It goes without saying, that since strong reflectivity is exhibited at the far-infrared range $L > 2~\mu m$ (low temperature rays), this would also apply to the near-infrared range $0.7~\mu m < L < 2~\mu m$. In other words, since the major portion of thermal rays are suppressed during summertime, this means that cooling equipment doesn't have to work nearly as hard to maintain the same temperature in the greenhouse.

2.2 Experimental Procedure and Results

A range of test panel sections were prepared, and their thermal properties measured using the thermal box method.

type \ flame	wood	iron
S		
S^2		
S^3		
$S^2 \cdot R$		
$S^2 \cdot A$		

Fig.3 Sample panels

For the first group of test panels (see Fig. 3), five film configurations were fabricated, and two kinds of frame materials were used; namely, wood and square metal tubing. In the figure, S designates rigid polyester transmissive film, R is a transmissive thermal reflecting film, and A is the complete reflecting sheet. The distance separating the two surfaces D is 20 mm, and the panel samples are 560 mm × 420 mm in size.

For the second group of samples, S^2R was used as the film for the entire group, but the separation D between the films was varied. Samples were constructed where D = 5, 10, 22, and 40 mm. This group of samples was also constructed in such a way that the gas inside the test sections could be changed to evaluate the effects of different gases such as Ar and CO_2.

The thermal box was constructed out of 100
-mm-thick insulating material, with the box
sealed at the bottom but left open at the
top. A heater, a radiating plate, and a
fan were then installed in the box, while
the test panels were fit into the top
opening of the box. Then, by operating the
box in an environmental test room, measured
values for the temperature inside and
outside the test samples, power consumption
of the heater, and so on could be obtained
under steady-state control conditions, to
derive thermal resistances for each test
panel.

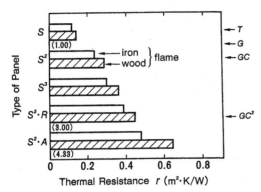

Fig.4 Thermal properties by type of panel

The thermal properties obtained for the
first group are shown in Fig. 4. It is
apparent that the thermal resistance
increases with the number of film layers,
but where the same number of layers is
involved—say, comparing films S^2R and S^3—
then the thermal resistance is increased by
employing thermal reflecting film R.
Notations on the right side of the figure
are to facilitate comparison between
different structures; here, T = tunnel house,
G = glass house, GC = glass with single
curtain, and GC^2 = glass with double curtain.
A thermal resistance of 0.45 $m^2 \cdot K/W$ (heat
transmission coefficient 2.22 $W/m^2 \cdot K$) for
the S^2R test panel with wooden frame
exceeded that for the glass greenhouse with
double curtain. It should be noted, however,
that the S^2R panel is always transmissive,
and thus the obtained value doesn't vary.
This is obviously not the case for the
curtained greenhouse, where thermal
resistance is maximized when the curtains
are pulled completely shut (GC^2) and no
light enters. By opening the curtain
halfway (GC), the wall is made transmissive,
but the thermal resistance is weak at 0.3
$m^2 \cdot K/W$. Then, when the curtain is
completely opened (G), thermal resistivity
falls to 0.17 $m^2 \cdot K/W$. It is thus apparent
that the light transmissive and thermally
insulated wall is always in a state that
conserves the maximum amount of energy.
The results for the second group of
samples are shown in Fig. 5. It can be
seen that the thermal resistance increases

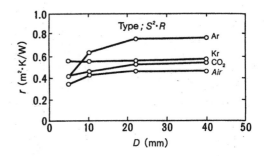

Fig.5 Thermal properties of the $S^2 \cdot$R-type panel

with increasing separation D between the
two surfaces.It was found that for air, the
thermal resistance saturates where D = 15
-20 mm, so this probably close to the
optimum range for practical applications.
As for the best type of gas sealed between
the films to enhance thermal resistance,
the order of efficiency is air < CO_2 < Kr
< Ar for D > 30 mm; and Kr and Ar are
reversed for a separation D of 5 mm, to
yield the order air < CO_2 < Ar < Kr. A
gas heat transfer suppression effect is
clearly evident, but whether such an
approach is economically viable will require
further study.

3. Solar Heat Collection Properties of a Greenhouse Constructed with LT^2I Wall Panels

It is desirable to enhance solar collection
efficiency of greenhouses as much as
possible. This way solar heat can be used
to maintain the optimum temperature during
the day, and also surplus heat can be
stored to maintain the temperature of the
greenhouse during the night.
Solar collection efficiency is evaluated as
the relation between quantity of solar
emissions I W/m^2 and the maximum temperature
differential inside and outside the
greenhouse ΔT_{1o}, such that $\Delta T_{1o}/I = Xc$
(maximum value of equivalent temperature
difference).
Employing this criterion, we investigated
the optimum greenhouse wall arrangement by
constructing 6 model greenhouses, as shown
in Fig. 6(a). All models were constructed
to the same dimensions: 800 mm wide ×
1,600 mm deep × 800 mm high, i.e., with a
wall surface to floor area ratio of 2.37.
The greenhouses were built on an East-West
axis to enhance solar collection efficiency
during winter. Floor were thermally
insulating and its surfaces were black.
Key to the symbols used in the figure is
as follows: S = rigid film; R = thermal
reflecting film; A = complete reflection
sheet; S^2 = two layers of S separated by 30
mm; S^3 = three layers of S; $S^2 \cdot R$ = two
layers of S plus one of R; $S^2 \cdot A$ = two
layers of S plus one of A. Types 1 through
4 were all constructed employing a single
type of panel, respectively, for the north
and south walls and for the roof surfaces.

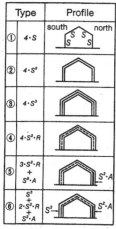
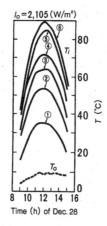

	Type	Profile
①	$4 \cdot S$	south S S north / S S S
②	$4 \cdot S^2$	
③	$4 \cdot S^3$	
④	$4 \cdot S^2 \cdot R$	
⑤	$3 \cdot S^2 \cdot R$ $+$ $S^2 \cdot A$	$S^2 \cdot A$
⑥	S^3 $+$ $2 \cdot S^2 \cdot R$ $+$ $S^2 \cdot A$	S^2 $S^2 \cdot A$

$I_o = 2,105 \, (W/m^2)$

(a) Type of Greenhouse   (b) Change of Temperature

Fig.6 Different greenhouse types and their
temperature change

For type 5, only the north wall was
constructed with $S^2 \cdot A$ panels; and for type
6, only the south wall was constructed with
S^3 panels.

The test results for changing diurnal
temperature conditions are shorn in Fig.6(b).
Increasing temperatures were observed in
the same order the types are numbered from
1 to 6. The only difference between the
two triple-film panel structures (types 5
and 6) was whether a sheet of R was
employed in the construction. Also,
regarding the two structures incorporating
a sheet of R film, (types 4 and 5), we
were trying the ascertain whether a
reflective north wall had any substantial
effect on incident rays.Finally, based on
the difference between types 5 and 6, we
were attempting to clarify whether the
south wall was solar energy transmissive or
not.

Calculated Xc values based on the results
are shown in Fig. 7. As expected, solar
heat collection efficiency Xc increased in
the same order as the types are numbered
from 1 to 6. As can be seen from the
comparative solar collection efficiencies
(indicated in parentheses in the figure),
the values obtained for type 6 was 2.71
times that of type 1.

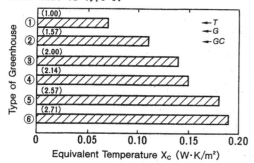

Fig.7 Solar heat collection efficiency and
greenhouse type

A conventional tunnel-house (T) provides
equivalent efficiency to type 1, and the
efficiency of a glass greenhouse (G)falls
somewhere between types 1 and 2. The double
-curtain greenhouse is on a par with type 2
when the curtain is half open (GC), and of
course Xc = 0 when the curtain is closed
(GC²)because the walls are no longer
transparent. It is thus apparent that the
solar collection efficiency of greenhouses
constructed with LT²I panels is
impressively high during winter months.

4. Thermal Shading Properties of a
Greenhouse Constructed with LT²I
Wall Panels

During summer months, especially during
the day, some means of suppressing thermal
rays entering the greenhouse is required.
Of course, light in the visible spectrum is
necessary for photosynthesis, but we want
to inhibit thermal rays. Thus, a
distinction must be made between thermal
shading as opposed to light shading.

As we observed earlier in Fig. 2 and
discussed at some length in Section 2.1,
transmissive thermal reflecting film does
possess superior thermal shading
capability. Figure 2 revealed that, while about 50%
of thermal rays (near-infrared rays and high-
temperature rays) are reflected, some 70%
of visible-light rays pass through, to yield
a solar energy transmissivity efficiency on
the order of 60%. With the $S^2 \cdot R$ wall
construction, since solar energy
transmissivity is 50%, the cooling load can
be cut in half with virtually no effort
because the walls do not require any
adjustment.

The ratio of visible-light transmissivity
τv to solar-energy transmissivity τs (i.e.,
$\rho \tau = \tau v / \tau s$) is termed the transmission
efficiency. This is one measure of thermal
shading efficiency. Almost all ordinary
films and light-shading materials provide a
transmission efficiency of $\rho \tau \leq 1$, where
a smaller proportion of visible light
passes through than thermal rays. Thermal
-reflecting and thermal-absorbing materials
both provide $\rho \tau > 1$. Absorbing materials,
however, tend to increase in temperature as
more heat is absorbed,and thus a heat
-release mechanism is required.

For thermal shading purposes, therefore,
thermal-reflecting materials represent a
better solution. For the present wall
construction, $\rho \tau$ values range from 1.1 to
1.3. Thus, while the light transmissive
and thermally insulated panels provide
excellent thermal shading efficiency, the
best structural layout is that of type 6 in
Fig. 6 where the panels are mainly applied
to the roof.

5. Application of the Panels to an
Actual Greenhouse

5.1 Solar Greenhouse
An experimental solar greenhouse was

constructed using the new LT²I panels in
the farming cooperative of Higashikawa,
Kamikawa District, Hokkaido. The
greenhouse measured 8 meters across the
front and 25 meters to the back, with a
floor space of 200 m². The structure was
equipped with a solar heating system
consisting of an underground heat converter
and solar collectors.

The panels were constructed according to
type 5 specifications (3·S²·R + S²·A)
and were applied to cover an area of about
300 m². A typical example of the S²R
panels is shown in Fig. 8. The panels are
1 m wide by 2 m in height. The panel
frames must be both strong and provide a
good insulating seal; two materials were
applied for the construction, 30-mm-square
wooden pieces and metal tubing of the same
dimension.

Fig.8 An external view of a typical panel

Testing continued over a three-year period,
during which energy efficiency and thermal
characteristics were constantly monitored
while cultivating a variety of fruits and
vegetables including melons, cucumbers, and
strawberries. The cultivation of melons is
especially noteworthy considering
Hokkaido's harsh, cold winters (See Fig.9,
and 10).

5.2 Additional Advantages

In addition to energy efficiency, a number
of other advantages associated with the
greenhouse construction became apparent or
were verified through the demonstration.
For example,(1) water droplets and mud do
not accumulate on the surface, so the
inside of the greenhouse stays much cleaner;
(2) there is no attendant work that needs
to be done to maintain the walls;(3) the
structure requires fewer internal supports,
which contributes to a lighter more open
interior; (4) the panels are exactly the
same thickness, so they stay in flush
alignment giving a neater appearance; (5)
the panels are light-weight and won't
fracture, so there is no attendant risk or
danger in handling them. Over three years,
the panels proved very sturdy even in the
face of hailstorms, thus providing more
than adequate safety.

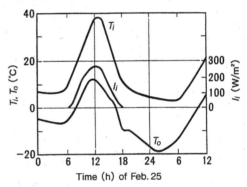

Fig.9 Solar greenhouse during winter
Upper figure shows an external view.
Lower figure shows melons under
cultivation.

Time (h) of Feb. 25

Fig.10 Temperature variation of the solar
greenhouse
I_i = volume of incident rays; T_0 = outside
temperature; T_i = inside temperature. The
temperature differential between inside
and outside the house at hight is 20℃
(without plants under cultivation or
external heating).

The panels also represent a good
investment in terms of natural resource
usage, because the rigid polyester film out
of which they are made has a long service
life (more than 10 years). What is more,
even if the panels are discarded and
incinerated, they emit no toxic gasses into
the atmosphere. And finally, the fact that
the panels not only have a long service
life but are also implemented as a fixed
installation, they create a greenhouse that
is essentially maintenance-free. Therefore,
to sum up this section, greenhouses
constructed with the new panels conserve

natural resources, they are clean, and they are energy efficient. Clearly, these very attributes will become more and more important to society in the years ahead.

5.3 Summary

In addition to the demonstration greenhouse described in the previous section,the panels have also been applied to a number of other structures, including a solar greenhouse and a horticultural factory.Given the excellent insulating properties of greenhouses incorporating the panels, such structures are perfectly suited for high-temperature forced cultivation in cold winter areas, and just as well suited for air-conditioned cultivation in summertime. In short, the panels provide an all-purpose year-round mediating mechanism for plants that require sunlight to grow and flourish.

It's not an exaggeration to say that the first requirement of a viable plant cultivation system is the walls out of which the system is constructed. With well -designed walls, all the beneficial energy available in sunlight reaching the greenhouse can be effectively utilized as a matter of course, so that the various types of equipment to control the interior environment of the system can function at their peak.

Where conservation of natural resources, energy efficiency, and cleanliness are important concerns, the LT^2I panels will have much to offer regardless of the specific system. The initial investment to construct a greenhouse using the panels may be marginally higher than in the past, but the long-term advantages of the panels more than out-weigh the initial cost.

6. Conclusions

A new type of airtight light transmissive and thermally insulated wall panels have been developed for horticultural structures, that are fabricated using transmissive film and thermal reflecting film. The thermal insulating resistance of the panels (horizontally-set, upward-thermalflow) is on the order of $0.45 m^2 \cdot K/W$ (heat transmission coefficient = $2.2 W/m^2 \cdot K$), which exceeds that for double-curtain glass greenhouses (i. e., when the curtains are closed).

A greenhouse constructed of the light transmissive and thermally insulated panels provides a solar heat collection efficiency (maximum value of equivalent temperature difference) during winter that is some 2.7 times greater than that provided by tunnel or glass greenhouses. At the same time, the thermal shading efficiency of the panels during summer is also quite impressive. With a near-infrared reflectivity of 50% and sunlight energy transmissivity also on the order of 50%, the transmission efficiency is 1.1 to 1.3.

The panels-and greenhouses constructed out of the panels-are not only highly energy efficient all year-round, they also offer a range of other attractive features: they represent a good investment in natural resources,they are labor-saving, and they are clean. A demonstration solar greenhouse was constructed with the panels in an area that experiences severe cold winters, and melons were successfully cultivated in the facility over a three-year period. The panels have also been applied to a number of other agricultural structures with similar success. The panels are certain to make a substantial contribution in the area of indoor horticultural cultivation.

Areas calling for further study include (1) accumulation of cultivation knowhow (regional differences, improvement of plant species, different cultivation regimes), (2) development of more sophisticated film materials for the panels (solar-light transmissive, but far-infrared reflective), and (3) fine-tuning to achieve greater optimization (reducing the weight of the panels and making them easier to install).

References

Abe, K., and Nara, M. (1987). Light Transmissive and Thermally Insulated Walls-Their effectiveness on Heat Collection and Shading of Non-visible Ray in a Greenhouse—. JOURNAL OF THE SOCIETY OF AGRICULTURAL STRUCTURES, JAPAN.,18,47-60.

Abe, K., and Nara, M. (1989). Long-term Performance of a Solar greenhouse. JOURNAL OF THE SOCIETY OF AGRICULTURAL STRUCTURES, JAPAN.,20,177-184.

Abe, K., Kashimura, M., and Nara, M. (1991). Development of Light Transmissive and Thermally Insulated Walls. The HITACHI HYORON., 73,6, Hitachi Hyoronsha, Tokyo.

Nara, M., and Abe, K, (1986). Study of Transparent and Adiabatic Walls. JOURNAL OF THE SOCIETY OF AGRICULTURAL STRUCTURES, JAPAN ., 17, 46-64.

Nara, M., Morinaga, M., Abe, K. and co-workers. (1988). Development of a Solar Greenhouse and its Thermal Analysis. JOURNAL OF THE SOCIETY OF AGRICULTURAL STRUCTURES, JAPAN ., 19, 81-97.

OPTIMAL GREENHOUSE TEMPERATURE TRAJECTORIES FOR A MULTI-STATE-VARIABLE TOMATO MODEL

I. Seginer

Dept. of Agricultural Engineering, Technion, Haifa 32000, Israel

Abstract. Present day greenhouse crop models contain many state variables. Obtaining an optimal environmental control trajectory for such crops by classical methods requires prohibitively large computational resources. In this study two shortcut methods were tried with some success: SCS, which is a simple version of model predictive control, and RMP, which is a reduced version of the Pontryagin maximum principle. The results indicate that the system may be represented for control purposes by just a fraction of the original number of state variables. In addition, the solution is relatively insensitive to future events.

Keywords. Agriculture; Greenhouse tomato; Environmental control; Temperature control; Optimal control.

INTRODUCTION

Temperature control in greenhouses affects several crop processes, most basic of which are photosynthesis and development. While it is normally desirable to increase photosynthesis rate, there may be situations where development must be retarded in order to meet timing requirements. An optimal solution to the temperature control problem must consider the biological and engineering properties of the crop-greenhouse system, as well as the weather and the economic environment. In principle, if a sufficiently accurate model of the system is available, and the exogenous variables (weather and prices) are known or assumed, an optimal solution can be obtained by the Pontryagin maximum principle (PMP; Pontryagin and others, 1962), or by dynamic programming (Bellman and Dreyfus, 1962). This approach involves, however, a prohibitive amount of computations when models with more than very few state variables are involved. The purpose of the present study is to explore computationally certain shortcuts, which, due to the particular properties of crop models, seem to produce acceptable results.

Computational studies of optimal control for greenhouses with simple, single-state-variable crops, showed that for every set of outside conditions there is a unique control, independent of the state of the crop (Challa and van de Vooren, 1980; Challa and Schapendonk, 1986; Seginer and others, 1986; Critten, 1991). When a two-state-variable lettuce model was considered, the optimal setpoints turned out to be a function of the state of the crop, as well as of the outside conditions (Marsh and others, 1987; Seginer and others, 1991). Seginer and McClendon (in preparation), who trained a neural network (NN) with optimal control results for the lettuce model, found that, indeed, information about the state of the crop improves considerably the performance of the NN as a control-generating device.

While the extension to general multi-state-variable models is expected to be difficult, an irrigation optimization study by Swaney and others (1983), involving a multi-state-variable

soybean model, seems to have obtained good results. Swaney used an optimization method similar to that of Marsh and others (1987), which may be classified as a simple version of model predictive control (MPC; Garcia and others, 1989), and which will be referred to here as sequential control search (SCS). The SCS cannot guarantee an optimal solution, since its control decisions are almost independent of future developments. Nevertheless, this rather primitive method gave surprisingly good results in the aforementioned studies. In particular, Seginer and McClendon (in preparation) showed that the control trajectory obtained by SCS was identical with the one obtained by the PMP.

The SCS apparently produces good results for multi-state-variable systems in which current control decisions are largely independent on future events. It does not, however, provide information about the relative importance of the various state variables of the system. Since this is expressed by the co-state variables, a modified PMP method was devised to supplement the information obtained by the SCS (Seginer and McClendon, in preparation). In the present study both methods were applied to a multi-state-variable tomato model (TOMGRO; Jones and others, 1991). Most of the computations were aimed to explore the behavior of a typical control trajectory and its response to different input information. Additional computations were used to evaluate the expected gain from optimal temperature control over a constant setpoint policy. The emphasis in this study is on the qualitative behavior of the system.

METHODS

The problem to be solved was formulated as follows: A tomato crop was assumed to be growing in a capacity-less greenhouse. The model representing the crop had over 50 state-variables, including 10 classes of leaves, stems and fruit, while the greenhouse model had none. The available control equipment consisted of a ventilation fan and a convective heater. An upper limit was placed on the capacity of the fan, while the capacity of the heater was unlimited. Weather conditions and prices were assumed to be known in advance for the whole season (deterministic problem).

An optimal daytime setpoint temperature was to be determined for each day separately. The nighttime setpoint was fixed in advance and was constant throughout the whole season. The temperature in the greenhouse was maintained at the setpoint by heating or ventilation. Ventilation was not applied during the night and sometimes was incapable of maintaining the required setpoint during daytime. The energy balance of the greenhouse did not consider evaporation explicitly. Heat was exchanged through the greenhouse glazing by convection, conduction, and radiation, all lumped into a single overall transfer coefficient. In addition, sensible heat and CO_2 were exchanged through ventilation and (constant) infiltration.

A suitable seasonal performance criterion for the problem (based on a unit floor area) was formulated:

$$J = \int_o^{t_f} k\{t\}f\{t\}dt - c_r t_f - c_h \int_o^{t_f} F\{t\}dt - c_v \int_o^{t_f} Q\{t\}dt, \qquad [1]$$

where the terms on the right are, respectively, the income from selling marketable fruit, cost of rent, cost of heating and cost of ventilation. Here t is time and t_f is termination time; f is the rate at which marketable fruit is produced; F is heating flux and Q is ventilation flux; k, c_r, c_h and c_v are the unit prices of fruit, rent, heat and ventilation. Note that the market price of tomato may vary with time.

As already mentioned, two methods were used to obtain the required daytime setpoints: the SCS and a modified maximum principle method. Both are described by Seginer and McClendon (in preparation), and the following is a brief abstract of that description.

The SCS assumes a reasonable sequence of future disturbances (weather) as well as a reasonable future control policy. At decision time, a number of control options are put to the test by running corresponding seasonal simulations of the system. The control decision which yields the best value of the performance criterion is selected as the one to be implemented.

The modified maximum principle method utilizes a reduced form of the Hamiltonian and a correspondingly reduced set of co-state equations. In this study the "reduced maximum principle" (RMP) utilized only two of the state variables, namely number of nodes, N, and total dry weight, W. The first is a measure of plant development, and responds mainly to temperature. The second is a measure of assimilation and responds mainly, but not solely, to solar radiation and CO_2 concentration. Their rates of change are functions of the state of the crop, the weather conditions and the control fluxes F and Q. The form of the reduced Hamiltonian is

$$H = k\{t\}f\{t\} - c_r - c_h F\{t\} - c_v Q\{t\} +$$

$$p_N\{t\}(dN/dt) + p_W\{t\}(dW/dt) \qquad [2]$$

where $p_N\{t\}$ and $p_W\{t\}$ are the co-state variables of N and W.

The numerical solution started with initial values of the state variables which correspond to transplanted seedlings. A guess of the initial co-state values $p_N\{0\}$ and $p_W\{0\}$ made it possible to search for the control fluxes which maximize the Hamiltonian (Eq. [2]). The state and co-state variables were then advanced by the state equations (the complete TOMGRO model) and the co-state equations for p_N and p_W. The latter

step required numerical evaluations of $\partial H/\partial N$ and $\partial H/\partial W$. The whole cycle was repeated until termination time has been reached, at which point the performance criterion (Eq. [1]) could be evaluated. An outer computational loop, searching for the pair $[p_N\{0\}, p_W\{0\}]$ which maximizes the performance criterion, completed the program. Note that the performance criterion is still based on a complete simulation of the system behavior. The approximation is restricted to the control decision mechanism.

RESULTS

The results can be divided into two unequal parts. The first part explores computationally some of the intrinsic properties of the system, such as sensitivity to assumptions about the future and relative importance of state variables. The second part, reported in the last sub-section, compares the performance obtainable with the optimal solution with the results of a constant setpoint regime.

Reference case

The somewhat arbitrary reference case considered a tomato crop transplanted on 1 January 1963 and grown for the next 240 days under Ohio weather. All prices, including the price of tomato, were fixed throughout the season as follows: $c_h = 0.02$ \$/MJ; $c_v = 2 \times 10^{-6}$ \$/m^3; k = 2 \$/kg. Expenditure for rent, which is proportional to the length of the growing season, was not charged, but can be properly corrected for whenever t_f is given in advance. A coarse setpoint scale was considered: 10 to 35°C with 5K intervals.

The reference solution to this problem was obtained with the SCS. Future temperature setpoints were taken to be constant at 30°C, since this was the temperature which produced the highest yield. It was also found to be the economically best constant setpoint for the reference situation.

The mean daytime outside temperature and the total solar radiation for each day are shown at the top of Fig. 1. Below these, the daytime temperature setpoints for the reference case, as obtained by the SCS, are shown (Fig. 1[a]). The results show a transition from low setpoints (mostly 15°C) during the first 50 days, to high setpoints (mostly 30°C) during the rest of the season. The excursions of the setpoints on individual days (upward from 15°C and downward from 30°C), seem to correlate positively with exceptional solar radiation levels.

The total marketable yield, total heating energy and total ventilation volume, as well as the resulting performance criterion, are shown in Table 1A, Line [a] (Columns 2 to 6). Deviations from the relevant reference performance criterion are also shown in the table (Column 7). It should be mentioned that the results do not reflect the actual performance of spring tomato crop in Ohio. There are important differences between the true and simulated cases in terms of variety, schedule, CO_2 enrichment (none in the simulation) and prices.

The results of a variation on the reference case, with a finer setpoint scale (10°C to 36°C at 2K intervals) are presented in Table 1A, Line [b]. There is an improvement of over 0.4 \$/m^2 in the performance of the system, but there is no qualitative change of the setpoint sequence (not shown). As a result, and in order to economize on computer time, the reference case was **not** replaced by case [b].

Table 1 Summary of results

Run	1 transp date Julian day	2 season length days	3 total yield kg/m²	4 heating energy MJ/m²	5 vent volume k(m³)/m²	6 perform criterion $/m²	7 relative loss* $/m²
A reference							
[a] reference	1	240	19.00	1019	219	17.173	0.000
[b] variation	1	240	19.02	1000	211	17.606	-0.433
B sensitivity to future							
[c] initially 10°C	1	240	18.91	1012	224	17.133	0.040
[d] 2nd iteration	1	240	19.10	1029	217	17.196	-0.023
[e] 1962 weather	1	240	18.97	1018	216	17.152	0.021
C RMP versus SCS							
[f] p_N	1	240	18.99	1046	187	16.682	0.491
[g] p_W	1	240	12.99	790	333	9.507	7.666
[h] p_N and p_W	1	240	18.94	1038	186	16.738	0.435
D Effect of state (uniform weather)							
[i] optimal		240	16.68	911	63	15.018	0.000
[j] constant 30°C		240	17.46	1061	36	13.611	1.407
E Various price sequences; standard timing							
[k] const price, opt.	1	240	19.00	1019	219	17.173	0.000
[l] " " , 30°C	1	240	20.41	1264	186	15.160	2.013
[m] high to low, opt.	1	240	19.49	1208	265	24.113	0.000
[n] " " , 30°C	1	240	20.41	1264	186	22.883	1.230
[o] low to high, opt.	1	240	12.22	824	305	19.312	0.000
[p] " " , 30°C	1	240	20.41	1264	186	7.438	11.874
F Various price sequences; best timing							
[q] const price, opt.	71	235	22.22	352	279	36.832	0.000
[r] " " , 30°C	71	235	22.31	389	239	36.366	0.466
[s] high to low, opt.	5	175	12.11	1146	102	24.612	0.000
[t] " " , 30°C	5	175	12.04	1164	87	23.924	0.688
[u] low to high, opt.	101	265	21.08	700	270	66.575	0.000
[v] " " , 30°C	101	265	21.58	900	233	64.638	1.937

* negative numbers in Column 7 represent gain relative to the relevant reference.

Sensitivity to assumptions about the future

The SCS makes at each decision time assumptions about the future values of the control actions, the weather and the prices. The sensitivity of the solution to the assumed future control sequence was examined first, by setting it to a constant 10°C rather than to the reference 30°C level. The resulting setpoint sequence (not shown) and the performance criterion (Table 1B, Line [c]) are only marginally different from those of the reference case.

Along the same vein, instead of an arbitrary initial setpoint sequence, the solution of the **reference** case could be used as the initial sequence in a second iteration. As Table 1B, Line [d] shows, the improvement obtained in this manner was marginal.

Having shown that the solution is hardly sensitive to what the control might be in the future, the effect of assumed future weather was explored next. Run [e] was started with the weather sequence of 1962, which was replaced day by simulation-day with that of 1963 (standard sequence). This process is an emulation of a real on-line operation with a perfect 24 hour forecast. The results in Table 1B Line [e] show again that precise knowledge of only the immediate future is required to make correct control decisions.

It should be clear that ridiculous assumptions about the control or weather of the far future may lead to a wrong solution. For example, if future weather is absurdly extreme (say no solar radiation), the current SCS decision would be to cut the losses to a minimum by never heating above the required minimum (10°C).

The effect of wrong assumptions about future prices (of tomato) will be considered in the last subsection.

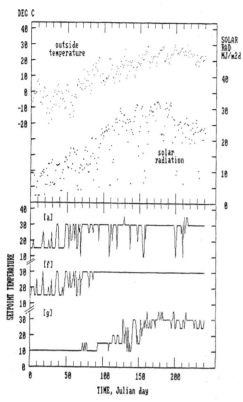

Fig. 1. Weather conditions and calculated optimal daytime setpoints for greenhouse tomato under 1963 Ohio weather. Top to bottom: Mean daytime outside temperature; Total daily solar radiation; [a] Optimal daytime setpoints with SCS; [f] RMP setpoints obtained with p_N; [g] RMP setpoints obtained with p_W.

Comparison of the SCS with the RMP

Based on previous experience (Seginer and McClendon, in preparation) and the tests just mentioned, it is likely that the SCS solution is in fact close to the global optimal solution. The next question, of how well does the RMP method do in comparison, could now be raised. Three runs were conducted to explore this question: In the first, [f], only p_N was retained in the Hamiltonian, in the second, [g], - only p_W, and in the third, [h], - both p_N and p_W.

The results, Table 1C, show the expediency of retaining p_N rather than p_W in the Hamiltonian. Compared to the reference solution, retaining just p_N (out of the 50 odd co-state variables of the system) resulted in a loss of less than 0.5 $/m² per season. Retaining only p_W resulted in a severe loss - over 7.6 $/m². Adding p_W as a second co-state variable to p_N, resulted in a negligible improvement. Other choices of co-state variables were not tried.

Figures 1[f] and 1[g] show the setpoints obtained with p_N and p_W respectively. Sequence [f] is similar to the reference solution [a], except that it clings to 30°C from about day 90 and on. Sequence [g] is quite different from [a] and [f], particularly in that the transition from low to high temperatures is postponed by roughly 80 days. This proved to be a bad strategy.

Figure 2 shows the trajectories $p_N\{t\}$ and $p_W\{t\}$ for Run [h]. The value of p_N is practically constant with time, while $p_W\{t\}$ first rises slightly and then dips considerably toward the end. This may be interpreted as indicating that promoting plant development is equally important throughout the season, while the production of dry matter becomes less important toward its end. This behavior may be an artifact of the truncated Hamiltonian, but probably not, since it makes sense: During the first part of the season promotion of development is presumably required in order to initiate production as early as possible, while toward the end of the season its main purpose is to ripen already formed green fruit. In contrast, there is not much point in creating new dry matter toward the end of the season, when most of it may not mature in time for marketing.

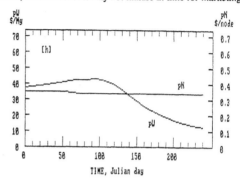

Fig. 2. Evolution of the co-state variables p_N and p_W with time along the season.

Effect of the state of the crop

All the results to this point were for essentially the same problem, and they all show how the chosen setpoints increase from the beginning of the season to its end. It is not clear, however, if this change is mainly related to the transition from winter to summer or to the changing state of the crop. To explore this question, an artificially uniform weather sequence was created, repeating a daily cycle typical of April (solar radiation 20.4 MJ/(m²d) and temperature range 1.7°C to 10.6°C). The setpoint results as obtained by the SCS are shown in Fig. 3. The season starts and ends with high (30°C) setpoints (presumably promoting earliness in the beginning of the season and fruit maturity toward the end), while around day 175 there is a pronounced minimum (15°C). Line [i] of Table 1D summarizes the results for this case in comparison with the results of a constant 30°C setpoint on Line [j]. A loss of about 1.4 $/m² is predicted as a result of maintaining a constant 30°C. This shows that the optimal setpoint trajectory depends to a measurable degree on the changing state of the crop.

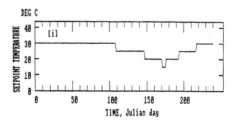

Fig. 3. Optimal setpoints for a uniform weather sequence.

Best timing of growing season

Given a particular climate and economic environment, one expects that there should be an optimal transplanting time and an optimal termination time. To explore this problem, three artificial tomato-price sequences were assumed:

(1) Constant price at 2 $/kg throughout the year (as was assumed for the reference case).

(2) Price of 4 $/kg for the first 180 days of the year and zero for the rest of the year.

(3) Zero value for the first 180 days and 4 $/kg for the rest of the year.

First, the optimal solution was calculated for the standard timing - transplanting on Julian day 1 and growing for 240 days. The results for the three price sequences are compared in Table 1E, lines [k] (which is a copy of [a]), [m] and [o], with the results of a constant 30°C setpoint, on lines [l], [n] and [p]. The optimal solutions for the three price sequences show an economic performance of the same order. The optimal solution is always significantly better than its constant 30°C counterpart, and remarkably better for price sequence (3). Note that the yield is always higher for the constant 30°C regime.

Figure 4 shows the optimal setpoints and the resulting accumulation of marketable fruit for the latter case, [o], both in comparison with the results for the reference solution [k] ($\equiv$[a]). It is clear that the control attempted to minimize the expenditure on heating during the first 110 days, and by doing so it effectively delayed the development of the crop. Marketable fruit only starts to appear on day 140, about 35 days later than for the constant price case [k]. Marketing starts on Julian day 180 (vertical segment in the figure), when 4 $/kg can finally be fetched for the fruit (note that the **rate** of production has just reached its maximum about day 180). This suggests that a better performance could be obtained by postponing the transplanting date, and by moving the termination date to the end of the year, when the price of tomato drops back to zero.

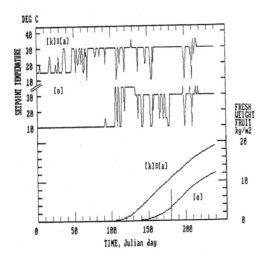

Fig. 4. Optimal setpoints and resulting accumulation of mature fruit for two price sequences. [k] $\equiv$ [a] uniform price at 2 $/kg; [o] zero price up to julian day 180 and 4 $/kg from then on.

The best transplanting and termination dates for the three price sequences and a constant 30°C regime (Lines [r], [t] and [v] in Table 1F) were obtained by trial and error. The improvement over the standard timing is relatively small for sequence (2), but otherwise it is very impressive. The improvement for sequence (1) (Line [r] compared to Line [l]) is due to shifting of the production period into the summer, resulting in a higher yield and lower expenses for energy. The small improvement for sequence (2) (Line [t] compared to Line [n]) is mainly due to termination on day 180, when the price of tomato drops to zero. As a result, the greenhouse is only occupied for 175 days. The improvement for sequence (3) (Line [v] compared to Line [p]) is dramatic and results from shifting and expanding the growing season, as suggested in the previous paragraph. Note that finding the optimal timing requires information about the cost of rent (here $c_r = 0$).

The optimal solutions for the new timing are summarized in Table 1F, Lines [q], [s] and [u]. The improvement over the constant temperature regime, Lines [r], [t] and [v], is now by an order of magnitude smaller than observed for the standard timing. This suggests that while proper timing produces much better results than arbitrary timing, the gain by optimization is less pronounced when the proper timing is selected.

DISCUSSION AND CONCLUSIONS

Optimal control of the greenhouse climate is often viewed as having a hierarchical structure, where the control rule (law, policy) is obtained off-line, perhaps once a year. This rule is then invoked on-line at frequent intervals, to generate specific control decisions. The present study deals mainly with the first stage of developing a control rule, namely finding optimal solutions to historic control problems. At a later stage, when solutions are available for the widest range of conditions that one can expect, these solutions may be used to form the control rule.

This study is just a preliminary exploration into the properties of multi-state-variable crop models from the point of view of greenhouse environmental control. Nevertheless, some tentative conclusions may be drawn from this limited evidence.

1. The SCS apparently could find optimal solutions under wrong (but not absurdly wrong) control and weather assumptions beyond the immediate future (Table 1B). This indicates that the control decisions of this and similar problems may be based on the current state of the system and on simple, short term, weather forecasts. When eventually the optimal solutions are distilled into a control rule, the required inputs could, therefore, be readily available. This may not be the case if the system includes an important operational storage, such as soil water capacity or temperature integration capability of the crop. The characteristic of an operational storage is that it is limited. As a result, its inclusion in a system leads to an optimization problem with constraints on the storage state variable (e.g., Bailey and Seginer, 1989). Intuitively, if the characteristic time of the operational storage is of the order of several days, a weather forecast for that duration will be required. In addition, the optimal control sequence for that whole period will have to be obtained at each decision time. Israeli and Lambert (1986) applied a similar approach to irrigation scheduling.

2. Unlike assumptions about future control and weather, assumptions about future prices have had a significant effect on the control decisions in this study (Table 1E; Fig. 4). The dramatic effect is, however, the result of comparing very different (even opposite) price sequences. The price difference

from one season to the next is in reality not nearly as drastic. Therefore, one may expect to manage quite well even if actual prices differ from the expected ones.

Figure 4 shows that in Run [o] heating started on Julian day 105, 75 days in advance of the price step. This is more than the time required for fruits to mature (30 to 50 days; not shown), since it was desirable to reach day 180 not just with mature fruit but with the maximum possible rate of production. This suggests that an initially unexpected steep rise in tomato price should be known almost two months in advance in order to take full advantage of it. On the other hand, if prices drop suddenly and are expected not to rise again (Run [m]), heating can be stopped on a short notice.

3. Experimentation with the RMP method (Table 1C, Fig. 2) showed that a properly chosen single state variable is capable of producing a satisfactory control trajectory. This gives some insight into the question which state variables are more important than others at a given stage of the crop's development. Originally, the single state-variable formulation of Seginer and others (1986) resulted in a constant value for the co-state variable. (This value was regarded as an expression of the intensity of cultivation by Seginer (1989).) Later, a study of the two state-variable lettuce model (Seginer and others, 1991) showed that one of the variables, namely total dry matter, is dominant, and that its importance grows steadily towards harvest time.

In the present study N is a very dominant state variable (Table 1C). The value of its co-state variable, p_N, is roughly constant throughout the season (Fig. 2). This means that to a first approximation the control, responding mainly to the changing weather, is largely independent of the state of the crop. The behavior of co-state variables depends, however, not only on the basic system equations, but also on the control available in the greenhouse, on the general nature of the disturbances (weather and prices) and on the performance criterion. For instance, Run [i] shows that under uniform weather the optimal control does depend to a significant degree on the state of the crop. As a result, at least one measure of the state of the crop, preferably number of nodes (or at least time from transplanting), will have to be included in the argument list of the control rule.

4. Optimal timing of the season is a critical factor, because it considerably affects the performance criterion, even when the price of produce is constant throughout the year (Run [q] compared with Run [a]≡[k]). In practice, however, the optimal timing requires only minor adjustments from one year to the next. Since TOMGRO tends to produce maximum yield at a constant 30°C daytime setpoint, optimal timing tends to match the growing season with the period when pushing for maximum yield is justified. As a result, when timing is optimal, further gain from day by day optimization is not large (Table 1F).

In summary, the SCS seems to produce close to optimal setpoint trajectories. This will probably hold true for similar problems, provided that the system contains no important operational storage, and as long as the uncertainty about the future is not substantial.

Further, despite the multi-state-variable structure of present-day crop models, the RMP results suggest that they may be replaced by much simpler models for control purposes. The equivalent models may contain just a few properly selected state variables, preferably measurable under commercial conditions. They may be explicitly determined or implicitly included in a control rule trained with optimal solutions based on the complete model. In the latter approach, the argument list of the control rule would include the selected state variables together with parameters describing the impending weather and the expected prices.

ACKNOWLEDGEMENT

I would like to thank Dr. J.W. Jones, the University of Florida at Gainesville, and Dr. R.W. McClendon, the University of Georgia at Athens, for fruitful discussions when this study was launched. Dr. W. Bauerle, the Ohio State University at Wooster, provided information about tomato practices in Ohio.

REFERENCES

Bailey B.J., Seginer I. 1989. Optimum control of greenhouse heating. Acta Hort. 245:512-518.

Bellman R.E., Dreyfus S.E. 1962. Applied Dynamic Programming. Princeton University Press.

Challa H., van de Vooren J. 1980. A strategy for climate control in greenhouses in early winter production. Acta Hort. 106:159-164.

Challa H., Schapendonk A.H.C.M. 1986. Dynamic optimalization of CO_2 concentration in relation to climate control in greenhouses. Chapter 9 in: Carbon dioxide enrichment of greenhouse crops, H.Z. Enoch and B.A. Kimball, Editors, Vol. I:147-160, CRC Press.

Critten D. 1991. Optimization of CO_2 concentration in greenhouses: A model analysis for the lettuce crop. J. Agric. Eng. Res. in press.

Garcia C.E., Prett D.M., Morari M. 1989. Model Predictive Control: Theory and practice - a survey. Automatica 25(3):335-348.

Israeli I., Lambert J. 1986. Irrigation scheduling of movable irrigation systems to maximize return. ASAE Paper No. 86-2073, 33 p.

Jones J.W., Dayan E., Allen H., van Keulen H., Challa H. 1991. A dynamic tomato growth and yield model (TOMGRO). Trans. ASAE, in press.

Marsh L.S., Albright L.D., Langhans R.W., McCulloch C.E. 1987. Economically optimum day temperatures for greenhouse hydroponic lettuce production. ASAE Paper No. 87-4023, 36 pp.

Pontryagin L.S., Boltyanskii V.G., Gamkrelidze R.V., Mishchenko E.F. 1962. The mathematical theory of optimal processes. John Wiley Interscience. 360 pp.

Seginer I., Angel A., Gal S., Kantz D. 1986. Optimal CO_2 enrichment strategy for greenhouses. A simulation study. J. Agric. Eng. Res. 34:285-304.

Seginer I. 1989. Optimal greenhouse production under economic constraints. Agricultural Systems 29:67-80.

Seginer I., Shina G., Albright L.D., Marsh L.S. 1991. Optimal temperature setpoints for greenhouse lettuce. J. Agric. Eng. Res. In press.

Seginer I., McClendon R. 1992? Methods for the optimal control of the greenhouse environment. In preparation.

Swaney D.P., Mishoe J.W., Jones J.W., Boggess. 1983. Using crop models for management: Impact of weather characteristics on irrigation decisions in soybeans. Transactions of the ASAE 26(6):1808-1814.

NOTATION

symbols		SI units
c_h	unit price of heating energy	$/J
c_r	unit price of rent	$/(m²[soil]s)
c_v	unit price of ventilation	$/(m³)
F	heat flux from heater	W/m²[soil]
f	rate at which marketable fruit is produced	kg/(m²[soil]s)
J	seasonal performance criterion	$/m²[soil]
k	unit market price for tomato	$/kg
N	number of nodes	[node]/m²
p_N	co-state variable of N	$/[node]
p_W	adjoint variable for W	$/kg[d.m.]
Q	ventilation rate	m³/(m²[soil]s)
t	time	s
t_f	termination time	s
W	total dry matter	kg[d.m.]/m²

FUNDAMENTAL STUDY OF PLANT FACTORIES

M. Takatsuji

School of High-Technology for Human Welfare, Tokai University, Numazu, Shizuoka, Japan

Abstract. Two basic conditions are indispensable to the creation of a plant factory: to standardize the relation between environmental conditions and plant growth and to promote plant growth by environmental control. In the present study, lettuce and sweet pepper were used as samples and the relations between light, temperature and CO_2 concentration and growth indices, such as photosynthesis and plant weight, were quantified in detail. In addition, the control of these environmental conditions achieved substantial growth promotion effects on the sample crops. Based on the data thus obtained, we conducted a feasibility study on plant factories, and technical problems concerning these factories were clarified.

Keywords. Plant factory; measurement of growth; environmental control; promotion of growth; lettuce.

INTRODUCTION

Plant factories are referred to as "the forth agriculture" as they are situated at the last stage of the development of cultivation technology, following cultivation in open fields, one in plastic green houses and hydroponics. While hydroponics eliminates the use of soil and thus enables us to control the intake of nutrients by the roots of plants artificially, plant factories apply an advanced environmental control to the portion of plants above the ground. By this techniques, we can attain a year-round, higher productivity in cultivation as the effects of natural environments are reduced.

There are two types of plant factories: one utilizing sunlight and totally controlled one that uses only artificial light. The latter is an ideal plant factory in the sense that we can achieve crop production with almost no reliance upon natural environments.

Thus, when we utilize a plant factory, we should know the relation between environmental conditions and plant growth far more precisely both from the viewpoint of plant physiology and of economic effects. But there was no attempt to quantify the

growth of higher plants from these angles in the past, although several plant factories have already been constructed.

Factory production has two main characters: standardization and high speed. Thus, if we are to materialize a plant factory, we have to (1) quantify the relation between environmental conditions and plant growth and (2) study the method of growth promotion by environmental control. This paper summarizes the results of our study regarding these two subjects which was continued for ten years from 1975 to 1985. More detailed descriptions of the study will be found in References given at the end.

MEASURING INSTRUMENTS

Plants conduct such basic physiological reactions as photosynthesis, respiration and transpiration, and translocate and distribute the substances they produce by biosynthesis to each of their organs, resulting in the growth of the entire plants. The growth of plants can be measured by such indices as weight, volume, water content and shape. We can culture plants in a growth chamber where different environmental conditions are given and study the state of their growth by measuring these indices.

Table 1 Selected Growth Indices and Methods of Their Measurement

		Photosynthesis →	Biosynthesis →	→	Plant		
Growth Indices	Leaf Vegetable (Lettuce)	Photosynthesis Respiration			Total Weight	Water Content	Shape
	Fruit Vegetable (Sweet Pepper)	Photosynthesis Respiration	Electric Potential Translocation		Volume		Partial Weight
Methods of Measurement	Leaf Vegetable (Lettuce)	Mass Spectrometer			Semiconductor Strain Gage Balance	Drying	Eye-measurement
	Fruit Vegetable (Sweet Pepper)	Mass Spectrometer	Electrode Mass Spectrometer ($^{13}CO_2$)		Immersion in Water		Resonance Method

The measurement should preferably be made on a non-destructive basis. Lettuces (butterhead type) were selected as a representative of leaf crops (vegetative growth plants) and sweet peppers, of fruit crops (reproductive growth plants) for the present study. Table 1 shows the measuring methods applied to these crops which were newly developed for the studies and applied to the major growth indices of the above-mentioned growth process.

A multi-component mass spectrometer was used to measure the gas exchange conducted by the plant bodies in photosynthesis, respiration and transpiration. The CO_2 absorbed by the leaves of sweet peppers and the CO_2 discharged as a result of the translocation of assimilates to their fruits were distinguished from each other using ^{13}C, and the $^{13}CO_2$ was measured also with the mass spectrometer. To monitor the growth conditions of sweet peppers, bioelectric potential was used. A pair of scales was used to measure the weight of the sample vegetables in general, but when non-destructive tests were desirable, an on-line measuring unit utilizing a semiconductor strain gauge was utilized. To conduct the non-destructive measurement of the weight of fruits' parts, forced oscillation was applied to the fruits, and the fruits' proper frequency was employed to calculate the weight. The volume of fruits can easily be measured by immersing them in water, while their water content and dry weight can be measured after drying them in a drying chamber. Needless to say, the latter measurement is conducted on a destructive basis. The shape of fruits was examined visually in almost all cases.

Growth Measurement of Lettuces

Quantification of growth

The relations between five environmental conditions (light intensity, day length, temperature, water temperature and CO_2 concentration) and four growth indices (photosynthesis, respiration, weight and water content) were measured in detail. Many years were needed to obtain the data which would be commonly applicable and recurrable since plant growth takes a long time and living plants show complex reactions to environmental stimuli, including memory effects, compound environmental effects, adaptation and transient responses.

The factor affecting plant growth effectively is the integral dry mass produced by photosynthesis per day (ΔM) rather than photosynthetic rate (P). This mass can be obtained by deducting the aggregate respiration volume during the dark period from the aggregate photosynthetic mass during the time when light is applied. The dependence of this ΔM on day length, temperature and CO_2 concentration were measured. Growth ratio ΔW (ratio of change in weight per day) is a direct index of the visual growth of plants. Thus, this ratio's dependences on day length, temperature and CO_2 concentration were also measured. From commercial viewpoint, quicker growth is by no means the most important element, plant shape and other qualitative characters are also significant factors. Water content Wc is a principal index that shows the quality of lettuces and other leaf crops. This is the ratio of wet weight to dry weight. To measure this ratio required much trouble since destructive tests had to be used, but its dependences on the three environmental factors were measured, too.

The outcome of the above measurements is summarized in Fig.1. It shows the growth locus of lettuce corresponding to changes of environments in the three dimensions, ΔM (relative value), ΔW (relative value) and Wc (absolute value).

The first curve is the one when the temperature Ta changes from 14°C to 30°C, while the three others show the growth locus when Ta is fixed at 20°C, the optimal photosynthetic temperature and when water temperature Tw changes from 20°C to 26°C, day length L from 6 to 24 hours and CO_2 concentration N from 400 to 1,200ppm. The growth state in other temperature conditions can be obtained almost precisely by moving the three curves in parallel along the Ta curve, although some deviations occur as a result of the compound effects of environments. The growth in any other combination of environmental conditions is also obtainable approximately by the parallel movement of the curves, but care should be taken in extreme conditions since the compound effects are great. When the water content is 23 or less, the quality of grown lettuce will be commercially acceptable.

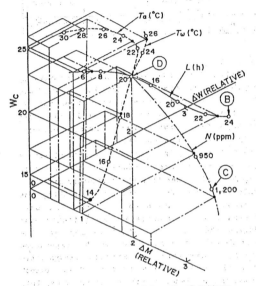

Fig.1. Quantification of Growth of Lettuce

Figure 1 tells us a number of interesting facts. Here, a brief description is made on the impact of each of the environmental conditions selected. ΔM is maximized at 20°C and ΔW at 22°C, with a small gap in the maximization temperature between them. Both air and water temperatures have a noticeable effect on water content, when the temperature is high, water content goes up, making the shape of leaves worse. Considering this, 20°C was selected as the optimal temperature for the growth of lettuce (Point D).

The total weight and dry weight (which increases in proportion to ΔM) and day length roughly maintain a proportional relation. They reach a peak when L is 24 hours (Point B). Day length promotes the growth of lettuces more than temperature. CO_2 also promotes growth greatly and at the same time reduces water content. When N is 1,200ppm, Wc drops to 16 (Point C). Day length and CO_2 concentration give only a small effect on leaf shape but a great influence on leaf color and thickness.

Figure 1 is the first that quantified the relation between environmental conditions and the growth of a higher plant almost perfectly. It enables us to

estimate at once the growth state of lettuce in a given environment. This method is applicable to other vegetatively growing plants, such as spinach, too.

Promotion of Growth

Figure 2 shows the actual growth curve of lettuce which grew from 25g to 200g under the several different combinations of environmental conditions given in Table 2. Conditions B,C and D in the table correspond to the respective symbols used in Fig.1. From this chart it becomes evident that the growth of lettuce is greatly affected by the three factors, day length, CO_2 concentration and temperature.

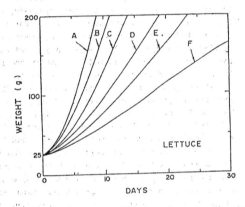

Fig.2. Growth curves of Lettuce Corresponding to Environmental Conditions given in Table 2

Table 2 Various Combinations of Environmental Factors used for Growth Experiment of Lettuce

Factors \ Conditions	A	B	C	D	E	F
Day Length (h)	24			12		
CO_2 Concentration(ppm)	1000	400	1200	400		
Temperature(°C)	20				26	14
Relative Humidity(%)	80 ~ 85					
Light Intensity(klx)	18					

GROWTH MEASUREMENT OF SWEET PEPPERS

Similar quantitative measurement was conducted on the relation between environmental conditions and the growth of sweet pepper, and large growth promotion by optimum environmental control was also observed.

The optimal photosynthetic temperature of the selected variety was 24°C and its optimal day length, 16 hours. Figure 3 shows the measurement results of the enlarging growth process of the fruits: the volume of the fruits under conditions A and B and change in the increase in the volume per day. Under condition A, the pace of growth was nearly 3 times quicker than that under average open-field conditions. Besides growth promotion, the artificial cultivation of fruit crops gives rise to great extension of the harvesting period, resulting in an enormous harvest.

Environmental Conditions	A	B
Temperature(Day/Night)(℃)	24/18	24/18
Day Length(h)	16	12
CO_2 Concentration(ppm)	1000	400

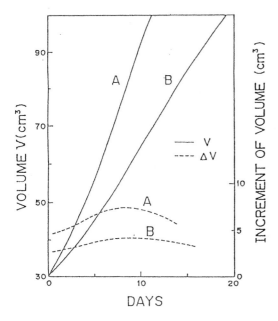

Fig.3. Acceleration of Growth of Sweet Pepper under Controlled Environments

FEASIBILITY STUDY

After the growth data of lettuces and sweet peppers were collected, a feasibility study was conducted on plant factory for these crops using the data obtained. First, a conceptual design of a lettuce factory was prepared, and its production scale and cost were estimated. An actual plant factory has a number of technical problems to be solved: among them are system design, illumination, air conditioning, dense planting, hydroponics, and selection of proper adiabators and coating materials and fast-growing varieties. In addition, there is a difference in the importance of these technical questions between the case where totally controlled artificial environments are adopted and the case where sunlight is utilized. To study these problems, an experimental factory of the totally controlled type and of the sunlight utilizing type was built. Utilizing the practical data collected from these factories, commercial plant factories have been constructed and technical guidance has been given.

We have designed and built several commercial plant factories. One example is the "Biofarm" constructed in The Daiei Inc., a leading supermarket chain in Japan, at one of its stores. This farm is located near the store's vegetable corner, and customers can see it and buy the vegetables raised there using no agricultural chemicals.

Studies of plant factories have been active in Japan in recent years, and the commercial operation of totally controlled factories is just around the corner. Some plant factories have already achieved profitability in the production of lettuce and other leaf crops. For example, Miura farm which I have directed from the beginning, succeeded to operate a profitable lettuce factory. The cultivated area is about 300m² and it can produce 500 bundles of high-quality lettuces everyday.

However, electric power still accounts for about 40% and depreciation expenses for production facilities 30% of the total production cost. This means the need of further efforts at saving energy and cutting cost. To realize more efficient plant factories, a number of problems are to be solved. Among them are the development of high-efficiency illuminating lamps, establishment of better hydroponics technology and breeding of new varieties whose growth is quicker.

REFERENCES

Takatsuji,M.(1979). Plant Factories. Kodansha,Tokyo

Takatsuji,M.(1986). Vegetable Factories. Maruzen, Tokyo

Takatsuji,M.(1987). Fundamental Theory and Practice of Plant Factories, 3rd ed. Gijutsu Joho Center. Osaka

Takatsuji,M.(1987). Introduction to Plant Factories. Ohmsha,Tokyo.

Takatsuji,M.(1990). The Birth of a Plant Factory. Nihonkogyo-shinbunsha, Tokyo.

Copyright © IFAC Mathematical and Control Applications
in Agriculture and Horticulture, Matsuyama, Japan 1991

PLANT FACTORY AND ITS PROSPECTS

S. Nakayama

Central Research Laboratory, Mitsubishi Electric Corporation, Amagasaki, Japan 661

Abstract. Plant factory which produces vegetables in full artificial environment is expected as one of the production methods
in agriculture even manufacturing industry especially in Japan.
By reviewing a series of investigations we conducted, the significance of artificial environment control and the prospects of
plant factory are discussed.

Keywords. Air conditioning; Agriculture; Biology; Computer
control applications; Environment control; Farming.

INTRODUCTION

Agriculture has been developed by concur-
ring inconvenient natural environment. The
environment controls such as watering and
dosing chemical fertilizers and improving
cognate species of plants have greatly
expanded the yield and the area of farm-
land. However, agriculture can not be
formed without using energy of sun light,
therefore, completely breaking away from
natural is impractical. The principle of
the production in suitable locations and
unstable and low land productivities will
essentially remain. In Japan, a decrease
of farming population and an advanced age
make rapid progress, a part of which is
considered to arise from the essential
problems.

Agriculture is one of basic industries, but
fields which get behind in the introduction
of industrial technologies. A part of
nonagricultural industries have become of
interest in plant factory as a new market.

Plant factory can be defined as a system
producing plants under artificially controlled environment by using industrial
technologies and ways of thinking. It is
characterized by using lamps instead of
sunlight. A glass house system using
sunlight together with supplementary lamps
can be also regarded as a plant factory,
but not considered in this article, because
the principle of the cultivation should be
basically different.

Several commercial plants have been in
operation. However, because of the higher
production cost, it was not widely generalized. Hereafter, investigations for improving plant factory in both the constituent and system technologies, and establishing the concepts for using it reasonably
must be performed. In order to argue about
this subject and others relating to plant
factory, a series of investigations we
carried out is presented.

PLANT FACTORY IN JAPAN

Much attention has been denoted to plant
factory in Japan. The Ministry of Agriculture, Forestry and Fisheries has been
investigating the potential of plant factory. Society of plant factory having more
than 600 memberships was established in
1989.

There have been several commercial and
demonstration plants in operation. Typical
examples in operation or under investigation are shown in Fig. 1 - 3. It can be
seen that the design of lighting or the
style of the cultivation bed including a
hydroponic method are characterized. But
almost all plants including others are
producing leaf vegetables, and few quantitative data for discussing the details are
published.

PRINCIPLE OF PLANT FACTORY

An important necessary condition going well
to plant factory is considered to accelerate the growth rate of harvest parts of
plants. First of all, let's discuss how
the condition relates on artificial environment control by reviewing growth characteristics of lettuce and turnip[1,2].

Exponential growth

Kimura presented a model of vegetative
growth of plants illustrated in Fig. 4[3].
C and F are the masses of nonphoto- and
photosynthetic organs, respectively. a is
the photosynthetic rate, by which gross
production rate (P_g) is given as aF. m is
the distributing ratio of P_g to the photosynthetic organ. r and k are the respiration rates for maintaining life and transforming the photosynthetic products to
organs.

Assuming validity of this model and parameters such as a, r and k as constant during
the growing stage, the weight of plant
(F+C) against the cultivation time is
represented by Equation 1[3]. The second

Fig. 1 Plant factory developed by Chubu Electric Power Co.,Inc and Mitsubishi Electric Co.

Fig. 2 Plant factory developed by Hitachi Co.

Fig. 3 Plant factory developed by Q.P. Co.

$$W = F_0(1+\gamma)e^{\alpha t} + (C_0 + \gamma F_0)e^{-\beta t}$$

$$\alpha = (ma - r_f)k_f, \ \beta = r_c k_c, \ \gamma = \frac{(1-m)ak_c}{\alpha+\beta} \qquad 1$$

term may be relatively small, so it could be said that plants grow with an exponential mode when assuming constant parameters. Getting exponential growth mode might be considered to be one of conditions for the rapid growth.

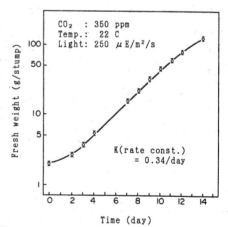

Fig. 4 Model of vegetative growth of plants

Fig. 5 shows semilog plots of fresh weight of an individual lettuce vs. cultivation time under constant artificial environment[1]. For keeping the space distribution of the irradiated light intensity at constant during the experiments, fluorescent lamps are arranged around the lettuce. It can be regarded that, except for the beginning and the latter terms, the exponential growth mode is realized. It could be reasonably assumed that plants are able to grow exponentially under constant or periodic environment, at least in a part of vegetative growth stage. This is one of importances of artificial environment control.

CO$_2$: 350 ppm
Temp.: 22 C
Light: 250 μE/m^2/s

K(rate const.) = 0.34/day

Fig. 5 Growth curve of lettuce under omni-directional irradiation (average of three experiments)

Growth rate

The slope of the part of the straight line of the semilog plots gives the growth rate constant (K). For getting rapid growth of plants, higher rate constant is also required. In general, the rate constant is said to become higher with increased light intensity and applied CO$_2$ concentration

Table 1 Effects of CO_2 concentration and lighting intensity on rate constant (K) of lettuce under omnidirectional irradiation

Lighting intensity ($\mu E/m^2/s$)	CO_2 concentration (ppm)	K (day^{-1})	Water content (%)
250	350	0.34	90.1
	950	0.31	90.3
	2000	0.32	87.2
	4000	0.33	86.4
164	350	0.26	91.0
	1200	0.31	91.3
	4000	0.27	89.2
85	350	0.21	92.2
	1200	0.29	94.1
	4000	0.27	89.5
50	1200	0.21	-

except for extreme cases. Table 1 shows the dependency of the rate constant on the light intensity and applied CO_2 concentration[1]. All experiments were performed under the omnidirectional irradiation and constant environment condition. This Table indicates that the dependency of the light intensity and CO_2 concentration on the rate constant are not coincide with that generally accepted. For example, at the stronger light intensity, the rate constant is seemed to be almost independent on CO_2 concentration. It is also noticeable that the high rate constant is obtained at low light intensity and high CO_2 concentration. These results suggest that other rate determining mechanism than light intensity and CO_2 concentration must be exist and the light intensity does not the rate determine factor. General principle for increasing the rate constant is not know, but partly discussed in next section.

Deviation from the exponential growth

As seen in Fig. 5, in the latter term of the growing stage, a deviation from the exponential growth is apparently observed. The rate constant seems likely to decrease gradually in the later term. In Fig. 6, CO_2 uptake rates (NPR) per one stump of lettuce are plotted against the existing fresh weight at fixed light intensity and several different photoperiods and CO_2 concentrations. As the photoperiod becomes longer, the uptake rate is apparently reduced and There seemed to be a marked tendency for the increase of the rate with increasing fresh weight to saturate. These correspond to the deviation mentioned above.

Fig. 7 shows the similar plots also for lettuce at fixed CO_2 concentration and several different photoperiods and light intensities. Under the continuous irradiation, the dependency of the CO_2 uptake rate on the light intensity becomes small. Fig. 8 shows the CO_2 uptake rate of lettuce before and after breaking the irradiation for 24 hours. The differences between the rates at before and after the dark period is greater at stronger light intensity.

As compared those results with those of experiments for turnip having larger tuber, remarkable differences in the characteristics are observed as shown in Fig. 9 and 10[2]. There is no difference of CO_2 uptake rate between photoperiods of 12 and 24 hrs, and the uptake rate increases with increased light intensity. The differences

in the CO_2 uptake rates coincide with the differences in the stomatal resistance of leaf as shown in Fig. 11 and 12.

Deviation from the exponential growth could be considered to attribute to an unbalance between photosynthetic rate and converting rate of the photosynthetic products from the leaves. Environmental control for balance is suggested to be important for reducing lighting power consumption.

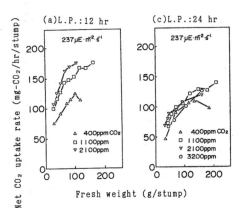

Fig. 6 Dependency of CO_2 uptake rate of lettuce on CO_2 concentration and lighting period

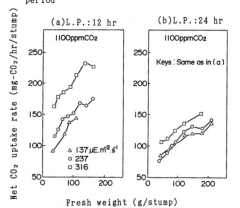

Fig. 7 Dependency of CO_2 uptake rate of lettuce on ligting intensity and period

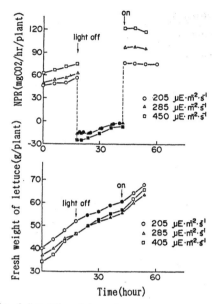

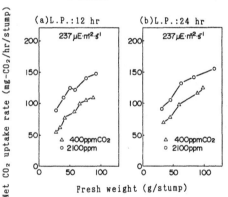

Fig. 8 Net CO_2 uptake rate of lettuce before and after dark period

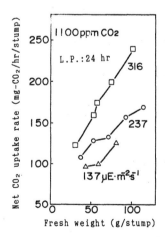

Fig. 9 Dependency of CO_2 uptake rate of turnip on CO_2 concentration and lighting period

Fig. 10 Dependency of CO_2 uptake rate of turnip on lighting intensity

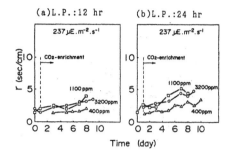

Fig. 11 Dependency of stomatal resistance of leaf of lettuce on CO_2 concentration and lighting time

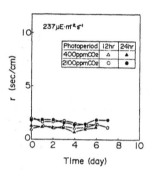

Fig. 12 Dependency of stomatal resistance of leaf of turnip on CO_2 concentration and lighting time

Environment control in plant factory

From above discussions, basic thinking way of reasonable environmental control in plant factory is proposed as follows;

(a) Environment control should be a means for an exponential growth.
(b) Environment control should be balanced with the rates of photosynthesis and the transforming the products.
(c) Designs of lighting for omnidirectional irradiation to the groups of leaves should be considered as a means of reducing the lighting power consumption.

We progressed a development of plant factory followed by this principle.

INVESTIGATION OF PLANT FACTORY

We have being on developing a plant factory in a collaboration with Chubu Electric Power Co. Inc. Fig. 2 and 12 show the plant factory. In this chapter, the outline of the investigations are presented.

Main features of the plant factory are as follows;

(a) Fluorescent lamps with relatively lower contents of green light are used and are arranged close to cultivation bed at the distance of 40 cm from the top of the panel.
(b) Main equipments including two stages of cultivation beds are arranged in a package.

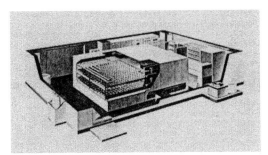

Fig. 13 Schematic diagram of plant factory developed by Chubu Electric Power Co.,Inc. and Mitsubishi Electric Co.

Basic investigation

Before constructing the plant factory, some basic investigations focusing to reduce the power consumption for lighting was performed[4].

Growth characteristics of groups of lettuce.
We considered that a basic principle for reducing the power consumption is to irradiate the light to the groups of plant leaves from omnidirection. In order to achieve this, fluorescent lamps were arranged close to plants. Fluorescent lamp is suitable for this close arrangement because of the lower content of heat radiation.

Growth characteristics of groups of lettuce were investigated by using a small growth chamber consisting of small rooms of 30 cm (width) x 120 cm (length) x 20 cm (height). The inner walls of the rooms are painted with white color with the reflective index of 80 %. The light intensity distribution in the space was preliminarily confirmed to be relatively homogeneous. Lettuces were planted in the rooms in a line at a regular interval.

Fig. 13 and 14 show semilog growth curves of groups of lettuce planted by 3 and 8 stumps per room at different light intensities. Applied CO_2 concentration was 4000 ppm. The dependency of the growth rate on light intensity is smaller in the coarser arrangement than that in the closer arrangement. This result indicates that the growth characteristics under the coarser arrangement are resemble to those under the omnidirectional irradiation. The light intensity dependence become larger at closer arrangement.

Fig. 15 shows relations between total daily increased fresh weight in a room against the whole existing fresh weight. With increasing the total weight, the increased weight is linearly increased, and passing through a maximum, and then gradually reduced. In the four cases the set numbers are different, the plots are on the same curve. The maximum growth rate and the corresponding the existing weight depend on light intensity as shown in Fig. 16.

In cultivation in groups, the combination of the planted density and irradiated light intensity is suggested to be important for efficient growth. By adjusting the spacing and the lighting intensity suitably, the maximum efficient yield will appear.

Assuming the maximum yield would obtained by a suitable spacing adjustment, required amounts of the lighting power for producing 100 gr of lettuce are estimated as Fig. 17, where 40 W of fluorescent lamps were used.

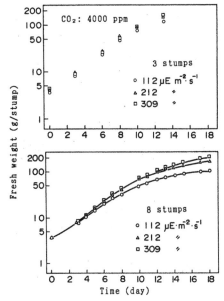

Fig. 14 Growth curves of groups of lettuce

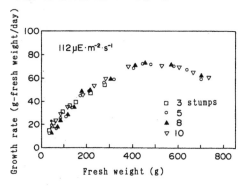

Fig. 15 Growth rate of group of lettuce against existing weight

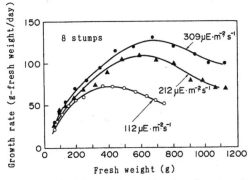

Fig. 16 Dependency of growth rate of group of lettuce against existing weight on lighting intensity

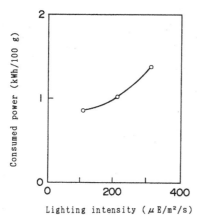

Fig. 17 Consumed power for lighting for producing 100 g of lettuce

Effect of light spectrum.

Spectrum of the light irradiated is said to connect with quantity and quality of plants harvested. Trial fluorescent lamps of different radiative intensity ratio of red, green and blue lights were evaluated by cultivating lettuce. The results are shown in Table 2, showing that the spectrum effects on the growth rate.

Table 2 Comparison of yield of lettuce with different lamps

Relative radiative power				Yield[*]
Red	Green	Blue	Total	
23	51	26	100	6.12(1.00)
38	24	38	99	6.72(1.10)
30	32	38	110	8.32(1.36)

[*]Yields of dry weight in 7 days from 30 gr of fresh weight

Outline of the plant factory[5]

The main specification of the plant factory we developed is summarized in Table 3. The main components and the functions are as follows:

Budding equipment.

This equipment is composed with three beds and produces infantile seedlings. Photoperiod, atmospheric temperature, humidity and temperature of hydroponic solution are automatically controlled. In the case of lettuce, 180/day of the seedlings (about 0.2 gr) are produced in 9 days.

Seedling equipment.

Seedlings to supply to main equipment are produced from infantile seedlings with the three beds equipment. The composition and the control mode are basically the same as the budding equipment. About 130 of seedling of lettuce (around 8 gr) are produced in 9 days.

Main cultivation equipment.

Seedlings from the seedling equipment are planted on the floating panels from the side of the double step beds. The planted panels are daily moved to another side of the beds where being harvested. A spacing mechanism of the plants is equipped only on the lower bed.

Heater, heat exchanger, humidifier and dehumidifier are located on the upper section of the facility. Air regulated by these components is distributed to the cultivation rooms and circulated by funs set at the sides of the rooms. Storing tanks of hydroponic solution are in the lower section, and the solution regulated was intermittently surprised to the beds and returned to the tanks. Main control of the temperatures of all components including the budding and seedling equipments are performed by controlling the flow rate of the chilled water produced outside of the equipment.

Atmospheric temperature, humidity CO_2, concentration, photoperiod, temperature, pH and electric conductivity (EC) of the solution are automatically controlled. Light intensity and the velocity of the wind are set manually. Control of these parameters are performed by a sequencer (MELSEC-A3, MITSUBISHI ELECTRIC). Results of the measurements including power consumption of each component and the trends are illustrated on CRT of a personal computer.

Results of operation

Daily variations of temperatures, humidity, CO_2 concentration, EC and pH of hydroponic solution are shown in Fig. 18 - 19. The light intensity at full power in main cultivation equipment is 650 uE/m2/s (45 klx) and the distribution of the intensity on the bed is in ±25 %.

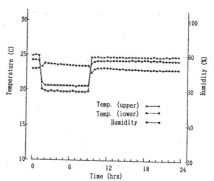

Fig. 18 Daily variations of temperature and humidity in main cultivation equipment

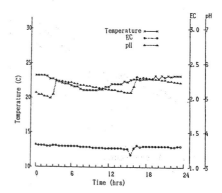

Fig. 19 Daily variation of temperature, EC and pH of circulated solution in main cultivation equipment

Fig. 2 shows growing lettuces in the main cultivation equipment. In the case of lettuce, 120 stumps/day (about 70 gr) are produced by the stay 9 days. Fig. 20 shows typical growth curves of lettuce, indicating near exponential growth mode. However, the growth rates of lettuce vary with breeds as shown in Fig. 21.

Power consumption of this plant factory is shown in Fig. 22. The daily average power consumption throughout a year was about 300 kWh. 45 % of the total consumption was consumed by lighting, 35 % by air conditioning and 20 % by others such as pumps.

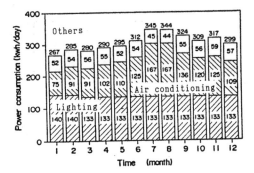

Fig. 22 Power consumptions of plant factory

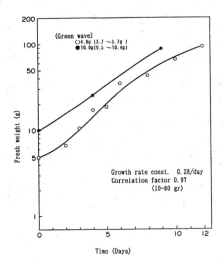

Fig. 20 Growth curve of lettuce in main cultivation equipment

PROSPECTS OF PLANT FACTORY

In this article, a series of investigations was discussed by focusing on those relating to the lighting power consumption of plant factory. At this point of time, it does not clear whether the evaluated power consumption for the lightning is practically acceptable or not.

Power consumption for lighting is indispensable to plant factory. The strong point of it arise from the lighting, so simple comparison of the cost of production with a cost consumed in an open field or a glass house is not constructive. What kinds of new value could it create and how should it be applied must be discussed.

Growth rate of lettuce was shown to be different from the species. As a matter of course, different kinds of plants will also be so. Therefore what kinds of plants should be cultivated is particularly important in to plant factory. Production of seedlings may be more close to a practical use. For its wider applications, researches for clearing the rate determining mechanism which is differ from plants and reasonable environment control conforming to the mechanism are required.

Under the existing circumstances, it could be seemed that enough industrial technologies conforming to special requirements of plant factory are not systematically applied. Air conditioning is one of them. By making use of a feature of free from the location, local energy should be applied.

I mentioned that plant factory defined here should be on different bases of cultivation principle from natural dependent cultivation. But as you see, there are few differences in the principle. Importance of the environment control should be realized again.

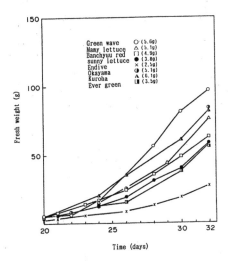

Fig. 21 Growth curve of lettuce in main cultivation equipment

Table 3 Main specification of plant factory

Productivity:	120 stumps/day (lettuce base)
Composition :	Budding, Seedling and Main cultivation equipment
Used lamp :	110 W fluorescent lamp
Area of beds:	Budding : 0.4 m² x 2
	Seedling: 2.6 m² x 3
	Main : 12.2 m² x 2

A part of this work was carried out in a joint research project with Chubu Electric Power Co., Inc..

REFERENCES

1)A. Ikeda, K. Ezaki and S. Nakayama, Envirn. Control in Biol., 22, 71-77 (1984).
2)A. Ikeda and S. Nakayama, Envirn. Control in Biol., 25, 1-5 (1987).
3)M. Kimura and T. Tozuka, "A Series of Lectures on Ecology 9, Production Process of plants", Kyoritu Syuppan (1973).
4)A. Ikeda, S. Nakayama, Y. Kitaya and K. Yabuki, Acta Horticulture, 229, 273-282 (1988).
5)A. Ikeda, Y. Tanimura, K. Ezaki, Y. Kawai, S. Nakayama, K. Iwao and H. Kageyama, Annual Meeting of Japan Society of Envirn. Control in Biol., (1990).

MEASUREMENT, MODELLING AND CONTROL PROBLEMS FOR BIOTECHNICAL PROCESSES

A. Munack

Institute of Biosystems Engineering, Federal Agricultural Research Centre, Braunschweig, Germany

Abstract. Measurement techniques for biotechnical systems are characterized by the fact, that the sampling frequency of important variables is low and that key variables may even be not measurable on-line. Starting from a discussion of possible sensoring techniques, the importance of models for estimation of non-measurable states is pointed out. However, in order to establish meaningful dynamic models and to use them on-line in model-based control algorithms, a thorough study of parameter identifiability should be performed. It is demonstrated that the evaluation of the information content in measurements and its optimization gives valuable results for the instrumentation of a process, and that calculations of the obtainable parameter estimation accuracy may be used to improve the performance of model-based predictive control.

Keywords. Biotechnology; sensors; identifiability; parameter estimation; adaptive control; predictive control.

INTRODUCTION

Also in the classical biotechnological processes of the food processing industry - like production of beer, wine, baker's yeast etc. - measuring and control techniques play a major role. In the past, these tasks were taken over by man almost entirely, e.g. by taking manual samples, by performing various sequential process steps, by suitable feeding of nutrients to the process or by stopping the process at the 'best' time instant. The rapidly increasing number and amount of biotechnologically produced products, the rising demands for increased product quality - also with regard to legal requirements (GMP) - as well as the permanent pressure to increase the process productivity have led to a continuously growing application of automation and control equipment for biotechnical processes. This trend is supported by the availability of inexpensive electronic components which enable to apply sophisticated algorithms of data processing and control, and facilitate the application of complex on-line procedures to medium-scale plants which in former times seemed to be economically feasible only for large production plants.

The situation in agricultural applications is quite similar. Here, however, the demand for cheap equipment is even stronger, since production units are small and most of the products have to be sold at low prices. The protection of the environment and legal restrictions demand for agricultural production processes with lowest possible inputs of fertilizers and agrochemicals. Having in mind, that - at least seen world-wide - the agricultural production must be maintained at the actual output quantity and should even be increased in proportion to the increasing world population, high productivity and low input can be obtained only by sophisticated production methods, using advanced control equipment. Thus, new technologies will find a high level of application in agriculture in the near future.

In this contribution, the emphasis is laid on applications to biotechnical processes. Various parallels will be evident with respect to agricultural applications. At first, the major measurement techniques for bioreactors are presented and discussed. In the more control-oriented part of the paper the use of dynamic process models and the specific aspect of the information content of measurement data for model generation will be considered. It is not the purpose of this paper, to give a complete overview on measurement and control of biotechnical processes (which can be found in textbooks, cf. e.g. Rehm and Reed (1991)); instead, some techniques will be addressed which are most likely to be relevant for control of agricultural processes, too. As application, a distributed-parameter system is chosen, since agricultural production processes also show transport and diffusion/convection phenomena in their distributed compartments soil and air.

MEASUREMENT TECHNIQUES FOR BIOTECHNICAL SYSTEMS

A peculiarity of measurements at biotechnical systems lies in the fact, that data from different sources reach the data-logger at different times. One can divide into three kinds of data:

- quasi continuous measurements, which are sampled at a high frequency and without any lag time, like pressure, temperature, pH, optical density, etc.;
- discontinuous measurements, which are sampled with a small lag time, like off-gas analyses and auto-analyzers for measurements of the broth concentrations;
- discontinuous measurements with long lag time, like manually processed analyses for the composition of biomass, and for concentration of substrate and product, in particular for cell-internal concentrations.

For process automation, one will rely on data of the first two groups, while data of the last group are mainly used for documentation and further calculations after the fermentation is finished. The use of these data during the actual cultivation is under research in some theoretically working groups. In any case, an effective data management is necessary in order to process the different data according to their temporal origin.

Direct measurements

In the following, only the first two groups are addressed; Table 1 gives an overview on the various possibilities for measurements at biotechnical processes. The list is by no means exhaustive, since the development of new sensors and analyzers shows a significant progress. However, a critical review exhibits, that many of the important measurement probes for automation of biotechnical processes have not reached the state of reliability and simplicity which is needed for industrial or at least pilot scale applications. Many of the recently developed sensors still show various cross-sensitivities, may only be used in restricted concentration ranges, are subject to disturbances and drift, or are not applicable in situ due to their physical construction. In the moment, automated analytical methods have attained a great deal of interest. Taking probes from the process and carrying out analyses outside the reactor, e.g. by FIA, GC, HPLC, or MS, enables to calibrate the measurement device as often as necessary; the cost of these devices,

Measured variable	measuring principle
volume	differential pressure ultrasonic devices
foam	contact electrode
stirrer speed	counter
temperature	PT100, thermocouple
gas flow	volumetric flow sensors
pH, redox potential	glass electrodes ion-sensitive FETs
biomass	filtration probe optical density fluorescence specific impedance
dissolved O_2	polarographic probe PTFE-probe + analyzer
dissolved CO_2	PTFE-probe + analyzer
off-gas O_2	paramagnetic detector ZiO_2-sensor mass spectrometer
off-gas CO_2	infrared detector mass spectrometer
ethanol	PTFE-probe + analyzer
glucose	enzyme electrode enzyme thermistor
L-lactate	enzyme thermistor
urea	optical biosensor
NADH in cells	fluorescence
morphological characterization	image processing
various substances	FIA, MS, HPLC

Table 1: Measured variables and measuring principles for biotechnical processes

however, is still too high for a wide-spread application, and the devices need a highly qualified staff in order to be operationable over long time periods.

Model-based measurements

The very restricted opportunities for direct measurements at biotechnical processes lead to the question whether the very efficient methods of control theory, like observers or filters, may also be applied to fermentation processes. Somewhat problematical in this respect is the fact, that in many cases reliable models for bioprocesses do not exist, since the dynamical equations are nonlinear, and the interdependencies of various major metabolic reactions inside the organisms are not yet fully understood. Nevertheless, efficient and general modelling techniques for bioprocesses are a very promising research subject for the future, since the formulation of a process model also enables to use sophisticated (predictive) control strategies. Thus, there is a need for off-line structural modelling as well as for on-line and off-line parameter identification.

The first industrial cultivation process driven by an observer was the Pekilo protein process, cf. Halme and Tiussa (1983). Further observer principles for bioprocesses were proposed by Stephanopoulos and San (1984) and by Dochain (1986). The latter author avoids to specify a growth kinetic for the process. This is a well-suited means for process observation, however it does not enable to calculate long-term predictions of the process.

MODEL BASED PREDICTIVE CONTROL

A control structure which seems to be well suited for biotechnical processes is the so-called Open-Loop Feedback Optimal Controller (OLFO). This algorithm, cf. Luttmann et al. (1985), relies heavily on a good (structural correct) process model, and is very similar to the recently established generalized predictive controller types, cf. Clarke et al. (1987). The structure is shown in Fig. 1. The algorithm solves the combined problem of system identification and process optimization separately. As in most adaptive algorithms, the time is divided into time slices, the adaptation intervals. In each interval, the unknown parameters of the system and - maybe - some deterministic disturbances are determined by minimizing an output least squares error criterion which weights the differences of measured system outputs and model outputs of the most recent adaptation interval(s). These identified parameters are extrapolated into the future and serve as a basis for calculation of future control actions, which control the process in an optimal way. Various criteria for optimization are possible. Then the control inputs are set to the calculated optimal ones, but only for the forthcoming adaptation interval. During this interval, the whole identification and optimization computations are carried out again, which may lead to a corrected control input for the following interval, and so on.

It can be stated, that this strategy should be well suited for processes with unknown or slowly time-varying parameters, if a structurally correct model is known and if the on-line estimation of parameters is possible by the available measurements. Therefore the evaluation of the information content of measurements plays a key role in this concept; a problem which is considered in detail in the following paragraphs.

INFORMATION CONTENT OF MEASUREMENTS

The problem of the information content is directly related to the precision of parameter estimates which may be obtained by using a certain measurement data set. This data set is dependent on (i) the experiment conditions under which the data were obtained, e.g. the process input functions, and (ii) the measurement devices, their precision and - for distributed-parameter systems - their location. The general theoretical framework which allows to deal with these problems is reported here shortly; for a more detailed discussion see Goodwin (1987) and the literature referenced therein.

Let the system be described by the nonlinear differential equation

$$\dot{\mathbf{x}}(t) = \mathbf{f}(\mathbf{x}, t, \mathbf{u}, \mathbf{P}) \; ; \quad \mathbf{x}(0) = \mathbf{x}_0(\mathbf{P}) , \qquad (1a)$$

and the system outputs be given by

$$\mathbf{y}(t_i) = \mathbf{g}(\mathbf{x}(t_i, \mathbf{u}, \mathbf{P}), t_i, \mathbf{P}), \qquad (1b)$$

where $\mathbf{x} \in \mathbb{R}^n$ denotes the state, $\mathbf{u} \in \mathbb{R}^m$ denotes the input, $\mathbf{y} \in \mathbb{R}^q$ is the (discrete time) input, and $\mathbf{P} \in \mathbb{R}^p$ is the vector of unknown system parameters. As pointed out above, we will treat the situation that a structural correct model is available, which allows for a formulation of the model equations identical to Eq. (1a, b), where only $\mathbf{P}$, $\mathbf{x}$, and $\mathbf{y}$ are replaced by the model parameters, states, and outputs $\hat{\mathbf{P}}$, $\hat{\mathbf{x}}$, and $\hat{\mathbf{y}}$, respectively.

With the difference between model and process outputs,

$$\mathbf{e}_i = \mathbf{e}(t_i, \mathbf{u}, \hat{\mathbf{P}}) = \hat{\mathbf{y}}(t_i, \mathbf{u}, \hat{\mathbf{P}}) - \mathbf{y}(t_i), \qquad (2)$$

the quadratic identification functional

$$J_I(\mathbf{u}, \hat{\mathbf{P}}) = \sum_{i=1}^{N} \mathbf{e}_i^T \mathbf{Q}_i \mathbf{e}_i \; ; \quad \mathbf{Q}_i \geq 0 \qquad (3)$$

is obtained. The optimization problem to be solved for parameter identification is now posed as follows:
"Find $\hat{\mathbf{P}}_{opt}$ such that $J_I(\mathbf{u}, \hat{\mathbf{P}}_{opt}) \leq J_I(\mathbf{u}, \hat{\mathbf{P}}) \; \forall \hat{\mathbf{P}} \in P_{ad}$ and $\hat{\mathbf{P}}_{opt} \in P_{ad}$, where P_{ad} is the set of physically, chemically, or biologically meaningful parameters."

Next we must consider that measurements on a system usually do not represent the undisturbed system's outputs, but will be corrupted by noise. For biotechnical processes, this noise is partly caused by the impreciseness of the measurement device, but most often a larger part is caused by the process itself with imperfect mixing, turbulent flow, gas bubbles, and other inhomogeneities in the reactor. This fact should be kept in mind when respresenting all noise characteristics by a zero-mean, Gaussian white noise process as follows:

$$\mathbf{y}^M(t_i) = \mathbf{y}(t_i) + \mathbf{w}(t_i); \qquad (4)$$

$$E\{\mathbf{w}(t_i)\} = 0 \quad , \; i = 1, \dots, N; \qquad (5a)$$

$$E\{\mathbf{w}(t_i)\mathbf{w}^T(t_j)\} = \delta_{ij} \mathbf{C}(t_i) \quad , \; i, j = 1, \dots, N. \qquad (5b)$$

Due to this noise, the result of the optimization problem, $\hat{\mathbf{P}}_{opt}$, becomes a random variable. If a bias-free estimate can be obtained (which will not be possible in most nonlinear situations by using the least squares criterion), one could state that, in the limit, the expectation for $\hat{\mathbf{P}}_{opt}$ could be equal to the true process parameters,

$$E\{\hat{\mathbf{P}}_{opt}\} = \mathbf{P}. \qquad (6)$$

Futhermore, it can be shown by using the Cramér-Rao inequality, that

$$E\{(\hat{\mathbf{P}}_{opt} - \mathbf{P})(\hat{\mathbf{P}}_{opt} - \mathbf{P})^T\} \geq \mathbf{S} = \mathbf{F}^{-1}, \qquad (7)$$

where $\mathbf{F}$ is the Fisher information matrix. Thus $\mathbf{S}$ gives a lower bound on the obtainable covariance of the parameter estimates.

The information matrix is computed by

$$\mathbf{F} = \sum_{i=1}^{N} \left[\mathbf{Y}_{\mathbf{P}}^T(t_i, \mathbf{u}, \mathbf{P}) \; \mathbf{C}^{-1}(t_i) \mathbf{Y}_{\mathbf{P}}(t_i, \mathbf{u}, \mathbf{P}) \right] , \qquad (8)$$

which shows that the optimal weighting of the errors in Eq. (3) is given by $\mathbf{Q}_i = \mathbf{C}^{-1}(t_i)$. Several criteria were proposed to give a measure for evaluation of the precision which is obtained; they may be applied to the information matrix $\mathbf{F}$ or to the covariance matrix $\mathbf{S}$. An overview is compiled in Table 2. The most simple criterion trace $(\mathbf{F}) \to$ max should not be applied since it may lead to non-informative experiments.

These well-known results are the basis for the application to a biotechnical process in the following paragraph.

criterion	definition	interpretation
A-criterion	min (trace $\mathbf{S}$) max (trace $\mathbf{F}$)	minimization of the variance in the arithmetical mean
D-criterion	min (det $\mathbf{S}$) max (det $\mathbf{F}$)	minimization of the variance in the geometrical mean
G-criterion E-criterion	min ($\lambda_{max}(\mathbf{S})$) max ($\lambda_{min}(\mathbf{F})$)	minimization of the largest variance of any parameter combination

Table 2: Criteria to evaluate and optimize the information content of measurements

APPLICATION TO A BIOTECHNICAL PROCESS IN A TOWER REACTOR

A simplified model of a biomass cultivation process is used to demonstrate the various aspects of utilization of the above formulated results. The detailed description of the plant and a derivation of a complete model can be found in the paper by Luttmann et al. (1985); calculations with a reduced model were already reported, cf. Munack (1985). The model equations used here are even more simplified; however, the main results remain the same. This shows, that the results are not heavily model-depending, an apprehension which is often expressed when treating nonlinear systems with the above (linearized) theory. The emphasis in this contribution is laid on a comparison of different criteria and the time-dependent availability of parameter estimates in different phases of the fermentation.

Description of the plant

The reactor used is a tower reactor, which is gassed from the bottom - a bubble column, cf. Fig. 2. It is filled with liquid, which contains the substrate and all other nutrients. The cells are submerged in the liquid phase;

no further mechanical agitation is needed, since the rising gas bubbles give a nice mixing. However, at higher cell concentrations, when the batch process reaches its end, an oxygen limitation is observed at the top of the reactor, since all available oxygen is consumed in the lower part of the reactor. Thus, distinct oxygen profiles can be observed, which lead to the necessity to describe the process by a distributed-parameter model.

The simple model used in the calculations is shown in <u>Table 3</u>. It consists of three balances: the first equation describes the dissolved oxygen in the liquid phase of the reactor, the second gives the oxygen molar fraction of the gaseous phase (bubbles), and the third equation gives the biomass concentration.

Measurements can be taken of the dissolved oxygen concentration by sensors, which can be installed at any spatial position in the column; furthermore, the biomass concentration can be measured (here the position is not critical, since the biomass is almost well mixed in the reactor), and an off-gas analysis gives the molar fraction of the oxygen contained in the exhaust gas, which may be used to calculate the integral oxygen uptake rate of the microorganisms.

Five parameters have turned out to be unknown. They are time-variant and varying from cultivation to cultivation, too. These are two fluiddynamical parameters, k_La^E and Δk_La, describing the oxygen transfer from gas phase into liquid phase and - roughly - its decrease towards the top of the reactor, as well as three biological parameters, μ_m, Y_{XO}, and K_O, the maximum specific growth rate, the yield coefficient, and the limitation constant in the Michaelis-Menten kinetics.

<u>Statement of the problems</u>

With respect to the instrumentation of the process, several questions must be solved. These are:
- Which are the best positions for the dissolved oxygen (DO) sensors to be placed in the reactor column?
- How many DO sensors are essential for a sufficient precision of the parameter estimates?
- Which types of the measurements - DO, cell mass, offgas - are essential or give the highest contribution to the parameter estimates?
- Are parameter estimates possible in all states of the cultivation, which means: Can the OLFO controller identify all parameters by using only measurement data from the preceding adaptation interval?
- Which are the parameters that are worst identified, and are these parameters of great influence on the course of the process (which would result in unreliable process predictions).

<u>Results concerning sensor type and allocation</u>

To answer these questions, the nominal trajectories of the process were computed as well as the associated sensitivities with respect to parameter changes. Then the information matrices were computed for various sensor configurations.

The results are summarized in <u>Fig. 3</u> to <u>Fig. 7</u>.

<u>Fig. 3</u> shows the evaluation of the E-criterion for the information matrix. When only one DO sensor is installed (1), it should be placed near the bottom of the fermenter, since the information content of the signal then is at its highest value. Adding a cell concentration

---

Dissolved oxygen in the liquid phase

$$\frac{\partial O_F}{\partial t} = D_F \frac{\partial^2 O_F}{\partial x^2} + k_La(x) \cdot (\frac{p}{H} X_{O,G} - O_F) - r_{OX}$$

$$\left.\frac{\partial O_F}{\partial x}\right|_{x=0} = 0; \quad \left.\frac{\partial O_F}{\partial x}\right|_{x=H_R} = 0; \quad O_F(t=0) = \frac{p \cdot X_{O,G}{}^E}{H}$$

Oxygen molar fraction in the gas phase

$$\frac{\partial X_{O,G}}{\partial t} = -v_G \frac{\partial X_{O,G}}{\partial x} - \frac{\epsilon_F}{\epsilon_G} k_La(x) \cdot \frac{RT}{p} (\frac{p}{H} X_{O,G} - O_F)$$

$$X_{O,G}(0) = 0.21 ; \quad X_{O,G}(t=0) = 0.21$$

Biomass concentration

$$\frac{\partial X}{\partial t} = D_F \frac{\partial^2 X}{\partial x^2} + r_{OX} Y_{XO} - r_D$$

$$\left.\frac{\partial X}{\partial x}\right|_{x=0} = 0; \quad \left.\frac{\partial X}{\partial x}\right|_{x=H_R} = 0;$$

where $k_La(x) = k_La^E - \Delta k_La \cdot \frac{x}{H_R}$

$$r_{OX} = \frac{\mu_m \cdot O_F}{Y_{XO}(K_O + O_F)} X ; \quad r_D = \mu_D \cdot X$$

<u>Table 3</u>: Model equations for the tower reactor

---

measurement (1+X) gives a result for λ_{min} (F) which is more than one decade higher, if the DO sensor is placed near the bottom. Note that this means that the standard deviation of the worst identified parameter combination is reduced by a factor of approx. 3.5. A further decade is gained by taking one DO sensor and the off-gas analysis (1+O), however, now the best position for the DO sensor is on the top of the fermenter. Best identification, of course, is possible by taking all three measurements (1+X+O). Fixing the first DO sensor on the top position, now a second DO probe is placed. This results in the curve (2+X+O), indicating that the best position of this sensor is at the bottom of the reactor. A third DO sensor (this curve is not shown) wouldn't add very much to the information content of the measurements.

One could object that the calculation for the two DO sensor placement problem is not complete, because it starts with the assumption that one sensor is fixed at the top of the reactor and then only the second position is sought. Therefore, the complete problem of allocating two sensors (in combination with cell concentration and off-gas analysis) was treated. The contour plot of the resulting λ_{min} (F) functional depending on the two sensor positions is shown in <u>Fig. 4</u>. It proves that in fact an allocation at the bottom and at the top gives best results. Relative maxima are obtained by allocating both sensors at the bottom or both sensors at the top of the reactor, while an allocation of both at 1.25 m of height would be the worst choice. It must be pointed out, that for cases where two sensors are mounted on the same height, there should be a sufficient (radial) distance between them in order to ensure the assumed statistical independence of the noise signals.

A comparison of the results gained by application of the E-criterion with the other criteria gives some further insights into the statements which are possible. Taking the determinant of **F** leads to quite comparable results. The application of the E- and D-criteria to **S** instead of **F** would, of course, give the same results. This is not the case when taking the A-criterion. Fig. 5 demonstrates that trace (**S**) gives quite comparable results to det (**F**) and λ_{min} (**F**); however, Fig. 6 shows that trace (**F**), the most simple criterion, does not provide much information.

The results may be summarized as follows:
- The off-gas analysis is the most informative measurement for parameter identification for the process; in fact, it can be shown that it is superior to eight DO measurements which are optimally placed.
- The best practical instrumentation would comprise cell concentration measurements, off-gas analysis and three DO sensors, which should be placed at the bottom, 40 % of the reactor height and at the top. Then one of the DO sensors could fail during the process, and the other two would still be sufficient.
- This result is "robust" in the sense, that it holds for the simplified model and for the more detailed model as well - indicating that a severe model dependence is not to be expected.

Identifiability of parameters in different stages of the cultivation

For application of the parameter identification in the OLFO control algorithm, one must ensure parameter identifiability also by inspection of small data sets, e.g. the preceding adaptation interval. This is of great interest, if the plant exhibits time variant parameters or if the model used is quite simple, such that some structural model inaccuracies are compensated by time variant model parameters.

The utilization of the eigenvector components of the information matrix in a supervising system for fermentations was first described by Posten (1990). Concerning the example treated here, Fig. 7 shows the relative standard deviations of the five unknown system parameters, and the standard deviation of the worst identified parameter combination ($\sqrt{\lambda_{max}}$ (**S**)), when only measurement data of the last four hours are used. It can be seen, that at the beginning of the cultivation, K_O is not identifiable - which is easy to understand, since the process is then running under non-limited conditions. During the central part of the cultivation, all parameters are nicely identified. At the end, when the DO profile goes down, identifiability conditions become worse again. Now the coalescence factor $\Delta k_L a$ becomes practically non-identifiable, and the standard deviation of K_O rises, too. This means, that at the start of the fermentation, one should calculate the process predictions with parameter values from preceding experiments, and at the end, one should not attempt to identify fluid dynamical parameters. Surprisingly, the problem of K_O identification is not as severe as it was reported for batch processes in stirred-tank reactors, cf. Holmberg (1982). This can be explained by the fact that in the middle stage of the cultivation there is always some part of the reactor where limiting conditions are present, while at the bottom unlimited growth conditions are to be found. This results in relatively nice identifiability conditions for the Michaelis-Menten kinetics. Thus predictive control may be very efficient, since the main parameters (the biological ones) are identified with sufficient precision.

CONCLUSIONS

Measurement facilities for biotechnical processes are somewhat restricted, in particular when relevant biological states are to be measured. Furthermore, biotechnical processes are quite complex and often not reproducibly run. Therefore, model-based measurement techniques, parameter identification techniques for formulation of models and their update, and adaptive predictive control algorithms are well-suited tools for control of these processes. In this contribution, the effects of suitable instrumentation of the plants with sensors and analytical devices were discussed, and the problem of identifiability of the parameters in different stages of the cultivation was addressed. Further research with respect to automatic evaluation of identifiability and automatic modification of input functions in order to guarantee (on-line) identifiability is needed.

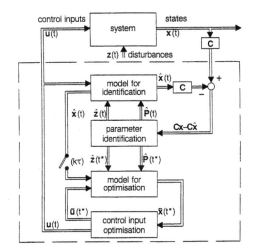

Fig. 1: Structure of the OLFO controller

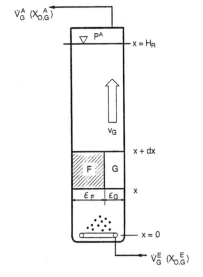

Fig. 2: Schematic diagram of the tower reactor

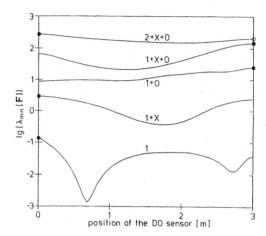

Fig. 3: Results of the sensor allocation problem (l_{min} (**F**))

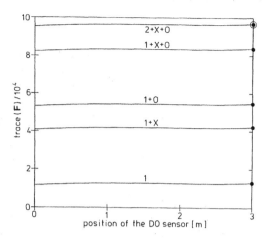

Fig. 6: Results of the sensor allocation problem (trace (**F**))

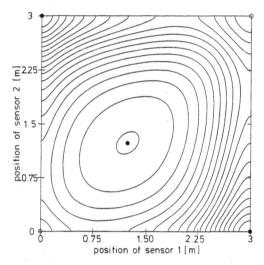

Fig. 4: Contour plot of l_{min} (**F**) for two DO sensors

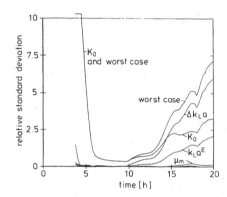

Fig. 7: Relative standard deviations of the unknown parameters during the course of the cultivation (identification horizon = 4 h)

REFERENCES

Clarke, D.W.; Mohtadi, C.; Tuffs, P.S. (1987). Generalized predictive control - Part I/II. Automatica, 23, pp. 137-148, 149-160.

Dochain, D. (1986). On-line parameter estimation, adaptive state estimation and adaptive control of fermentation processes. Dissertation, Université Catholique de Louvain.

Goodwin, G.C. (1987). Identification: Experiment Design. In: System & Control Encyclopedia, Pergamon, Oxford, Vol. 4, pp. 2257-2264.

Halme, A.; Tiussa, E. (1983). Use of computers in Pekilo-protein production. Proc. 1st IFAC Workshop on Modelling and Control of Biotechnical Processes, Pergamon, Oxford. pp. 267-272.

Holmberg, A. (1982). On the practical identifiability of microbial growth models incorporating Michaelis-Menten type nonlinearities. Math. Biosc., 62, pp. 23-43.

Luttmann, R.; Munack, A.; Thoma, M. (1985). Mathematical modelling, parameter identification and adaptive control of single cell protein processes in tower loop reactors. Advances in Biochemical Engineering, Vol. 32, Springer, Berlin. pp. 95-205.

Munack, A. (1985). On parameter identification for complex biotechnical systems. 1st IFAC Symp. on Modelling and Control of Biotechnological Processes, Pergamon, Oxford, pp. 139-145.

Posten, C.; Munack, A. (1990). On line application of parameter estimation accuracy to biotechnical processes. Proc. of the ACC, San Diego, Vol. 3, pp. 2181-2186.

Rehm, H.J.; Reed, G. (1991). Biotechnology - A Multi-Volume Comprehensive Treatise. 2nd ed., Vol. 4.

Stephanopoulos, G.; San, K.-Y. (1984). Studies on on-line bioreactor identification. Biotech. and Bioeng., 26, pp. 1176-1218.

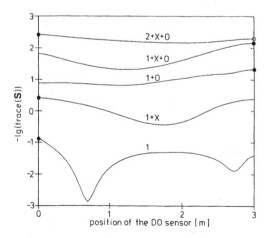

Fig. 5: Results of the sensor allocation problem (trace (**S**))

Copyright © IFAC Mathematical and Control Applications
in Agriculture and Horticulture, Matsuyama, Japan 1991

ENVIRONMENTAL CONTROL IN PLANT TISSUE CULTURE AND ITS APPLICATION FOR MICROPROPAGATION

T. Kozai

Faculty of Horticulture, Chiba University, Matsudo, Chiba 271, Japan

Abstract. Environmental factors in plant tissue culture or micropropagation are discussed. The effect of physical environment in plant tissue culture on the growth and development of plantlets in vitro during multiplication and rooting stages is then discussed. Advantages of photoautotrophic micropropagation over conventional (heterotrophic) micropropagation are finally discussed in terms of reduction in production costs and automation.

Keywords Biotechnology, Environment control, Environmental engineering, Micropropagation, Physiological models, Tissue culture.

INTRODUCTION

Micropropagation of plants by means of plant tissue culture or simply tissue culture has many advantages over conventional vegetative propagation and its commercial use in horticulture, agriculture and forestry is currently expanding worldwide (DeBergh and Zimmerman, 1990). However, its widespread commercial use for major crops and forestry trees is still restricted as a result of its relatively high production costs. The high production costs in conventional micropropagation are mainly due to high labor costs, limited rates of growth during multiplication stage, poor rooting, and low survival rates of the plantlets during acclimatization stage.

Development of automated systems for reducing manual operation and for control of the environment in plant tissue culture is essential to achieve a drastic reduction in production costs for a large-scale micropropagation process.

Recently, research has been conducted extensively on automation and/or robotization for reducing manual operation in micropropagation, including automation of liquid medium preparation and feeding, plant image recognition and processing, microcutting, transplanting, etc. The reader is referred to the reviews by Aitken-Christie (1990) and Kozai (1990) for the automation/robotization.

On the other hand, there is not much research on the environmental effects and controls in micropropagation or tissue culture. This is partly because conventional culture vessels such as test tubes and flasks are generally so small and airtight that automation in measurement and control of the environment is difficult to conduct. In fact, even off-line measurement of the environment is often difficult to conduct.

An excellent review has been prepared on environmental effects in conventional micropropagation with special attention to the environmental effects during stages preparation of plant material and initiation of culture (Read, 1990). The reader is also referred to the reviews by Dunstan and Turner (1984), Preece and Sutter (1990) and Kozai (1991a) for the environmental effects on growth and development of micropropagated plants during acclimatization

stage. However, there is not much research on the effects of the physical environment on the growth and development of plantlets in vitro during stages multiplication and rooting. In this article, the features of in vitro environment and the environmental effect on the photosynthesis and growth of plantlets in vitro are discussed.

Definitions of technical terms in plant tissue culture and micropropagation used in this article are mostly based upon Donnelly and Vidaver (1988) and Debergh and Zimmerman (1990). This article is a revised and reduced version of Kozai (1991c) and the reader is referred to it for further details.

ENVIRONMENTAL FACTORS IN PLANT TISSUE CULTURE

The mechanism of environmental changes in a culture vessel is similar to that of the environmental changes in a greenhouse. That is, the interrelationship among environments inside and outside the culture vessel and plants in the culture vessel is similar to the interrelationship among environments inside and outside the greenhouse and crops in the greenhouse. In fact, a culture vessel is a miniature greenhouse or a small growth chamber in a sense. Also, an explant to be cultured in vitro may be thought of as a miniature cutting in conventional vegetative propagation (Read, 1990).

Hence, ecological, ecophysical, ecophysiological, environmental physiological and environmental engineering studies on micropropagation processes should be important (Huges, 1981). In reality, however, micropropagation research has too long ignored physical environmental factors and has too long relied upon pharmaceutical levels of exogenous plant growth regulators application in huge factorial experiments (Wilkens, 1988).

There are much research on modeling and dynamic simulation of growth for horticultural and agricultural crops as a function of environmental variables. There is, however, few research on modeling and dynamic simulation of growth for plantlets in vitro. Development of a growth model for plantlets in vitro should be easier than that for greenhouse or field-grown crops because plantlets in vitro are grown in a culture room where air temperature and light intensity are mostly kept constant with time during each of photo- and dark periods.

Environmental control during micropropaga-tion is considered to be important from a practical point of view for the reduction of production costs. Because, as will be shown later, it promotes growth and development (increases in fresh and dry weights, number of nodes, total leaf area, in vitro rooting, in vitro branching, etc.). it also reduces morphological and physiological disorders such as vitrification (also called glassiness; Ziv, 1990) and thin cuticular wax formation on leaves, and loss of plants in culture due to biological contamination (a rapid and accidental growth of undesirable bacterial, fungal or algal microorganism in culture vessel).

Furthermore, It increases uniformity of growth and development and reduces excess application of exogenous plant growth regulating substances. Furthermore, environmental control in culture may help plants grow more rapidly and vigorously during acclimatization stage.

In reality, however, the physical environment in vitro is not well maintained in conventional tissue culture and the in vitro environment which is very different from the environment in the greenhouse or in the field often brings about undesirable phenomena as indicated by DeBergh and Maene (1984).

The main objective of multiplication should be to get a large number of genetically identical plants in a shorter time period and at a lower production cost, which have a higher photosynthetic or photoautotrophic ability (getting carbon dioxide as carbon source from the air) under stressful environmental conditions. Traditionally, however, the plantlets in vitro are cultured under predominantly heterotrophic or photomixotrophic conditions relying on sugar in the medium as the main carbon source.

Explants and regenerated shoots in culture have been considered to have little photosynthetic ability and require sugar (carbohydrates) as a carbon and energy source. Recent research, however, has revealed that chlorophyllous explants/shoots, in general, have relatively high photosynthetic ability and that they may grow faster in some cases under photoautotrophic conditions than under heterotrophic and photomixotrophic conditions, provided that the physical and chemical environments in the culture are properly controlled for photosynthesis (Kozai, 1990).

Research on photosynthesis and transpira-tion of plantlets in vitro is important to develop a better micropropagation system.

SOME CONSIDERATIONS FOR MEASUREMENT AND CONTROL OF THE ENVIRONMENT

For the measurement of the in vitro environment, many of the plant environmental sensors widely used in the fields of agronomy, horticulture, agricultural meteorology and ecology can be used. Measurement and control of the plant environments are well described, for example, by Langhans (1978). For the measurement of the in vitro environment, sensors and/or the amount of sampling material should be as small as possible so as not to disturb the environment.

Light

Spectral distribution of light from different light sources significantly differs from each other. Fluorescent lamps have been the primary light source used in micropropagation. The spectrum generally matches the requirements of cultures. When the lamps are placed close together, they generally give a relatively uniform horizontal distribution of photosynthetic photon flux density (PPFD) on the culture shelf. However, there is a large difference in PPFD between the inside and outside of the vessels, and the distribution of PPFD in culture vessels on the shelf is largely dependent upon the closure type, vessel type, and their arrangement on the shelf (Fujiwara et al., 1989).

The light source is almost always installed above the vessels and the tissues/plants in the vessels generally receive downward light. In this case, as the plants grow, more light energy is intercepted by the upper part of the plants and less light energy by the lower parts. In future micropropagation systems, it might be better to direct light from the sides by means of optical fibers or other thin or tiny light sources (Kozai, 1990).

With this lateral or sidewards lighting method, plants may receive more light evenly with less electricity consumption for lighting. The lateral lighting promotes better growth and gives better plant shape (Kozai et al., 1991a). For control of plant photomorphogenesis with a low level of blue, red and far-red lights, different types of light emitting diodes (LED), which emit either blue, red or far-red light separately, can be used at a relatively low cost. Application of LED for growing plants would also be practical (Bula, 1991).

Morini et al. (1990) tested the effect of light-dark cycles different from the largely adopted 16 hours light- 8 hours dark, without altering the amount of radiation supplied to the cultures. They found that the growth of peach tree shoots was much greater with 4 hours light - 2 hours dark compared to 16 hours light- 8 hours dark. The method of lighting (light-dark cycles, light source, spatial arrangement of light sources, etc.) is an area for further research.

Gas exchange characteristics of vessels

It has been shown that the type of vessel closure affects the gaseous composition as well as the light environment, and hence vitrification and growth of plants in culture; the looser types being better than the tighter ones for reducing vitrification of plantlets in vitro (Hakkaart and Versluijs, 1983; Dillen and Buysens) and for promoting the growth of strawberry plantlets in vitro (Kozai and Sekimoto, 1988).

The air exchange characteristics of the vessel are best expressed by the number of air changes (infiltration) per hour of the vessel, which is defined as the hourly air change rate of the vessel divided by the air volume of the vessel (Kozai et al., 1986). The number of natural air changes for a flat bottom glass test tube (air volume: 45 ml) closed with an aluminum foil cap, plastic formed cap and silicon foam rubber plug are 0.18, 1.5 and 0.6 per hour respectively (Kozai et al., 1986). The number can be increased up to 3-6 times per hour by using a gas permeable microporous polypropylene film as part of the vessel closure (Kozai and Sekimoto, 1988).

The number of air exchanges per hour, E, is a sort of physical property of the vessel and is basically constant with time. The gas concentration in the vessel containing shoots or plantlets and medium varies with the gas concentration outside the vessel, the E value and gas production/absorption characteristics of the plants/medium in the vessel. The mathematical relationship among the factors previously mentioned is given by Fujiwara et al. (1987). The E can be defined similarly for forced air changes in the vessel.

Relative humidity and water potential

The interchange of water in the gaseous and/or liquid phase among plants, aerial and root zone environments in the vessel and the air outside the vessel plays an important role in plant growth and development. Directions and rates of water flow are determined by the spatial distribution of water potentials inside and outside of the vessel. The spatial gradient of water potentials in the vessel is almost always very small, resulting in little water movement in the vessel.

Relative humidity is normally high in the vessel. The high relative humidity in the vessel can be rather accurately measured with a high polymer sensor (impedance type) if calibration is conducted using saturated salts at a regular time interval. The compact cylinder type sensor (8 mm in diameter and 35 mm in length) is commercially available at a reasonable cost. Most humidity sensors are attached with a thermistor for temperature measurement.

Kozai et al. (1990) showed the dependence of relative humidity in the vessel upon the number of air changes per hour, relative humidity in the culture room and total leaf area of the plantlets in the vessel, based upon the relative humidity measurements using the high-polymer sensors.

The water potential of a culture medium can be estimated based upon the nutrient composition of minerals and carbohydrates (Kozai et al., 1986).

Temperature

Although the setpont of air temperature in culture room is not changed throughout the day in most cases, the temperature distribution is somewhat uneven in space and time. The actual temperature is approximately 1 $^{\circ}C$ higher during the photoperiod than during the dark period. The temperature inside the vessel is approximately 1 $^{\circ}C$ higher than that outside the vessel during the photoperiod. The temperature is often higher at the lower shelf than at the upper shelf in the culture room. Uraban and Jaffrin (1990) developed a model for heat and mass transfers in the vessels and applied to various physical conditions to predict thermal phenomena in the vessel.

Kozai et al. (1991b) conducted an experiment on the effect of the difference in air temperature between photoperiod and dark period on morphogenesis and growth of potato plantlets in vitro under photoautotrophic and CO_2 enriched conditions. They showed that the plant height is shorter and the number of unfolded leaves is a little greater with decreasing the difference in temperature. The dry and fresh weight of the plantlets is almost the same when the daily average of the temperature is the same regardless of the difference in temperature between the photoperiod and dark period.

CO_2, O_2, and ethylene

Concentrations of CO_2, O_2 and ethylene (C_2H_4) can be measured simultaneously using a gas chromatograph with a FID (flame ionized detector). CO_2 concentration is also measured with an infrared type CO_2 analyzer. This is convenient for a continuous measurement but the analyzer requires a relatively large amount of sample gas for the continuous measurement and the sampling may disturb the gaseous environment.

There are several ways of modifying the gaseous components in the vessel. They include: a) addition of agents absorbing or producing a particular gas component in the vessel, b) control of the gaseous environment in the culture room and use of a gas permeable film as closures, and c) use of a forced ventilation or air circulation system with a controller. For details, the reader is referred to Kozai (1991c).

CO_2 concentration in the vessel during the photoperiod

The decrease in CO_2 concentration in the culture vessel during the photoperiod (C_{in}) was first reported by Ando (1978) for orchids being cultured in vitro at the rooting stage.

The C_{in} in relatively air-tight vessels containing ornamental plantlets decreased to 70-80 ppm within 2-3 hours after the start of the photoperiod (Fujiwara et al., 1987; Infante et al., 1989). The C_{in} was as low as the CO_2 compensation point of C_3 plants and was about 250 ppm lower than the normal atmospheric CO_2 concentration (ca. 340 ppm).

Experimental results suggest the following for chlorophyllous explants/plantlets being cultured in vitro by conventional micropropagation: (1) The explants/plantlets have a photosynthetic ability because the C_{in} decreases sharply with time at the onset of the photoperiod. (2) Their photosynthesis is restricted by low CO_2 concentrations during most of the photoperiod because there is insufficient CO_2 from the outside air through the vessel caps. (3) They are forced to develop hetero- or photomixotrophy and a higher PPFD will not increase net photosynthetic rates under such low CO_2 conditions. (4) They may develop photoautotrophy and grow faster under photoautotrophic, high CO_2 and high PPFD conditions than under hetero- or photomixotrophic conditions. (5) The initial growth rate is greater for an explant with a larger area of highly chlorophyllous tissue (or green leaves).

Net photosynthetic rate

Estimation of the net photosynthetic rate of tissue cultured explants/plantlets had been mostly conducted using a small assimilation chamber with a forced air mixing system. CO_2 measurements are made with an infrared-type CO_2 analyzer which requires a significant volume of sample gas (e.g., Grout and Ashton, 1978; Donnelly and Vidaver, 1984; Reuther, 1988; Capellades, 1989). In this case, the net photosynthetic rate per plantlet is estimated by multiplying the forced air flow rate though the chamber by the difference in CO_2 concentration between the inlet and outlet of the chamber. The difference is relatively small even at a low air flow rate because the chrolophyllous parts of the explants/plantlets and hence CO_2 uptake in the chamber is small.

Furthermore, with the assimilation chamber method, physical parameters such as the gas diffusion coefficients adjacent to the plantlets and the physiological conditions such as water relations of the explants/plantlets in the chamber may differ from those in vitro and in situ (in the vessel). The net photosynthetic rates of in vitro explants/plantlets in situ may then differ from those in the chamber for identical PPFD, CO_2 and temperature levels. We must be careful, therefore, to interpret the results obtained with assimilation chambers.

Fujiwara et al. (1987) showed a numerical method for estimating the parameter values of photosynthetic response curves of in vitro plantlets in situ using the data on the diurnal courses of CO_2 concentration inside and outside the vessel. Kozai and Iwanami (1988) estimated the steady-state net photosynthetic rate of in vitro plantlets in situ by multiplying the difference in CO_2 concentration between the inside and outside of the vessel by the natural air ventilation rate; hourly natural ventilation rate being a product of the number of air changes per hour of the vessel, the air volume of the vessel and a unit conversion factor. In these two methods, CO_2 measurements are made using a gas chromatograph with a negligible disturbance allowing the net photosynthetic rate of in vitro plants can be estimated in situ. A similar method to be employed for estimating the dark respiration rate, transpiration rate, etc.

ENVIRONMENTAL EFFECTS ON PHOTOSYNTHETIC GROWTH AND DEVELOPMENT (Kozai, 1991c)

In photoautotrophic micropropagation, the photosynthesis, growth and development of in vitro explants/plantlets are largely influenced by the physical environmental factors which include light, CO_2, humidity, air flow speed, temperature, O_2 etc. Research of the environmental effect on photoautotrophic micropropagation has been extensively reviewed by Kozai (1991a, 1991b, and

1991c). In the following sections, the environmental effect on the net photosynthetic rate, growth and development of explants/plantlets in vitro will be briefly reviewed.

Photosynthetic response curves of explants/plantlets in vitro

Kozai et al. (1990b) showed the photosynthetic response curves for in vitro Cymbidium plantlets in situ as affected by CO_2, PPFD and temperature. They showed that the curves did not differ largely from those of plants grown under shade in the greenhouse.

The net photosynthetic rate of in vitro Primula malacoides (C_3 plant) plantlets in 1% O_2 and 10% O_2 were, respectively, about 3 times and 1.5 times larger than that in 21% O_2, at a CO_2 concentration of approximately 200 ppm, due to a reduced photorespiration at lower O_2 concentrations (Shimada et al., 1988).

The net photosynthetic rate of in vitro rose plantlets was greater when cultured on the medium with a lower sucrose concentration (Capellades, 1989). A similar result was obtained for potato plantlets in vitro (Nakayama et al., 1991). The starch content in leaves of plantlets was higher when cultured on the medium with a higher sucrose concentration while a higher starch content in the leaves brought a lower net photosynthetic rate (Capellades, 1989). She showed that the rate was also affected by the relative humidity.

The net photosynthetic rate of strawberry plantlets was greater when cultured in a vessel with a forced ventilation system than when cultured in a normal vessel with natural ventilation (Kozai et al., 1989). This is probably because CO_2 diffusion into the stomatal cavities of the plantlets in vitro is restricted by the stagnant air in the normal vessel; the air movement in the vessel being caused by natural convection.

The net photosynthetic rates of leaves of plantlets and seedlings in vitro measured under saturated PPFD, 340 ppm CO_2 and a leaf temperature of 20°C were similar, nevertheless the shape of the photosynthetic response curves relating the rate and CO_2 concentration indicated some differences in their photosynthetic parameters (Pospisilova et al., 1987).

Research on photosynthetic responses of in vitro explants/plantlets in situ as affected by the in vitro physical environment has only been recently conducted and there are many problems that still need to be solved.

CO_2 enrichment under high PPFD

Based upon the data given above, one can expect an increase in net photosynthetic rate and hence growth of plantlets in vitro by increasing the CO_2 concentration in the vessel during the photoperiod. There are several practical ways of increasing the concentration, as shown below.

a) Gas permeable film in closures

There are more than several reports indicating the positive effects of use of gas permeable film as closures under high PPFD on the increases in net photosynthetic rate and growth of plantlets in vitro (Kozai, 1990). When leafy single node cuttings were used as explants, the growth of plantlets of some species is greater when cultured on medium without sugar in the vessel closed with the film than when cultured on medium with sugar in the relatively air-tight vessel. With this approach, the effects of the passive CO_2 enrichment under high PPFD on growth in vitro can be verified with a minimum change in the existing micropropagation facility. The percentage of vitrification is also decreased with the use of the film possibly due to lower humidity and increased gas exchange in the vessel and drying out of the medium.

b) CO_2 enrichment in the culture room

CO_2 enrichment under high PPFD (100-200 μmol m^{-2}s^{-1}) was effective for promoting the chlorophyllous explant/plantlet growth of tobacco (Mosseau, 1986), Cymbidium (Kozai et al., 1987), carnation (Kozai and Iwanami, 1988), potato (Kozai et al., 1988), etc., when cultured on the medium both with and without sugar.

c) A larger vessel with a CO_2 supply system

Forced ventilation with atmospheric air or nitrogen-oxygen-carbon dioxide mixtures resulted in reduced propagule weights and shoot numbers for the multiplication stage of Rhododendron in the vessels with 400 ml air volume when cultured under a PPFD of 39 μmol m^{-2}s^{-1} (Walker et al., 1988).

The dry weight and net photosynthetic rate of strawberry plantlets cultured with sugar free liquid medium were promoted when cultured in a large vessel with a forced ventilation system under a PPFD of 96 μmol m^{-2}s^{-1}, compared with those cultured by a conventional method (Fujiwara et al., 1988).

With the systems given above, not only CO_2 concentration but also relative humidity, ethylene, gas diffusivity, etc. in the vessel are modified. Therefore, the changes in growth of explants/plantlets in vitro resulting from the use of these systems cannot be attributed only to CO_2 enrichment. However, the changes are considered to be brought primarily by the CO_2 enrichment in most cases. More detailed studies on the effect of gaseous environment on the photosynthetic growth and development is an area for further study. Much more research should be done in future on the effects of forced ventilation on the growth of explants/plantlets.

ADVANTAGES OF PHOTOAUTOTROPHIC MICROPROPAGATION

Hetero- or photomixotrophic micropropagation has many advantages over conventional vegetative propagation. However, its widespread commercial use is still restricted as a result of its relatively high production costs. Some disadvantages of hetero- and photomixotrophic micropropagation are shown in Table 1.

Photoautotrophic micropropagation seems to have many advantages for reducing the production costs and for increasing productivity and plantlet

Table 1 Some disadvantages of hetero- and photomixotrophic micropropagation
----
1. Sugar in the medium may cause biological contamination.
2. Air-tight, small vessels must then be used to reduce a loss due to contamination.
3. Automated and computerized micropropagation system is thus difficult to develop.
4. The inside air is almost saturated with water vapor.
5. Besides, CO_2 and ethylene concentrations may become abnormal.
6. Thus, high PPFD is not effective for the growth promotion.
7. Growth regulators are often necessary for regeneration.
8. The abnormal environment may induce physiological/morphological disorders, retardation of plantlet growth, and mutation.
9. The disorders may result in high percentage of death during the acclimatization stage.
----

quality (See Table 2). Some of the advantages shown in Table 2 have been proven in experiments but many have not been and require additional research, especially on the environmental effects on photosynthetic growth of explants/plantlets in vitro.

However, potential of photoautotrophic

Table 2 Some advantages of photoautotrophic
micropropagation
----
1. Growth of plantlets are promoted.
2. Physiological/morphological disorders and
   plantlet quality is improved.
3. Procedures for rooting and acclimatization
   are simplified or even eliminated.
4. Application of growth regulators and other
   organic matter can be minimized.
5. A loss of plantlets due to contamination
   is reduced.
6. A larger vessel can be used with minimum
   contamination.
7. Then, the environmental control of the vessel
   is easier.
8. The control of growth and development by
   means of environmental control is easier.
9. Automation and computerization are easier.
----

microprogation using chlorophyllous shoots/nodal
cuttings, whether automated or not, seems
promising.

REFERENCES

Aitken-Christie, J. (1990) Automation. In:
Micropropagation: Technology and
Application. eds. DeBergh, P. and
Zimmerman, R. Kluwer Academic Publishers,
Dordrecht, 363-388.

Ando, T. (1978) Gaseous environment in the
airtight culture vessel containing
orchids. Abstr. Annual Autumn Meet. Jap.
Soc. Hort. Sci., 368-369. (in Japanese).

Bula, R.J., R.C. Morrow, T.W. Tibbitts, B.J.
Barta, R.W. Ignatius, and T.S. Martin.
(1991) Light-emitting diode as a
radiation source for plants. HortScience
26, 203-205.

Capellades, M.Q. (1989) Histological and
ecophysical study of the changes occurring
during the acclimatization of in vitro
cultures. Dissertation. Gent University,
Belgium, pp.98.

Cuello, J.L., P.N. Walker, and C.W. Heuser
(1989) Effects of ventilated CO_2, light
and sucrose levels on stage II
micropropagation of Buddleia alternifolla
(Butterfly Bush). ASAE Paper No. 89-6091.
(ASAE/CSAE presentation paper).

Debergh, P., and L. Maene. (1984) Pathological
and physiological problems related to the
in vitro culture of plants. Parasitica 40,
69-75.

Debergh, P., and R. Zimmerman. (1990)
Micropropagation: Technology and
Application. Kluwer Academic Publishers,
Dordrecht. pp.469.

Dillen, W., and S. Buysens (1989) A simple
technique to overcome vitrification in
Gypsophila paniculata L. Plant Cell,
Tissue and Organ Culture 19, 181-188.

Donnelly, D.J., and W.E. Vidaver. (1984)
Pigment content and gas exchange of red
raspberry in vitro and ex vitro. J. Amer.
Soc. Hort. Sci. 109, 177-181.

Donnelly, D.J., and W.E. Vidaver. (1988)
Glossary of plant tissue culture. Belhaven
Press, London. pp.141.

Dunstan, D.I., and K.E. Turner. (1984) The
acclimatization of micropropagated plants.
In: Cell Culture and Somatic Cell Genetics
of Plants. Vol.1. Ed. Vasil, I.. 123-129.
Academic Press.

Fujiwara, K., T. Kozai, and I. Watanabe.
(1987) Measurements of carbon dioxide gas
concentration in closed vessels containing
tissue cultured plantlets and estimates of

net photosynthetic rates of the plantlets.
J. Agric. Meteorol. 43, 21-30. (in
Japanese with English summary).

Fujiwara, K., T. Kozai, Y. Nakajo, and I.
Watanabe. (1989) Effects of closures and
vessels on light intensities in plant
tissue culture vessels. J. Agric.
Meteorol. 45 (3), 143-149 (in Japanese
with English summary).

Grout, B.W.W., and M.J. Ashton. (1978)
Transplanting of cauliflower plants
regenerated from meristem culture. II.
Carbon dioxide fixation and the
development of photosynthetic ability.
Hort. Res. 17. 65-17.

Infante, R., E. Magnanini, and B. Righetti (1989)
The role of light and CO_2 in optimizing
the conditions for shoot proliferation of
Actinidia deliciosa in vitro. Physiol.
Plantarum. 77. 191-195.

Hakkaart, F.A., and J.A. Versluijs. (1883) Some
factors affecting glassiness in carnation
meristem tip cultures. Neth. J. Pl. Path.
89. 47-53.

Hughes, K. W. (1981) In vitro ecology:
Exogenous factors affecting growth and
morphogenesis in plant culture systems.
Environ. and Experimental Botany. 21, 281-
288.

Kozai, T., K. Fujiwara, and I. Watanabe (1986)
Relation between the culture medium
composition and water potential of liquid
culture media. J. Agric. Meteorol. 42, 1-
6. (in Japanese with English summary).

Kozai, T., K. Fujiwara, and I. Watanabe.
(1986) Effects of stoppers and vessels on
gas change rates between inside and
outside of vessels closed with stoppers.
J. Agric. Meteorol. 42, 119-127. (in
Japanese with English summary).

Kozai, T., H. Oki, and K. Fujiwara (1987)
Effects of CO_2 enrichment and sucrose
concentration under high photosynthetic
photon fluxes on growth of tissue-cultured
Cymbidium plantlets during the preparation
stage. Proc. Symposium Floriezel 87 Arlon
- Belgium. 135-141.

Kozai, T., and K. Sekimoto. (1988) Effects of
the number of air changes per hour of the
closed vessel and the photosynthetic
photon flux on the carbon dioxide
concentration inside the vessel and the
growth of strawberry plantlets in vitro.
Environ. Control in Biol. 26, 21-29.

Kozai, T., and Y. Iwanami (1988) Effects of CO_2
enrichment and sucrose concentration under
high photon fluxes on plantlet growth of
carnation (Dianthus caryophyllus L.) in
tissue culture during the preparation
stage. J. Japan. Soc. Hort. Sci. 57(2),
279-288.

Kozai, T., Y. Koyama, and I. Watanabe (1988)
Multiplication of potato plantlets in
vitro with sugar free medium under high
photosynthetic photon flux. Acta
Horticulturae 230, 121-127.

Kozai, T., C. Kubota, and M. Nakayama (1989)
Net photosynthetic rates of plantlets in
vitro under natural and forced ventilation
conditions. J. Jap. Soc. Hort. Sci. 58.
(special issue 1). 250-251.

Kozai, T. (1990) Environmental control and
automation in micropropagation. Proc. of
the 4th Toyota Conference on Automation in
Biotechnology, Aichi-ken, Japan. 201-227.

Kozai, T. (1990) Micropropagation under
photoautotrophic conditions. In:
Micropropagation: Technology and
Application. eds. P. Debergh, and R.
Zimmerman, Kluwer Academic Publishers,
Dordrecht. 447-469.

Kozai, T., K. Tanaka, I. Watanabe, and M.

Hayashi (1990a) Measurements of relative humidity and transpiration characteristics of plantlets in vitro. Abstr. of Annual Meet. of Agric. Meteorol., 24-25.

Kozai, T., H.Oki, and K. Fujiwara (1990b) Photosynthetic characteristics of Cymbidium plantlet in vitro. Plant Cell, Tissue and Organ Culture. 22, 205-211.

Kozai, T. (1991) Photoautotrophic micropropagation, IN VITRO Cellular & Developmental Biology, 27P, (in press)

Kozai, T., M. Hayashi, and M. Ochiai. (1991a) Effect of the sidewards lighting on the growth and morphology of potato plantlets in vitro. J. Japan. Soc. Hort. Sci. 60 (special issue 1).

Kozai, T., S. Kushihashi, and C. Kubota. (1991b) Effect of the difference in air temperature between photoperiod and dark period on morphogenesis and growth of potato plantlets in vitro under photoautotrophic and CO_2 enriched conditions. J. Japan. Soc. Hort. Sci. 60 (special issue 1).

Kozai, T. (1991a) Acclimatization of micropropagated plants. In: Biotechnology in agriculture and forestry. Vol. 17. Ed. Bajaj, Y. Springer-Verlag. (in press).

Kozai, T. (1991b) Autotrophic micropropagation. In: Biotechnology in agriculture and forestry. Vol. 17. Ed. Bajaj, Y. Springer-Verlag. (in press).

Kozai, T. (1991c) Controlled Environments in conventional and automated micropropagation. In: Cell Culture and Somatic Cell Genetics of Plants, Vol. 8. 213-230.

Langhans, R.W.(Ed.) (1978) A growth chamber manual. Comstock Pub., Ithaca. pp.222.

Moe, R. (1988) Effect of stock plant environment on lateral branching and rooting. Acta Horticulturae 226, 431-444.

Morini, S., P. Fortuna, R. Sciutti, and R. Muleo. (1990) Effect of different light-dark cycles on growth of fruit tree shoots cultured in vitro. Advances in Hort. Sci. 4. 163-166.

Mousseau, M. (1986) CO_2 enrichment in vitro. Effect on autotrophic and heterotrophic cultures of Nicotiana tabacum. Photosynthesis Research 8, 187-191.

Nakayama, M., T. Kozai, and K. Watanabe. (1991) Effect of Presence/absence of sugar in the medium and natural/forced ventilation on the net photosynthetic rates of potato explants in vitro. submitted to Plant Tissue Culture Let. (in Japanese with English summary)

Pospisilova, J., J.Catsky, J. Solarova, and I.Ticha (1987) Photosynthesis of plant regenerants. Specificity of in vitro conditions and plantlets response. Biologia plantrum (Praha). 29, 415-421.

Pospisilova, J., J.Solarova, J.Catsky, M.Ondrej, and Z.Opatrny. (1988) The photosynthetic characteristics during the micropropagation of tobacco and potato plants. Photosynthetica 22, 205-213.

Preece, J.E. and E. Sutter. (1990) Acclimatization of micropropagated plants to the greenhouse and field. In: Micropropagation: Technology and Application. Kluwer Academic Publishers. Dordrecht. 71-94.

Read, P.E. (1989) Environmental effects in micropropagation. In: Handbook of plant cell culture. Vol.5. Eds. Ammirato, P.V. et al., McGraw-Hill Pub., 95-125.

Reuther, G. (1988) Comparative anatomical and physiological studies with ornamental plants under in vitro and greenhouse conditions. Acta Horticulturae 226, 91-98.

Urban, L. and A. Jaffrin. (1990) Steady state thermal conditions inside plant tissue culture vessels submitted to a constant level of irradiation. Biotronics. 19,71-81.

Walker, P.N., C.W. Heuser, and P.H. Heinemann (1988) Micropropagation: Studies of gaseous environments. Acta Horticulturae 230. 145-151.

Wilkins, H.F. (1988) Techniques to maximize cutting production. Acta Horticulturae 226, 137-143.

Ziv, M. (1990) Vitrification: morphological and physiological disorders of in vitro plants. In: Micropropagation: Technology and Application. Kluwer Academic Publishers. Dordrecht. 45-70.

COMPUTER INTEGRATED PLANT GROWTH FACTORY FOR AGRICULTURE AND HORTICULTURE

Y. Hashimoto

Dept. of Biomechanical Systems. Ehime University, Tarumi, Matsuyama 790, Japan

Abstract. Management and control for greenhouse production of vegetable and flower based on computer applications have been remarkably developed in northern European countries as the leader in the world. Furthermore, vegetable could be cultivated in the factory where such artificial environmental factors as lamps are controlled just like "process automation(PA)" in chemical process industries. These factories have been developed mainly by Japanese industrial company and are so called "Plant Growth Factory", though its original system appeared in Denmark , about 30 years ago. In the system, many computers are used for environmental control , nutrient control, and management of cultivation. Recently, it seems evident that a seed bed and a nursery of young plant should be improved from the system in the frame of the plant growth factory. Of course, the artificial intelligence is also introduced to the expert system for control and diagnosis of the cultivating crops. On the other hand, progress in automated mechanization for seeding and transplanting has made "greenhouse automation" fit for practical use just like "factory automation(FA)" in industries. Now, process industries are rationalized based on the concept of so called "computer integrated manufacture (CIM)" after the technological development involved in PA followed by FA. Therefore, such the system in the agricultural production as the plant growth factory including vegetable factory and nursery factory as well as the advanced greenhouse, should also be considered based on the concept of CIM. That is the "computer integrated agricultural production (CIAP)", proposed in our previous papers.
In this paper, the CIAP is examined from much more wide and generalized point of view, which may be called "computer integrated plant growth factory for agriculture and horticulture".
It might be noted that the newly proposed concept based on CIAP discussed in this paper should be expected as the most effective system in the coming generation.

Keywords. Computer applications; Artificial intelligence; Control applications; System identification; Agriculture

INTRODUCTION

It seems very difficult to control the environment factors strictly in solar greenhouse from the engineering point of view, because of the great disturbance from the time-variable sunlight . As for the environmental (namely climate) control of greenhouse, great progress in physics model and control system has been made for these ten years (Bot,1983; Takakura, 1988; Tantau, 1981; Udink ten Cate, 1981, 1983). In the environment control of greenhouse, it has been proved valid that optimal control is much more important than strictly constant control based on the setpoint (Tantau, 1990; Udink ten Cate, 1985). Furthermore, wide applications of computers are introduced to the management of greenhouse production as well as the environment control (Cros,1990).

On the contrary, under these controlled environment in the greenhouse, horticultural crops show the different behaviour from the traditional one which have been acquired through a long history of agriculture and horticulture, cultivated in the field. Especially, it might be interesting problem that the crop response is closely related to the control strategies in greenhouse (Challa,1991). Optimal conditions for the horticultural crops should be examined both from the horticultural operation(Vogel, 1989) based on physiological plant ecology as well as plant and cell physiology (Nonami, 1991a, 1991b) and from the management strategies for economic yield, resulting the decision of the strategies for the setpoints of the control in the greenhouse (Reinisch, 1989).

Recently, plant growth factory becomes prosperous. There are two types in the plant factory system. These are perfectly controlled system with artificial light and solar system. The former control system is similar to that in the growth chamber. On the other hand, the latter one is similar to that in the greenhouse. Comparing with the greenhouse, plant factory is not so familiar to our readers. Therefore, we would like to describe it in more detail. When plant is limited for producing vegetable only, it is sometimes called vegetable factory. If we are asked to find the origin of the plant factory, we should go back to thirty years ago when the first plant factory in the world was born in Denmark. Therefore the history is not so long. In the original system, the cultivation stage is treated as a manufacturing process. Several cultivating processes are combined into a batch process and controlled as assembling line. This made a great change in the concept of agriculture, and had an impact on the image of the traditional cultivation. Since then, several plant factories have been developed in the world. Now, the plant factory noticed in Japan (Takatsuji,1991) has environmental control system like a large growth chamber. In this system, the emphasis is laid on the process control rather than on manufacturing process. It is the system with standardized cultivating process, with which the optimum production can be realized in the shortest time (Iwao,1990). In detail, the system can make plants growing three to five times faster than in the field cultivation by supplying the twenty four daily illumination hours with artificial lighting system, controlling the CO_2 concentration in several times of that in the natural environment, and regulating the temperature, pH, EC in the optimal state. This system is the factory with process control. Moreover, in the factory, computer, all the sensor possible for use and some robots are invested to reduce the personnel expenses and to get the maximum harvest in the shortest time. In other words, this system seems to be the factory automation in agriculture and horticulture. Now the concept and system of the plant factory are going to be introduced to a seed bed and a nursery of young plant. Although it is hardly to say that system for seed bed and nursery are not different from cultivating systems for various crops, from the process control or the system automation point of view, they may be much the same.

In such a symbolized system as plant factory including advanced greenhouse, several types of computers and its networks as well as almost all kinds of high technologies are introduced. Therefore, it might be necessary for us to examine these computer integrated system, which we proposed as "computer integrated agricultural production (CIAP)" in our previous paper (Hatou, 1990b).

PROCESS CONTROL OF PLANT GROWTH FACTORY (GREENHOUSE CLIMATE CONTROL)

From the aspect of computer applications , the routine of the process control is established based on micro-computer (micro-processor). Control algorithm is also ready in PID algorithm based on feedback control system. Computer calculates the deviation between the set-point for the environment and the data monitored from the change of the environment in each sampling interval, and decides the manipulating signal by operating PID action on the deviation. Environmental factors including the light, temperature, relative humidity and CO_2 concentration can be divided into several sub-systems when those factors are controlled respectively. Then, several minutes of sampling interval may be sufficient enough for controlling the environmental factors. Furthermore, only very short time is needed for processing sampled data in PID control algorithm. Therefore, it is not difficult to combine all the sub-systems by one micro-computer based on a time sharing method. Nutrient control, however, has a big dead time. Therefore, it might be necessary that the system should be examined based on feedforward control algorithm as well as feedback one. Furthermore, EC and pH could not identify the state of nutrient control. Ion sensors are introduced to the system (Hashimoto, 1988c). As for the control of light, intensity and spectrum of the light have strong effect on plant growth. It seems interesting that the effect of frequency involved in the on-off interval of fluorescent lamp is not meaningless (Hashimoto, 1987b, 1988b).

To control the environment more flexibly, it is necessary to change the set point of the environment and the system parameters disturbed by the sudden changes in the outside. This belongs to one of the adaptive control problems (Tantau, 1990). To reduce the cost to the minimum in non-biological system, the optimal strategy is needed. This can be solved with the optimal control theory. The optimal control is much more effective for biological system, which is proposed in another section. Though the advanced control algorithm is rather complicated , it could be improved partly from the fundamental PID algorithm.

For the development of the better control system, the system identification and the parameter estimation are inevitable (Young, 1985; Morimoto, 1985, 1991a). From this point of view, control system design is proposed for greenhouse system (Young, 1991). On the other hand, fuzzy logic and AI are also necessary for the realization of better control system (Morimoto, 1991b).

These technology of process control is now introduced to process of micropropagation of tissue culture and nursery system (Kozai, 1991).

PLANT GROWTH FACTORY AUTOMATION (GREENHOUSE AUTOMATION)

Human labor has been used necessarily in seeding,

transplanting, harvesting and spreading farm chemicals, even in the process control system. Some robots and automated mechanizations begin to take place of these human labour in the agricultural and horticultural field (Su, 1989; Dario, 1990; Yamashita, 1991). In England, we can find an interesting example in commercial grower, where several types of anutomated mechanization substitute farmers as a pioneer of greenhouse automation.

Furthermore, clean and micro-robot are used in the LSI (Large Scale Integrated circuit of semi-conductor) factory, which could get over the unable barrier by human labour. Circumstance of micro-propagation is closely similar to that of LSI industry. The robots for micropropagation and transplanting in tissue culture are developed (Smith,1989; Fujiwara, 1991; Kinase, 1991), which will greatly improve nursery industry in agriculture and horticulture.

In the near future, robots, containing intelligent robot and intelligent tele robot, will substitute the human labor in plant growth factory including tissue culture. With these robots and mechanization, the plant growth factory may be called "Plant Growth Factory Automation" just like FA (factory automation) in the industrial manufacture (Hashimoto, 1989b).

SPEAKING PLANT APPROACH TO CULTIVATING PROCESS CONTROL

In order to promote the process optimization of plant growth factory, beyond the mere process control without biological elements, the identification of the system in the relationship between the environmental physics and plant responses to the environmental stresses and the control concomitant with the advanced algorithm should be indispensable. In another word, it is important to improve the environmental control from the fundamental level without biological element to the advanced level including biological elements.

Optimal condition should be obtained based on the deep understanding of physiological plant ecology as well as the mathematical model of the environmental physics. Describing in more detail, the optimal cultivating condition could not be identified without measuring the plant responses, while only environmental variations are measured in the process control .

From the measurement point of view, there are many kinds of data obtained from the plant responses, from long-term responses about growth to short-term responses about initial process of photosynthesis. In biological process of cultivating crop, knowledge processing is still important rather than mathematical processing. The knowledge processing means the expert system based on the knowledge base. Therefore, in the advanced control including biological elements, its success may be at the mercy of measurement , and knowledge processing as well as system theory.

This is the concept that should be defined as the speaking plant approach(SPA) to the process (or environmental) control of plant growth factory or greenhouse (Hashimoto, 1979, 1981, 1985a, 1989a).

Approach from Cultivating Technology and Plant Responses

Increasing the quality and quantity of farm products means cultivating the variety with good inheritance and supplying an optimal environment condition. The optimal environmental condition varies with plant variety and also varies with growing stage. Furthermore, as a result of the evolution for long years, plant growth is related to seasonal change affected by the altitude in the earth. Plant have the adaptability to daily environmental change existing in anywhere. All those show that the optimal environmental condition could not be simple.

In general, agricultural products with high added-value

could not be yielded with the environment only promoting the speed of the growth except when only the bio-mass is the harvesting object. Optimal environmental condition does not always mean the environment promoting plant growth, sometimes, the environmental stress that retards the growth but improves the quality is also concluded in the optimal condition. Environmental stress is necessary to the plant according to the growing period. This kind of cultivation can only be realized based on the knowledge of the expert who is praised to an excellent farmer. The knowledge of the expert has become possible to be input into the AI system (computer with artificial intelligence), but the plant stress such as water stress by withering can not be identified by the untrained people or the computer. This should be analyzed scientifically.

Responses of plants to environment are dealt scientifically in physiological ecology (Hashimoto, 1987a, 1990a). Growth is promoted by photosynthesis. The quantity of photosynthesis is evaluated with photosynthetic rate. It is mainly affected by quantum flux density and CO_2 concentration. Photosynthesis is usually measured in net CO_2 uptake with the light intensity (Parkinson, 1990). Photosynthesis is carried on the existence of CO_2 in the air surrounding the plant. Plant absorbs CO_2 from the stomata distributing in the leaf. Behavior of stomata (Omasa, 1985) is not negligible for plant growth, because the photosynthesis will stop when the stomata are closed. Stoma is controlled by the water status in the leaf, which can be quantified by water potential (Boyer, 1990). Quantitative analysis of stomata is carried out with transpiration and stomata conductance. Photosynthesis can not go smoothly even when the stomata are open, and the quantum flux density and CO_2 are sufficient if the photosynthate is accumulated in the leaf. Therefore, translocation in the leaf should be grasped. In Duke University, dynamics of translocation is measured by using ^{11}C, and the measurement system is a large one integrated by the computer (Strain, 1990). Its application to the problem discussed here would be difficult at present.

The realization of optimal environment control should be done based on the premise of grasping plant responses to the environmental stress in relation to the academic knowledge of physiological ecology, and making full use of the knowledge by farmers engaged in the actual cultivation.

Approach from Information and Knowledge Processing

Now, almost all kinds of data can be processed in real time by the aid of computer system. Information processing, including analog electric signal, could be easily done using computer-soft for digital signal processing. Especially, application of image processing is owing to Fourier transform , one of the most popular algorithm among digital signal processing. Image processing technique has elucidated plant responses in two dimension (Eguchi, 1990; Hashimoto, 1982, 1984c, 1990b; Omasa, 1982, 1990). NMR-CT used in medical diagnosis begins to be used in physiological plant ecology (Kramer, 1990). This may never be considered if without fast Fourier transform. Sensor with processing ability (Yamasaki, 1990) is also expected to get plant responses much more.

Computer processing the human's knowledge and being capable to make decision in stead of digital data comes into practical use. Because this system can make decision according to the input data, it is called AI (artificial intelligence) or knowledge processing. Fuzzy and neural networks is also found effective (Hirafuji, 1991; Morimoto, 1991b). These processing should be introduced into speaking plant approach.

Approach from System Identification and Control Engineering

Identifying the environment-plant system based on physiological plant ecology should be investigated, though identification for the physical system without biological elements has already been established. Some results have been reported (Hashimoto, 1984a, 1984b, 1985b, 1988a; Morimoto,1988; Tantau, 1985), but at present, there are no established methodology. It may be noted that identification should be examined in the combination of the dynamics of physiological plant ecology and mechanical or physical environment. Approach from this aspect is also important to make full use of the computer.

As shown by the history of the engineering for industrial process, feedback control is just a starting point. Now, the control engineering is advanced from DDC (Direct Digital Control) to adaptive control, optimal control, further to the complicated system based on fuzzy, neural, and AI. Also, the control system is extended from single computer to computers system with various functions. Furthermore, the control system is combined into LAN (Local Area Network), the ability of which is greatly increased.

COMPUTER SUPPORT SYSTEM FOR CONTROL, HORTI-CULTURAL OPERATION AND MANAGEMENT

Application of computer becomes more broad (Day, 1991). Computer is introduced to decision support system in horticultural operation and management as well as to real time control system. Furthermore AI (artificial intelligence) is now used for the expert and diagnosis system.

The tomato cultivating support system, based on the program of horticultural operation, was designed and examined (Hatou, 1990a, 1991b), system of which had the computer network composed of both usual micro-computer for environmental control of crops and the special computer for the artificial intelligence. The support system is found to be effective for large scale plant growth factory.

On the other hand, "ES (expert system)" is more biological and use of AI computer is inevitable. Expert system for disease and pest diagnosis for tomato was examined (Hoshi, 1988). Expert system for control system as well as adjustment of set-point based on horticultural operation, is also examined (Hatou, 1991a). In the important stage through the whole cultivating term, the status of the crop is able to be discriminated based on the AI computer. Thus, the adequate set-point of the environmental control could be decided based on the status of the crop.

As the next stage, expert system for the management of plant growth factory based on some strategies may be expected, which is so called CIM in agriculture, in other words, the high levelled decision system from the point of management view. This problem should be examined in "computer integrated agricultural production" rather than in the support system.

COMPUTER INTEGRATED PLANT GROWTH FACTORY

In industrial manufacture, application of robot has graded up the automation from process automation (PA) to factory automation (FA). Furthermore, the computer has been introduced into the factory management as called computer integrated manufacture (CIM).
Referring the circumstance in the industries, computer integrated system in agriculture and horticulture may be proposed as shown in figure 1.

Figure 1 shows the recent computer integrated system. The system has the computer network composed of host computer, usual micro-computer for process control and factory automation, and the special computer for knowledge processing based on the artificial intelligence. Necessary menu is chosen through the conversation between operator and computer. Six main menus are prepared , these are "LAN", "Process Control", "Factory Automation", "SPA(speaking plant approach)", "Support System", and "Knowledge Processing" as shown in the left part of the

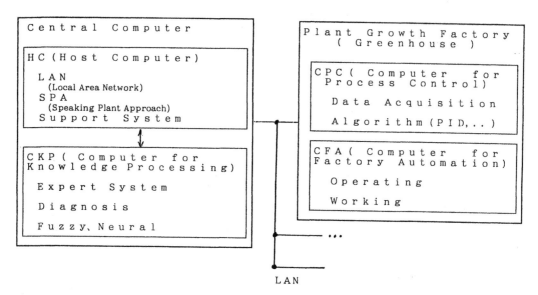

Figure 1 Computer integrated system(CIS)

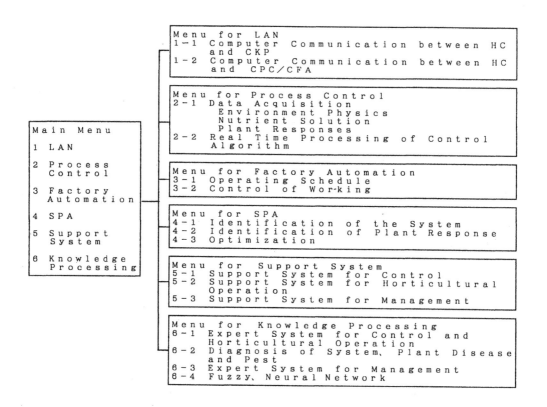

Figure 2 Main and sub-menu of CIS

108

figure 2. Each menu is used for the purpose described in the right part of the figure. The languages used in the menu are "BASIC", "C" and "PROLOG".

Menu for "LAN" deals with computer communication with proper protocol. The system is programmed with about several hundred steps.

Menu for "Process Control" is always used for monitoring and real time control, depending on mainly mathematical and physical method.

Menu for "Factory Automation" is ready for robot and mechanization system in primitive control.

On the other hand, menu for "SPA" is more complicated for the purpose of identification and optimization of total and sub-system based on physiological plant ecology.

Menu for "Support System" is prepared for the decision support of horticultural operation and management.

Finally, menu for "Knowledge Processing" is developed for expert system and diagnosis system based on the artificial intelligence.

CONCLUSION

Plant growth factory has been applied both to cultivating process and to micropropagation process. The value of them increases enormously, if computer integrated system could be introduced into these system. As PA (process automation), FA (factory automation), and CIM (computer integrated manufacture) in chemical process industry have much improved the productivity, greenhouse and plant growth factory which are considered to be closely similar to them among agriculture and horticulture, are also expected to improve the productivity.

Computer integrated system, composed of " computer for communication and network (LAN)", "computer for environmental control", "computer for nutrient control", "computer for factory automation", "computer for identification and optimization", "computer for support system of horticultural operation and management", and , "computer for knowledge processing in expert and diagnosis system", is inevitably necessary for the industrialization of agricultural production. It seems evident that computer integrated system proposed in this paper should be the most reasonable system.

REFERENCES

Bot,G.P.A. (1983). Greenhouse climate: from physical processes to a dynamic model. pp.240, Agricultural University, Wageningen, The Netherlands.

Boyer,J.S., and H.Nonami (1990). Water Potential and Its Components in Growing Tissues. In Y.Hashimoto et al.(Ed), Measurement Techniques in Plant Science, Academic Press, San Diego.Chap.2,pp.101 ～ 112

Challa,H. (1991). Reflections about Optimal Control in Greenhouse Cultivation. Proc. IFAC/ISHS 1st Workshop on Mathematical and Control Applications in Agriculture and Horticulture (in press) Pergamon Press, Oxford

Cros,M.-J., and R.Martin-Clouaire (1990). Management of Agricultural Production Systems. Proc. 3rd International Congress for Computer Technology 263 ～ 287

Dario,P. (1990). Research on Sensors for Advanced Robotics. Italy-Japan Joint Workshop on "Advanced Robotics" (Organized by P.Dario et al.)

Day,W. (1991). Computer Applications in Agriculture and Horticulture. Proc. IFAC/ISHS 1st Workshop on Mathematical and Control Applications in Agriculture and Horticulture (in press) Pergamon Press, Oxford

Eguchi, H. (1990). Digital Processing of Plant Images Selected by Spectral Characteristics of Reflectance for Evaluation of Growth. In Y.Hashimoto et al.(Ed), Measurement Techniques in Plant Science, Academic Press, San Diego. Chap.6, pp.361 ～ 372

Fujiwara, H. (1991). Discriminating robot system for carnation seedlings with fuzzy logic. Proc. IFAC/ISHS 1st Workshop on Mathematical and Control Applications in Agriculture and Horticulture (in press) Pergamon Press, Oxford

Hashimoto,Y. (1979). Computer control of short term plant growth by monitoring leaf temperature. Acta Horticulturae 106, 139 ～ 146

Hashimoto,Y.,T.Morimoto,S.Funada and J.Sugi(1981).Optimal Control of Greenhouse Climate by the Identification of Water Deficiency and photosynthesis in short-term Plant Growth. Proc. IFAC 8th World Congress 3621 ～ 3626, Pergamon Press, Oxford

Hashimoto,Y., T.Morimoto and S.Funada(1982). Image processing of Plant Information in the Relation between Leaf Temperature and Stomatal Aperture. In Technological and Methodological Advances in Measurement. (Proc. IMEKO 9th World Congress) Vol.3:313 ～ 320 North Holland Publ.Co., Amsterdam

Hashimoto,Y., and T.Morimoto (1984a). System identification of plant response for optimal cultivation in greenhouses. Proc. IFAC 9th World Congress, Vol.7:2039 ～ 2044, Pergamon Press, Oxford.

Hashimoto,Y., B.R.Strain and T.Ino (1984b). Dynamic behaviour of CO_2 uptake as affected by light. Oecologia 63, 159 ～ 165

Hashimoto,Y., T.Ino, P.J.Kramer, A.W.Naylor and B.R.Strain (1984c) Dynamic analysis of water stress of sunflower leaves by means of a thermal image processing system. Plant Physiology 76:266 ～ 269

Hashimoto,Y., T.Morimoto and T.Fukuyama (1985a). Some speaking plant approach to the synthesis of control system in the greenhouse. Acta Horticulturae 174, 219 ～ 226.

Hashimoto,Y., and T.Morimoto (1985b). Identification of water relation and CO2 uptake in physiological ecological processes in a controlled environment. Proc. IFAC 7th Symposium on Identification and System Parameter Estimation: 1677 ～ 1681, Pergamon Press, Oxford.

Hashimoto,Y., Y.Yi, T.Morimoto, F.Nyunoya, H.Nishina and Y.Nakane (1987a). Pilot chamber for the identification of the growth process in a vegetable factory. Proc. IFAC 10th World Congress, Vol.2:338 ～ 343, Pergamon Press, Oxford.

Hashimoto,Y., Y.Yi, F.Nyunoya, Y.Anzai, H.Yamazaki, S.Nakayama, A.Ikeda (1987b). Vegetable growth as affected by on-off light intensity developed for vegetable factory. Acta Horticulturae 229, 259 ～ 264

Hashimoto,Y., and Y.Yi (1988a). Dynamic model of CO_2 uptake based on system identification. Acta Horticulturae 248, 295 ～ 300

Hashimoto,Y., and Y.Yi (1988b). Control model of plant as affected by the pulsed light illumination. Acta Horticulturae 245, 448 ～ 454

Hashimoto,Y.,T.Morimoto,T.Fukuyama,H.Watake,S.Yamaguchi , and H.Kikuchi (1988c). Identification and control of hydroponic system using ion sensors. Acta Horticulturae 245, 490 ～ 497

Hashimoto,Y. (1989a). Recent Strategies of Optimal Growth Regulation by the Speaking Plant Concept. Acta Horticulturae 260, 115 ～ 121

Hashimoto,Y. (1989b). Dynamical Approach to Plant Factory. Proc. AGROTIQUE 89, 111 ～ 118, Bordeaux

Hashimoto,Y., and H.Nonami (1990a). Overview of Current Measurement Techniques from Aspects of Plant Science. In Y.Hashimoto et al.(Ed), Measurement Techniques in Plant Science, Academic Press, San Diego. Chap.1, pp7 ～ 24

Hashimoto,Y.(1990b). Leaf Temperature based on Image Processing. In Y.Hashimoto et al.(Ed), Measurement Techniques in Plant Science, Academic Press, San Diego. Chap. 6, pp372 ～ 386

Hatou,K., H.Nishina, and Y.Hashimoto (1990a). Tomato Cultivating Support System Based on Computer

Integrated Agricultural Production. Proc. 3rd International Congress for Computer Technology 100 ～ 107

Hatou, K., H. Nishina, and Y. Hashimoto (1990b). Computer Integrated Agricultural Production. Proc. IFAC 11th World Congress, Vol(11) 306 ～ 310, Pergamon Press, Oxford.

Hatou, K., Y, Kamio, and Y. Hashimoto (1991a). Computer Integrated Plant Factory based on AI. Proc. IFAC 9th Symposium on Identification and System Parameter Estimation (in press), Pergamon Press, Oxford.

Hatou, K., H. Nonami, M. Itou, I. Tanaka, and Y. Hashimoto (1991b). Computer Support System for Tomato Cultivation in Plant Factory. Proc. IFAC/ISHS 1st Workshop on Mathematical and Control Applications in Agriculture and Horticulture (in press) Pergamon Press, Oxford

Hirafuji, M. (1991). A Plant Growth Model by Neural Networks and L Systems. Proc. IFAC 9th Symposium on Identification and System Parameter Estimation (in press), Pergamon Press, Oxford.

Hoshi, T., and T. Kozai (1988). Disease and pest diagnosis for tomato. Proc. 2nd International Congress for Computer Technology 457 ～ 472.

Iwao, K., A. Ikeda, and S. Nakayama (1990). Control Application for Plant Factory. Proc. IFAC 11th World Congress, Vol(11) 295 ～ 301, Pergamon Press, Oxford

Kinase, A., and H. Watake (1991). Robot for masspropagation. Proc. IFAC/ISHS 1st Workshop on Mathematical and Control Applications in Agriculture and Horticulture (in press) Pergamon Press, Oxford

Kozai, T. (1991). Environmental Control and Automation in Micropropagation. Proc. IFAC/ISHS 1st Workshop on Mathematical and Control Applications in Agriculture and Horticulture (in press) Pergamon Press, Oxford

Kramer, P. J., J. N. Siedow, and J. S. MacFall (1990). Nuclear Magnetic Resonance Research on Plants. In Y. Hashimoto et al. (Ed), Measurement Techniques in Plant Science, Academic Press, San Diego. Chap. 6, pp403 ～ 431

Morimoto, T., Y. Hashimoto, and T. Fukuyama (1985) Identification and control of hydroponic system in greenhouses. In Identification and System Parameter Estimation (IFAC 7th Symposium on Identification) 1689 ～ 1693, Pergamon Press, Oxford

Morimoto, T., T. Fukuyama, Y. Yi, and Y. Hashimoto (1988). Identification of physiological dynamics in hydroponics. Proc. IFAC 8th Symposium on Identification Vol. 3 :1736 ～ 1741, Pergamon Press, Oxford.

Morimoto, T., I. Cho, and Y. Hashimoto (1991a). Identification of Hydroponics in an Advanced Control System of the Greenhouse. Proc. IFAC 9th Symposium on Identification and System Parameter Estimation (in press), Pergamon Press, Oxford.

Morimoto, T., and Y. Hashimoto (1991b). Application of Fuzzy Logic and Neural Network to the Process Control of Solution pH in Deep Hydroponic Culture. Proc. IFAC/ISHS 1st Workshop on Mathematical and Control Applications in Agriculture and Horticulture (in press) Pergamon Press, Oxford

Nonami, H., Y. Hashimoto, and J. S. Boyer (1991a). Environmental Control for Plant Growth in Plant Factory Operation and Greenhouse Management from Physiological Viewpoint. Proc. IFAC/ISHS 1st Workshop on Mathematical and Control Applications in Agriculture and Horticulture (in press) Pergamon Press, Oxford

Nonami, H., T. Fukuyama, I. H. Cho, and Y. Hashimoto (1991b). Growth Regulation of Tomato Plants by Nutrient Supply Management in Greenhouses. Proc. IFAC/ISHS 1st Workshop on Mathematical and Control Applications in Agriculture and Horticulture (in press) Pergamon Press, Oxford

Omasa, K., I. Aiga, and Y. Hashimoto (1982). Image Instrumentation for Evaluating the Effects of Air Pollutants on Plants. In Technological and Methodological Advances in Measurement (IMEKO 9th World Congress). Vol. 3:303 ～ 312, North Holland Publ. Co., Amsterdam

Omasa, K., Y. Hashimoto, P. J. Kramer, B. R. Strain, I. Aiga, and J. Kondo (1985) Direct Observation of Reversible and Irreversible Stomatal Responses of Attached Sunflower

Leaves to SO_2. Plant Physiology 79: 153 ～ 158

Omasa, K., and K. Shimazaki (1990). Image Analysis of Chlorophyl Fluorescence in Leaves. In Y. Hashimoto et al. (Ed), Measurement Techniques in Plant Science, Academic Press, San Diego. Chap. 6, pp837 ～ 401

Parkinson, K. J., and W. Day (1990). Design and Testing of Leaf Cuvettes for Use in Measuring Photosynthesis and Transpiration. In Y. Hashimoto et al. (Ed), Measurement Techniques in Plant Science, Academic Press, San Diego. Chap. 3, pp207 ～ 228

Reinisch, K., E. Arnold, A. Markert, and H. Puta (1989). Development of Strategies for Temperature and CO2 Control in the Greenhouse Production of Cucumbers and Tomatoes Based on Model building and Optimization. Acta Horticulturae 260, 67-76

Smith, G. P. (1989). The Application of robotics to Horticultural Micropropagation. AGROTIQUE 89. 179-194

Strain, B. R., J. D. Goeschl, Y. Fares, C. E. Magnuson, and C. E. Jaeger (1990). In Y. Hashimoto et al. (Ed), Measurement Techniques in Plant Science, Academic Press, San Diego. Chap. 4, pp265 ～ 276

Su, W. M., and J. C. Guinot (1989). Analyse des Niveaux de Robotisation d'un Engin de Travail du Sol. Proc. AGROTIQUE 89. 119 ～ 136, Bordeaux

Takakura, T. (1988). Protected Cultivation in Japan. Acta Horticulturae 230, 29 ～ 37

Takatsuji, M. (1991). Fundamental Study of Plant Factory. Proc. IFAC/ISHS 1st Workshop on Mathematical and Control Applications in Agriculture and Horticulture (in press) Pergamon Press, Oxford

Tantau, H.-J. (1981). The ITG digital greenhouse climate control system for energy conservation. Proc. IFAC 8th World Congress, Vol. 7:3617 ～ 3620, Pergamon Press, Oxford.

Tantau, H.-J. (1985). Measurement of Stomata Opening Using Energy Balances. In Identification and System Parameter Estimation (IFAC 7th Symposium on Identification) 1983 ～ 1986, Pergamon Press, Oxford

Tantau, H.-J. (1990). Automatic Control Application in Greenhouse. Proc. IFAC 11th World Congress, Vol. 11 :302 ～ 305, Pergamon Press, Oxford.

Udink ten Cate, A. J., and J. van de Vooren (1981). Adaptive Systems in Greenhouse Climate Control. Proc. IFAC 8th World Congress, Vol. 7:3609 ～ 3616, Pergamon Press, Oxford.

Udink ten Cate, A. J. (1983). Modelling and (adaptive) control of greenhouse climate. pp159, Agricultural University, Wageningen, The Netherlands.

Udink ten Cate, A. J. (1985). Simulation Models for Greenhouse Climate Control. In Identification and System Parameter Estimation (IFAC 7th Symposium on Identification) 1683 ～ 1688, Pergamon Press, Oxford

Vogel, G., A. Heissner, and P. Klaering (1989). Ways and Possibilities for Automatic Control of Plant Growth and Yield/Formation Processes and Computer Aided Soil and Crop Management in Vegetable Production. Acta Horticulturae 260, 25-66

Yamasaki, H. (1990). From Sensor Devices to Intelligent Sensing Systems. In Y. Hashimoto et al. (Ed), Measurement Techniques in Plant Science, Academic Press, San Diego. Chap. 1, pp25 ～ 42

Yamashita, J., T. Abe, K. Satou, T. Imoto, and M. Hikita (1991). Development and Travelling Performance of Unmanned Automatic Transport Vehicle Running in Parallel with Ridge Side and Row Spacing for Use in the Greenhouse. Proc. IFAC/ISHS 1st Workshop on Mathematical and Control Applications in Agriculture and Horticulture (in press) Pergamon Press, Oxford

Young, P. C. (1985). A Practical Approach to Identification and System Parameter Estimation. In P. C. Young et al. (Ed), Identification and System Parameter Estimation (Proc. IFAC 7th Symposium on Identification), Pergamon Press, Oxford, Vol. 1, 1 ～ 15

Young, P. C. (1991) Recursive Identification and Estimation in Computer Aided Control System Design for Glasshouse System. Proc. IFAC/ISHS 1st Workshop on Mathematical and Control Applications in Agriculture and Horticulture (in press) Pergamon Press, Oxford

INDUCED LIGNIFICATION AND ELICITORS - A CASE OF PLANT HOST-PARASITE INTERACTIONS

Y. Asada

Ehime University, Matsuyama 790, Japan

Abstract. Formation and induction of lignins in fungus-infected plants is one of the results of host-parasite interactions. A lignification-inducing factor (LIF) bound to cell walls in an inactive form is liberated in active form from cell walls by an infection or other stimuli. The LIF induced formation of "disease lignin" in the infected tissues, by activating the metabolic pathways of lignin biosynthesis. LIF was partially purified and presumptively identified as a glycopeptide. It is considered to be an endogenous elicitor which is a constitutive component of plant cells and associated with disease resistance. That derived from one plant can initiate lignification in other plant species. Lignin formation is, however, not initiated only by the application of the LIF, but requires the active participation of an infecting organism. LIF can transfer into the upper leaves and activate lignin formation upon infection.

Keywords. Biology; biocontrol; biological model; lignification-inducing factor; host-parasite interactions.

INTRODUCTION

Biological regulation systems have been evolved since life was formed on the earth. It is known that regulations and controls in biological metabolisms are most complicated and efficient, although they have not been understood completely. One of well-studied examples of such biological systems is lignin formation in plants when plants are infected by fungi. In this paper, mechanisms of formation and induction of lignins in fungus-infected plants are presented.

BIOLOGICAL MODEL OF DISEASE CONTROL

We have been interested in the formation and induction of lignins in fungus-infected plants, because lignin formation or lignification of diseased parenchyma cell walls is one of the typical host resistance responses induced by host-parasite interactions. For our studies, we used mainly the Japanese radish root as a susceptible organ, because the plant is easily obtained and also useful because of the absence of chlorophyll and its adaptability to experimental treatment. As an infecting fungus, we used mainly Peronospora parasitica, a downy mildew fungus, because of a strong pathogenicity. The cell walls of infected parenchyma tissue are red in color, a positive reaction for the presence of lignin. Of course, in healthy parenchyma tissue, there is no histochemical evidence for the presence of lignin here and in the cell walls of vessels positive reaction for the presence of lignin. Then, we have chemically isolated lignins from the healthy and diseased tissues and obtained the empirical formulae of both the isolated lignins as follows (Asada and Matsumoto, 1972):

Healthy lignin: $C_9H_{10.90}O_{2.20}(OCH_3)_{1.16}$
Disease lignin: $C_9H_{8.93}O_{2.80}(OCH_3)_{0.75}$

Figure 1 shows probable pathway of lignin biosynthesis in the plant. Lignin seems to be produced from p-coumaryl, coniferyl and sinapyl alcohols by the catalytic reaction of peroxidases. We assumed that in the diseased tissue either the pathway for the synthesis of sinapyl alcohol or the dehydrogenation polymerization was inhibited, because sinapyl alcohol has 2 methoxyl groups. If o-methyltransferase was completely inhibited then no ferulate, hence no coniferyl alcohol will be produced unless some methyltransferase isozymes present in the tissues. Thus the inhibition or activation of dehydrogenation polymerization are more likely responsible for the drift in lignin constituents in the diseased tissues. Since in the dehydrogenation polymerization of p-hydroxycinnamyl alcohols, peroxidase was thought to be involved, isoperoxidases in both the healthy and diseased tissues were analyzed. From assay for the substrate specificity of these isozymes, we have found that the synthesis of guaiacyl lignin in the diseased tissues might result from the decreased concentration of the isozymes which catalyze polymerization of sinapyl alcohol and also from the increased concentration of basic isozymes which catalyze polymerization of p-coumaryl and coniferyl alcohols (Ohguchi and Asada, 1974). To supply enough amounts of p-hydroxycinnamyl alcohol, we thought it would be necessary to increase the shikimic acid pathway activity. Qualitative analysis of intermediates in the pathway revealed, the presence of quinate, prephenate, phenylpyruvate, trans-cinnamate, p-coumarate and caffeate by gas-liquid chromatography of the extract from the infected root (Asada et al., 1972). These phenolic acids could not be detected from healthy tissue when we used an equal amount of samples. Generally, the hexose monophosphate shunt seems to be the predominant pathway for glucose catabolism in diseased plants. For such activation of the shikimic acid pathway in the roots, enzymes which catalyze these changes should be activated. In fact,

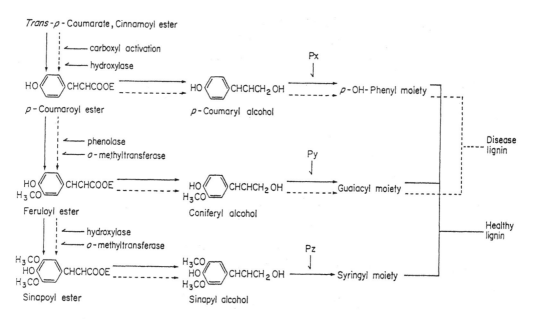

Fig. 1. A possible pathway for lignin biosynthesis in healthy (———→) and diseased
(- - - -→) plants. P_x, P_y, P_z: Peroxidase isozymes x, y, z.

increase in L-phenylalanine ammonia-lyase activity was observed in the infected root and
this was shown to be due to de novo synthesis of
the enzyme.

Lignin was formed not only in cell walls of root
tissues infected with the fungus, but also in
those of healthy tissues infiltrated with the
homogenate of the infected tissues. After the
treatment with the homogenate, lignin began to be
formed in cell walls in about 12 hours.
Lignification of the tissue occurred much earlier
than those infected by the pathogen, and
completely inhibited by blasticidin S, which is
an inhibitor of protein synthesis. It is likely
that some type of inducer is involved in the
homogenate for the enhanced lignification. The
inducer may either directly originated from the
pathogen or indirectly produced by the host
tissues as a response to the infection.

The homogenate-induced lignification was
completely inhibited by inhibitors of nucleic
acid synthesis, such as cordycepin, ethidium
bromide, chromomycin A₃ and actinomycin D when
they were applied within 30 min following the
homogenate infiltration. When these inhibitors
were applied within 2 hours after homogenate
infiltration, the lignification proceeded only to
a limited extent. These inhibitors exerted no
effect on tissue lignification when applied 3
hours or more after homogenate infiltration.
Blasticidin S, a protein synthesis inhibitor,
exerted a similar effect. It inhibited
lignification completely when it was applied
within 2 hours after homogenate infiltration and
to a limited extent when applied from 2 to 6
hours after homogenate infiltration. These
results may suggest that enzyme proteins

responsible for the lignification was
synthesized in the tissue at least 7 hours after
homogenate treatment. Lignification of root
tissues was also blocked when they were exposed
to a temperature of 50 degree C for 1 min in
water or for 1 hour in air prior to homogenate
infiltration. All cells in the tissues survived
these heat treatments as evidenced by their
ability to plasmolyze. The effect of these heat
treatment is reversible, as the lignification
could be induced when the heat-treated tissue was
subjected to the homogenate infiltration after
incubating at 20 degree C for 24 hours following
the heat treatment.

In order to know the origin of the
lignification-inducing factor, LIF, in the
homogenate, slices in which lignification had
been induced by infiltration with the homogenate
of diseased tissues were homogenized and fil-
trated. The filtrate was used as the second
homogenate. Similarly, the filtrate obtained from
the fresh tissues treated with the second
homogenate was used as the third homogenate.
Lignification-inducing ability of fresh tissues
was assayed by using the second and third
consecutive homogenate. The lignification-inducing factor, LIF, should be a substance which is
produced in the root tissues in response to the
treatment, because it was not only withstood
massive dilution but also increased in activity
through the sequential repetition of homogenate
infiltration. Therefore, we have isolated LIF
from the homogenate. A single spot was obtained
on thin layer chromatography. From the UV absorption spectrum of this compound the LIF have a
peptide bond, and also since this compound was
qualitatively anthrone positive, the content of
carbohydrate and protein in this substance were

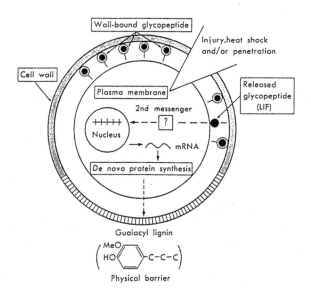

Fig. 2. A speculative diagram of lignin formation in diseased and/or stressed plant tissues. Recognition between LIF and receptors may possibly activate metabolic pathways of lignification in cells through mediation by a second messenger.

estimated. The ratio of carbohydrate and protein contents was 1 to 1.31, suggesting that the LIF is a glycopeptide. The results of dialysis through visking tubes and Sephadex column chromatography showed that the molecular weight of this compound seems to be in the order of several thousands. Treatments with pronase and sodium periodate caused the loss of biological activity of the LIF, indicating that both carbohydrate and protein moieties are essential for the biological activity of the LIF. The activity also decreased when this was kept at 60 degree C for 20 min or at pH 2 or 12. These physico-chemical properties suggest that the LIF is a glycopeptide and its activity depends on a complex structure of carbohydrate and protein.

The LIF is not the product of pathogens, but is a constitutive substance of plants which are able to form disease lignins. This is indicated by the fact that even the homogenate of tissues exposed to high temperatures of over 40 degree C is capable of inducing lignin formation in healthy tissues. It is thus suggested that the LIF may possibly be produced by degradation of the constitutive substances or liberated in active form from plant cells under various stresses. Therefore, experiments were conducted to ascertain the localization of the LIF in the cells. The homogenate of the uninoculated root was centrifuged at a low speed. Both the supernatant and pellet were treated at 120 degree C for 20 min and each fraction was assayed for lignin formation. The supernatant did not show any capacity for lignin formation, but the pellet fraction did. In the homogenate of the diseased root, however, active LIF was found in both the supernatant and pellet, suggesting liberation of LIF from cell walls. In the homogenate of the root tissues exposed to high temperature, the activity was found only in the supernatant. Lignin induction was also found in the autoclaved

homogenate of the root tissues which had been treated with blasticidin S. These results indicate that the LIF located in the cell walls was activated by its liberation. The idea that the LIF is located in cell walls also supported by the fact that lignification was induced in the homogenate of root tissues which had previously been treated with blasticidin S to block de novo synthesis within the cells. To confirm this idea, we have prepared a cell wall fraction from the root. The wall preparation was suspended in a phosphate buffer and autoclaved at 120 degree C for 15 min. This extract induced the lignification, therefore, the LIF is located in the cell walls.

The LIF was very active in lignin induction not only in radish root, but also in radish leaves, turnip roots, leaves of cabbage and Chinese cabbage, cauliflower, sweet potato tuber and cucumber fruit. However, it was not active in white potato tuber and onion bulbs which do not form wound lignins. It is of interest that the LIF from radish root was very active in lignin induction in all the plants which have the basic capability to form wound lignin, including turnip root. Thus, the LIF produced in tissues of these plants could be the same, and it may have a generalized lignin induction activity. The LIF produced in turnip root was also found to induce lignification in cucumber fruit as well as radish root. It is thus plausible that the LIF should prevent the invasion or damage caused by a pathogen in the plant tissues, if the lignified walls could block fungal invasion. The LIF was sprayed on the first and second leaves of cucumber, and the treated leaves and untreated leaves were then inoculated with conidia of Colletotrichum lagenarium. The intensity of the disease symptoms in the second and third leaves was estimated by the number and size of lesions. A large reduction in the number and size of lesions was observed not only in the LIF-treated second leaves, but also in the untreated third leaves. Given the high inoculum density, 10^7 per ml, of the conidia, the reduction of the lesion number and inhibition of the size development were significant. Germination of the conidia in LIF solution was better than that in water, indicating that LIF is not a germination inhibitor. Yet lesion number and size was inhibited when LIF was applied to cucumber leaves. It is particularly of interest that the LIF was active in resistance induction in the upper untreated leaves, suggesting that resistance induction by the LIF is transferred systemically to the upper leaves. However, the LIF which elicits lignification in host cell walls is naturally present in cell walls as a constitutive and inactive form. Thus, the LIF may be classified as a kind of endogenous elicitor since it is activated by its liberation from the plant cell walls. As mentioned earlier, the LIF can induce lignification in all the plants that are capable of lignification in response to cut-injury. Based on the results mentioned so far, a speculative model for lignification in the diseased tissues is illustrated in Fig. 2. However, if lignification in the diseased tissues is induced by the LIF which was activated after release from cell walls, then why does not lignification occur right after spraying the activated form of the LIF? Another factor which would probably be induced by alien stimuli and functions in releasing the LIF from cell walls may be involved. Preliminary experiments showed that the LIF was applied with 2-chloroethyl phosphate, which become an ethylene producer within the cells, lignification occurred extensively, so that the induction of lignification may be regulated in part by a plant hormone.

CONCLUSION

A lignification-inducing factor (LIF) bound to cell walls in an inactive form is liberated in active form from cell walls by an infection or other stimuli. The LIF induced formation of "disease lignin" in the infected tissues, by activating the metabolic pathways of lignin biosynthesis. LIF was partially purified and presumptively identified as a glycopeptide. It is considered to be an endogenous elicitor which is a constitutive component of plant cells and associated with disease resistance. That derived from one plant can initiate lignification in other plant species. Lignin formation, however, not initiated only by the application of the LIF, but requires the active participation of an infecting organism. LIF can transfer into upper leaves and activate lignin formation upon infection.

REFERENCES

Asada, Y., and I. Matsumoto (1972) The nature of lignin obtained from downy mildew-infected Japanese radish root. Phytopathol. Z., 73, 208-214.

Asada, Y., I. Matsumoto, and T. Tashiro (1972) The formation of phenolic acids in the root of downy mildew-infected Japanese radish. Ann. Phytopathol. Soc. Japan, 38, 405-409.

Ohguchi, T., and Y. Asada (1974) Dehydrogenation polymerization products of p-phydroxycinnamyl alcohols by isoperoxidases obtained from the downy mildew-infected roots of Japanese radish (Raphanus sativus). Physiol. Plant Pathol., 5, 183-192.

WATER AND THE PHYSIOLOGICAL REGULATION OF GROWTH IN CONTROLLED ENVIRONMENTS

J. S. Boyer* and H.Nonami**

*College of Marine Studies, University of Delaware, Lewes, DE, USA
**Dept. of Biomechanical Systems, Ehime University, Tarumi, Matsuyama, Japan

Abstract. The process of photosynthesis imposes water use on the plant that is far larger than the needs for growth and metabolism. Most of the water is lost by evaporation (transpiration) and can cause high and variable humidity in controlled environments. Different humidities result in differential development and quality of various plant parts. As a consequence, humidity control with large capacity dehumidification is required for reproducible plant growth. The differential development at various humidities can be understood with a mass budget showing that transpiration and growth compete for the same water supply. Gradients in water potential (Ψ_w) affect the competition. Growth requires a growth-induced Ψ_w field around the vascular supply to provide water at the required rate. Water is transported along the field to all the cells, moving from cell to cell and encountering significant frictional resistances. Water for transpiration also probably requires a potential field but little is known about it because the flow path is not well understood.

The competition for vascular water forces growth to operate with a lower potential field when transpiration is faster. The lower field is supported by a lower osmotic potential because of accumulated solute that also acts to maintain pressures inside the growing cells. The quality of the tissue is affected by the increased solute. Thus, depending on the objectives, plant development can be altered in controlled environments by changing humidity to favor rapid growth that maximizes the size of the product or by slowing growth, which increases the content of sugars and amino acids and thus alters the quality of the product.

Keywords. Water potential, growth-induced water potential field, osmotic potential, humidity control, photosynthesis, transpiration.

INTRODUCTION

Most plants possess the unique ability to carry out photosynthesis. Using the products of the process, all the organic constituents of the cells can be derived as long as CO_2 is available from the air, and H_2O and a few salts are available from the soil. The CO_2 and H_2O combine to form sugars that together with the elements from the soil are the building blocks for the other cell constituents.

The CO_2 enters the cells by dissolving in water at the cell surfaces, and thus the cells must expose wet surfaces to the air inside the leaf. This causes a large water loss (transpiration) from the plant because water tends to evaporate from the wet cell surfaces. The water vapor diffuses out of the leaf and into the surrounding air.

The H_2O entering the air causes special problems for controlled environments intended for plants. In a confined space, the evaporation causes humidities to rise. The more plants there are, the higher the humidity. High humidities affect the quantity and quality of the plant. Leaves and fruits tend to be large and low in structural material. Soils are often waterlogged, and salt uptake may diminish. Root development may be inhibited. Because the humidity depends on how many plants there are in the chamber, the plants vary in these characteristics according to plant density.

Reproducible plant growth is thus difficult to obtain without humidity control.

The amount of water lost to the air depends mostly on the amount of radiation striking the leaves. The radiation causes leaf stomata to open. The greater the radiation input, the more open the stomata. The diffusive resistance to water loss is thus decreased by high radiation, and transpiration increases. In addition to these diffusive effects, the incoming radiation is partitioned between transpiration and sensible heat according to the humidity, windspeed, and temperature of the leaf and air. The lower the humidity, the greater is the fraction of the incoming radiation devoted to transpiration. Leaf temperatures decrease as transpiration becomes more rapid and they may fall below air temperature.

The effect of these interacting factors is to cause humidity control to become more difficult as humidities are brought down to those approaching the external environment. To deal effectively with this problem, it is not enough to achieve humidity control in an empty chamber. Moreover, one cannot fill a chamber with plants and determine their water use at unspecified humidities. Design must be aimed at chambers fully laden with plants exposed to the required humidities.

PHYSIOLOGICAL CONTROL OF WATER
MOVEMENT THROUGH THE PLANT

Because humidity can have significant effects on plant growth, it is important to consider the principles that are involved. The amount of water lost by transpiration is much larger than the amount required as a substrate for photosynthesis and cell growth, and this imposes larger requirements on water extraction from the soil than would otherwise occur. The situation can best be understood by putting the water lost from the plant into a mass budget that includes all the water lost and gained by the plant (Boyer, 1985). The budget would have the form:

$$A + T + G - H \qquad (1)$$

where the water taken in is the absorption flux A, water loss is the transpiration flux T, the water used in all growth processes is the growth flux G, and the water stored in the leaf is the storage or hydration flux H. Each flux has units of moles·m^{-2}·s^{-1} and is based on the same unit of area, usually the projected leaf area, but the area is arbitrary and could be chosen anywhere in the plant as long as it is the same for each flux.

The flux A is toward the plant and accordingly is a positive flux whereas the flux T is a loss and thus a negative flux. Because the flux G involves water used in cell enlargement and metabolism, it also is negative. The hydration flux affects the water content of the tissue and depends on the balance of A, T, and G. Thus, H is positive when A exceeds T plus G, or negative when A is less than T plus G. In many cases H is zero because the other fluxes balance each other.

This budget includes G because cell enlargement and metabolism are essential plant functions even though the flux may be only 1/100 of T. The physical requirements are different for G and T, and conditions favoring T may inhibit G and vice versa. In particular, rapid T can dehydrate the shoot tissues and decrease G.

Water absorption for A depends to a large extent on the development of tensions (negative pressures) in the plant that extract water from the soil (Boyer, 1985). When T plus G exceed A, H becomes negative and the cells dehydrate. This dehydration is common early in the day and causes tensions to develop that are applied to water in the plant and ultimately the soil. As sufficient water is being absorbed to equal the requirements for T plus G, further dehydration is prevented and H becomes zero. In this condition, the shoot is dehydrated but stable. The process reverses as transpiration diminishes late in the day. Water enters faster than T plus G require. The H is positive, the shoot rehydrates, and tensions become less until T plus G are again balanced by lower A. This sequence of events indicates that absorption always lags transpiration, and it illustrates what has been called the absorption lag (Kramer, 1937; 1938).

The tensions that form as a result of dehydration are determined by the tissue water potential (Ψ_W, i.e., the chemical potential of water expressed in units of pressure). The Ψ_W varies with tissue water content and is determined mostly by solute (Ψ_S) and pressure (Ψ_P) components according to:

$$\Psi_W - \Psi_S + \Psi_P \qquad (2)$$

Water potentials develop in cells because photosynthesis generates solute at concentrations usually much higher than in the solutions surrounding the cells. The high concentrations lower the Ψ_W of the cell solution by the amount Ψ_S which causes water to move into the cells (Fig. 1). As water enters, pressures develop because the increased volume is opposed by the rigidity of the cell walls. The increased pressure raises the Ψ_W by the amount Ψ_P (Fig. 1).

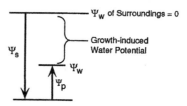

Figure 1. Potential diagram illustrating the osmotic potential (Ψ_S), pressure potential (Ψ_P), and water potential (Ψ_W) of a cell in a growing tissue. The tissue is in contact with water having a Ψ_W of zero. In this case, the growth-induced Ψ_W forms because Ψ_P is not as high as it would be in the absence of cell enlargement. The Ψ_P is suppressed by the yielding of the cell walls.

Because solute concentrations are generally low outside the cell (Scholander et al., 1965; Boyer, 1967; Klepper and Kaufmann, 1966; Nonami and Boyer, 1987), the Ψ_W is transmitted to the solution surrounding the cells mostly as a tension, which moves water into the cells and ultimately from the soil. When water is optimally available to plants, tensions tend to run in the 0.2 to 2.0 MPa range (Scholander et al., 1965; Boyer, 1967).

As tissues dehydrate (H is negative), their Ψ_W becomes lower and greater force is exerted on the surrounding water. Tensions as high as 5 to 6 MPa (50 to 60 bars) have been measured in dehydrated tissues (Oechel et al., 1972). The increased tension increases A and balances T plus G. In effect, the balancing occurs because Ψ_W changes in a corrective direction whenever an unbalance occurs and H is not zero.

The concept that the mass budget is balanced by the effects of Ψ_W has large implications for plant growth in controlled environments. Leaves and indeed all plant organs enlarge in a coordinated fashion by absorbing water and thus water for the growth process must enter all the cells (Molz and Boyer, 1978). As a result, cells closest to the water supply must transmit water at rates many times their own needs. Despite the fact that they may have a relatively high water conductivity, the additional flow necessary for the other cells causes substantial Ψ_W gradients to form. For growth, Ψ_W in the outermost cells can be 0.2 to 0.4 MPa lower than in the inner cells (Molz and Boyer, 1978).

Similar principles apply to the water lost by transpiration except that the loss is not uniform for all the cells, and the gradients are thus different from those required for growth (Boyer, 1974). The external leaf cells (epidermis) are coated with waxes and there are waxes on the surfaces of internal cells close to the stomatal pores (Boyer, 1985; Nonami et al., 1991). The waxy layers retard evaporation and it appears that transpiration occurs mostly from cells close to the veins. Thus, the cells close to the leaf surface are bypassed by the water vapor diffusing in the intercellular air spaces (Boyer, 1974; Boyer, 1985; Nonami and Schulze, 1989; Nonami et al., 1991). It also has been suggested that water moves readily to the epidermis and evaporates from the underside close to the stomatal pores (Meidner, 1976; Sheriff and Meidner, 1974). Whatever the fact, water appears to bypass many of the leaf cells (Rayan and Matsuda, 1988) in contrast to the situation for growth. Little is known about the Ψ_W gradients associated with transpiration except that they are substantial over large distances (e.g., from root to shoot) and they are detectable in individual leaves (Nonami and Schulze, 1989; Nonami et al., 1991).

GROWTH-INDUCED WATER POTENTIALS

The Ψ_W associated with growth can be studied if plants are grown in the dark in saturating humidity where T is zero. Under steady conditions, H also is zero and Equation 1 becomes:

$$A = -G \qquad (3)$$

The Ψ_W of the tissue is then devoted entirely to absorbing water for the growth process.

Figure 2 shows Ψ_W along the stem of a soybean seedling grown in this way (Cavalieri and Boyer, 1982; Boyer, 1988). The Ψ_W are averages and are about 0.15 MPa lower in the rapidly growing (rapidly elongating) tissue at the top of the stem compared with the slowly growing (nonelongating) tissue at the base. In contrast, the nonelongating tissue is in near-equilibrium with the Ψ_W of the root medium because there is no water uptake by the cells (A, T, G, and H are zero in this tissue). Thus, the Ψ_W in the vascular supply also equilibrates with the Ψ_W of these cells.

On the other hand, because the Ψ_W of the elongating tissue is lower than in the nonelongating tissue, the Ψ_W must be lower than in the vascular supply. The amount by which it is lower represents the force applied to the water in the vascular system by the enlarging cells. The result is that water moves out of the vascular system at a rate sufficient to feed the enlargement process.

Although these potentials are averages, they indicate that a three dimensional gradient exists around the vascular system (Fig. 2, middle graphs). This gradient has a higher potential close to the vascular tissue and decreases radially as water moves outward from the vascular system (Molz and Boyer, 1978). It also decreases as water moves inward from the vascular tissue. Because of its three-dimensionality, it is most properly considered a potential field arrayed around the vascular system (Nonami and Boyer, 1987; Nonami and Boyer, 1989).

The potential field is responsible for moving water into each cell in a coordinated fashion so that the entire stem enlarges. Because the cells closest to the vascular supply must transmit water for the outlying cells, the potential field is steepest there. The cells farthest from the system require only water for their own needs and the field becomes fairly flat at this position (Fig. 2).

ORIGIN OF GROWTH-INDUCED
WATER POTENTIALS

The potential field associated with growth is a fundamental property of all growing plant tissues (Boyer, 1985; Nonami and Boyer, 1987). It was suggested that the field forms from the growth activity of the cells because the enlarging cell walls would yield to internal pressures, preventing them from becoming as high as they otherwise would if the walls were rigid (Boyer, 1968). The effect would be to create a Ψ_W lower than in the vascular system (Fig. 1). A different view is that the cell protoplasts could be surrounded by a concentrated solution in the cell walls sufficient to cause Ψ_W to be low in the cells (Cosgrove and Cleland, 1983). Nonami and Boyer (1987) found that high concentrations were not present, however. They observed instead a significant tension on the water surrounding the cells. Thus, the original concept based on wall yielding is the most likely, and the tension in the wall solution probably arises by transmission of the low potential of the cell protoplasts to the surrounding solution in the usual way.

Water entering the enlarging cells probably flows simultaneously through the cells and around them in the cell walls (apoplast) according to the relative resistances of each path (Boyer, 1985; Molz and Ferrier, 1982; Steudle, 1989; Steudle and Frensch, 1989; Steudle and Jeschke, 1983). The magnitude of the tensions supporting this flow indicates that enlarging tissues have a significant frictional resistance. This probably occurs because the vascular supply is not well developed in enlarging tissues. Also, there are undifferentiated tissues throughout the enlarging region (Steudle and Boyer, 1985).

The presence of the field has two consequences. First, if its shape is modified it will alter the flow of water to the outer tissues for growth. For example, if the Ψ_W decreases in the vascular supply, the initial effect is to invert the field immediately adjacent to the vascular system (Fig. 3). This inversion prevents water from moving out of the vascular system and blocks water transport to all the outlying tissue. However, the outlying tissues are not initially changed in their Ψ_W and most of the potential field is still intact (Fig. 3).

Such a block rapidly inhibits the enlargement process (Nonami and Boyer, 1989; Nonami and Boyer, 1990), probably within seconds, because there is a change in Ψ_W in only a few of the cells. Rapid inhibitions have been observed under conditions that would change the field (Acevedo et al., 1971; Nonami and Boyer, 1990). In time, the potential field of the outlying cells adjusts, a new field is established, and growth resumes gradually (Nonami and Boyer, 1990). The gradual resumption is in contrast to the rapid inhibition initiated by the vascular system. The difference is probably attributable to the number of cells involved.

Thus, growth competes with transpiration for the water in the vascular system, and growth is sensitive to the rate of transpiration (Westgate and Boyer, 1984; Fiscus et al., 1983). In effect, transpiration lowers the potential in the vascular supply and the growth of the cells needs to compete for water at a correspondingly lower potential.

The growing cells respond to vascular conditions by adjusting their Ψ_S (Westgate and Boyer, 1984) which maintains the Ψ_P and the growth-induced Ψ_W (Fig. 1). The Ψ_S determines the maximum force that can be applied for the growth process. The Ψ_S is lowered by solute entering the cells faster than it is utilized. The solute is derived mostly from photosynthesis (or from reserves previously accumulated by photosynthesis) and consists of sugars and amino acids (Meyer and Boyer, 1981; Morgan, 1984). It is used in biosynthesis for growth and will accumulate if it is unused in the growth process. There is evidence that growth slows as the solution in the vascular system decreases in potential, and solute accumulation results (Meyer and Boyer, 1981). It has been shown (Meyer and Boyer, 1981; Michelena and Boyer, 1982; Sharp and Davies, 1979; Westgate and Boyer, 1985) that the accumulation is particularly large in the growing regions. Because the accumulation increases the quantity of sugars and amino acids, the quality of the tissue is changed.

The field provides a way to explain how growth responds to the effects of transpiration and variations in the water supply around the roots, and why the quality of the tissues also changes. It is important to note that, because the potential of the growing cells is not in equilibrium with the surroundings, a change in potential in the vascular supply will not cause the same change in the growing cells. Initially, the Ψ_W of most of the cells is unchanged and later Ψ_W adjusts to a new Ψ_W. Thus, growth experiments require the Ψ_W to be measured. If the growing tissue is excised, the Ψ_W also will change. Excision removes the external water supply but solute transport cannot occur, the Ψ_S cannot adjust, and enlargement continues until Ψ_P decreases to a threshold value at which enlargement is no longer possible (Cosgrove et al., 1984; Boyer et al., 1985). The decreased Ψ_P results from cell wall relaxation that occurs as the walls yield but no water enters. The effects are quite small (Ψ_W decreases about 0.5 to 1.0 MPa) and are delayed by attached mature tissue which can act as an internal water supply (Matyssek et al., 1988). For some growth experiments, however, these changes will need to be taken into account.

CONCLUSIONS

The competition between transpiration and growth results in a differential development of plant organs. The effects are expressed not only as variations in the size of the organs but also variations in solute content that can affect the quality of the tissues. The competition is mediated by a potential field in the enlarging tissue that interacts with the water potential of the vascular tissue supplying water for both transpiration and growth. Humidity control is thus essential for reproducible plant development in controlled environments and requires a large capacity for dehumidification because of large rates of transpiration.

REFERENCES

Acevedo, E., Hsiao, T.C. & Henderson, D.W. (1971). Immediate and subsequent growth responses of maize leaves to changes in water status. *Plant Physiol.*, 48, 631-636.

Boyer, J.S. (1967). Leaf water potentials measured with a pressure chamber. *Plant Physiol.*, 42, 133-137.

Boyer, J.S. (1968). Relationship of water potential to growth of leaves. *Plant Physiol.*, 43, 1056-1062.

Boyer, J.S. (1974). Water transport in plants: mechanism of apparent changes in resistance during absorption. *Planta*, 117, 187-207.

Boyer, J.S. (1985). Water transport in plants. *Annu. Rev. Plant Physiol.*, 36, 473-516.

Boyer, J.S., Cavalieri, A.J. & Schulze, E.-D. (1985). Control of cell enlargement: effects of excision, wall relaxation, and growth-induced water potentials. *Planta*, 163, 527-543.

Boyer, J. S. (1988). Cell enlargement and growth-induced water potentials. *Physiol. Plantarum*, 73, 311-316.

Cavalieri, A.J. & Boyer, J.S. (1982). Water potentials induced by growth in soybean hypocotyls. *Plant Physiol.*, 69, 492-496.

Cosgrove, D.J. & Cleland, R.E. (1983). Solutes in the free space of growing stem tissues. *Plant Physiol.*, 72, 326-331.

Cosgrove, D.J., Van Volkenburgh, E. & Cleland, R.E. (1984). Stress relaxation of cell walls and the yield threshold for growth: demonstration and measurement by micropressure probe and psychrometer techniques. *Planta*, 162, 46-54.

Fiscus, E.L., Klute, A. & Kaufmann, M.R. (1983). An interpretation of some whole plant water transport phenomena. *Plant Physiol.*, 71, 810-817.

Klepper, B. & Kaufmann, M.R. (1966). Removal of salt from xylem sap by leaves and stems of guttating plants. *Plant Physiol.*, 41, 1743-1747.

Kramer, P.J. (1937). The relation between rate of transpiration and rate of absorption of water in plants. *Amer. J. Bot.*, 24, 10-15.

Kramer, P.J. (1938). Root resistance as a cause of the absorption lag. *Amer. J. Bot.*, 25, 110-113.

Matyssek, R., Maruyama, S. & Boyer, J.S. (1988). Rapid wall relaxation in elongating tissues. *Plant Physiol.*, 86, 1163-1167.

Meidner, H. (1976). Water vapour loss from a physical model of a substomatal cavity. *J. Exp. Bot.*, 27, 691-694.

Meyer, R.F. & Boyer, J.S. (1981). Osmoregulation, solute distribution, and growth in soybean seedlings having low water potentials. *Planta*, 151, 482-489.

Michelena, V.A. & Boyer, J.S. (1982). Complete turgor maintenance at low water potentials in the elongating region of maize leaves. *Plant Physiol.*, 69, 1145-1149.

Molz, F.J. & Boyer, J.S. (1978). Growth-induced water potentials in plant cells and tissues. *Plant Physiol.*, 62, 423-429.

Molz, F.J. & Ferrier, J.M. (1982). Mathematical treatment of water movement in plant cells and tissues: a review. *Plant Cell Environ.*, 5, 191-206.

Morgan, J.M. (1984). Osmoregulation and water stress in higher plants. *Annu. Rev. Plant Physiol.*, 35, 299-319.

Nonami, H. & Boyer, J.S. (1987). Origin of growth-induced water potential: solute concentration is low in apoplast of enlarging tissues. *Plant Physiol.*, 83, 596-601.

Nonami, H. & Boyer, J.S. (1989). Turgor and growth at low water potentials. *Plant Physiol.*, 89, 798-804.

Nonami, H. & Boyer, J.S. (1990). Primary events regulating stem growth at low water potentials. *Plant Physiol.*, 93, 1601-1609.

Nonami, H. & Schulze, E.-D. (1989). Cell water potential, osmotic potential, and turgor in the epidermis and mesophyll of transpiring leaves. *Planta*, 177, 35-46.

Nonami, H., Schulze, E.-D. and Ziegler, H. (1991). Mechanisms of stomatal movement in response to air humidity, irradiance and xylem water potential. *Planta*, 183, 57-64.

Oechel, W.C., Strain, B.R. & Odening, W.R. (1972). Tissue water potential, photosynthesis, ^{14}C-labeled photosynthate utilization, and growth in the desert shrub *Larrea divaricata* Cav. *Ecol. Monogr.*, 42, 127-141.

Rayan, A. & Matsuda, K. (1988). The relation of anatomy to water movement and cellular response in young barley leaves. *Plant Physiol.*, 87, 853-858.

Scholander, P.F., Hammel, H.T., Bradstreet, E.D. & Hemmingsen, E.A. (1965). Sap pressure in vascular plants. *Science*, 148, 339-346.

Sharp, R.E. & Davies, W.J. (1979). Solute regulation and growth by roots and shoots of water-stressed maize plants. *Planta*, 147, 43-49.

Sheriff, D.W. & Meidner, H. (1974). Water pathways in leaves of *Hedera helix* L. and *Tradescantia virginiana* L. *J. Exp. Bot.*, 25, 1147-1156.

Steudle, E. (1989). Water flow in plants and its coupling to other processes: an overview. In S. and B. Fleischer, (Eds.), *Methods in Enzymology*, Volume 174, Biomembranes Part U, Cellular and Subcellular Transport: Eukaryotic (Nonepithelial) Cells. Academic Press, NY. Pp. 183-225.

Steudle, E. & Boyer, J.S. (1985). Hydraulic resistance to water flow in growing hypocotyl of soybean measured by a new pressure-perfusion technique. *Planta*, 164, 189-200.

Steudle, E. & Frensch, J. (1989). Osmotic responses of maize roots. Water and solute relations. *Planta*, 177, 281-295.

Steudle, E. & Jeschke, W.D. (1983). Water transport in barley roots. *Planta*, 158, 237-248.

Westgate, M.E. & Boyer, J.S. (1984). Transpiration- and growth-induced water potentials in maize. *Plant Physiol.*, 74, 882-889.

Westgate, M.E. & Boyer, J.S. (1985). Osmotic adjustment and the sensitivity of leaf, root, stem, and silk growth to low water potentials in maize. *Planta*, 164, 540-549.

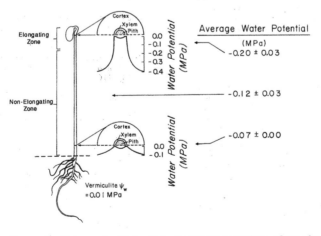

Figure 2. Water potentials (Ψ_H) at various positions along the stem (hypocotyl) of soybean seedlings growing rapidly in saturating humidities in the dark. The average water potential is shown (right-hand table) together with the profile of water potential along the diameter of the stem (center graphs). Note that the profile is steep next to the vascular supply (xylem) and flat at the outer surface of the stem. The profile, when rotated around the longitudinal axis of the stem, gives the 3-dimensional potential field.

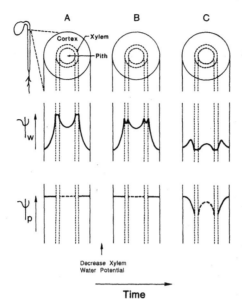

Figure 3. Changes in the growth-induced water potential (Ψ_W)
profile after a decrease in the Ψ_W of the vascular supply. A)
Tissue grows rapidly before decrease and a Ψ_W field is present
that moves water out of the xylem. The pressure potential (Ψ_P)
is uniform in all the cells. B) Growth is inhibited
immediately after xylem Ψ_W begins to decrease. C) Ψ_W field
collapses after 1-2 h, and Ψ_P decreases in cells close to
vascular system. Note that water movement and cell enlargement
are inhibited at B and C because profile is inverted next to
the xylem. For growth to resume, profile must move downward to
re-establish a favorable Ψ_W field.

PHYSIOLOGICAL AND GENETIC REGULATION IN PLANTS

A. W. Naylor

Dept. of Botany, Duke University, Durham, North Carolina 27706, USA

Abstract. We are now in a position to make a major thrust in controlling the productivity of plants. For a long time it has been known that all living things are the product of the interplay of the environment and their genetics. Until the past 60 years the environmental component of the genetics-environmental equation has received the most attention. Advances in biochemistry and knowledge of steps in metabolic pathways together with information on the division of labor in organelles, membranes and compartments together with discoveries of how physical conditions are sensed and how this information is transmitted to specific genes have revealed many control points. Such knowledge allows for reasoned human intervention. Examples are presented showing how environmental factors such as light and temperature are perceived and how this information is transmitted through the cytoplasm, into the nucleus and targeted to a specific gene or group of genes. Response of these activated or repressed genes is crucial in the production of peptides - the chief products of the genetic code. At this point the peptides come predominantly under the control of the chemical environment of the cell. Regulation of enzymes through the production of isoenzymes, their allosteric properties and feed-back control are considered. How protection against the destructive effects of light generated free-radicals to the photosynthetic apparatus is described. A model of how a hormone such as indole acetic acid brings about genetic activity is presented.

Keywords. Light control; temperature control; gene regulation; enzyme regulation, bio-protection against free-radicals; metabolic pathway regulation; indole acetic acid action.

INTRODUCTION

All organisms are the vectorial resultant of the interplay of their genetics and the environment. There is a bewildering array of control points. These are ever present at the environmental, genetic and biochemical levels. Without these controls we could not have distinctly different forms of life - only amorphous protoplasm, there could be no division of labor we associate with cell walls, membranes, nucleus, cytoskeleton, chloroplasts, mitochondria, golgi apparatus, and vacuoles. Differentiation at the cellular level had to take place before there could be roots, stems, leaves, flowers, fruits and seeds. Obviously, an understanding of the differentiation process is necessary if we are to rationally control it.

ENVIRONMENTAL FACTORS AS SIGNALS

Within limits, environmental factors can be looked upon as sets of signals. The question then becomes one of how are the signals perceived, how does transduction occur and what are the consequences? The principal environmental factors impinging on plants are light, temperature, water, minerals, pH and gravity. How does the plant perceive these signals? Because of time limitations, only the factors of light and temperature can be considered.

Light is perceived by at least three different pigment systems. One of these, chlorophyll, is an energy converter the other two, phytochrome and cryptochrome, are especially important in photomorphogenesis. Chlorophyll makes it possible for the energy of blue and red light to be converted to chemical energy - the high energy compounds being NADPH and ATP. It is the energy stored in these compounds that drives all the life processes of multicellular plants and animals. Phytochrome is activated by red light and transformed to a molecular form that has a multitude of actions such as initiation of germination, promotion of formation of leaf primordia, or chlorophyll accumulation depending upon the molecular environment in which it finds itself. This is thought to be done by a transducer of small Ca^{2+} concentration changes into major physiological responses. Activated calmodulin can bind and regulate a wide range of regulatory enzymes in cells. More than six of these are already known. Much less is known about how cryptochrome transduces the energy it receives. Action spectra, however, reveal that this pigment is most sensitive to blue light, is present in membranes and exerts its effects through the activation of a kinase. Perhaps in the process using a second messenger system similar to phytochrome's.

Temperature is sensed by every chemical and chemical reaction in the cell. Ions, atoms, molecules, and supra-molecules are all affected. Life exists, generally speaking, in unstable equilibrium with the environment within the narrow temperature range of 0° to 85°C. Two basic concepts: "rate effects" and "weak bond" effects are keys to controlling metabolic pathways and the complex biological molecules on which those pathways and reactions are dependent.

From the standpoint of rate effects, changes in the average kinetic energy of the atoms and molecules of the organism, that is, the temperature of the organism, will be translated into changes in the rates at which the chemical reactions comprising metabolism occur. This would include respiration, photosynthesis, nucleic acid synthesis, protein synthesis, lipid synthesis and the vast array of other reactions required in making and maintaining the multitude of constituents of the new and old protoplasm. Since all of these diverse reactions with their differing rate maxima must be coordinated in a highly structured time frame, it is readily understandable why the temperature window for life is so narrow.

The cause of the narrowness of this window becomes even clearer when the importance of weak bonds, meaning van der Waal forces, hydrogen bonds, ionic bonds and hydrophobic interactions in living organisms are taken into consideration. It is at the molecular and macromolecular levels that the crucial relation between structure and function can be most readily visualized. Virtually all of the higher orders of biochemical structure, that is, the 3° and 4° structure of proteins, membrane structure, nucleic acid structure and most of the biochemical interactions demanding a high degree of stereochemical specificity (e.g. the binding of substrates of enzymes) are highly, if not entirely, dependent on weak bonds. Many enzymes are made up of two or more subunits. An example of this is ribulose bisphosphate carboxylase oxygenase, the enzyme of paramount importance in CO_2 fixation. This enzyme consists of eight large and eight small subunits. These subunits must be precisely arranged if maximum photosynthesis is to be achieved. These sixteen subunits are assembled through the aid of a protein called chaperonin (Gatenby and Ellis, 1990) and held together in a single macromolecule by means of multiple weak bonds without the aid of a single covalent bond. Multienzyme complexes, involved in some other metabolic sequences, are even larger and less stable.

Proteins play major roles in cells and tissues. Some function as storage reservoirs; others are of primary importance from a structural standpoint as in the cytoskeleton; and very large numbers are enzymes - catalytic agents, that accelerate reaction rates without shifting equilibria. The general concept - "one gene-one polypeptide" introduced in the 1940s has had an almost overwhelming impact on the direction of biological research for the past 50 years. All of the genetic information in the DNA strands is made available on a highly selected basis in the form of peptides and proteins, most of which have an enzymatic function.

REGULATION OF PROTEIN SYNTHESIS

Different cell types synthesize different sets of proteins although the gene sets are the same for each cell. Basically, adjacent cells such as those of xylem and phloem, although they start out looking alike become different from one another, because, early on, they synthesize and accumulate different sets of proteins. Ultimately, these difference are made manifest in terms of form and function of these cells. There are at least six control points for a given protein's formation between the genetic code for it in a DNA strand and the finished protein. Antibiotics block many of these steps. Let us consider each of the six steps.

Any or all of the steps leading from DNA to the functioning protein could be involved in establishing differences in population densities for a given protein in adjacent cells. Protein differences can arise by 1) regulation of when a given gene is transcribed, i.e. transcriptional control; 2) regulation of the processing of the primary RNA transcript, i.e., processing control; regulation of which of the many completed mRNAs in the cell's nucleus are exported to the cytoplasm, i.e., transport control; 4) determining which mRNAs in the cytoplasm will be translated by ribosomes, i.e. translational control; 5) selectively stabilizing specific mRNAs against degradation by RNAses, i.e., mRNA degradation control; and 6) control of the activity, particularly if the protein is an enzyme. It is at this point that "fine tuning" of enzyme activity takes place through interation with metals, nucleotides, amino acids, carbohydrates and sometimes end products of enzyme actions. More will be said later about enzyme regulation. In this hierarchy of control points, however, transcriptional control and protein activity control predominate. Light, high and low temperatures, certain small molecules, hormones, specialized proteins can all indirectly target specific genes causing them to be active or silent.

Gene regulatory proteins can either activate or repress gene transcription. This is done in higher plants by a large set of specialized sequence-specific DNA binding proteins, the gene regulatory proteins. Each of them recognizes specific DNA sequences about 8-15 nucleotides long. The binding of these proteins to DNA sequences can either facilitate (positively regulate) or inhibit (negatively regulate) transcription of an adjacent gene. Of particular significance is the finding different cell types in a multicellular organism have different sets of genes to be transcribed by each cell type. This results in the development of tissues such as xylem, phloem, cambium, cortex and epidermis.

There are two main categories of enzymes: constitutive and inducible. The constitutive enzymes are those that are present in all cells of the organism while the inducible enzymes are evanescent and are produced in response to internal and external signals. The constitutive enzymes can be thought of as housekeeping enzymes and the genes for them are turned-on all the time. The inducible enzymes are those that are most subject to environmental control. Some of the inducible enzymes have a common function but differ slightly in molecular weight, pH and temperature optima and other characteristics. These are known as isoenzymes. A given set of isoenzymes may be induced in toto, differentially or not at all in a given tissue. The causes of such variable expression tissue by tissue, organ by organ are not known.

INTERMEDIARY METABOLISM

Intermediary metabolism in the living organism involves so many synthetic, degradative, branching and linking processes that it almost defies the imagination. Remarkably, however, this complex network of activities is maintained under control matching metabolic activity to metabolic requirements - a state called homeostasis. A fine example of oscillatory control involving regulation of the central respiratory pathway, the tricarboxylic acid cycle (TCA), and the use of NH_3 in amino acid synthesis is found in the synthesis of glutamic acid. α-ketoglutaric acid is at a crucial branch point. One path leads to continued operation of the TCA cycle and making energy available in the cell while the other, in the presence of ammonia, can lead to the formation of glutamic acid, the precursor of many amino acids and glutamine. The allosteric control properties of α-ketoglutaric acid dehydrogenase determines which path will make use of α-ketoglutaric acid's carbon skeleton. α-Ketoglutarate dehydrogenase is markedly stimulated in its activity, leading to the production of succinic acid and indirectly adenosine triphosphate (ATP) by binding adenosine diphosphate (ADP) or adenosine monophosphate (AMP) to its surface away from the active site. Binding of these compounds results in alteration of the shape of the enzyme causing it to bind more tightly to its substrate, α-ketoglutaric acid, thereby increasing the enzyme's efficiency. As ATP accumulates and ADP and AMP are correspondingly used up, α-ketoglutarate dehydrogenase becomes less active and α-ketoglutarate tends to accumulate. In the presence of NH_3 and NADH the α-ketoglutaric acid is converted to glutamic acid.

Meanwhile, ADP and AMP once again become available causing α-ketoglutarate dehydrogenase to become more active. Thus the system oscillates back and forth between energy production and glutamate synthesis. This system is only one of perhaps hundreds of similar branch point control loci producing in the cell the integrated chemical flux called metabolism.

Almost all metabolic processes are regulated to some degree. The widespread occurrence of feedback regulation of biosynthetic pathways is well documented. But at least as important is the control of thermodynamically "irreversible" reactions which must operate in both directions. Such control is achieved by employing different reaction mechanisms under separate regulatory control; for example, glycogen phosphorylase and glycogen synthestase catalyze the same reaction in opposing directions by very different mechanisms and are subject to different controls. Such dual reactions provide much of the flexibility of intermediary metabolism.

CONTROL OF PHOTODESTRUCTIVE REACTIONS

A major problem faced by photosynthesizing plants is the dissipation of excess electrons generated as a result of the relative slowness of the carbon fixation phase of photosynthesis in constrast with the speed of the light phase. Free radicals formed under such conditions can quickly become lethal. Differential abilities to deal with this problem leads to what is known as shade or light loving plants. Although $NADP^+$ is by far the most efficient electron acceptor in the chloroplast, and required in CO_2 reduction, if it were the only electron acceptor in the chloroplast the system would soon be overwhelmed by electrons, especially in bright light. However, other electron acceptors are present. Among them are oxygen, glutathione, ascorbate, α-tocopherol, peroxidases and violaxanthin. Violoxanthin appears to be the primary alternative electron acceptor from photosystem I while O_2 serves this role for PSII. When O_2 accepts an extra electron superoxide ($O_2^{\cdot-}$), a destructive free-radical, is formed which must be quickly inactivated or serious damage to the photosynthetic apparatus can occur. Superoxide dismutase, violoxanthin, ascorbate, glutathione and NADPH all appear to be important in this process. Christine Foyer, et al (1990) have assembled the information on this point into a plausible model for the minimization of free radicals. Free radicals, however, are never completely eliminated and some destruction regularly takes place. Sun loving plants would be expected to be richest in the several electron quenching components while shade loving plants may be deficient in one or more of them. Conceivably, genetic engineering could correct for single or even multiple deficiencies and convert shade loving to high light plants.

HORMONAL REGULATION

One last point. Plant hormones have powerful regulatory functions. Although much effort has gone into determining at what level hormones such as indole-acetic acid (IAA) exert growth regulatory functions, much remains to be determined. How a cell with a rigid cell wall manages to elongate presents many puzzles. Indole-acetic acid has been known for a long time to play an important role in the elongation process and so has turgor pressure. But knowledge of the intermediate steps has been slow to accumulate. A current view on the cascade of events following introduction of IAA to the cell has been provided by Brummell and Hall (1987).

The first consequence of IAA binding to its receptor in the plasma membrane is to cause a second messenger, inositol triphosphate, acting in the cytoplasm to cause the endoplasmic reticulum to secrete Ca^{2+} ions into the cytosol. The rise in Ca^{2+} level is transient, and acts as a signal activating Ca^{2+}/H^2 antiporters in the vacuole membrane. As the vacuole takes up Ca^{2+}, the Ca^{2+} level in the cytosol drops and as exchange for protons occurs there is a lowering of cytosolic pH which stimulates electrogenic ATP-driven H^+ extrusion into the cell wall resulting in plasticization, allowing elongation to take place. This is the "fast response" and is observable within 10-20

minutes after application of IAA. Thus the
first growth response occurs as a consequence
of raised Ca^{2+} concentration. The increased

level of Ca^{2+} exerts an additional and longer
term effect and with or without the
involvement of calmodulin stimulates the
activity of protein kinase bringing about
protein phosphorylation. In this way, a gene
regulatory protein, already present in the
cytosol, becomes activated and capable of
binding IAA. The regulatory protein-IAA
complex moves into the nucleus where it
interacts with the genome repressing certain
genes and activating the transcription of
specific mRNA sequences. The first genes
transcribed seem to code for polysaccharide
synthesis. These genes promote the synthesis of
cell wall components that are added to the
cell wall as elongation takes place. This
can be observed within 30 minutes following
presentation of IAA to the surface of the
epidermis. Since cortical cells do not have
plasma membranes with auxin receptors it is
only the epidermal cells that respond to
auxin as just described.

REFERENCES

Brummel, D.A. and J.L. Hall. (1987. Rapid
    cellular responses to auxin and the
    regulation of growth. Plant, Cell and
    Environment, 10, 523-543.
Foyer, C., R. Furbank, J. Harbinson and P.
    Horton. (1990). The mechanisms
    contributing to photosynthetic control
    of electron transport by carbon
    assimilation in leaves. Photosynthesis
    Research, 25, 83-100.
Gatenby, A.A. and R.J. Ellis (1990).
    Chaperone function: the assembly of
    ribulosebisphosphate carboxylase-
    oxygenase. Ann. Rev. Cell Biology, 6,
    125-149.

ENVIRONMENTAL CONTROL FOR PLANT GROWTH IN PLANT FACTORY OPERATION AND GREENHOUSE MANAGEMENT FROM PHYSIOLOGICAL VIEWPOINT

H. Nonami*, Y. Hashimoto* and J. S. Boyer**

*Dept. of Biomechanical Systems, Ehime University, Tarumi, Matsuyama 790, Japan
**University of Delaware, College of Marine Studies, Lewes, Delaware 19958, USA

Abstract. The enlarging plant tissue displayed significant water potential disequilibria between the water source and the enlarging cells. This disequilibria formed a water potential field (i.e., potential field of chemical potential of water) in the enlarging tissue, which is called a growth-induced water potential field. Theoretically, the growth-induced water potential field can be expressed by an equation derived from a combination of Fick's law and the conservation of mass, and can be used to determine the direction of water flow in the expanding tissues. Experimentally, cell water potentials of cells in the zone of elongation were measured with combinations of a cell pressure probe and a nanoliter osmometer. Both measured and theoretically predicted water potentials coincided well, showing that the growth-induced water potential field existed in the elongating tissue. The average magnitude of the growth-induced water potential field was estimated from the water potential difference between the water source and the expanding cells by using psychrometry. Prior to growth recovery by acclimation to environmental stress by plants, the growth-induced water potential recovered. Thus, we found that if the growth-induced water potential was measured, both growth recovery and inhibition could be predicted. Because changes in water status in plants occur prior to changes in cell expansion, if the system for water status measurements is established in plant factories and greenhouses, it may be possible to regulate crop growth more effectively.

Keywords. Plant biology; agriculture; plant factory; greenhouse; plant growth; environment control; water stress.

INTRODUCTION

Because plant growth is one of the most sensitive physiological mechanisms of plants to environmental stresses, slight changes in environmental conditions in plant factories and greenhouses cause changes in growth of crops. In order to cultivate high quality crops under a controlled environment, methods for adjustment of environmental conditions have been studied. However, most methods that have been developed for regulating climatic conditions for crops have not included a means for evaluating the physiological status of crops. Growth of crops is most commonly measured in terms of dry matter accumulation. However, measurements of dry matter accumulation are not suitable for use as a parameter for control in order to regulate growth of crops in plant factories and greenhouses, as such measurements must be evaluated over a long period of time such as days and weeks. Significant changes in cell volume expansion, on the other hand, occur in a matter of hours in most plants, and thus, are the most appropriate parameter for growth regulation. Because water occupies more than 90% of the total molecules in expanding cells of most plants, water movement into cells regulates cell expansion rates in most cases. Thus, we have studied plant-water relations of crops under controlled environments, and will suggest new types of growth regulation methods for crops grown under controlled

environments. Because physiological principles for plant growth should be applicable to any kind and any size of plants, physiological regulation for growth should be applicable for cells under tissue culture and also for cells of crops grown in greenhouse conditions.

THEORY

The theory of water movement through an isotropic tissue is based on the hypothesis that the rate of transfer of a diffusing substance through a unit area of a section is proportional to the concentration gradient measured normal to the section (Fick's law). Because the water content of plant tissue is related to the tissue water potential and because water flux must be proportional to the water potential gradient, the relation between water flux (J) and water potential (Ψw) can be written as follows;

$$J = -D \text{ grad } \Psi w \qquad (1)$$

where D is the diffusion coefficient. It is not necessary to know the microscopic pathway for water movement in order to apply this concept. However, it is likely that water moves from cell to cell along gradients in potentials that are in local equilibrium between the cell wall and the protoplast (Molz, 1976a; Molz and Ikenberry, 1974).

Water moving into plant tissue must either pass out of the tissue or remain within the tissue. This principle, from the law of conservation of mass, can be stated as the divergence theorem according to;

$$\frac{\partial \Psi w}{\partial t} = -\text{div } J \qquad (2)$$

where $\partial \Psi w / \partial t$ indicates the water potential increase per unit time within the cells (hydration) and div J is the difference in influx and efflux per unit volume.

If in addition some portion of the influx is used for cell enlargement, we represent this process by a sink term in the conservation law. The equation can be modified to

$$\frac{\partial \Psi w}{\partial t} = -\text{div } J - S \qquad (3)$$

where S is a sink term used to indicate the increase in cell volume due to growth. The sink term is negative because growth removes a fraction of the water from the component undergoing strictly influx, efflux, and hydration changes. Combining Eqs. 1 and 3 yields;

$$\frac{\partial \Psi w}{\partial t} = D \text{ div grad } \Psi w - S \qquad (4)$$

which states that the change in hydration of a tissue will be determined by the ability of the tissue to transmit water, represented by the diffusivity, and the water potential gradients in the growing tissue.

In the case of modeling stem elongation of plants, it is assumed that stems have a cylindrical shape in the present study. Thus, cylindrical coordinates can be used to simplify the analysis. Expressing Eq. 4 in these coordinates;

$$\frac{\partial \Psi w}{\partial t} = D \frac{\partial^2 \Psi w}{\partial r^2} + \frac{D}{r} \frac{\partial \Psi w}{\partial r} - S \qquad (5)$$

Equation 5 describes the growth-induced water potential field in the zone of elongation of a cylindrical tissue.

In order to apply Eq. 5, it is necessary to obtain a reasonable estimate for D. This can be done by evaluating the theoretical expression for D in terms of protoplast and cell wall hydraulic and elastic properties. One may utilize the expression given by Molz (1976a, 1976b) and Molz and Ikenberry (1974);

$$D = \frac{\Delta x \ (Pa + K\Delta x A/2)}{W_v s + V_o/(\varepsilon + \pi_o)} \qquad (6)$$

where A = cross-sectional area of vacuolar pathway, a = cross-sectional area of cell wall pathway, π_o = osmotic pressure at zero turgor, Δx = diameter of cell, K = permeability of cytoplasmic complex separating cell wall from vacuole, V_o = cell volume at zero turgor pressure, W_v = volume of cell wall/cell, P = hydraulic conductivity of the cell wall, and s = specific water capacity of the cell wall material.
Molz and Boyer (1978) theoretically derived S, assuming local equilibrium between the protoplast and cell wall. Defining G = $(1/V_o) \partial V_o/\partial t$, S is given by

$$S = (\varepsilon + \pi_o)G \qquad (7)$$

Although D and S in Eq. 5 are given, analytical solutions for Eq. 5 are still difficult to obtain. In order to simplify Eq. 5 further, we assume that changes in water potentials with respect to time are much slower than changes in water potentials with respect to position within the elongating stem tissue. This is very likely under natural conditions because influences of changes in the surrounding water status first appear in the root region and gradually move to the elongating region of the stem. If so, $\partial \Psi w / \partial t$ in Eq. 5 can be determined directly from slopes of a curve plotted with water potential changes with respect to time. Thus, Eq. 5 can be modified to

$$\psi_w(t) = D \frac{\partial^2 \Psi w}{\partial r^2} + \frac{D}{r} \frac{\partial \Psi w}{\partial r} - S \qquad (8)$$

where $\psi_w(t)$ is changes in water potential at a given time and can be replaced by the measured values. If one defines the constant α as follows;

$$\alpha = (\psi_w(t) + S)/D,$$

then Eq. 8 becomes;

$$\frac{d}{dr}\left(\frac{d\Psi w}{dr}\right) + \frac{1}{r}\frac{d\Psi w}{dr} = \alpha \qquad (9)$$

Substituting M = $d\Psi w/dr$ reduces Eq. 9 to a linear first order equation in M. Defining M/r = N and differentiating this relationship with respect to r allows further rearrangement of Eq. 9 to give;

$$r \frac{dN}{dr} + 2N = \alpha$$

which can be integrated by standard techniques to give the general solution;

$$\Psi w = \frac{\alpha}{4} r^2 - \frac{C_1}{2} \ln r + C_2 \qquad (10)$$

where C_1 and C_2 are constants of integration. First, we will solve this equation in the cortex where it is assumed that $d\Psi w/dr = 0$ at the epidermal surface of the stem and Ψw is equal to the xylem water potential at the region which contains the xylem. For the pith, the approach is similar, but the boundary conditions are $d\Psi w/dr = 0$ at the center of the stem.

MATERIALS AND METHODS

Plant Material

Soybean (Glycine max (L.) Merr. cv. Williams) and kidney bean (Phaseolus vulgaris L. cv. Shiraginugasa) seedlings were grown from seeds disinfected in a 1% solution of NaOCl and rinsed with flowing water for 1h.

Each soybean seed was sown in a 200 ml beaker containing vermiculite with adequate water (5.0 ml of 0.1mM CaCl$_2$/g of vermiculite; water potential of -0.01 MPa). The soybean seedlings were grown at 29 $\pm$ 0.5 ℃ and 100% RH in the dark. After 60 h, each seedling was subjected to water stress by being transplanted to a 200 ml beaker containing vermiculite having -0.28 MPa of water potential.

Kidney bean seedlings were grown hydroponically in nutrient solution (in mol m⁻³: 0.33 NH_4^+, 4.0 NO_3^-, 0.27 PO_4^{3-}, 4.0 K^+, 0.14 Mg^{2+}, 3.5 Ca^{2+}, 0.62 SO_4^{2-}, 0.014 Fe^{3+}, 0.009 Mn^{2+}: EC = 0.1 S·m⁻¹, water potential = −0.07 MPa) at 29 ± 0.5 ℃, 40 ± 10% RH and 150 μ mol·m⁻²·s⁻¹ of photosynthetically active photon flux density. Water stress was applied to the seedlings by increasing concentrations of the culture solution by 5-fold (−0.35 MPa of water potential) or 7.3-fold (−0.51 MPa of water potential).

Growth Measurements

The stem length of 8 plants was measured with a ruler for each measurement. Detailed measurements of growth rate were obtained with a radial displacement transducer clamped to the upper part of the stem and a rigid reference bar clamped to the lower stem. The length between the transducer and rigid bar was recorded electrically. The relative growth rate was calculated from the changing length as a function of time divided by the length of the zone of elongation, assuming that the stem diameter was constant during each measurement interval.

Tissue Water Potential Determinations

The water potential was measured by the isopiestic technique (Boyer and Knipling, 1965). A thermocouple chamber was coated with melted and resolidified petrolatum (Boyer, 1967) and loaded with stem segments about 15 mm long. Water potential of the elongation zone and the mature zone was measured. Also, water potentials of vermiculite and nutrient solution were measured similarly.

Cell Water Potential Determinations

Cell water potentials were measured in the elongating region of soybean hypocotyls. Prior to the measurement, the seedling was coated with petrolatum and further covered with wet tissue paper on the petrolatum to minimize water loss from its surface. Cell turgor was measured with a cell pressure probe according to the procedure described by Nonami et al. (1987). Immediately after cell turgor determination, cell solution was extracted from the same cell with the probe, and transferred to a nanoliter osmometer (Clifton Technical Physics, Hartford, N.Y. U.S.A.) in order to determine cell osmotic potential according to Nonami and Schulze (1989). Cell water potential was obtained by algebraic summation of the turgor and osmotic potential in the same cell (Nonami and Schulze, 1989). All manipulations were conducted under green safelight.

RESULTS

Growth Inhibition and Water Status

When soybean seedlings were transplanted from vermiculite having −0.01 MPa to −0.28 MPa, stem elongation was inhibited immediately (Fig. 1A). The xylem water potential was estimated from the water potential of the mature region, because water potential of the xylem is equilibrated with that of the mature region as shown by Nonami and Boyer (1990a). Immediately after application of

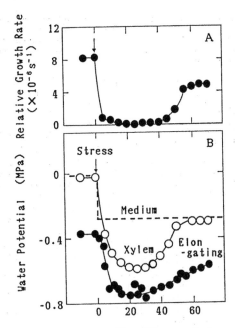

Fig. 1. Relative growth rate (A) and water potentials (B) of vermiculite (Medium), the xylem (Open circles) and the zone of elongation (Closed circles).

water stress by transplanting, the xylem water potential decreased and became much lower than water potential of the growing medium (i.e., vermiculite having −0.28 MPa of water potential) (Fig. 1B). Water potential of the zone of elongation decreased after water stress was applied (Fig. 1B). However, decreases in water potential of the elongation zone were slightly slower than decreases in the xylem water potential (Fig. 1B).

Eventually, the xylem water potential equilibrated with the water potential of the medium (Fig. 1B), and then, growth recovery was initiated (Fig. 1A).

Growth-induced Water Potential Fields

When the water potential of the elongation zone was measured with the psychrometry, the water potential indicated the average water potential of cells contained in the zone of elongation. If changes in water potential measured with psychrometry are considered to be equivalent to $\psi_w(t)$ in Eq. 8, water potential fields associated with growth can be determined by applying the boundary conditions and using Eq. 10. The xylem water potential was measured in Fig. 1B and $\psi_w(t)$ was determined from the slope of the curve of the water potential of the elongating region in Fig. 1B. Values of biological parameters to calculate D were adopted from Molz and Boyer (1978).

Figure 2 shows water potential fields formed in the elongating region at a given time after application of water stress. Before applying water stress, water potential was lowest next to

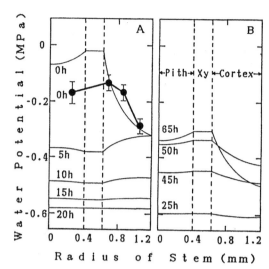

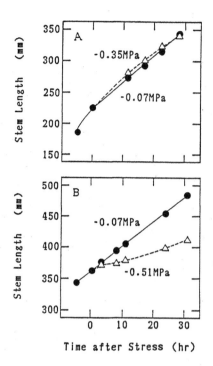

Fig. 2. Water potential profiles across the stem of soybean seedlings in the zone of elongation during growth inhibition (A) and growth recovery (B). Water potential profiles are shown in the pith region (Pith), the xylem region (Xy) and the cortical region (Cortex). Potential profiles shown in thin lines were calculated, and time was indicated hours after the stress application. Cell water potentials (closed circles) were measured directly, and the vertical bars indicate 95% confidence intervals.

Fig. 3. Stem length of kidney bean seedlings when the nutrient solution water potential was increased from -0.07 MPa to -0.35 MPa (A) and from -0.07 MPa to -0.51 MPa (B).

the epidermal surface and the highest in the xylem region (see a thin line marked with Oh in Fig. 2A). Such a potential field associated with actively growing tissue is called a growth-induced water potential field. Immediately after application of water stress, the xylem water potential decreased, but the water potential near the epidermal surface was unchanged (see a line marked with 5h in Fig. 2A). Progressively, water potential fields flattened out across the stem, and at the 20th h after stress, water potential gradients disappeared completely across the stem (Fig. 2A). Recovery of water potential gradients was initiated by the rise in the xylem water potential. When the growth began to recover at the 45th h after stress (Fig. 1A), the growth-induced water potential field was gradually formed (Fig. 2B). At the 65th h, the growth-induced water potential field was formed to support the partial recovery of growth after stress (Figs. 1A and 2B).

In order to prove the existence of the growth-induced water potential field, the cell water potential across the stem was measured in actively growing soybean seedlings before applying water stress (closed circles marked with Oh in Fig. 2A). In the cortex region, measured cell water potentials were similar to water potential fields predicted theoretically (Oh in Fig. 2A). This indicates that a growth-induced water potential field existed and was extending from the xylem toward the epidermal surface in order to support synchronized cell expansion across the stem. Slight deviations between measured values and theoretically-determined values were observed in the pith region (Fig. 2A).

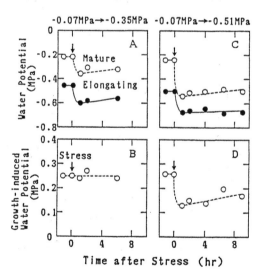

Fig. 4. Water potentials of the mature region and the elongation region (A, C) and the growth-induced water potential (B, D) when the nutrient solution water potential was increased from -0.07 MPa to -0.35 MPa (A, B) and from -0.07 MPa to -0.51 MPa (C, D). The growth-induced water potential was estimated by water potential difference between the elongation region and the mature region.

Application of the Growth-induced Water Potential to Hydroponic Culture

Because the growth-induced water potential field was initiated from the xylem water potential and formed by cell water potentials in the growing tissue, water potential difference between the elongating region and xylem may be a good estimate of the magnitude of the growth-induced water potential field. In order to test this hypothesis, kidney bean seedlings were grown hydroponically, and growth and water potentials of the elongating region and mature region were measured simultaneously. Water stress was applied by increasing the concentration of nutrient solution.

When water potential of the culture solution was decreased from -0.07 MPa to -0.35 MPa, growth of kidney bean seedlings was slightly promoted (Fig. 3A). However, immediately after water potential of the culture solution was decreased from -0.07 MPa to -0.51 MPa, growth of kidney bean seedlings was inhibited (Fig. 3B). After 10 h of the water stress, gradual recovery of growth of the seedlings was observed (Fig. 3B).

In order to understand the mechanism of growth inhibition by water stress created by increasing the concentration of the nutrient solution, the growth-induced water potential was estimated from the differences between water potentials of the elongation region and mature region. In plants which did not have growth inhibition under -0.35 MPa of the culture solution, the growth-induced water potential was unchanged (Fig. 4B). In plants whose growth was inhibited under -0.51 MPa of the culture solution, the growth-induced water potential decreased immediately after the stress, and began to recover prior to growth recovery (Fig. 4D). It is evident that growth of plants is regulated by the growth-induced water potential field.

DISCUSSION

In the present study, dark-grown soybean seedlings were used to demonstrate the concept of a growth-induced water potential field when the seedlings were subjected to water stress. Because dark-grown soybean seedlings have a restricted growing region in their stems and because the stems have a cylindrical shape, the theoretical analysis could be applied easily. Furthermore, because dark-grown soybean seedlings had a relatively large cell size, water potentials of individual cells could be measured directly and compared with theoretically predicted values. We found that both predicted and measured values coincided well in the cortical cells. Deviation of a water potential field in the pith region between theoretically predicted and measured values is probably caused by the assumption that all cells in the elongation zone had the same water permeability, i.e., K in Eq. 6, when the field was calculated theoretically. Because the measured cell water potentials in the pith region were lower than the theoretically predicted values, cells in the pith region must have lower water permeability than the value used in the present study (i.e., the value of K which was taken from the work of Molz and Boyer (1978)).

The xylem water potential cannot be measured with the combination of a pressure probe and a nanoliter osmometer, because the xylem does not

have confined cell membranes. Thus, actual xylem potential in Fig. 2 should be represented by values shown in thin lines in the xylem region (Xy in Fig. 2), whose values were measured directly with psychrometry and used as the boundary conditions in calculation. Thus, it is evident that the steepest water potential gradient existed surrounding the xylem in the growth-induced water potential field when growth was occurring actively (Oh in Fig. 2).

When a growth-induced water potential field is considered in plants under water stress, physiological phenomena observed under stress can be explained logically. Nonami and Boyer (1989) observed a decrease in turgor of cells located near the xylem immediately after water stress, but no decrease in turgor of cells located near the stem surface. Because their experimental setups were similar to our theoretical analysis in the soybean stem, both results can be compared. When the water stress was applied to the root region of soybean seedlings, water potentials of the xylem and its surrounding cells decreased but not in cells located near the stem surface as shown in Fig. 2A (see the line marked with 5h). Because accumulation of solute in cells does not seem to occur drastically within such short intervals, osmotic potential (Ψs) of cells can be considered to be constant. Relations of cell water status can be written as follows;

$$\Psi w = \Psi p + \Psi s$$

where Ψw and Ψp indicate cell water potential and cell turgor, respectively. This relation was experimentally confirmed by works of Nonami et al. (1987) and Nonami and Schulze (1989). Thus, when the water potential in the root region was altered, the initial changes in water potentials of cells in the elongation zone of the stem must coincide with changes in cell turgor. Because a drop in water potential of cells located near the xylem was about 0.2 MPa within 5h after stress according to Fig. 2A, a decrease in turgor of those cells must be about 0.2 MPa. Nonami and Boyer (1989) observed about 0.2 MPa of decrease in turgor of cells located near the xylem, supporting our theoretical analysis.

Although we have restricted our calculations and measurements of a growth-induced water potential field in the zone of elongation of dark-grown soybean seedlings, the same concept must be applicable to any growing tissue. Because the expanding cells must absorb water from the water source in order to increase their volume, the growing tissue must form a growth-induced water potential field. For practical purposes, the magnitude of a growth-induced water potential can be estimated from water potential differences between the elongating tissue and the water source. When the growth-induced water potential was measured by using hydroponically grown kidney bean seedlings, it was found that growth of the seedlings was correlated with changes in the growth-induced water potential.

Nonami and Boyer (1990a) studied the kinetics of physiological parameters regulating plant growth under water stress by using a newly developed guillotine psychrometer. They found that the growth-induced water potential was the first of the physiological parameters to decrease to a growth-limiting level (Nonami and Boyer, 1990a). The decreased water potential gradient within the growing tissue was caused by a decreased water potential of the xylem. This was followed by a similar decrease in cell wall extensibility and tissue conductance for water (Nonami and Boyer, 1990a).

In order to check the observations which indicate that the inhibition of stem growth at low water potentials is accompanied by decreases in cell wall extensibility and tissue hydraulic conductance to water, Nonami and Boyer (1990b) developed a theory to relate rheological properties of cell walls measured with an extensiometer and thermodynamical properties of cell walls measured with psychrometry. They found that the plastic properties of the cell walls determined cell wall extensibility (Nonami and Boyer, 1990b).

Both cell wall extensibility and hydraulic conductance are under metabolic control, and their low levels under water stress indicate that substantial metabolic change occur in the enlarging cells (Nonami and Boyer, 1990a). Wall extensibility is probably determined by the action of wall enzymes on the polymerization of wall constituents and the cross-linking of structural proteins with wall constituents (Cassab and Varner, 1988; Fry, 1986; Theologis, 1986). In seedlings similar to those used here, polyribosome and mRNA levels decreased a few hours after water stress (Mason et al., 1988). Increased amounts of proteins were extractable from the walls and a 28 kD protein present in the cytoplasm accumulated in the walls (Bozarth et al., 1987).

Growth recovery occurred when the xylem water potential was equilibrated with the medium (Fig. 1). This was caused by osmotic adjustment in cells (Nonami and Boyer, 1989, 1990a). Prior to the growth recovery, recovery of the growth-induced water potential was observed (Figs. 2 and 4, Nonami and Boyer, 1990a). The recovery of the growth-induced water potential induced metabolic changes in cells, resulting in increases in wall extensibility and hydraulic conductance (Nonami and Boyer, 1990a, 1990b). Afterward, growth was resumed. Thus, the growth-induced water potential can be considered to be a primary physiological factor regulating growth under water stress.

To apply the concept of crops grown in plant factories and greenhouses, we grew kidney bean seedlings hydroponically and estimated the magnitude of the growth-induced water potential from water potential differences between the mature region and elongating region. Because growth conditions of these kidney bean seedlings simulate most crops grown under controlled environment and because the estimated growth-induced water potential agreed well with growth of kidney bean seedlings, it is evident that the concept of the growth-induced water potential can be applicable for the regulation of crop growth cultivated in plant factories and greenhouses.

CONCLUSION

We have introduced a concept of a growth-induced water potential field which governs water influx to the expanding cells. In growing tissue, the xylem had the highest water potential and cells located near the epidermal surface had the lowest water potential. Theoretically predicted values coincided with the experimental measurements.

When plants were subjected to environmental stresses, the growth-induced water potential field collapsed, resulting in growth inhibition. Practically speaking, the magnitude of a growth-induced water potential field can be estimated from differences between water potentials of the elongating region and the water source. Because the growth-induced water potential field recovers prior to growth recovery, if the growth-induced water potential field can be determined appropriately, growth behavior of crops can be predicted under controlled environments.

REFERENCES

Boyer, J.S. (1967) Leaf water potentials measured with a pressure chamber. Plant Physiol., 42, 133-137.

Boyer, J.S., and E.B. Knipling (1965) Isopiestic technique for measuring leaf water potentials with a thermocouple psychrometer. Proc. Natl Acad. Sci., 54, 1044-1051.

Bozarth, C.S., J.E. Mullet, and J.S. Boyer (1987) Cell wall proteins at low water potentials. Plant Physiol., 85, 261-267.

Cassab, G.I., and J.E. Varner (1988) Cell wall proteins. Annu. Rev. Plant Physiol. Plant Mol. Biol., 39, 321-353.

Fry, S.C. (1986) Cross-linking of matrix polymers in the growing cell walls of angiosperms. Annu. Rev. Plant Physiol., 37, 165-186.

Mason, H.S., J.E. Mullet, and J.S. Boyer (1988) Polysomes, messenger RNA and growth in soybean stems during development and water deficit. Plant Physiol., 86, 725-733.

Molz, F.J. (1976a) Water transport through plant tissue: the apoplasm and symplasm pathways. J. Theor. Biol., 59, 277-292.

Molz, F.J. (1976b) Water transport in the soil-root system: transient analysis. Water Resource Res., 12, 805-808.

Molz, F.J., and J.S. Boyer (1978) Growth-induced water potentials in plant cells and tissues. Plant Physiol., 62, 423-429.

Molz, F.J., and E. Ikenberry (1974) Water transport through plant cells and cell walls: theoretical development. Soil Sci. Soc. Am. J., 38, 699-704.

Nonami, H., and J.S. Boyer (1989) Turgor and growth at low water potentials. Plant Physiol., 89, 798-804.

Nonami, H., and J.S. Boyer (1990a) Primary events regulating stem growth at low water potentials. Plant Physiol., 93, 1601-1609.

Nonami, H., and J.S. Boyer (1990b) Wall extensibility and cell hydraulic conductivity decrease in enlarging stem tissues at low water potentials. Plant Physiol., 93, 1610-1619.

Nonami, H., J.S. Boyer, and E. Steudle (1987) Pressure probe and isopiestic psychrometer measure similar turgor. Plant Physiol., 83, 592-595.

Nonami, H., and E.-D. Schulze (1989) Cell water potential, osmotic potential, and turgor in the epidermis and mesophyll of transpiring leaves: Combined measurements with the cell pressure probe and nanoliter osmometer. Planta, 177, 35-46.

Theologis, A. (1986) Rapid gene regulation by auxin. Annu. Rev. Plant Physiol., 37, 407-438.

MICROPROPAGATION BY NUTRIENT MIST SUPPLY: EFFECTS OF A MIST SUPPLY PERIOD ON POTATO PLANTLET GROWTH

Y. Ibaraki and K. Kurata

Dept. of Agricultural Engineering, University of Tokyo, Yayoi 1-1-1, Bunkyo-ku, Tokyo 113, Japan

Abstract. Since nutrient mist culture enables one to widely control the micro environment around cultured tissues, precise understanding of the effects of control factors such as mist supply period and mist wafting region is very important. In order to investigate effects of the mist supply period on the potato plantlet growth, different mist supply periods were tested. These included three different mist supply periods which were constant through the culture period and mist supply variations which involved changing the supply period at different plantlet growth stage during the culture period. The average growth parameters (fresh and dry weights, shoot length, and the number of nods per potato plantlet), were observed to be increased by each mist culture in comparison with the respective control cultures (an agar medium). Longer mist supply, however, caused root development inhibition. The shortest mist supply period, i.e., nutrient mist supply for 1 min. per a 10 min. cycle (hereafter treatment 1/10), brought about the largest increase of fresh and dry weights and the best root development among the treatments with the constant mist supply periods. The experiment on the mist supply variations showed promising results for better plantlet growth, i.e., plantlet shoot growth was more promoted than in the treatment 1/10, although root growth was suppressed. It is expected that further optimization of the mist supply variations may bring about more efficinet propagation.

Keywords. Biotechnology; nutrient mist; micropropagation; environmental control; ultrasonic transducer; computer control.

INTRODUCTION

To achieve higher quality seedlings and faster propagation rates, micropropagation techniques have recently been applied to seed breeding, although the large labor requirements for these techniques have limited their applications to only highly profitable species. Significant progress, however, has been made in mechanizing micropropagation processes (Tisserat and Vandercook, 1985, Aitken-Christie and Davies, 1988), and efficient propagation systems have been developed (Akita and Takayama, 1988).

A novel one of these systems is a micropropagation system by nutrient mist supply (Weathers and Giles, 1988, Kurata, Ibaraki and Goto, 1991) which supplies a medium in a form of small particles, i.e., a nutrient mist. Nutrient mist cultures are expected to offset the disadvantages of conventional medium supply methods and also to reduce labor.

One of the biggest advantages of these cultures is that culture environmental conditions can easily be controlled, i.e., alteration of controllable factors such as mist supply period and mist wafting region in the culture space enable one to change the culture conditions related to nutrient absorption and gas exchange of plant tissues. Furthermore, many control parameters can be nondestructively changed according to cultured tissue growth. It should be noted, however, that if control parameters are not properly set, an unsuitable culture environment may easily result due to a wide range of possible conditions, hence both a precise understanding of the effects of these parameters on the cultured tissues and their optimization are necessary.

In the nutrient mist culture of potato plantlets, the mist wafting region in the culture space, i.e., the nutrient mist supply area, is one of the most important factors involved in controlling culture conditions. In fact, it was reported that when a growth chamber was filled with nutrient mists, potato plantlet axially shoot growth was actually inhibited, resulting in abnormal plantlets (Kurata, Ibaraki and Goto, 1991). This problem was then solved by supplying nutrient mists to only the potato plantlet's root zone, with plantlet shoot growth being higher as compared to those in an agar medium culture (Kurata, Ibaraki and Goto, 1991). To obtain the best environmental growing condition for cultured explants, mist culture control parameters must be optimized.

Considering this, the presented study focused on the mist supply period, i.e., the nutrient mist supply time during one control cycle. Previous experiments supplied nutrient mists for 9 minutes out of a 10 minute control cycle (indicated as 9/10), and even though shoot growth was enhanced, the older explant leaves died, with the mist supply period not being effectively optimized (Kurata,Ibaraki and Goto, 1991). This led to the presented investigation which examined the effects of different mist supply periods on potato plantlet growth and confirmed that culture conditions can be improved by alteration of the supply periods, leading to the advantage of not having to perform transplanting operations.

MATERIALS AND METHODS

Nutrient Mist Supply System

A previously described nutrient mist culture system was used (Kurata, Ibaraki and Goto, 1991), in which the liquid medium was converted into a nutrient mist by an ultrasonic transducer placed in the bottom of a growth chamber (8 L glass cylinder, diameter: 190mm, height: 300mm). Figure 1 shows the growth chamber, with explants being set on a plant supporter (stainless steal mesh, pore size: 25mm). Nutrient mist was supplied only to the potato plantlet root areas located underneath the plant supporter. A temperature controlled water bath was used around the growth chamber to remove the heat produced by the ultrasonic transducer, with the water temperature being experimentally determined so as to maintain a constant growth chamber air temperature for each mist supply period. Air was humidified by sending it through a water column, and was passed through a 0.2um membrane filter (Millipore, Millex-FG 50), in order to prevent the growth chamber's contamination. The operation of the ultrasonic transducer and photo period were controlled using a personal computer (NEC, PC-9801).

Explants

Explants (node cuttings having one leaf, stem length: around 1 cm) were excised from potato plantlets (Solanum tuberosum, cultivar; Benimaru) which were cultured in vitro using a Murashige and Skoog (1962) basal medium with 0.8% agar and 3% sucrose.

Culture Medium

A half strength Murashige and Skoog (1962) basal

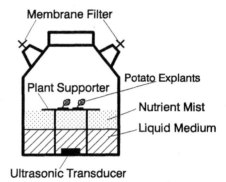

Fig. 1. Mist culture system growth chamber.

medium with 3 % sucrose (initial condition, pH: 5.8) was used in the mist culture because there was a possibility of medium condensation caused by evaporation during the culture period (Kurata,Ibaraki and Goto, 1991). The medium volume dispensed into the growth chamber was determined in a way which sufficiently supported mist generation, being determined to be 1800 mL. In turn, in control culture using a testtube (diameter 25mm, height 150mm), a full strength Murashige and Skoog medium with 3% sucrose and 0.8% agar (initial condition: pH 5.8, dispensed volume 20 mL), was used as a conventional medium.

Culture Conditions

Culture conditions for both the mist and control cultures are given in Table 1. All experiments were conducted in a temperature-controlled clean room (temperature; 22 °C). Photosynthetic photon flux (PPF) in the mist culture was measured inside the growth chamber with no plantlet present, and fluorescent lamps were used as the light source.

Experiment 1: Mist Supply Period Effects on Potato Plantlet Growth

In order to investigate the effects of the mist supply period, in addition to the previous experiment (mist supply period; 9/10, culture period; 3 weeks), two different mist cultures with shorter mist supply periods

TABLE 1 Culture Conditions

		Msit Culture	Agar Culture (Control)
PPF	[μmol/m^2]	50	50
Photo Period	[h/d]	16	16
Outside Air Temperature	[°C]	22	22
Inside Air Temperature	[°C]	24	24
Approximate amount of Outside CO_2 Concentration	[ppm]	400	400
Density of Explants	[no./cm^2]	1~12/68	1/20

"Outside" means clean room in which culture vessels were set, while "inside" means inside the culture vessels.

were conducted, with plantlet growth of all these mist cultures being compared. A two week culture period, was used in these experiments, with being considered to allow a good comparison between mist and control cultures. Table 2 shows the mist supply period of each mist culture.

TABLE 2 Mist Supply Periods in Experiment 1

Treatments	Mist Supply Period for a 10 min Control Cycle
9/10 [a]	9 [min]
5/10	5
1/10	1

a) Previous experiment

The ultrasonic transducer operation time was considered as the nutrient mist supply period. The chamber's air supply flow rate and flow period were respectively maintained at 0.8 L/min and 9 minutes out of a 10 minute control cycle, with the control culture (agar medium) being simultaneously supplied with a corresponding mist culture.

Experiment 2: Mist Supply Variation Effects on Potato Plantlet Growth

Mist supply variations involved changing the mist supply period at different growth stages during the culture period. Controlling the mist supply variations enabled different tissue growth requirements to be met during these different stages. In order to observe these effects an experiment was arranged in the following manner. Nutrient mist was supplied 9 minutes out of a 10 minute control cycle for the first three days, when planted tissue starts to heal the wounds received in the planting cutting operation and 1 minutes out of the same cycle for the next 11 days, when the shoot and root formation occurs (Table 3).

TABLE 3 Mist Supply Variations in Experiment 2

Parameter	0-3 days	4-7 days	8-14 days
Mist Supply Period [min per 10 min]	9	1	1
Air Flow Period [min per 10 min]	9	9	9
Air Flow Rate [L/min]	0.5	0.5	2

RESULTS AND DISCUSSION

Experiment 1

Table 4 shows the average of mist/control culture growth parameters (number of nods, shoot length, and fresh/dry weight) at the end of the culture. Potato plantlet growth was promoted by each mist culture, with the ratio of mist culture growth parameters to the respective control culture ones, subsequently termed the "growth ratio," being greater than 1.

Mist culture treatments were independently conducted at different times because there was only one available mist culture system, thus the explant materials were not completely homogenized between different mist culture treatments. Since different growths were therefore obtained in each control culture, all treatments were compared using their growth ratios.

Growth ratios of the fresh/dry weight of the 1/10 treatment were the largest of all treatments.

The ratio of fresh root weight to fresh total weight, termed "root weight ratio," decreased with respect to an increase in the mist supply period, i.e., the 9/10 treatment ratio was smaller than its control culture. It is believed that a long mist supply period may inhibit root growth.

TABLE 4 Experiment 1 Average Growth Parameters after Culturing

Treatment		Culture Period [day]	Numbers of Nods	Shoot Length [mm]	Fresh Weight [g]	Dry Weight [g]	Root Weight Ratio
9/10 [a]	Mist	21	8.9	111.0	.567	.038	.21
(11) [b]	Control		7.1	67.3	.333	.025	.32
	Growth Ratio		1.25	1.65	1.70	1.52	0.64
5/10	Mist	14	6.5	87.9	.599	.040	.28
(12)	Control		5.7	69.0	.283	.202	.21
	Growth Ratio		1.15	1.27	2.12	2.00	1.34
1/10	Mist	14	7.0	87.7	.505	.031	.35
(11)	Control		5.8	57.1	.198	.014	.24
	Growth Ratio		1.21	1.54	2.55	2.25	1.46

a) Previous experiment
b) Plantlet number

TABLE 5 Experiment 2 Average Growth Parameters for a Mist Culture
with Mist Supply Variations[a]

	Numbers of Nods	Shoot Length [mm]	Shoot Fresh Weight [g]	Total Fresh Weight [g]	Dry Weight [g]	Root Weight Ratio	Leaf Aera [cm²]
Mist	7.0	89.8	.362	.418	.029	.13	4.8(3.6)[b]
Control	4.9	58.4	.123	.157	.012	.22	1.8(0.8)

a) 12 plantlets grown in a 14 day culture period
b) Area of new leaves

The older explant leaves which were observed to die in the 9/10 treatment did not appeared in either the 5/10 or 1/10 treatment.   A few vitrificated leaves, however, to be seen in liquid cultures were observed in 5/10 treatment.

These results suggest that the mist supply period is an important factor in mist cultures, and that altering this period changes the culture environment of potato explants, thus causing different effects on their growth.

Plantlets are normally used in order to obtain high quality seedlings, thus plantlets which have adequate acclimatization characteristic are favorable. From this point of view, a shorter mist supply period was considered to be better due to a larger increase in both the fresh and dry weight, and also better root development. It was, however, difficult to collect the roots under the plant supporter because they were tangled, thus improvements of the supporter are necessary.

Experiment 2

Table 5 shows the average of growth parameters 14 days after planting, with all values except the root weight ratio being larger than in control agar culture. As shown Fig. 2., the growth ratios concerned with shoot growth such as number of nodes, fresh shoot weight, and dry weight were larger than in the 1/10 treatment of Experiment 1.

These results suggest that a longer mist supply period immediately after planting enhanced shoot growth, although it simultaneously inhibited root growth.   The potato explants used in these experiments   were   stem   cuttings   which   were immediately excised from potato plantlets prior to transplanting, thereby causing wounds which allowed both nutrient absorption and water evaporation to occur.   Because the nutrient mist was abundantly supplied to explants, which   were continuously surrounded by a fresh medium, favorable conditions are considered to have occurred, thereby enabling the explant to easily absorb nutrients and also making it difficult to release water from wounds' surface. This condition is thought, however, to be detrimental for root growth, hence it is important how to reduce the mist supply period from the initially long one. The mist supply variations of Experiment 2, especially the period having an initially longer mist supply, was determined without being based on preliminary optimization

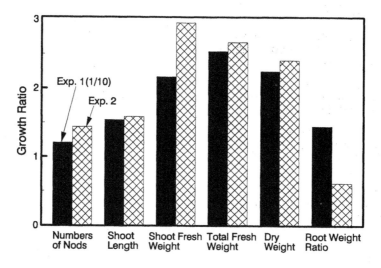

Fig. 2. Growth comparison using mist supply variations.

experiments, thus additional investigations are required to determine the best way to alter the mist supply period.

Throughout the experiment it was observed that traces of mist which entered into the shoot regions escaped from the root regions. This may cause undesirable growth characteristics due to a resulting high humidity, thus suggesting that the shoot and root regions should be partitioned into two independent control regions. In this condition, nutrient mist cultures will have a wider range of controllability for the environmental conditions surrounding the tissues, thereby obtaining a culture system that minimizes the number of required transplants, which should subsequently results in both more efficient propagation and saving labor.

CONCLUSION

Nutrient mist culture of potato plantlets were conducted in order to optimize the mist supply period so as to obtain the best growing conditions. The shortest mist supply period (1/10) gave the best results, i.e., the largest increase in fresh and dry weights, and also better root development.

Mist supply period variations were also conducted during plantlet growth, and showed promising results for obtaining better plantlet growth. By altering these variations, it is believed that micropropagation of potato explants can be further optimized.

REFERENCES

Aitken-Christie, J.and H. E. Davies (1988). Development of a semi-automated micropropagation system. Acta Hort. 230, 153-158

Akita, M. and S. Takayama (1988). Mass propagation of potato tubers using jar fermentor techniques. Acta Hort. 230, 55-61

Kurata, K., Y. Ibaraki, and E. Goto (1991). System for micropropagation by nutrient mist supply. Trans. of Am. Soc. of Agri. Engi. in press

Murashige, T. and F. Skoog (1962). A revised medium for rapid growth and bioassays with tobacco tissues cultures. Physiol. Plant. 15, 473-479

Tisserat, B. and C. Vandercook (1985). Development of an automated plant culture system. Plant Cell Tissue Organ Culture. 5, 107-117

Weathers, P. J. and K. L. Giles (1988). Regeneration of plants using nutrient mist culture. In Vitro Cellular. 24, 727-732

"ROOTS DIVIDING METHOD" FOR NUTRIENT
SUPPLY IN SOILLESS CULTURE

K. Kurata

Dept. of Agricultural Engineering, University of Tokyo, Yayoi 1-1-1, Bunkyo-ku, Tokyo 113, Japan

Abstract. A new method of nutrient supply in hydroponics was tested, in which plant roots were divided into several groups and each group was located in a different solution containing one salt, except for the group for micronutrients. In this method, change of nutrient elements balance in solution over time, which sometimes hinders normal crop growth in a conventional method, is no more a problem. Concentrations of each element can be easily monitored with an EC meter and can be easily adjusted. A preliminary experiment suggested the effectiveness of this method, although it was not statistically confirmed. The experiment also showed that the technical problem in transplanting exists in this method.

Keywords. Biocontrol; control system design; hydroponics; nutrient supply; sensors

INTRODUCTION

In factory-style crop production systems, crop can grow twice or three times faster than outside. This shows that the crop has such a potential, which is not realized under normal conditions (uncontrolled or less controlled conditions). However, constant monitoring and precise control of crop environment are required to allow the crop to exhibit its high potential. For these purposes, much of hardware and software, including precise understanding of crop interaction with the environment, is needed. One of the future directions of the research on crop production systems is pursuing such a system installed with many devices for monitoring and control.

However, such a system is not the only possible view of the future crop production systems. Such a system requires not only many devices but also much energy to operate and much labor as well. There is another way to pursue in which we rely more on the wide adaptability of the crop. Crop adaptability is one of the major factors we should consider in designing agricultural or horticultural production systems. There are possibilities we can reduce amount of control (actions required for control the crop environment) or we can reduce required devices, if we utilize crop adaptability in a proper way. Thus, investigating the crop adaptability and its application in production systems are also important subjects when considering the future crop production systems.

This report shows an example of ideas of utilizing crop adaptability in a crop production system and the results of a preliminary experiment based on the idea. In hydroponics, nutrient control is one of the major concerns. In commercial applications, nutrition is controlled based on the measured electrical conductivity (EC) and pH of the solution. However, problems often arise, because EC readings give only the total salt concentration, not the concentrations of individual elements. Many have reported (e.g. Son and Takakura, 1987) that the ratios of

individual elements change over time due to plant absorption and thus the "effectiveness" of the solution gradually reduces. Most growers restore nutrient balance by either routinely dumping the entire solution or leaking the solution from the reservoir. It goes without saying that constant monitoring the concentrations of individual elements and adjusting them based on the measured values are desired. For this purpose, public service of analyzing the concentrations of individual elements is in some area provided. Another effort to overcome this problem has been made to develop cheap reliable sensors for individual elements, so that the growers can easily monitor the individual elements concentrations.

In this way, in hydroponics more sensors are required for monitoring and more control is needed to keep the elements balance at the desirable level to achieve high productivity. However, thinking of the fact that change of individual elements ratio occurs because different salts are mixed in the solution, we think of another possibility to overcome this problem: different salts are not mixed and are supplied to the crop independently. Roots of each plant are divided into several groups and each group is placed in a different solution which contains only one salt. In other words, each group of roots is allocated a role to absorb only one salt.

We name this culture method a "roots dividing method". The roots dividing method can be expected to have the advantages over the conventional method that change of nutrient elements balance over time does not become a problem, and that the concentrations of individual elements can be easily measured by an EC meter because one solution contains only one salt. Therefore, adjustment of the concentrations of each salt at the desired level can be easily achieved independently from other salts. Dumping of the solution is not necessary. Moreover, interference of excessive amount of one ion with the plant uptake of other ions (e.g. Gerber, 1985) can be expected not to occur.

These arguments are based on the assumption that the crop can grow in the roots dividing method. Particularly, whether roots can grow in a solution with one kind of salt, and whether roots can uptake nutrients in such a situation are open questions. As far as the author knows, there have been no attempts to test the possibility of this method. The expected advantages of this method discussed above also assume an ideal situation. They assume that the resource water contains no nutrients, that the roots discharge no organic matter into the solutions, and that pH adjustment of each solution by acid or alkali is not necessary. In real situations, these factors will be obstacles to maintaining each solution containing only one salt. However, we can expect that the concentrations of ions appearing in a solution with the resource water, from the roots, or from pH adjustment are much less than that of the planned salt and the roots dividing method is still effective, if the crop has the adaptability to this method. We can also foresee technical problems of limited number of roots groups (number of solutions) and/or a trouble involved in dividing the roots.

The objectives of this study are to test the possibility of the roots dividing method (whether the crop has the adaptability to this method) and to test technical problems involved in this method. For these objectives we conducted a preliminary experiment.

MATERIALS AND METHODS

Lettuce (*Lactuna sativa* L.; cultivar Okayama) of butterhead type was used in the experiment, which was conducted in January, 1987 in the campus of University of Tokyo, Japan. In a greenhouse three culture vessels (2 m long) were constructed in parallel in east-west orientation. The cross-section of the vessels are illustrated in Fig. 1. The vessels were named from north to south A, C, and B. The vessels A and B were for the conventional culture method, while the vessel C, which was located at the center, was for the roots dividing method. Experiments for the conventional method were conducted for comparison with the roots dividing method. As illustrated in this figure, a method of DFT was applied. The vessel C consisted of four channels (number 3 to 6), while other vessels consisted of only one channel. Thus, in this experiment the roots of lettuce on the vessel C was divided into four groups. The channel 1 and 2 (the vessels A and B) contained 78 L solution, while the channel 3 and 6 had 16.9 L, and the channel 4 and 5, 22.1 L. Each channel was connected to the reservoirs, the volumes of which were 150 L for the channels 1 and 2 and 70 L for the channels 3 to 6. Each solution was continuously circulated through the channel and the reservoir by a pump, to avoid DO deficiency. The air temperature of the greenhouse and the solution temperature were controlled at 15 to 20 °C. DO was not controlled but was maintained during the experiment at 5 to 7 mg/L. EC was not controlled, while pH was adjusted at 5.0 to 7.5.

The initial nutrients concentrations in each solution were determined to be those listed in Table 1. These were determined considering the number of channels in the vessel C and required macronutrients. As a result, ammonium form nitrate (NH_4^+) was not given in this experi-

ment. The concentrations of each salt were determined to be the same for all the vessels, except that the nutrients were supplied separately in the vessel C. Micronutrients were mixed also in the vessel C and were supplied in one channel.

The experiment began on January 15th, 1987 using 30 days old seedlings. Tap water was used for the culture, which was not completely free from the nutrients. On the vessel C roots were so divided into four groups, that each group had approximately the same amount of roots. Ten seedlings were transplanted onto each vessel and they were harvested on January 29th, when fresh and dry weights of root and shoot parts and the root length of each plant were measured.

RESULTS

Figure 2 shows lettuce roots on the vessel C. Divided roots into four groups can be recognized. Figure 3 shows an example of a harvested lettuce from the vessel C. Note that growth of the roots was different in different channels. The channels of the root groups on this figure are, from the right to the left, 3, 4, 6, and 5. The roots in the channels 5 and 6 had brownish color, while the rests were whiter. An extremely suppressed growth of the roots in the channel 6 can be recognized.

Figure 4 shows the plants on the vessels. The plants indicated by 1, 2 and 3 are on the vessel C. The plant 1 had yellowish leaves and the growth was suppressed. This was caused by the damage on the roots which had been given at the time of transplanting. However, the plants 2 and 3 showed a normal growth compared to those on the vessels A and B. Unfortunately, in this experiment a larger part of the plants was not free from damage at the time of transplanting. This could be recognized in the suppressed growth of the shoot part shown in Table 2. This table shows, however, that the root part growth in the vessel C was comparable to that in the vessels A and B. Another point which is noteworthy in this table is that the water content of the plants on the vessel C was less than that on the vessels A and B.

DISCUSSION

The roots dividing method tested in this study involves very troublesome work in dividing roots. The roots were often damaged during this process. Therefore, we could not obtain statistical confirmation of the effectiveness of this method.

However, as some plants as the plants 2 and 3 on Fig. 4 showed, plants can grow as normally in the roots dividing method as those in the conventional method, if they were free from the damage. Therefore, development of transplanting method without giving damage to the roots is required in order to have this method effective. We have already developed a seedling culture method, in which roots are automatically divided into several groups. Transplanting such seedlings involves much less trouble than this experiment.

If we pursue the way to control more precisely the plant environment to achieve higher productivity, more devices are needed for monitoring and control. As a result, the plant production system will become more and more heavily installed and energy consuming. Another possibility is investigating the plant adaptability to the environment and the method of utilizing it in a production system. The roots dividing method reported here is an example.

CONCLUSION

A new method of supplying nutrients in hydroponics were tested, in which important salts were supplied separately to the roots. This method eliminates problems of ion balance in the solution and requires no ion selective sensors. An experiment showed a strong suggestions for

the effectiveness of this method, although it was not statistically confirmed. The experiment also revealed a technical problem of transplanting involved in this method.

REFERENCES

Gerber, J.M. (1985). Plant growth and nutrient formulas. In A.J.Savage (Ed.), *Hydroponics Worldwide: State of the Art in Soilless Crop Production*, International Center for Special Studies, Honolulu. pp58-69.

Son, J.E., and T. Takakura (1987). A study on automatic control of nutrient solutions in hydroponics. *Journal of Agricultural Meteorology*, *43*, 147-151.

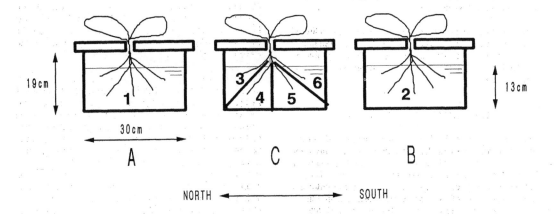

Fig. 1 Cross-section of culture vessels. Numerals refer to the channel number.

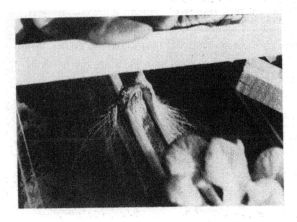

Fig. 2 Lettuce roots in the roots dividing method.

Fig. 3 An example of lettuces grown
by the roots dividing method.

TABLE 1   Initial Nutrient Concentrations

Vessel	Channel	Elements	Conc.
A	1	KNO_3	8 me/L
		$MgSO_4$ $7H_2O$	2 me/L
		$Ca(H_2PO_4)_2$ H_2O	6 me/L
B	2	micronutrients	
		(Otsuka No.5)	40 mg/L
	3	micronutrients	
		(Otsuka No.5)	40 mg/L
C	4	KNO_3	8 me/L
	5	$MgSO_4$ $7H_2O$	2 me/L
	6	$Ca(H_2PO_4)_2$ H_2O	6 me/L

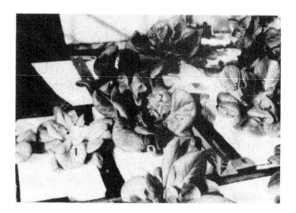

Fig. 4 Lettuces on the culture vessels.
Lettuces numbered 1, 2, and 3 are
grown by the roots dividing method.

TABLE 2 Growth Data of Lettuce (Average of Ten Plants)

Items	Vessels	Values
Fresh Weght	A	45.59 g
	B	32.27 g
	C	18.67 g
Max. Root Lenght	A	33.75 cm
	B	31.75 cm
	C	32.79 cm
Dry Weight (Whole Plant)	A	2.26 g
	B	1.78 g
	C	1.54 g
Dry Weight of Shoot Part	A	1.87 g
	B	1.40 g
	C	1.17 g
Dry Weight of Root Part	A	0.40 g
	B	0.38 g
	C	0.36 g
Water Content	A	0.95
	B	0.94
	C	0.92

140

Copyright © IFAC Mathematical and Control Applications
in Agriculture and Horticulture, Matsuyama, Japan 1991

AN ENVIRONMENTAL CONTROL SYSTEM FOR GROWING PLANTS UNDER LOW TOTAL PRESSURES

E. Goto, K. Iwabuchi and T. Takakura

*Dept. of Agricultural Engineering, Faculty of Agriculture, University of Tokyo, Yayoi 1-1-1,
Bunkyo-ku, Tokyo 113, Japan*

Abstract. An environmental control system including a growth chamber was constructed to study the feasibility of growing plants under low total pressures relating to crop production in space and CELSS. The system could control temperature, relative humidity, light intensity, and total and partial pressures. The chamber was cylindrical, and 63.5cm in diameter and 50cm in height. Thirty samples of leaf vegetables such as spinach could be grown in the chamber by hydroponics under artificial lights. This system could control total pressure, O_2 and CO_2 partial pressures separately, and consequently maintained them at constant during a growth experiment of spinach at a total pressure of 50kPa. Spinach(Spinacia oleracea L.), which was grown under the atmospheric pressure(101kPa), was transplanted in the chamber and exposed to various pressure conditions, and its photosynthetic rate was measured. Photosynthetic photon flux density(PPFD) was maintained at $120\mu molm^{-2}s^{-1}$ and the light period was set at 14h. Volume of nutrient solution was 9ℓ, and pH was 6.0 and electric conductivity was 2.2mS/cm. The setting values of temperature and relative humidity(RH) were 20°C and 73 to 75%, respectively. The result indicated that photosynthetic rates of spinach at 50 and 75kPa were almost the same as that at 100kPa.

Keywords. Environmental control; low pressure; plant growth; photosynthetic rate; spinach; pressure control; humidity control; computer control.

INTRODUCTION

In the past several years, there has been increasing interest in growing plants in extraterrestrial space (Tibbitts and Bula, 1989), and CELSS (Controlled Ecological Life Support System) has been proposed as a system in space and has been studied (Averner, 1989; Olson et al., 1988). A crop production component is considered to be one of the components comprising CELLS and is expected to play an important role in this system. On the other hand, the question whether plant can grow in space has not been solved, because there was no experiment on plant growth in space, except several experiments made in spaceships and space stations. The feasibility of plant growth in space has, however, been gradually identified from those experiments.

Gravity and gas composition in space including on the moon and Mars are different from those on the earth. Many researchers have considered that zero gravity or microgravity would affect plant growth. In fact, some researchers have been studying the effect of gravity on the growth. On the other hand, gas composition and concentrations such as O_2 and CO_2, which affect plant growth, have been already studied relating to crop production in controlled environments. The previous work, thus, can be useful in space under the condition where a total pressure is maintained at the same as that on the earth. However, total pressure can not be maintained easily in space at the level on the earth, because outside pressure is zero or very close to zero. In addition, Andre and Massimino (1991) showed, in their report, that the advantage of structures whose total pressure was reduced. We, therefore, do not consider that total pressure in a crop production component is kept necessarily at the level of a resident component.

Experiments on plant growth and development under low total pressures have been already made at some institutes. Lind (1971) constructed an experimental system simulating the conditions within the NASA Skylab, and NASA developed the Biomass Production Chamber as a central focus of NASA's CELLS project (Prince and Knott, 1989). Daunicht and Brinkjans (1991) and Andre and Massimino (1991) constructed growth chambers to study plant growth and photosynthesis under low total pressures. In these previous studies, the experiments were carried out under various total pressure conditions. However, there was no case where total, N_2, O_2, and CO_2 pressures were controlled separately.

The final purpose of our research was to study the feasibility of growing plants under various conditions of total and partial pressures. Consequently, an new experimental growth chamber was necessary to perform our research. In this report, first, we described an environmental control system including a growth chamber which was intended to grow plants under various conditions of total and partial pressures. Next, we discussed some technical problems resulting from the operating test, and last, showed the result of measurement of photosynthetic rate of spinach with this system. Growth experiments of spinach under different total pressures were carried out following those preliminary experiments. The details of these experiments will be reported in our next paper.

CONSTRUCTION OF AN ENVIRONMENTAL CONTROL SYSTEM

Characteristic of the System

As mentioned earlier, the final purpose of this research was to study the feasibility of growing plants under low total pressures. It seemed best to make experiments where the pressure difference between outside and inside a chamber was zero or close to zero. However, there are very few places on the earth suitable for above condition except mountains of extremely high altitudes, and if there are, the places are not easy for researchers to work in. In addition, there is no place on the earth where total pressure is lower than 30kPa. On the other hand, it seemed not easy to maintain pressure inside a chamber alone at lower pressure in an atmospheric pressure room, because the pressure difference between inside and outside complicated gas and humidity control as well as the maintenance of the total system. It was, therefore, considered that a chamber designed for use at the atmospheric pressure could not useful for our research. As a result, a new environmental control system including a growth chamber was designed and constructed to carry out growth experiments under low total pressures (Fig.1). The system was expected to realize the following five conditions.
(1) To maintain total pressures down to 40kPa.
(2) Independent pressure control of total, O_2 and CO_2.
(3) To grow plants by hydroponics.
(4) Continuous measurement of photosynthetic rate.
(5) Computer control of environmental factors.

Detail of the System

The body of the chamber was cylindrical and its side and base were made from stainless. The body was 50cm in height, 63.5cm in inner diameter, and 2mm in thickness. The chamber's cover was a transparent acrylic resin plate of 2cm thickness with a round stainless frame. To keep the chamber highly airtight, a silicon seat of 4cm in width and 3mm in thickness was put between the body top edge and the cover, and seal grease was painted on each seat's surface of the cover and the body top edge. Sixteen Swagelok female screw holes of 1/4" PT size were made in the side body. The pipes and tubes for gas or liquid exchange between the inside and the outside could be fitted in the screws with male connectors.

Electrical wires were put through the screws with bored-through type of male connectors, in order to supply power sources and to monitor output of sensors in the chamber. A plant bed made from vinyl chloride resin was set in the chamber for hydroponics. The plant bed was 40cm wide, 40cm long and 10cm high. A styrenes form plate with 30 holes was set for the plant bed. Nutrient solution was continuously aerated by an air pump placed in the chamber. A gas supply system was comprised of three lines for N_2, O_2 and CO_2. Each line had a gas bottle, a pressure regulator, a flow meter, and an electromagnetic valve. The supply of three kinds of gases were done separately by on-off control. For temperature control in the chamber, the chamber was placed in another larger chamber which had heaters, a refrigerator, and five 100W metal halides lamps. A 16-bit microcomputer with A/D converters and relays was used to monitor and control environmental factors.

Measurement of air temperature and air humidity was made by a wet and dry bulb thermometer with C-C thermocouples, which was placed in the chamber. A pressure sensor (MANOACE 30, Sayama Trading Co., Japan) was used to measure pressure inside the chamber. The measurable range and accuracy of its transducer (280E, Setra Systems Inc., USA) was between 0 and 100kPa and $\pm 0.5\%$F.S., respectively. Measurement of O_2 concentration was made by an O_2 analyzer (RO101Y1, Iijima Seimitsu Kogyo Co., Japan) which could measure concentration of 0-100% at the accuracy within 0.1%F.S. The measurement was carefully done at the atmospheric pressure, because its sensor was a galvanic cell type and could not operate in low pressures. Measurement of CO_2 concentration was made by an infrared gas analyzer (ZFP6, Fuji Electronics Co., Japan) placed in the chamber. The measurable range covered between 0 and 2000ppm at the atmospheric pressure, and its accuracy was within 0.5%F.S. Since the analyzer's output was affected by a total pressure, a conversion curve from output to pressure was necessary to be corrected at each total pressure.

CONTROLS OF ENVIRONMENTAL FACTORS

Air Humidity

The system was designed as an airtight one to minimize the loss of supplied gases. As a result, the airtightness caused a rapid increase in air humidity, because of transpiration from leaves and evaporation from nutrient solution. Moreover, diffusion rate of water vapor is inversely proportional to total pressure, which indicates that an increase in humidity is faster at lower pressures than at 100kPa. Therefore, the system seemed to require a dehumidifier which had a higher performance to maintain air humidity at a reasonable value for plant growth. For this reason, two dehumidifiers (DH-109 and DH-209, Komatsu Electronics Co., Japan), thermoelectric cooling types, were used and placed outside the chamber. Small vacuum vessels were used as drain tanks, and were

142

connected to the dehumidifiers through a rubber tube. The air between the chamber and the dehumidifiers was circulated with air pumps and rubber tubes.

Humidification of inside air was done by an ultrasonic humidifier which had a transducer driver and an ultrasonic transducer (NB-510S, TDK, Japan). The driver was placed outside the chamber, and the transducer was set on the bottom of a water tank placed inside the chamber. Water was led from an outside tank automatically with a float switch and an electromagnetic valve. Relative humidity in the chamber was controlled by on-off control of the humidifier while the dehumidifiers were continuously operated.

Total Pressure

Total pressure was controlled with the pressure sensor and a diaphragm vacuum pump. Air infiltration from outside into the chamber occurred due to pressure difference, because there were unavoidable gaps in the system. The infiltration increased total pressure in the chamber, and increase rate of the pressure became higher when a setting value of the inside pressure was lower.

O_2 Partial Pressure

Since the O_2 sensor was galvanic cell type and could not operate at low pressures, O_2 measurement was done with the following subsystem. The subsystem consisted of a pressurized box where the O_2 sensor was set, an diaphragm vacuum pump, an air pump, a pressure switch, and electromagnetic valves (Fig. 1). The air in the pressurized box was usually isolated from the air in the chamber except during the period of monitoring O_2 concentration at a certain interval. The interval was varied according to a total pressure, and was five minutes at 75kPa and eight minutes at 50kPa. First, the air to be measured was sent into the box with the vacuum pump and a pressure switch until the pressure in the subsystem reached the atmospheric pressure. Second, after the subsystem and the chamber were isolated by the valve, O_2 concentration was continuously measured while the air in the subsystem was being circulated. Last, the air in the box was isolated by the valves again, and the air both in the chamber and in the subsystem, were mixed down to an equilibrium pressure.

CO_2 Partial Pressure

The output of CO_2 analyzer was affected by a total pressure. Calibrations were, therefore, made with a vacuum desiccator which was provided with a pressure sensor and a vacuum pump. First, standard gas including N_2 and CO_2, whose concentration was already-known, was filled up the desiccator, and output was determined as the value of CO_2 concentration at 100kPa. Next, the gas was exhausted down to a certain total pressure, and then output was determined as the value of CO_2 concentration at the pressure. CO_2 concentration used as standard were 350, 599, 1488, 2015 and 3000ppm at 100kPa. Fig. 2 shows the relations

between CO_2 partial pressures and outputs of the infrared gas analyzer. The result indicated that a conversion curve from output to partial pressures varied according to total pressures. Therefore, in the experiments mentioned later, CO_2 partial pressure was estimated from the calibration curves.

OPERATING TEST

With the system, growth experiments of spinach (Spinacia oleracea L.) were carried out under various conditions of total and partial pressures. After germination, spinach was grown by hydroponics for approximately 20 days at the atmospheric pressure. Then, spinach with 2 to 4 true leaves was transplanted in the chamber and grown for five days. The number of samples was 30 in an experiment. Growth conditions were as follows. Photosynthetic photon flux density (PPFD) was $120 \mu molm^{-2}s^{-1}$ and the light period was 14h. Volume of nutrient solution was 9ℓ, and pH was 6.0 and electric conductivity was 2.2mS/cm. The setting values of temperature and relative humidity (RH) were 20°C and 73 to 75%, respectively.

Fig. 3 shows an example of the time changes in total pressure, temperature and RH. Temperature could be maintained at accuracy within 0.5°C. Mean total pressure was kept at about 50kPa, nevertheless, the values higher than 60kPa were sometimes observed. This higher values occurred when the operation of the vacuum pump stopped owing to overload. After a few minutes, however, this pump started to operate and to exhaust gas again. Therefore, this problem could be dissolved by using a higher performance pump.

RH increased up to 85% in the latter period, approximately after the third day from the beginning of the experiment. However, this higher RH was considered to make no significant difference in plant growth. The increase in RH indicated that generation rate of water vapor from transpiration and evaporation exceeded dehumidification rate. In this system, most of dehumidification was done by two dehumidifiers and the rest of it was dependent on gas exchanges. Dehumidification was one of significant problems to be solved when growing plants in a closed system. In this system, gas exchange was made at a certain interval to maintain O_2 or CO_2 partial pressure, and the change in RH was larger than the changes in the partial pressures. Therefore, the system must have become an open type system and have expended plenty of dry pure gases if dehumidification had been done by gas exchange alone. It seemed that the capacity of dehumidification should be determined by taking into account the maximum generation rate of water vapor, especially transpiration rate.

Fig. 4 shows an example of the time changes in total and partial pressures when O_2 partial pressure was lowered at 3kPa in the light period to study the effect of low O_2 on spinach growth. Total pressure as well as partial pressures could be maintained almost at constant and was not affected by O_2 and CO_2 partial pressures. This successful control was due to total pressure control independent of partial pressures.

MEASUREMENT OF PHOTOSYNTHETIC RATE OF SPINACH

Measurement of photosynthetic rate of spinach was made with the assimilation chamber technique. There were two practical ways to measure photosynthetic rate in this system. One was to measure continuously supply rate of pure CO_2 while CO_2 pressure was kept at constant. This way was useful for the measurement at the atmospheric pressure, because there was little infiltration and leak of gases in the system at the pressure. However, the way could not give precise measurement at lower total pressures than 75kPa, because air in the chamber was exhausted at a certain interval to keep total pressure at constant. The other way was to measure the decrease in CO_2 partial pressure for a certain period, approximately for 10 to 30min, after CO_2 supply stopped. This way could be useful at the lower total pressures. CO_2 partial pressure was measured continuously down to 80% of initial pressure in absence of operation of the vacuum pump, and CO_2 absorption rate by plants was estimated from the pressure gradient. In this way, air infiltration into the system was necessary to be taken into account when measurement was done under lower total pressures. CO_2 inflow accompanied with the air infiltration was estimated from the increase in total pressure during the measurement.

TABLE 1 Photosynthetic Rates of Spinach
under Different Total pressures

Pressure			Photosynthetic Rate
Total (kPa)	O_2 (kPa)	CO_2* (Pa)	$(mgCO_2dm^{-2}h^{-1})$
100	21	100	9.22
75	21	100	9.06
50	21	100	8.55

*Initial value at the beginning of measurement.

Table 1 shows photosynthetic rates under different total pressures. The number of spinach used in this measurement were 10 samples of 5g each in fresh weight. After spinach was transplanted and acclimatized to the condition in the chamber for 30min, measurement started in the way described above. Photosynthetic rate in the result was lower at 50kPa than at 75 and 100kPa. Initial CO_2 pressure at the beginning of the measurement was intended to be 100Pa in each measurement. Initial CO_2 pressure at 50kPa was, however, 850ppm and roughly 85% of those at other experiments. It was considered that at PPFD of $120\mu molm^{-2}s^{-1}$, in the range of CO_2 pressure between 0 to 100Pa, photosynthetic rate was affected by CO_2 pressure, and that the rate at 50kPa could

increase if initial CO_2 pressure was set at 100Pa. It can, therefore, be concluded that photosynthetic rates of spinach at the two lower total pressures were almost the same as that at 100kPa. In addition, this indicated that spinach could grow under low total pressures.

CONCLUSION

1. An environmental control system including a growth chamber was made to study plant growth under various conditions of total and partial pressures.

2. Total pressure, O_2 and CO_2 partial pressures, temperature and relative humidity in the chamber could be maintained at the intended values at total pressures down to 50kPa.

3. Photosynthetic rate of spinach was measured and the rates at total pressures of 50 and 75kPa were almost the same as that at 100kPa.

REFERENCES

Andre M and D. Massimino(1991). Growth of Plants at Reduced Pressures: Experiments in Wheat-Technological Advantages and Constraints. Proceeding of 28th COSPAR, July, 1990. Adv. Space Res(in press).

Averner M. M.(1989). Controlled Ecological Life Support System. Lunar Base Agriculture: Soils For Plant Growth. ASA-CSSA-SSSA, Madison. pp. 145-153.

Daunicht H. -J. and H. -J. Brinkjans(1991). Gas Exchange and Growth of Plant under Reduced Air Pressure. Proceeding of 28th COSPAR, July 1990. Adv. Space Res(in press).

Lind C. T.(1971). Germination and Growth of Selected Higher Plants in a Simulated Space Cabin Environment. AMRL-TR-70-121.

Olson R. L., M. W. Oleson and T. J. Slavin(1988). CELSS for Advanced Manned Mission. HortScience, 23, 275-293.

Prince R. P. and W. M. Knott, III(1989). CELSS Breadboard Project at the Kennedy Space Center. Lunar Base Agriculture: Soils For Plant Growth. ASA-CSSA-SSSA, Madison. pp. 155-163.

Tibbitts T. W. and R. J. Bula(1989). Growing Plant in Space. Chronica Hort., 29, 53-55.

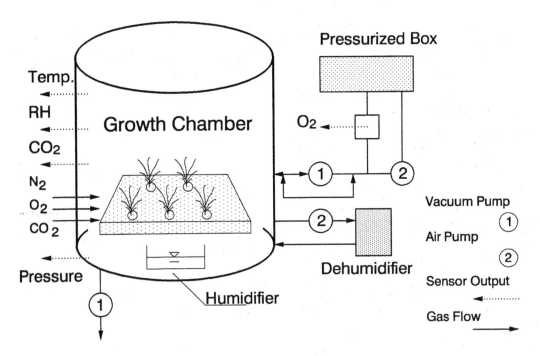

Fig.1. Schematic diagram of the environmental control system
for growing plants under low total pressures.

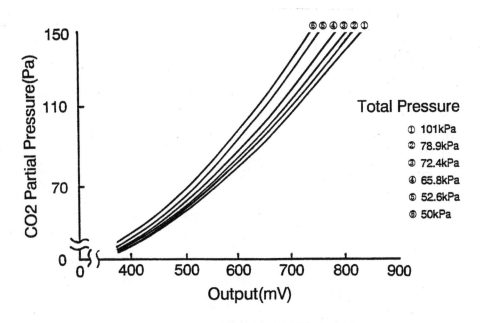

Fig.2. Relations between CO_2 partial pressure and the output
of the infrared gas analyzer under different total
pressures.

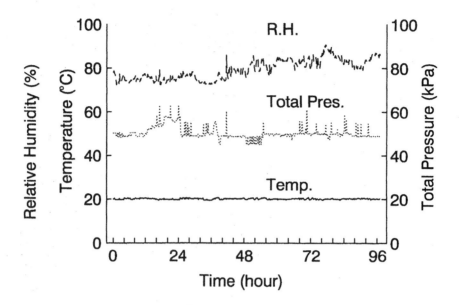

Fig.3. Time changes in total pressure, temperature, and
relative humidity at a setting total pressure
of 50kPa.

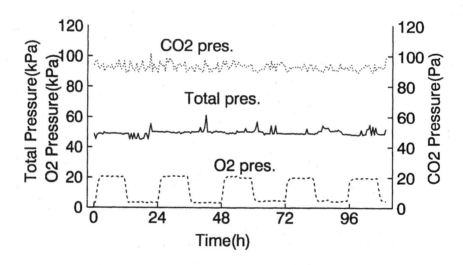

Fig.4. Time changes in total pressure, and O_2 and CO_2
partial pressures. O_2 pressure in the light
period was lowered down to 3kPa, to study the
effect of photorespiration inhibition.
O_2 pressure in the dark was kept at 21kPa.

APPLICATION OF FUZZY LOGIC AND NEURAL NETWORK TO THE PROCESS CONTROL OF SOLUTION pH IN DEEP HYDROPONIC CULTURE

T. Morimoto and Y. Hashimoto

*Dept. of Biomechanical Systems, College of Agriculture, Ehime University, Tarumi,
Matsuyama 790, Japan*

Abstract. Crop growth control systems in a practical cultivation are characterized by complexity and fuzziness. Fuzzy logic and neural network approach can be used to solve such complex ill-defined problems. They have high potential to deal with complicated systems with their own high flexible control and learning capability. In this paper, a new adaptive control method combined feedback control based on a fuzzy logic and feedforward control based on a neural network is applied to the control system of pH of the nutrient solution in deep hydroponic culture and then the efficiency of this method is discussed. Applying the fuzzy logic to the feedback control part of the pH control increased the flexibility and smoothness of the control performance. Furthermore, the operational actions show the same pattern as a skilled human operator's technique. On the other hand, the neural network composed of 3 layers was applied to the construction of inverse dynamic model taking the solution pH as the input and the acid or alkali supply as the output. Applying the inverse dynamic model to the feedforward control part, more speedy control performances were obtained. All the control results were superior in the flexibility, stability and deadbeat time to that of traditional control. This shows that smooth, safety and automatic adjustment of pH of the nutrient solution would be able to realize with a computer control.

Keywords. Fuzzy logic; Neural network; hydroponic automation; pH control; deep hydroponic system; nutrient solution

INTRODUCTION

Recently, hydroponic cultivation made remarkable developments in the fields of protected cultivation and plant factories. This is because the hydroponic system have great advantages for the automation of crop production. Following these developments, in order to control the hydroponic crops efficiently, we have tried to make clear its physiological dynamics as affected by environmental factors using traditional identification methods (Morimoto, 1988; Hashimoto, 1985). However, it is difficult to achieve the hydroponic automation by means of these traditional precise mathematical methods. Because as the hydroponic factors such as EC (Electric conductivity), pH, dissolved oxygen, dissolved CO_2 nutrient composition and temperature of the nutrient solution in the culture medium are remarkably influenced by biological and chemical reactions and climate conditions, the hydroponic control processes are characterized by complexity and fuzziness (Epstein, 1990; Wild, 1987).

The pH of the nutrient solution in the hydroponic culture is kept constant in the range of 5.5 to 6.5 by the addition of a acid or alkali. However, the solution pH is easy to change by the reasons of the imbalance of anion and cation uptake by plant roots, release and hydrolysis of CO_2, the efflux of H^+ as a result of excess cation over anion uptake and the efflux of organic acid from the roots (Mugwire, 1977; Ponnamperuma, 1966; Bangerth, 1979). This is almost caused by a low buffering capacity of the nutrient solution in the culture medium. Changes in pH of the nutrient solution are remarkably able to affect the ion uptake by plant roots (Hatch, 1984). Furthermore, Physiological disorders caused by excess high pH level are also many observed (Morimoto, 1991). It is therefore necessary to control the solution pH effectively. In a deep hydroponic culture

system, however, flexible and precise control of solution pH is difficult because the dynamics of solution pH as affected by an acid or alkali shows significant nonlinear characteristics. Furthermore, marked time delay of pH reaction occurs because the nutrient solution in the deep hydroponic system is recirculated between storage tank and culture vessel at a small rate. Moreover, marked distribution of pH arises in the culture vessel because the nutrient solution flow in it is significantly restricted by the amount of developing root systems. As mentioned above, the control system of the solution pH in a deep hydroponic culture system is also characterized by complexity and fuzziness.

It is known that fuzzy logic and neural network approaches are able to deal with real complex problems such as nonlinearity, time delay and time-varying parameters (Antsaklis, 1990; Lee, 1990). They have the potential to solve the difficult problems that cannot be handled by traditional precise mathematical methods. The neural network is able to identify a class of unknown nonlinear system with their own high learning capability (Bhat et al., 1990 ; Chu et al., 1990). This implies that a model predictive or adaptive control for hydroponic automation become possible by using the neural network model obtained by the identification (Chen, 1990; Psaltis, 1988). On the other hand, the fuzzy logic is effective in the control system where the human experience and knowledge play an important roles. This shows that flexible and smooth control of hydroponic factors may be realized using the fuzzy logic (Morimoto, 1990). It may be considered that we should utilize the fuzzy logic and neural network approaches as an active controller of hydroponic system to attain growth optimization of the hydroponic plants.

In this paper, we report on some results from an

initial study of the application of fuzzy logic and neural network in the adaptive control of solution pH in the deep hydroponic culture. The control method is mainly composed of two parts which are feedback control based on fuzzy logic and feedforward control based on neural network. Then the merits of this method is discussed.

MATERIALS AND METHODS

Control system of solution pH

Figure 1 shows schematic diagram of deep hydroponic culture and the pH control system of nutrient solution. The hydroponic system under consideration consists of two vessels, which are storage tank and culture vessel containing 8 and 5 liters of nutrient solution, respectively. The nutrient solution is continuously pumped back to the top ends of the culture vessel and is circulated between the storage tank and the culture vessel at a rate of 300 ml/min. The pH of the circulating nutrient solution is controlled by the addition of acid, 0.3 M H_3PO_4 or alkali, 0.2 M KOH. In this experiments, the acid or alkali liquid were added to the storage tank only in order to ensure the safety of plant roots. The supply rate of acid or alkali is 60 ml/min. As for other condition in this experiment, EC (Electric conductivity) of the nutrient solution was adjusted to 0.08 S/m, and solution temperature was held constant at approximately 15 degree C during the experiment.

Application of fuzzy logic and neural network to feedback and feedforward control of solution pH

In order to attain the crop growth optimization of the plant, we should learn high techniques of a skilled human operator. It is known that the control policies of a human operator are firstly determined based on a feedback control procedure, but the essential control rule gradually changes a feedforward control procedure as learning of its system movements proceeds (Kawato et al., 1987; Psaltis, 1988). In this paper, we utilized these procedures in order to solve the complex control systems. Figure 2 shows block diagram of the control system of the solution pH based on both feedback and feedforward control. In the figure, u(k) is the input variable, y(k+1) the output variable and e(k) is the error between desired and output value at time-step k. The input variable u(k) to the control object is given as the sum of outputs from the feedback and feedforward controller. Let's apply a fuzzy logic and neural network to the pH control system. Figure 3 shows block diagram of the control system in the case of applying the fuzzy logic to the feedback control part and the neural network to the feedforward control part. In this paper, such new control procedure using the fuzzy logic and neural network is proposed.

Neural network approach for feedforward control

In the figure 3, the purpose of using the neural network is to learn the nonlinear characteristics of the solution pH and then to make the inverse dynamic model for an adaptive control (Psaltis, 1988). Figure 4 shows the structure of the neural network architecture proposed in the previous paper (Morimoto, 1991). The network used in the present study consists of three layers: input, hidden, and output layer. Each layer have many interconnected processing elements called neuron. It is noted that signals passed through the neurons in the hidden and output layers are transformed based on a nonlinear (sigmoid) function but in the input layer, it is transformed based on a linear function. The strengths of interconnections between the neurons are called weights. The learning procedure used for the construction of the inverse dynamic model is the

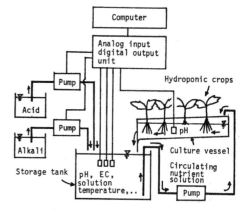

Fig. 1 Schematic diagram of the control system of solution pH in the deep hydroponic culture system composed of storage tank and culture vessel. Nutrient solution is circulated between storage tank and culture vessel.

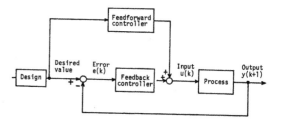

Fig. 2 Block diagram of general control scheme based on feedback and feedforward control.

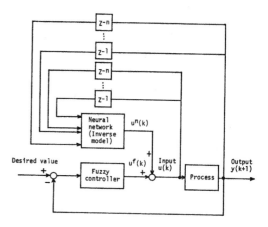

Fig. 3 Block diagram of the control system based on fuzzy logic and neural network.

error back propagation (Rumelhart, et al., 1986), which method gradually tunes the weights in the neural network so as to minimize the squared error between desired and actual responses of the solution pH. In this case, optimal amount of acid added to the tank is obtained from the inverse dynamic model taking the solution pH as the input and the amount of acid as the output of the neural network. The tuning procedure is accomplished when the squared error becomes smaller. The tuning sequence is mainly constructed with two processes which are forward calculation in order to obtain the network output and backward calculation in order to obtain the renewed weights and biases. The forward calculation is described as follows:

$$OT^s_j = \sum_{i=1} O^{s-1}_i \, W^s_{ij} + H^s_j \quad \ldots\ldots\ldots\ldots (1)$$

$$O^s_{kj} = \frac{1}{1 + \exp(-OT^s_j)} \quad \ldots\ldots\ldots (2)$$

where OT^s_j represents the total input to j-th neuron in the s-layer, which shows the weighted sum of the output O^{s-1}_i of i-th neuron in the (s-1)-layer. Furthermore, W^s_{ij} is the weight between i-th neuron in the (s-1)-layer and j-th neuron in the s-layer, and H^s_j is a bias (threshold level) of j-th neuron in the s-layer. On the other hand, O^s_{kj} represents the output of j-th neuron in the s-layer when the k sorts of input-output pair is applied to the neural network. As shown in Eq.(2), the output from the neuron is obtained based on the nonlinear transformation of sigmoid function. Next, backward calculation for tuning is carried out. Suppose the total error E followed by the k sorts of patterns. The values of E are obtained from the sum of the squared error between the estimated output O^{s+1}_{kj} of the network and desired (actual) output $u_j(k)$ which shows the amount of acid supply. The E is described as follows:

$$E(W_{ij}) = 1/2 \sum_{k=1} \sum_{j=1} (O^{s+1}_{kj} - u_j(k))^2 \quad \ldots\ldots (3)$$

Tuning of the weights and biases is carried out so that the total error E given by Eq.(3) is minimized and its procedure is based on a gradient descent algorithm. The error signal e^{s+1}_{kj} of j-th neuron in the output layer ((s+1)-layer) is described as follows:

$$e^{s+1}_{kj} = (O^{s+1}_{kj} - u_j(k)) O^{s+1}_{kj} (1 - O^{s+1}_{kj}) \quad \ldots\ldots (4)$$

The tuning of weights and biases in the output layer are carried out based on the error give by Eq.(4) as follows:

$$W^{s+1}_{ij}(p+1) = -a \cdot e^{s+1}_{kj} O^{s+1}_{kj} + b \cdot W^{s+1}_{ij}(p) \quad \ldots (5)$$

$$H^{s+1}_j(p+1) = a \cdot e^{s+1}_{kj} + b \cdot H^{s+1}_j(p) \quad \ldots\ldots\ldots\ldots (6)$$

where p represents iteration number for the convergence of the total error. The coefficient a is the learning rate, and b is the momentum factor. Furthermore, the error signal e^s_{kj} of j-th neuron in the hidden layer (s-layer) is expressed as follows:

$$e^s_{ki} = \sum_{j=1} e^{s+1}_{kj} \cdot W_{ji} \cdot O_{ki} (1 - O_{ki}) \quad \ldots\ldots\ldots (7)$$

Tuning forms of the weight and biases in the hidden layer is also expressed with the same forms as Eq.(5) and Eq.(6). These procedures are repeated until the total error given in Eq.(3) becomes small enough. Then, we can obtain the optimal values of weights and biases in the neural network. Initial weights and biases were selected randomly between -0.5 and 0.5. The values of the learning rate and the momentum factors used in this calculation are 0.02 and 0.8, respectively. Iterative back propagation computation for the convergence of the estimated error stopped when the iteration number is equal to 2000. In this study, we defined the root mean square error (RMSE) between the estimated and actual responses as a criterion of the estimated accuracy. The values of RMSE are equal to the square root of the total error described in Eq.(3).

Fuzzy logic approach for the feedback control
Fuzzy control systems are characterized by the

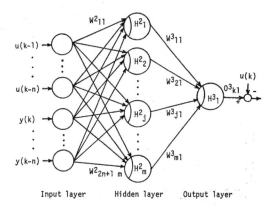

Fig. 4 Structure of neural network for inverse dynamic model construction.

linguistic statements which qualitatively expressed by "if-then" rules based on skilled operator's experience (Lee, 1990a). In this experiments, in order to achieve a flexibility, smoothness and robustness of the control performance, we utilized the fuzzy logic. These control performances using the fuzzy logic are completely characterized by membership functions. In general, it becomes a serious problem how the membership function are made up. In recent study, however, neural network approach for constructing membership functions is reported (Horikawa, 1991). In this study, the membership functions are determined by means of the method of trials and errors based on the skilled human operator's techniques. The control (input) variable, amount of adding a acid to the storage tank, was determined based on a fuzzy reasoning system which is constructed with two input variables, the error between desired and observed responses and the changing rate of the error (Morimoto, 1990). The method of fuzzy reasoning was based on the mamdani's minimum operation rule (Lee, 1990b).

RESULTS AND DISCUSSION

Behavior of solution pH in the deep hydroponics
Figure 5 shows time course changes in pH of the nutrient solution as affected by the addition of acid. From the upper to bottom figures, on-off operational pattern of acid supply, acid amount added to the storage tank, solution pH of the storage tank and solution pH of the culture vessel. The On-off operation of acid supply was performed based on a pseudo-random numbers (M-series). In this experiment, it is noted that the acid was added into the storage tank only, and the plants were not placed. As shown in the figure, significant nonlinear characteristics of solution pH are observed in both vessel. Furthermore, remarkable time delay properties of the pH response are also measured in the culture vessel part. This is caused by the recirculation of small amount of nutrient solution in the deep hydroponic culture system. Therefore, the reaction of solution pH of the storage tank is more smooth than that of the storage tank. If many plants were placed in the culture vessel, the reaction of solution pH may shows more complicated reactions.

Membership function for the fuzzy control
Figure 6 shows the membership functions of fuzzy sets. (a) is the cases of error with three sets: A_1=small , A_2=medium and A_3=big. (b) is the cases of changing rate of the error with three sets: B_1

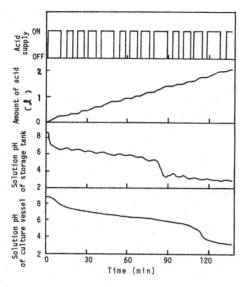

Fig.5 Time course changes of solution pH in storage tank and culture vessel as affected by acid addition followed by pseudo-random numbers.

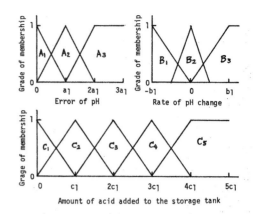

Fig. 6 Membership functions of fuzzy sets.
(a) Error of pH with three sets: A1=small, A2=medium, A3=big. (b) Rate of pH change with three sets: B1=negative big, B2= negative and positive small (zero), B3= positive big. (c) Amount of acid added into the storage tank with five sets: C1=small small (zero), C2=small, C3=small medium, C4=medium, C5=large.

= negative big, B2=negative or positive small and B3=positive big. (c) is the cases of amount of acid added in the tank with five sets: C1=small small (zero), C2=small, C3=small medium, C4=medium and C5=large. As mentioned above, these forms are tuned by the method of trials and errors so as to minimize the error between the desired and observed responses in the control performances. On the other hand, the fuzzy control rules have nine linguistic statements which are also determined by the methods of trials and errors using the membership functions based on the skilled operator's techniques. The fuzzy control for the feedback control was conducted according to this fuzzy control rules.

Control results based on fuzzy logic
First, we'll show the control results based on the fuzzy logic only. The control was carried out based on the membership functions shown in Fig. 6. Figures 7 and 8 show the control performances of solution pH by means of fuzzy logic. Figure 7 shows the pH responses of storage tank in the case when the pH was controlled by monitoring the storage tank's pH values only. On the other hand, the responses in Fig. 8 are the case when the pH was controlled by monitoring the culture vessel's pH values only. The solid lines and dotted lines are the cases of the fuzzy control and traditional control. The traditional control used in these experiments is the case when the fixed amount of acid was added to the storage tank at regular intervals. Comparing the both responses in the storage tank as shown in Fig. 7, more smooth and stable control trajectory than the method of traditional control is observed in the fuzzy control performance. On the other hand, as shown in Fig. 8, there was no major effect of fuzzy logic on the control performance. This is caused by the very slow reaction of solution pH in the culture vessel. In addition, the same control actions of acid supply as a skilled operator's technique were observed in both fuzzy control performances.

Inverse model construction by learning procedure
The inverse dynamic models used for the feedforward control were estimated by learning the

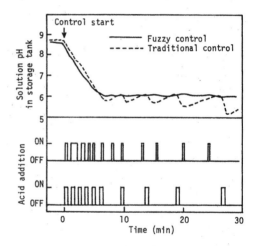

Fig. 7 Comparison of control performances of solution pH in the storage tank by means of fuzzy and traditional control.

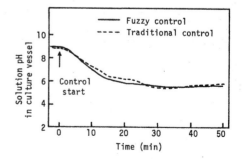

Fig. 8 Comparison of control performances of solution pH in the culture vessel by means of fuzzy and traditional control.

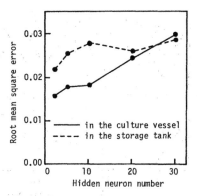

Fig. 9 Relation between hidden neuron number and root mean square error (RMSE).

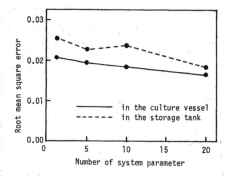

Fig. 10 Relation between system parameter number and root mean square error (RMSE)..

input (solution pH) and output (amount of acid added to the storage tank) data shown in Figs. 6, 7 and 8. Figure 9 shows the behavior of root mean square error (RMSE) as a function of neuron number in the hidden layer. The solid line represents the case when the solution pH of the storage tank was taken as the input variable, and the dotted line represents the case when that of the culture vessel was taken as the input variable. Iterative computation for the convergence of RMSE stopped when the iteration number is equal to 2000. In the case when the input variable is the solution pH of the storage tank, no significant differences in RMSE were observed over 10th neuron numbers. It is noted that the smallest RMSE value is observed when the neuron number is equal to 2. On the other hand, in the case of the culture vessel, the values of RMSE became larger as the neuron number increases. Figure 10 shows the behavior of RMSE as influenced by the number of system parameter (system order). As shown in the figure, slight reduction in RMSE was observed as the number of system parameter increases. This result is similar to the result shown in the previous paper (Morimoto, 1991). In this study, from a view of calculating time saving, optimal values of the number of the hidden neuron and the system parameter for the neural network calculation were selected 10th and 10th, respectively.

Control result based on fuzzy logic and neural network

Lastly, let's apply the fuzzy logic and neural network procedures to the control process of the solution pH. Figure 11 shows the control perform- ances based on the combination of fuzzy logic and neural network and the traditional control. The essential control action combining fuzzy logic and neural network was firstly given by the feedforward control based on the inverse model but after that time, it gradually changed to the combination of feedback control based on fuzzy logic and feedforward control. As shown in the figure, shorter deadbeat time was obtained by the new method than by the traditional control and the fuzzy logic only shown in Fig.7. It is clear that this faster response is caused by the feedforward control action. Furthermore, the control trajec- tory based on the combination of fuzzy logic and neural network is superior in the stability, smoothness and deadbeat time to that of traditional control.

CONCLUSION

In a deep hydroponic culture system, flexible and

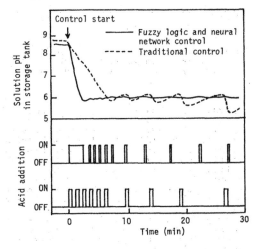

Fig. 11 Control performances of solution pH in the storage tank by means of the combination of fuzzy logic and neural network and the traditional control.

precise control of the solution pH is difficult because the dynamic properties of the solution pH as a function of acid or alkali show a significant nonlinear characteristics and time delay of circulating response. In this study, for the first step of the application of fuzzy logic and neural network procedures to hydroponic auto- mation, we applied these new methods to the pH control system in the deep hydroponic system. Applying the fuzzy logic to the feedback control part in the control system increased the smoothness and stability of control performance. Furthermore, applying the inverse dynamic model obtained by the neural network procedure to the feedforward part decreased the time required to reach the desired level. The satisfiable results of these two applications for the automatic control of solution pH were obtained in all experiments. This imply that flexible, smooth and safety automatic adjustment of solution pH would be able to realized with a computer control. It is concluded that the new control algorithm using the fuzzy logic and neural network would be a useful and practical way of adjusting the pH of the nutrient solution in a deep hydroponic culture. In addition, in order to confirm the better utility of the fuzzy logic and neural network procedures, we should apply these ones to more complicated control systems.

REFERENCES

Bangerth, F. (1979). Calcium-related physiological disorders of plants. Annual Review of Phytopathology. 17, 97-122.

Bhat, N.V., P.A. Minderman, Jr., T.M. McAvoy, and N.S. Wang. (1990). Modeling chemical process systems via neural computation. IEEE Control System Magazine. 10(3), 24-30.

Chen, F.C. (1990). Back-propagation neural networks for nonlinear self-tuning adaptive control. IEEE Control System Magazine. 10(3), 44-48.

Epstein, E. (1990). Roots : new ways to study their function in plant nutrition. In: Measurement techniques in plant science. (Hashimoto, Y., P. Kramer, H. Nonami and B. Strain, eds.), 291-318, Academic Press, Inc.

Hashimoto, Y. and T. Morimoto. (1985). Identification of water relations and CO_2 uptake in physiological ecological processes in a controlled environment (I) The time domain approach to CO_2 uptake. Proc. 7th IFAC Symposium on Identification and System Parameter Estimation. Vol.2, 1677-1681, Pergamon Press, Oxford.

Hatch, D.J. and R.J. Canaway. (1984). Control of pH in a system of flowing solution culture using a microcomputer. Journal of Experimental Botany. 35(161), 1860-1868.

Haynes, R.J. (1990). Active ion uptake and maintenance of cation-anion balance: A critical examination of their role in regulating rhizosphere pH. Plant and soil. 126, 247-264.

Horikawa, S., T. Furuhashi, S. Okuma and Y. Uchikawa. (1991). A leaning fuzzy controller using a neural network. Society of Instrument and Control Engineers. 27(2), 208-215, (in Japanese).

Kawato, M., K. Furukawa, and R. Suzuki. (1987). A hierarchical neural-network model for control and learning of voluntary movement. Biological Cybernetics. 57, 169-185.

Lee, C.C.(1990a). Fuzzy logic in control systems: Fuzzy logic controller - Part I. IEEE Transactions on Systems, Man, and Cybernetics. 20(2), 404-418.

Lee, C.C. (1990b). Fuzzy logic in control systems: Fuzzy logic controller, part II. IEEE Transactions on Systems, Man, and Cybernetics. 20(2), 419-435.

Morimoto, T., Y. Hashimoto and T. Fukuyama. (1985). Identification and control of hydroponic system in greenhouses. Proc. 7th IFAC Symposium on Identification and System Parameter Estimation. Vol.2, 1677-1681, Pergamon Press, Oxford.

Morimoto, T., T. Fukuyama, Y. Yi and Y. Hashimoto. (1988). Identification of physiological dynamics in hydroponics. Preprints of 8th IFAC Symposium on Identification and System Parameter Estimation. Vol.3, 1736-1741, Pergamon Press.

Morimoto, T. (1990). Application of fuzzy logic to the process control of pH in deep hydroponic system. Environ. Control in Biol.. 28(4), 177-178, (in Japanese).

Morimoto, T., I. Cho and Y. Hashimoto. (1991). Identification of hydroponics in an advanced control system of the greenhouse. Preprints of 9th IFAC Symposium on Identification and System Parameter Estimation (in press).

Mugwire, L.M. and S.U. Patel. (1977). Root zone pH changes and ion uptake imbalance by triticale, wheat, and rye. Agronomy Journal. 69, 719-722.

Ponnamperuma, F.N., E. Martinez, and T. Loy. (1966). Influence of redox potential and partial pressure of carbon dioxide on pH values and the suspension effect of flooded soils. Soil Science. 101(6), 421-431.

Psaltis, D., A. Sideris, and A.A. Yamamoto. (1988). A multilayered neural network controller. IEEE Control System Magazine. 8(2), 17-21.

Rumelhart, D.E., G.E. Hinton and R.J. Williams. (1986). Learning representation by back-propagation error. Nature. 323(9), 533-536.

Wild, A., L.H.P. Jones, and J.H. Macduff. (1987). Uptake of mineral nutrients and crop growth: The use of flowing nutrient solutions. Advances in Agronomy. Vol.41, 171-219.

Copyright © IFAC Mathematical and Control Applications
in Agriculture and Horticulture, Matsuyama, Japan 1991

ROOT TEMPERATURE EFFECT ON HYDRAULIC CHARACTERISTICS OF ROOTS IN HYDROPONICS

S. Yoshida and H. Eguchi

Biotron Institute, Kyushu University 12, Fukuoka 812, Japan

ABSTRACT. Water uptake and gas exchange in cucumber roots (*Cucumis sativus* L. cv Chojitsu-Ochiai) were examined in a root temperature control system of air-tightened hydroponics. The effect of root temperature on water uptake rate was found in a sigmoidal pattern: The water uptake rate was reduced at root temperatures lower than 12°C. The responses of O_2 decrease rate and CO_2 increase rate to the root temperatures appeared almost parallel to the pattern of water uptake rate. Furthermore, the effect of root temperature on total root resistance was examined in a detached root system by applying suction to the proximal end. The total root resistance became higher at lower root temperatures. Thus, it could be suggested that at the lower root temperatures, the increase in root hydraulic resistance in relation to the inhibited root respiration results in the decrease in water uptake in intact plants. On the other hand, the total root resistance at lower root temperatures in figleaf gourd plant (*Cucurbita ficifolia* B.) which is more tolerant of low temperature were kept higher as compared with that of cucumber plant. This fact indicates that low temperature tolerance in plant growth closely relates to root hydraulic characteristics at the lower root temperatures.

Keywords. Root hydraulic resistance; root gas exchange; temperature control; agriculture; biology; environment control; on-line operation; sensors; vacuum control.

INTRODUCTION

Rhizosphere environment is responsible for plant water relations which are essential for growth and production in plants. It is well known that lower soil temperatures reduce water absorption in roots and induce a water deficit in shoots (Kramer, 1940). Kramer (1942) has found the species specific differences in the water absorption by intact plants responding to lower soil temperatures between chilling sensitive plants and chilling tolerant plants, and Eguchi and Koutaki (1986) have reported that the transpiration rate in a cucumber plant grafted on a figleaf gourd stock is kept higher even in the case that the transpiration rate in an intact cucumber plant is reduced at low soil temperatures of 10°C. These facts indicate that the soil temperatures affect the water relations in whole plants through hydraulic properties of roots.

On the other hand, Lawrence and Oechel (1983) and Szaniawski and Kielkiewicz (1982) have reported the inhibition of root respiration at lower root temperatures, and Everard and Drew (1987) have found that water transport in roots is sensitive to oxygen deficiency in hydroponic solution. It is further necessary to examine the relationships among root water uptake, root gas exchange and root temperature, aiming at better understanding of root temperature effect on hydraulic characteristics in plants.

The present paper deals with analysis of the root temperature effects on root water uptake, root gas exchange and root hydraulic resistance in cucumber plant by on-line measurements in root temperature control system of hydroponics.

MATERIALS AND METHODS

Control and Measurement System

Figure 1 shows the schematic diagram of root temperature control system provided with measurements of water uptake and gas exchange in intact roots in air-tightened hydroponics. Root temperature in a stainless steel pot (3.7 ℓ) was controlled from 8 to 32(±0.1)°C by a water bath method: The pot was placed in the water bath where temperature-controlled water was circulated, and the upper surface of the pot was covered with heat insulating materials of styrofoam. The pot was filled with aerated nutrient solution which was slowly stirred.

Water uptake rate in a root system was measured automatically by a potometer, where the decrease of the solution was detected by a float connected to a potentiometer. (The solution surface in the potometer was sealed with paraffin liquid layer to prevent diffusion of air into the solution). A polarographic O_2-meter (UD-1, Central Kagaku Co., Ltd.), a potentiometric electrode CO_2-meter (CGP-1, Toa Electronics Ltd.), and pH-meter (HM-7E, Toa Electronics Ltd.) were employed for on-line measurements of dissolved O_2 and CO_2 concentrations. The respective sensor signals were transmitted to CPU through interfaces.

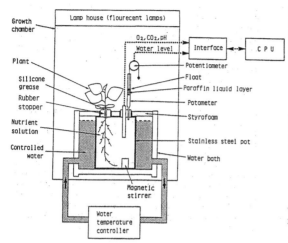

Fig. 1. Schematic diagram of a root temperature control system provided with the measurements of water uptake and gas exchange in an intact plant.

A material plant was transplanted to the root temperature control system in a growth chamber (Matsui and colleagues, 1971). The water uptake rate and the gas exchange in roots were measured for 4 days at an air temperature of 25°C and a relative humidity of 40% and continuous light with a intensity of 200 $\mu mol\ m^{-2}s^{-1}$ (fluorescent lamps; FLR110-EHW/A, Toshiba Corp.).

Measurement of CO_2 Concentration

Dissolved CO_2 is composed of unionized CO_2 (CO_2 and H_2CO_3) and ionized CO_2 (HCO_3^- and CO_3^{2-}) which are in equilibrium, as described by Helder (1988).
The molarity of total inorganic carbon (ΣCO_2) can be obtained by summing the molarities of unionized CO_2 and ionized CO_2 as follows,

$$[\Sigma CO_2] = [CO_2 + H_2CO_3] + [HCO_3^-] + [CO_3^{2-}] \tag{1}$$

The molarity ($[CO_2 + H_2CO_3]$) of unionized CO_2 was directly measured by the CO_2-meter, and the molarity ($[HCO_3^-] + [CO_3^{2-}]$) of ionized CO_2 was evaluated by using measured $[CO_2 + H_2CO_3]$ and pH from Henderson-Hasselbach equations:

$$\log[HCO_3^-] = \log[CO_2 + H_2CO_3] + pH + \log K_{a1} \tag{2}$$

$$\log[CO_3^{2-}] = \log[HCO_3^-] + pH + \log K_{a2} \tag{3}$$

where K_{a1} and K_{a2} are equilibrium constants for the ionization of dissolved CO_2 in the solution, and can be calculated at a given temperature of T(K) by Eqs. (4) and (5):

$$\ln K_{a1} = -14554.21T^{-1} + 290.9097 - 45.0575 \ln T \tag{4}$$

$$\ln K_{a2} = -11843.79T^{-1} + 207.6548 - 33.6485 \ln T \tag{5}$$

Thus, the CO_2 concentration was evaluated as the molarity of total inorganic carbon.

Measurement of Root Hydraulic Resistance

Measurement system. Figure 2 shows the schematic diagram of root temperature control system provided with the measurements of gas exchange and hydraulic resistance in a detached root system. For the root temperature control and the on-line measurements, the instruments shown in Fig. 1 were also employed. The healthy root system of a material plant was detached at the stem base in $CaCl_2$ solution of 20 mmol ℓ^{-1}. The proximal end (cut end of the root system) was connected to a glass tube, and the tube was sealed with silicone rubber. The root system was set in the pot, and the pot was air-tightened with rubber stoppers and silicone grease. The glass tube was connected to a digital vacuum controller (VC-20S, Okano Works Ltd.) with a vacuum pump through a water trap and a buffering tank. The pressure at the proximal end was controlled at an external atmospheric pressure minus 80 kPa, so that the suction of 80 kPa was applied to the root system.

Calculation of total root resistance. Hydraulic resistance in a root system was defined as total root resistance (total hydraulic resistance per

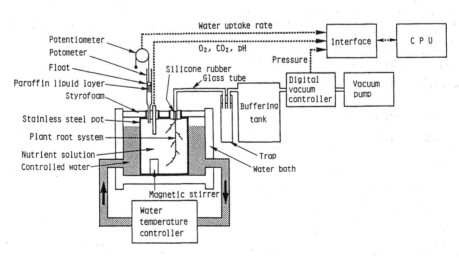

Fig. 2. Schematic diagram of a root temperature control system provided with the measurements of hydraulic resistance and gas exchange in a detached root system where suction is applied to the proximal end.

whole root system) in this experiment. The total root resistance can be expressed by

$$R = \frac{\Psi_1 - \Psi_2}{F} \qquad (6)$$

where R is the total root resistance, Ψ_1 is the water potential of the nutrient solution, Ψ_2 is the water potential at the proximal end of the root system, and F is the water uptake rate. Ψ_1 is composed of the external atmospheric pressure (Ψ_{p1}) and the osmotic potential of the nutrient solution ($\Psi_{\kappa1}$), and Ψ_2 is composed of the pressure at the proximal end (Ψ_{p2}) and the osmotic potential of the xylem sap ($\Psi_{\kappa2}$). Therefore, R can be rewritten by using the pressure potential difference ($\Delta\Psi_p$) and the osmotic potential difference ($\Delta\Psi_\kappa$) as follows,

$$R = \frac{(\Psi_{p1} + \Psi_{\kappa1}) - (\Psi_{p2} + \Psi_{\kappa2})}{F} = \frac{\Delta\Psi_p + \Delta\Psi_\kappa}{F} \qquad (7)$$

For evaluation of osmotic potential, xylem sap exuded from the root system was sampled, and the osmotic potential was measured by the method of Campbell and colleagues (1973): The sample was placed in a sample chamber of a thermocouple psychrometer (HR-33T, Wescor Inc.). The measurement was performed after 30 min of water vapor equilibrium in the sample chamber. The output of the thermocouple was converted to osmotic potential by using NaCl standards of known osmotic potentials. The osmotic potential of the nutrient solution was also measured in the same manner.

Thus, measured osmotic potentials of the nutrient solution ($\Psi_{\kappa1}$ =80 kPa) and of xylem sap ($\Psi_{\kappa2}$ =100±10 kPa) were obtained. Therefore, the osmotic potential difference ($\Delta\Psi_\kappa$) was 20 kPa. So that, the total root resistance was calculated from Eq. (7) by using the water potential difference ($\Delta\Psi_p + \Delta\Psi_\kappa$) of 100 kPa and measured water uptake rate (F).

Plant Materials

Cucumber plants (*Cucumis sativus* L. cv Chojitsu-Ochiai) and figleaf gourd plants (*Cucurbita ficifolia* B. cv Kurodane) were used. The plants were grown in fully aerated hydroponics at an air temperature of 23℃ and a relative humidity of 70%. The whole root system of the 3 leaf stage plant of healthy growth was used as a specimen. The dry weight of the whole root system used was 0.12±0.03 g in cucumber plant and 0.14±0.03 g in figleaf gourd plant (there was no significant difference between those two species at 5% level).

RESULTS AND DISCUSSION

Figure 3 shows examples of the time course patterns of water uptake rate in roots and dissolved O_2 and CO_2 concentrations in nutrient solution at respective root temperatures of 12, 20 and 28℃. Water uptake rate at the root temperature of 12℃ was kept remarkably lower than those at the higher root temperatures: The water uptake rate at the root temperatures of 20 and 28℃ became higher during 30 h after the start of the measurements and gradually decreased (Fig. 3a). The O_2 concentration, which was initially 0.25±0.01 mmol ℓ^{-1}, decreased to 0.01 mmol ℓ^{-1} in 38 h at 12℃, in 18 h at 20℃ and in 8 h at 28℃ (Fig. 3b). The velocity of the O_2 decrease was enhanced with the higher root temperatures. On

the other hand, the initial CO_2 concentration was 0.06±0.01 mmol ℓ^{-1}. At the root temperature of 28℃, the CO_2 concentration reached to about 3.1 mmol ℓ^{-1} in 80 h, and thereafter it was kept steady-state. At the root temperatures of 12 and 20℃, the CO_2 concentration continued to increase in the time course of 96 h (Fig. 3c). The velocity of the CO_2 increase was enhanced with higher root temperatures as well as the velocity of the O_2 decrease. Thus, CO_2 concentration continued to increase even after the time when O_2 concentration decreased to 0.01 mmol ℓ^{-1}. From these results, it became clear that the CO_2 release by roots can be caused even in the O_2 deficit solution. So, it was difficult to calculate the root respiratory quotient from the balance of O_2 uptake and CO_2 release in the nutrient solution. Thus, it was obvious that both gas exchange and water uptake in roots are inhibited with lower root temperatures and are enhanced with higher root temperatures (Yoshida and Eguchi, 1989).

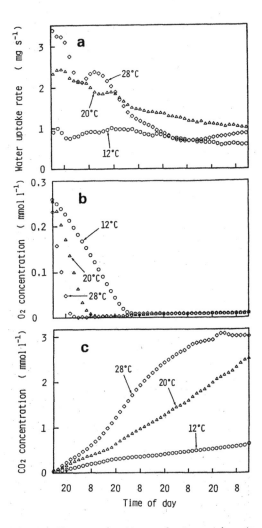

Fig. 3. Time course patterns of water uptake rate in intact roots (a) and O_2 (b) and CO_2 (c) concentrations in nutrient solution at respective root temperatures of 12, 20 and 28℃.

To examine root temperature effects on water uptake and gas exchange in an intact root system of cucumber plant, water uptake rate and O_2 decrease rate and CO_2 increase rate were evaluated by using the mean of the measured values in 4 days. Figure 4 shows distributions of the water uptake rate (a), the O_2 decrease rate (b) and the CO_2 increase rate (c) on the root temperatures in the temperature region of 8 to 28℃. Their response curves appeared almost parallel with each other in the sigmoidal pattern, where the response rates decreased at root temperatures lower than 12℃ and increased at root temperatures higher than 16℃. These patterns agreed with the results which have been obtained in water uptake rate in response to root temperatures (Markhart and colleagues, 1979; Running and Reid, 1980). Thus, both of the water uptake and the gas exchange in intact roots clearly responded to the root temperatures (Yoshida and Eguchi, 1989).

Boyer (1971) has reported that the hydraulic resistance in a root system is larger than 50% of that in a whole plant. In order to analyze the root temperature effect on water uptake rate in intact root systems, hydraulic resistance in a whole root system in response to root temperatures was examined. In general, hydraulic conductance and resistance in roots have been calculated in detached roots under hydrostatic pressure of 200 to 500 kPa applied to external root medium in a pressure chamber, in order to obtain the amount of water flow in intact roots of a transpiring plant (Fiscus, 1975, 1977). However, the high pressure may influence the hydraulic properties of roots grown at normal pressure (Koide, 1985; Nobel, Schulte and North, 1990; Salim and Pitman, 1984a 1984b). In this experiment, the suction of 80 kPa, which was applied to the proximal end of the detached root system, was used for the measurement of the total root resistance.

Figure 5 shows an example of the time course patterns of O_2 and CO_2 concentrations and water uptake rate in the cucumber root system, where the pressure was applied to the proximal end at the root temperature of 24℃. At the start of the suction ($t=0$ min), O_2 concentration was 0.24 ± 0.02 mmol ℓ^{-1} and CO_2 concentration was 0.02 ± 0.003 mmol ℓ^{-1}. During 120 min, O_2 and CO_2 concentrations became 0.17 mmol ℓ^{-1} and 0.04 mmol ℓ^{-1}, respectively, even when the concentrations changed most rapidly at the highest root temperature of 32℃. These small changes in O_2 and CO_2 concentrations did not influence on the water uptake (Yoshida and Eguchi, 1988). Just after the start of suction ($t=0$ to 15 min), the water uptake rate rapidly increased, and thereafter became steady-state. Therefore, the means of the measured values from 30 to 90 min were used as F in Eq. (7) for the evaluation of the total root resistance.

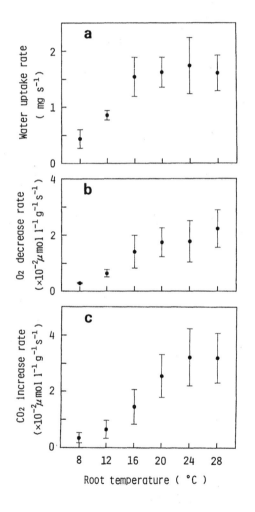

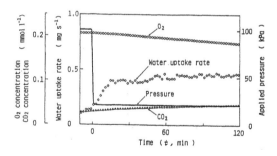

Fig. 5. Time course patterns of water uptake rate ($\diamondsuit$) in a cucumber root system at the pressure (—) applied to the proximal end of the root systems, O_2 concentration ($\bigcirc$), and CO_2 concentration ($\triangle$) in nutrient solution at 24℃.

Fig. 4. Distributions of water uptake rate (a), O_2 decrease rate (b) and CO_2 increase rate (c) on root temperatures, where the means of values measured in 4 plants are plotted with 95% confidence intervals.

Figure 6 shows the distributions of water uptake rates (a), O_2 decrease rates (b), and CO_2 increase rates (c) on root temperature in cucumber and figleaf gourd root systems. Water uptake rate was clearly reduced at lower root temperatures in those two species. The O_2 decrease rate and the CO_2 increase rate were also reduced at lower root temperatures in those two species. This result agreed with the characteristics observed in the root gas exchange of intact plants (Fig. 4).

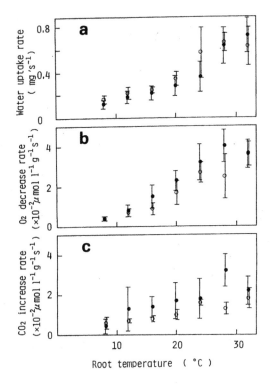

Fig. 6. Distributions of water uptake rates (a), O_2 decrease rates (b) and CO_2 increase rates (c) on root temperature in cucumber (●) and figleaf gourd (○) root systems under the suction of 80 kPa, where the means of values measured in 6 plants are plotted with 95% confidence intervals.

Figure 7 shows distributions of the total root resistances on root temperature. The total root resistances distributed in a region of about 0.2×10^9 to about 1.0×10^9 MPa s m^{-3} in those two species: The total root resistances became higher at lower root temperatures and reduced to about 0.2×10^9 MPa s m^{-3} at higher root temperatures. This fact suggests that decrease in water uptake at lower root temperatures is due to increase in the total root resistance at lower root temperatures (Yoshida and Eguchi, 1990).

The total root resistance consists of the radial resistance from the external solution to the root xylem and the axial resistance in the root xylem (Frensch and Steudle, 1989). Nagano and Ishida (1984) have reported that the radial resistance (absorptive resistance) constitutes 80-85% of the total root resistance in a soybean root system. Therefore, the total root resistance was influenced by the root temperature mainly through the radial resistance, and it could be conceivable that the temperature effect on total root resistance is associated with the permeability in the root cell membrane in response to the root temperature.

Furthermore, species specific difference in total root resistance was found between those two species at lower root temperatures: At the root temperature of 8°C, the total root resistance in the cucumber root system was higher than that in the figleaf gourd root system, and the difference in total root resistance between those two species was significant at 5% level. At higher root temperatures, however, the total root resistances were kept lower, and any appreciable differences were not found between those two species. In general, the cucumber plant grafted on the stock of the figleaf gourd plant has been used for cucumber production, because this grafted plant becomes relatively tolerant of lower soil

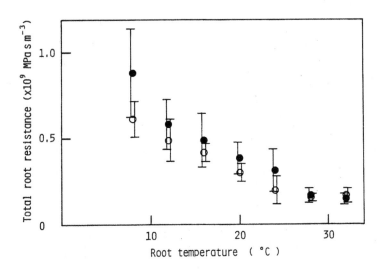

Fig. 7. Distributions of total root resistances on root temperature in cucumber (●) and figleaf gourd (○) root systems at the water potential difference of 100 kPa, where the means of values measured in 6 plants of respective species are plotted with 95% confidence intervals.

temperatures in the growing process (Tachibana, 1982). Eguchi and Koutaki (1986) have found the leaf transpiration in figleaf gourd and the grafted cucumber plants is more active than that in cucumber plant at the root temperatures lower than 10°C. From these facts, the low temperature tolerance in figleaf gourd plant could be considered to be brought by lower hydraulic resistance in a root system at lower root temperatures.

CONCLUSIONS

Newly developed root temperature control systems provided with instrumentations of root water uptake, gas exchange and hydraulic resistance in air-tightened hydroponics were reliably useful for exact analyses of temperature effects on the hydraulic characteristics and the respiration in roots. In this system, it became clear that at lower root temperatures, water uptake and gas exchange in the intact roots decrease, and hydraulic resistance in the detached roots increases. These facts suggest that root hydraulic resistance closely relates to root activity as observed in root gas exchange, and the higher hydraulic resistance at the lower root temperatures results in decrease in water uptake in roots.

From species specific difference in root hydraulic resistance between cucumber and figleaf gourd plants, it could be conceivable that lower hydraulic resistance at lower root temperatures is responsible for low temperature tolerance in plant growth.

REFERENCES

Boyer J. S. (1971) Resistances to water transport in soybean, bean, and sunflower. *Crop Sci.*, 11, 403-407.

Campbell E. C., Campbell G. S. and Barlow W. K. (1973) A dewpoint hygrometer for water potential measurement. *Agric. Meteorol.*, 12, 113-121.

Eguchi H. and Koutaki M. (1986) Analysis of soil temperature effect on transpiration by leaf heat balance in cucumber, cucurbit and their grafted plants. *Biotronics* 15, 45-54.

Everard J. D. and Drew M. C. (1987) Mechanisms of inhibition of water movement in anaerobically treated roots of *Zea mays* L. *J. Exp. Bot.*, 38, 1154-1165.

Fiscus E. L. (1975) The interaction between osmotic- and pressure-induced water flow in plant roots. *Plant Physiol.* 55, 917-922.

Fiscus E. L. (1977) Determination of hydraulic and osmotic properties of soybean root systems. *Plant Physiol.*, 59, 1013-1020.

Frensch J. and Steudle E. (1989) Axial and radial hydraulic resistance to roots of maize (*Zea mays* L.). *Plant Physiol.*, 91, 719-726.

Helder R. J. (1988) A quantitative approach to the inorganic carbon system in aqueous media used in biological research: dilute solutions isolated from the atmosphere. *Plant, Cell Environ.*, 11, 211-230.

Koide R. (1985) The nature and location of variable hydraulic resistance in *Helianthus annuus* L. (sunflower). *J. Exp. Bot.*, 36, 1430-1440.

Kramer P. J. (1940) Root resistance as a cause of decreased water absorption by plants at low temperatures. *Plant Physiol.*, 15, 63-79.

Kramer P. J. (1942) Species differences with respect to water absorption at low soil temperatures. *Am. J. Bot.*, 29, 828-832.

Kramer P. J. (1983) Factors affecting the absorption of water. In *Water Relations of Plants.*, Academic Press, New York. pp. 235-261.

Lawrence W. T. and Oechel W. C. (1983) Effects of soil temperature on the carbon exchange of taiga seedlings. I. Root respiration. *Can. J. For. Res.*, 13, 840-849.

Markhart A. H., Fiscus E. L., Naylor A. W. and Kramer P. J. (1979) Effect of temperature on water and ion transport in soybean and broccoli systems. *Plant Physiol.*, 64, 83-87.

Matsui T., Eguchi H., Hanami Y., Handa S. and Terajima T. (1971) A growth cabinet for the study on biotronics. I. Design and performance. *Environ. Control in Biol.*, 9, 37-46.

Nagano T. and Ishida T. (1984) Internal Plant-water-status and its control. III. The root resistance to water movement through the soybean root. *J. Agr. Met.*, 40, 229-233. (In Japanese with English summary)

Nobel P. S., Schulte P. J., and North G. B. (1990) Water influx characteristics and hydraulic conductivity for roots of *Agave deserti* Engelm. *J. Exp. Bot.*, 41, 409-415.

Running S. W. and Reid C. P. (1980) Soil temperature influences on root resistance of *Pinus contorta* seedlings. *Plant Physiol.*, 65, 635-640.

Szaniawski R. K. and Kielkiewicz M. (1982) Maintenance and growth respiration in shoots and roots of sunflower plants grown at different root temperatures. *Physiol. Plant.*, 54, 500-504.

Salim M. and Pitman M. G. (1984a) Pressure-induced water and solute flow through plant roots. *J. Exp. Bot.*, 35, 869-881.

Salim M. and Pitman M. G. (1984b) Water and solute flow through mung bean roots under applied pressure. *Physiol. Plant.*, 61, 263-270.

Tachibana S. (1982) Comparison of effects of root temperature on the growth and mineral nutrition of cucumber cultivars and figleaf gourd. *J. Japan. Soc. Hort. Sci.*, 51, 299-308.

Yoshida S. and Eguchi H. (1988) Relationship between gas exchanges in intact roots and water uptake in response to leaf transpiration in hydroponics. *Biotronics*, 17, 59-68.

Yoshida S. and Eguchi H. (1989) Effect of root temperature on gas exchange and water uptake in intact roots of cucumber plants (*Cucumis sativus* L.) in hydroponics. *Biotronics*, 18, 15-21.

Yoshida S. and Eguchi H. (1990) Root temperature effect on root hydraulic resistance in cucumber (*Cucumis sativus* L.) and figleaf gourd (*Cucurbita ficifolia* B.) plants. *Biotronics*, 19, 121-127.

DEVELOPMENT OF TS-STYLE PLANT FACTORY

S. Akagi*, M. Kiyosawa* and T. Kitamoto**

*Laboratory of Horticulture, QP Corporation, Tokyo, Japan
**Research and Development Dept., Bridgestone Corporation, Tokyo, Japan

ABSTRACT

It is a dream of everyone that constant quantities of plants with constant quality
are produced through the year, without being affected by season or weather. This is
essential for a stable supply of food.
With improvements in technology, various agricultural methods have been
developed and recently those with environmental control are becoming popular.
Especially plant/vegetable factory systems, a vegetable production system by
operating an optimal environment for plant growing, are expected to be used in
future agriculture. Hereunder, the theory and practice of the first commercialized
plant/vegetable factory systems "TS Farm" is outlined.

KEYWORDS

Agriculture; Biotechnology; Environmental control; Industrial production system;
Optimal condition; Pesticide-free; Quality control; Stability

INTRODUCTION

To establish a stable system for vegetable
production without pesticide was a
common dream of human beings.
Technology of agriculture is developed
according to an improvement of culturing
system such as open field culturing to
covered culturing and to hydroculturing.
Plant/vegetable factory is most advanced
technology of this field equipped with
environmental control including artificial
light in order to realize stable production.
Research on plant factory has been started
in Europe 1960's, in USA 1970's, and in
Japan 1980's. Although some systems are
installed for commercial production, most
of them were not successful due to its
profitability.
The presenters believe that most important
in establish plant/vegetable factory system
as an indus:ry is its productivity; and
started research from 1983 aiming at high
quality of final products, suitable farming
methods, proper hardware structure,
optimal environmental control, and
minimizing production costs. A slope type
vegetable factory, "TS Farm" was
announced in 1986, then sale of the system
started from 1988.
The system adopts airoponic culture using
artificial light, which is three-dimensional
culture in a triangular prism formed by
using panels joined at the top. This system
enhances productivity and reduces the
energy-costs with the TS Farm allowing
commercial production. Regarding the
name of "TS Farm", T stands for triangle
panels as a feature of the system, and S,

spray-misting adopted for supplying the
culture solution.
This report describes the structure,
performance, etc. of the TS Farm in
reference to the culture of butter-head
lettuce.
Furthermore, research has been conducted
on the production of vegetables, herbs,
edible flowers, etc. using the system.

STRUCTURE OF SYSTEM

The basic structure of the TS Farm is
comprised of modular rectangular panels
which form a series of three-sided pyramids
together with a nutrition-supplying sprayer
on the back side of the panels as shown in
Fig. 1.

Fig. 1. Structure of TS Farm System

1. Materials of the Structure

The main materials used for seedling
rearing and growing rooms are GRP
(glass-fiber reinforced plastic) panels which

are most suitable as structural material because of freedom from corrosion deterioration, low- heat conductivity and sufficient strength.

Furthermore, since the panels are provided as modules, they can be simply assembled.

2. Rearing Board

Rearing boards made of polystyrene formed in triangular prisms are arranged in the rearing room. Each of the boards is 900 mm x 1,800 mm and is wheeled. During transplanting, fix-planting and harvesting, they are removed, and during culture, they are moved on a line according to schedule. Plants are plugged in holes of the board with the roots hanging down from the board exposed to air. Then the roots can be absorb oxygen freely together with nutrient.

3. Nutrition Supply Sprayer

Under the rearing boards, i.e., in the underground portion, the culture-solution is sprayed on the roots by pump. The temperature of the culture-solution is kept at 2 to 3°C lower than the atmospheric temperature of the leaves and stems without being controlled. The culture-solution is circulated, and the fertilizer concentrations and pH are automatically adjusted.

4. Environmental Control

A computerized environment control system controls the air temperature, humidity, carbonic acid gas concentration and artificial sunshine. (Fig. 2)The light is a specially developed high-pressure sodium lamp to provide light equally on V-shaped rearing boards.

All environmental conditions can be monitored so that when something happens minimum countermeasures can be taken.

Accordingly, the rearing room and nursery is isolated from the outside. Although it is recommendable to have an outer structure basically, it is not required to be installed in another building or warehouse, and can be used in any existing prefabricated house.

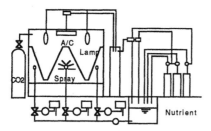

Fig. 2. Environmental Control System

CULTURING PROCEDURE

An outline of the culturing of butter head lettuce is shown in Fig. 3.

1. Germination Stage

Sterilized seeds of a special variety for the TS Farm are sown into 23 mm square foamed polyurethane cubes, and one day after the forcing of germination in darkness, the seedlings are reared under fluorescent lamps. To maintain high productivity, to obtain high germination and equal quality seed is important and it makes saving of energy cost.

2. Seedling Stage

After germination, the seedlings are grown for 7 days by the flat-flooding method. (i.e. seedling stage 1). Then, the seedlings are transplanted onto rearing boards (holes spaced 67mm x 67mm) and cultured under high pressure sodium lamps and sprayed for 12 days (i.e. seedling stage 2).

3. Rearing Stage to Harvesting

A total of 20 days after sowing, the seedlings are planted on rearing boards (holes spaced 138mm x 142mm and 84 holes per board), and on the 12th day after the fix-planting, they are harvested.

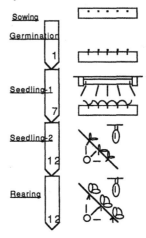

Fig. 3. Culturing Procedure
(Numbers indicate periods for each stage.)

The culture environmental condition is shown in Table 1. The composition of the culture- solution is specially mixed by the presenters.

TABLE 1	Environmental Conditions		
	S-1*	S-2*	Rearing
Daylength(Hr)	18	16	18
Illuminance(μmol/Sm^2)	105	225	250
Temperature(°C)	20±3	21±3	20±3
Humidity(%)	85	85	80
CO2(μl/l)	800	800	1000
Nutrition EC(S/m)	0.15	0.20	0.20
Nutrition pH	5.5-6.3	5.5-6.3	5.5-6.3

S-1*: Seedling Stage 1
S-2*: Seedling Stage 2

EVALUATION OF FINAL PRODUCT

1. Method of Evaluation

After harvesting, the number of leaves, fresh and dry-matter weight per board were measured. The leaf color was measured at the central portion of the largest leaf using the color difference meter, CR-200 (Minolta).

For analysis of ingredients, beta-carotene was determined by petroleum ether extraction, and vitamins, by thiochrome lumiflavin fluorophotometry.

For the quantitative analysis of bacteria, 10g was taken from the external leaf portions and central leaf portions of five samples, and subjected to a stomacker using sterilized salt solution. The solution obtained was measured as to the general viable cell number, the number of coliform group, Enterococcus, and Escherichiacoli.

2. Results and discussion

The results from evaluating the final product are shown in Table 2 and 3. The number of leaves was 18.9 , and the fresh weight of the aerial part was 75.1g with dry-weight of 2.86g. The system used for the tests has been operating since 1986. Since the beginning of the system operation, the vegetables produced here have been delivered to markets. As a result of market research, it was found that in the case of the butter-head type, fresh weights of 60 g or more seem to be accepted in the market. This shows that more than 90% of the products can be delivered as suitable. With regard to the size, it is to be noted that the butterhead lettuce in Japan is harvested earlier before head-formation unlike the practice in Europe to harvest after head-formation.

In the quantitative analysis of bacteria, both of the coliform groups and Enterococcus were negative (Table 2)while conventional products show 10$^{3-5}$ and 10 respectively. This is attributable to the system being perfectly enclosed without using soil at all, even though it has been continuously operated for 5 years and the culture-management has been intensive.

TABLE 2	Analysis of Bacteria	
	Outer Leaf	Core Part
SPC(#/g)	2.4x10^3	5.2x10^2
Coliform group(#/g)	<10	<10
Enterococcus(MPN/g)	<3	<3
Escherichia Coli	<3	<3

To compare results of ingredient analysis with the values in the Japanese Standard Food Ingredient Table (4th edition issued in 1990), although there were a little difference (Table 3), common products also show more differences according to area, season and variety and the results indicate the TS Farm- made product is equivalent to common products. The products of the system are being delivered still today (5 items), and are favorably accepted by consumer in color, shape and taste together with feature of pesticide-free.

TABLE 3 Ingredients Analysis		
	TS Farm	ISFIT*
Protein (g)	1.4	1.5
Fat (g)	0.2	0.2
Carbohydrate		
Fiber (g)	0.3	0.4
Sugar (g)	2.4	1.6
Ash (g)	0.6	0.9
Minerals		
Calcium (mg)	49	50
Phosphorus (mg)	38	44
Iron (mg)	0.7	2.2
Natrium (mg)	5	5
Pottasium (mg)	222	370
Vitamin		
Retinol potency (IU)	1520	780
Thiamine (mg)	0.03	0.05
Riboflavin (mg)	0.10	0.12
Ascorbic acid (mg)	16	13
(Ingredient numbers per 100g)		

ISFIT* :Japanese Standard Food Ingredient Table (4th edition issued in 1990)

PRESENT CONDITIONS

1. Plants which can be cultured

Plants which can be cultured by the TS Farm are listed in Table 4, and they have been confirmed to allow mass production. Other vegetables than these can also be produced, but in view of profitability, leafy vegetables high in growth-rate seem more suitable. Vegetables with head formation are slow in growth and high in the rate of non-edible portion among vegetable fruits, and so not very profitable.

The consumption of herbs has grown rapidly of late,but selling prices are high. In ordinary culture, it is difficult to always obtain high quality herbs due to influence of air temperature, daylength, etc., but this can be achieved by the TS Farm. Edible flowers also attracting attention recently. They are used for salads and toppings, and those cultured without pesticides are established as an independent commodity.

TABLE 4 Cultured variety
Leaf Vegetables:
Butterhead Lettuce, Leaf Lettuce, Garland
Chrysanthemum, Water Cress, Celery,
Japanese Honewort, Welsh Onion, Spring
Onion, Senpousai
Herbs:
Marjoram, Coriander,Thyme, Tarragon,
Chervil, Basil, Peppermint, Lemon Balm,
Rosemary, Chamomile, Borage
Edible Flowers:
Viola,Pansy, Dianthus, Cornflower,
Impatiens, Linaria, Snapdragon, Marigold,
Verbena, Lobelia

2. Supply Record

In practical application of the TS Farm,
seven systems were delivered in Japan up
to now (1990), including those for testing
and research (Table 5 shows that for
commercial production only).
The commercial production facilities using
the TS Farm are now producing several
species of leafy vegetables including lettuce,
herbs and edible flowers.

TABLE 5 Commercial Systems
Model	Location	Size	Products
TS-240	Osaka	65m^2	Lettuce
TS-600	Shizuoka	203m^2	Lettuce,Herbs
TS-1250	Kobe	320m^2	Lettuce,Herbs

(Model Number indicates approx quantity
daily produced in head of lettuce)

3. Production Costs

Production costs from the TS Farm system
cannot be simply stated, since it is
influenced by the kind of the product, the
scale of the system, depreciation method,
labor cost, electric power charge, etc.
However, an ideal production cost from
TS-1250, a standard production system, can
be calculated as shown in Table 6.

TABLE 6 Typical Production costs
Electricity	19.8	(22%)
Labor	17.0	(19%)
Consumption	12.0	(13%)
Depreciation	42.0	(46%)
Total	90.8	

(Numbers in Japanese Yen per head of
Butterhead lettuce in case of model TS-1250)

Comparing the productivity and electricity
consumption above mentioned with
conventional systems are shown in Table 7.

TABLE 7 Comparison of Production Costs
	TS Farm	System-A	Sys-B	Sys-C
Space(m^2)	390	430	333	250
Production	1344	1188	664	464
Productivity	3.5	2.8	2.0	1.9
Elec. Cost	19.8	46.1	31.6	35.4

Note: System-A: Gutter Culture
       System-B: NFT
       System-C: Flat Flood

CONCLUSION

Feature of the TS Farm
The features of the TS Farm described
above can be summarized as follows:

1.   Clean vegetables using no pesticides:
Since the system is perfectly enclosed,
culture without pesticides can be executed,
and clean vegetables with a minimum of
bacteria can be obtained. Less bacteria mean
an advantage in product distribution since
the deterioration of the products can be
retarded.

2.   Constant-quantity production: Since the
system is perfectly controlled, plants with
the same quality can be produced in
constant quantities through out the year.

3.   High productivity: Since the system is
three-dimensional, the available space is
double that of a flat system.

4.   Energy saving: Since the system is
three-dimensional, the costs for light,
temperature control and carbonic acid gas
are low compared with conventional
systems.

5.   Space saving: Since the system is
perfectly controlled, place for construction
is not limited. This allows reduction in
distribution costs and supply of fresh
products .

Future problems and development

1.   Hardware: According to the Table 6,
electricity and labor costs occupy 22% and
19% respectively less than those for
conventional systems. Those are most
important for cost savings together with
that for the system itself which depreciation
occupies 46% of production costs. This
means reasonable costs for commercial
production.

2.   Software: Developing faster growing
varieties will help to save costs by means of
saving on running costs and increasing the
number of harvestable items, including a
highly functioning plant for enriching,
which will help expansion in the market.

3.   Conditions for commercialization: The
prices indicated in Table 7 are 30 to 40%
higher than those for conventional
products, and so it will be necessary to
develop sales routes by taking advantage of
the stable supply, safety (perfectly free from
pesticides) and cleanliness (less bacteria).
As a result of the market survey, it was
found that the products from the TS Farm
are highly evaluated by the food service
industries such as the household dish
industry and food processing industry.
This reflects the above-mentioned
advantages and the high yield in commercialization (low loss) and furthermore

production costs can be reduced by shortening the distribution route (intermediate distributors are not required) and reducing the delivery costs (simplified packaging style).

Presenters believe that plant factory is essential technology for future considering increases of world population, worldwide food shortage, global environmental pollution, and climatic change. In order to realize this dream, multiple approach to the plant factory development is necessary and presenters' co-work will be the first step.

Presenters express their appreciation to Professors Hashimoto and Nonami of E'hime University together with all customers and sub-venders of the TS Farm system.

REFERENCES

1. Hashimoto, Y. and Takatsuji, M. (1986). Plant Factory. <u>Agriculture and Horticulture No.61.</u> Japanese pp169-173.
2. Hashimoto, Y. and Takatsuji, M. (1987). Recent Plant Factory. <u>Agriculture and Horticulture No.62.</u> Japanese pp86-90
3. Takatsuji, M. (1987). <u>Fundamental theory and practice of plant factories,</u> 3rd edition.Japanese. Technical Information Center.121-147p
4. Standard Tables of Composition in Japan. Rev.,4 (1990).
5. Itoh, T. (1987), Practical approach of factory vegetable production. <u>Agriculture and Horticulture No.61.</u> Japanese pp174-180
6. All about Plant Factory, Japan Horticulture Association(1987), pp130.
7. Abe,Y. (1989). M-type Hydroculture. <u>The Journal of Electrical Installation Engineers of Japan vol.9</u> 134-140pp.

Copyright © IFAC Mathematical and Control Applications
in Agriculture and Horticulture, Matsuyama, Japan 1991

CLASSIFICATION OF APPLES WITH A NEURAL NETWORK BASED CLASSIFIER

U. Ben-Hanan, P.-O. Gutman and K. Peleg

Faculty of Agricultural Engineering, Technion, Haifa 32000, Israel

Abstract: 352 apples were classified as green or red by three
human experts. Each apple got a probabilistic (fuzzy) membership
in the red and green sets according to the number of votes. The
majority decided the actual classification. This fuzzy data, and
features such as color hue, saturation, and intensity, and the ratio
between red and green areas, for the first 30 (or 60) apples were
used to train a Boltzmann Perceptron Network (BPN). The remaining
fruits were then classified by the BPN. It was found that hue is the
dominant feature. The misclassification was less than 15%, similar
to a Bayesian classifier trained on the same data. The advantages of
the BPN seem to be that it allows fuzzy input data, and many
features.

Keywords: Classification; agriculture; neural nets; optimization;
fuzzy sets

1. INTRODUCTION

The problem of automatically sorting
agricultural produce poses two main
problems:
1. A sufficient number of features has to
be identified, and extracted by the
automatic sorting machine, in order to
emulate a human classifier;
2. The statistical properties of the sorted
produce may vary, due to the origin of the
produce. A human clasifier has an overview
of the incoming produce, and adapts his
classification. An automatic sorting
machine must be as easy to retrain.

Consequently, it is desirable to find a
classifier that is able to deal with many
input features, and output classes, and
that is efficiently retrainable with a small
training set. It should be able to handle
"objective" features, such as the area of
blemishes, and be able to model
"subjective" features, such as color or
shape, by a combination of measurable
features (spectral contents, surface color
distribution; curvature, principal axes
ratio).

A major problem is to determine the
correct combination of measurable features
for a given subjective one. Typically, this

is done in a training session, where one or
more human classifiers sort the produce. In
classical classification theory Peleg (1985),
the sorted classes are then correlated to
the measured features, and decision
boundaries are determined using various
principles, such as clustering, minimum
misclassification, etc.

However, the human classification may be
ambiguous. Sometimes, the human is not
sure about the correct classification;
sometimes different humans classify
differently. While the classification
decision itself is easily resolved by voting,
it seems beneficial to be able to provide
the automatic classifier with the "soft" or
fuzzy classification during the training
session, in order to enhance the
performance for borderline cases.
Likewise, an automatic classifier that can
be tuned to provide various levels of "soft"
or fuzzy classification decisions during
operation, in the form of class probabilites,
may be efficient in indicating those items
for which a firm decision cannot be made,
enabling the formulation of an intermediate
class.

Neural networks seem to offer some
advantages for real life classification: Due
to their architecture, they can be built to
operate very fast. They are non-

parametric with respect to the feature space, and do not rely on explicit statistical properties of the produce. They can be easily retrained via corrected samples reclassified by a human expert, without the need to measure the corrected features.

It is true that the classification boundaries found by a neural network may also be found in other ways and approximated by a piecewise linear classifier. However, a neural network is particularly easy to train in a multidimensional feature space, and because of its non-linear internal structure probably needs fewer parameters to describe complicated classification boundaries than a piecewise linear classifier. Neural networks seem therefore well suited to serve as extractors of relevant features, even if another realization of the classifier is subsequently chosen.

In this paper, the Bolzmann Perceptron Network (BPN), a deterministic, feedforward network, that was defined in Yair and Gersho (1990) is applied to an apple sorting example. The BPN has the desired soft classification properties, and seems to be able to be trained well with small trainig sets. It has previously been applied only to a few synthetic examples.

The paper is organized as follows: In section 2, the classification problem and our method to collect the raw data are described. Section 3 discusses the color features we used. Our choice of parameters for the BPN is presented in section 4. The results are found in section 5. A short discussion (section 6) and a list of references conclude the paper.

2. THE CLASSIFICATION PROBLEM

The color of a fruit is one of its major features. The desired colors for different grades are defined by standards, including pictures of representative fruits. Lately, a lot of research has been done on the identification of the color of agricultural produce. Davenel et al (1988) described an automatic system to detect color and blemishes of apples . Upchurch et al (1988) described a method to sort peaches by maturity by measuring two characteristic wave lengths: one to detect the green area, and the other to detect the red area, Green indicates maturity, while red was used to compensate for the intensity of the reflected light.

In packing houses the grading by color is usually done by a human classifier or by a machine measuring the total light reflectance from the fruit surface. The first alternative is costly. The second method operates under the assumption that integration of the color over the whole fruit surface is sufficient.

In this paper a simple experiment was conducted to classify a batch of 352 Orlians apples into a Red and a Green class. Note that different individuals might grade a given apple differently, even though they were trained with the same set of standard color pictures. We therefore let three expert classifiers, look at the fruits. Each apple was thus assigned a fuzzy class membership, with probabilites 0, 1/3, 2/3, or 1, but was unambiguously classified by majority voting.

Each apple was video photographed, and the picture was transferred to an image processing system in which relevant color features were extracted. With some of it serving as a training set, the data was fed into a BPN. We studied the influence of different color features, and the neural network complexity, on the resulting classification boundaries, and the amount of misclassification.

3. THE COLOR FEATURES

In the video camera, color images are represented in terms of Red, Green, and Blue (RGB) components, with a frame buffer for each. See fig. 1. In the image processor, we computed a mean RGB vector from the respective histograms. The mean RGB vector for each apple was then transformed to a mean Hue, Saturation, and Intensity (HSI) vector (Smith, 1978). The motivations for this transformation is that the HSI representation is very similar to the human perception of color.

While the mean HSI representation gives a correct measure of the average apple color, it might not reflect the human impression of the apple. For instance, we hypothesized that an apple with an intensely green spot, and a large, but pale red area, might on an average be greenish, while a human will classify it as red. Therefore, in addition to the mean HSI-components, we also used as a feature the ratio of the "green" area to the "red" area (g/r): a pixel with its G-component larger than the R-component was said to belong to the "green" area, and vice versa.

4. THE NEURAL NETWORK CLASSIFIER

The Bolzmann Perceptron Network (BPN) is fully documented in Yair and Gersho (1989), and Yair and Gersho (1990). In this section we define the parameters we used for the different computational experiments.

The BPN is characterized by I inputs, J hidden units, and M output units. During operation, the input to the BPN is, for each apple, a feature vector with I components, consisting of some or all features described in section 3. The output is the computed class membership that may be either definite, or fuzzy, depending on a user chosen parameter (β). If the output was fuzzy, we classified according to maximum probability.

During the training phase, a set of N apples was used. For each apple, the BPN was provided with the actual feature vector, and a probabalistic class membership (section 2). The result is a set of BPN link weights, whose values have been optimized to fit the input-output data.

Yair and Gersho (1990) describe a conjugate gradient method for the optimization of the training phase; we chose to use the canned IMSL routine UMCGG. We had to supply the gradient, an initial guess of the link weights, the size of the initial gradient step, the maximum step size, and stopping criterion.

5. RESULTS

The classification decisions of a tested BPN gave rise to a de facto classification boundary. The amount of misclassification was computed with respect to the human classification that is assumed to be "correct". The BPN boundary and misclassification were compared with a Hue-based one-parameter minimum misclassification (Bayesian) classifier derived from the human classifier data (see fig. 2), Peleg (1985).

The *a posteriori* optimal thresholding for a Bayesian classifier is obtained by the Hue value where the two histograms in fig. 2 cross each other. The value in our example is 22.5. The total misclassification with this threshold would have been 41 out of 322 apples of the test set (12.7%), with 21 of the green apples classified as red, and 17 of the red were classified as green.

The *a priori* Basyesian classifier based on a

training set of the first 30 apples gives a threshold at 18, or any value between 22 and 25. When this classifier is used to grade all the apples, the misclassification is at best as the *a posteriori* described above, and at worst with 53 apples misclassified.

The results of the classification with a BPN, trained on the N=30 first apples, using Hue as the only input feature (I=1), with two hidden units (J=2), two output classes (M=2) and initial weights=0.2, are found in the first row of Table 1. This BPN was first trained with only one hidden unit but the optimization did not converge.

Table 1: Misclassification related to the BPN classification features.

Feature	Red in Green	Green in Red	Total Contam.	%
H	11	35	46	14.3
H+g/r	11	35	46	14.3
H+g/r+I	12	36	48	14.9

In Table 1 two more classification results with different combinations of features are presented.

The test set (322 fruits) with the human classification is depicted in fig. 3a with respect to Hue and g/r. In fig. 3b the "H+g/r" BPN classifier results are depicted. The parameters of this BPN classifier were:

$$R=\begin{bmatrix}0.5325 & 0.2034\\0.5325 & 0.2034\end{bmatrix} \quad W=\begin{bmatrix}0.1077 & 0.2317\\0.2923 & 0.1683\end{bmatrix}$$

$$Q=\begin{bmatrix}0.7108 & -0.0165\\0.7108 & -0.0165\end{bmatrix} \quad C=[0.4943 \ 0.4943]$$

$$S=[1.2176 \ -0.8176].$$

For the H+g/r+I combination three hidden units were implemented. For the other combinations, two hidden units were used.

When the classifier was trained on the first N=60 apples, the total misclassification on test set of 292 fruit with H and H+g/r as features was 14.4%. When using the H+g/r+I features the misclassification was 15%.

167

6. DISCUSSION

Although this study is extremely limited, indeed, it seems possible to draw a few tentative conclusions: 1) The average hue of the apple color is the feature that enables emulation of the human classification decision; 2) The BPN classifier is an efficient means to select significant features (see Table 1); and 3) a small data set seems sufficient to train the BPN to achieve results close to a human classifier.

It should also be noted that humans also commit classification errors. The apples that the humans "misclassified" with respect to the BPN decision based on Hue (see fig. 3) were checked. It was found that in some cases the reflected light intensity of the fruit was misleading; e.g. a dark red fruit was classified as green, and a bright greenish apple was graded red. Indeed, in such cases it is not desirable that a machine classifier emulates the human sensitivity to reflected light.

Analog sensing of the green and red reflections, which is currently in commersial use in apple packinghouses, can be translated into a mean hue measurement of the apple surface. The observations in this study indicate that the threshold of such a system could be periodically updated by a BPN, and a human classifier.

REFERENCES

Davenel, A., Ch. Guizaed, T. Labarre, F. Sevila. 1988. Auotomatic Detection of Surface Defects on Fruit by Using a Vision System. J. Agric. Engng. Res. 41,1–9.

Peleg, K. 1985. Produce Handling Packaging and Distribution. AVI Publishig Corp. Westport Conneticut.

Smith, A. R. 1978. Colour Gamut Transform Pairs. SIGGRAPH Conference, pp. 12–19.

Upchurch, B. L., M. J. Delwiche, and D. L. Peterson. 1988. Evaluation of Optical Measurments for Peach Maturity Sorting. ASAE Paper No. 88-6024. St. Joseph, MI 49085-9659.

Yair, E., and A. Gersho. 1989. Maximum A Posteriori Decision and Evaluation of Class Probabilities by Boltzmann Perceptron Classifiers. Submitted to the Proceedings of the IEEE for the special issue on neural networks.

Yair, E. and A. Gersho. 1990. The Boltzmann Perceptron Network: A Soft Classifier. Neural Networks, Vol. 3, pp. 203–221.

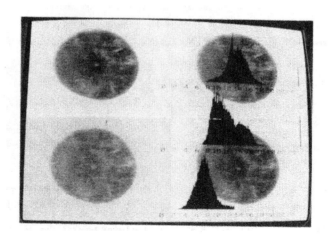

Fig 1: A picture of the video screen during image processing. The upper left picture shows one face of the apple. The lower left corner depicts a thresholding operation to exclude the background, together with the other face of the apple. On the right we see the R, G, and B histograms of the total apple surface.

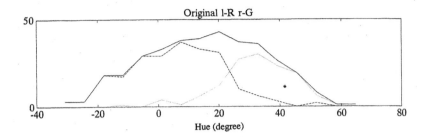

Fig 2. The hue histogram of 352 apples (solid line). The left (right) histogram represents the human classification of the red (green) apples.

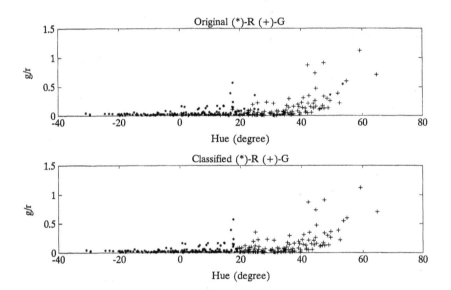

Fig 3a. Hue and g/r features for the 322 apples in the test set. Apples classified as red (green) by the majority vote of the human experts are marked * (+).

   b. Hue and g/r features for the 322 apples in the test set. Apples classified as red (green) by the BPN classifier (I=2,J=2,M=2) are marked * (+).

CONTROL OF INTEGRATED FARM ENERGY
AND TECHNOLOGY SYSTEM

I. Farkas

Institute for Mathematics and Computer Science, University of Agricultural Sciences, H-2103,
Gödöllő, Hungary

Abstract. Control of operation modes of an integrated farm energy/tech-
nology system is studied. The solar, the oil/gas burners and electrical
energy are considered as a source of energy. The main energy consumers
are a crop drier, service hot water making and other technological processes. A control strategy based on a hierarchy for energy distribution
among the consumers has been carried out. For these purposes a block oriented simulation model has been developed including the submodel of technological processes and the control algorithms. The model was applied to
designe the control of operation modes on a middle scale farm.

Keywords. Agriculture, computer simulation, energy, heat consumers,
identification, integrated system, modeling, operation modes.

INTRODUCTION

Due to the current energy crisis, including
oil price exploision and the rigorous world
market quality requirements, the optimal
consumption of energy is a key question on
a farm. It is needed to integrate the all
available energy sources including also the
renewable ones and to distribute them
optimally between the different technological consumers, such as greenhouses, animal buildings, crop driers, service water
heating, hot water storage, etc.

To set up a solar, preheating system can be
economically justified if the gathered
solar energy is used all over the year
round. In the present conditions the integration of solar energy into the farm
energy system seems to be a possible way of
economical application.

At the same time it is required to solve
the control problem of a coupled energy
system to provide the optimal operation
under any working condition. This problem
consists of two main parts. One is the
control of energy distribution by setting
up the operation mode for the actual source
of energy, and the second one is selfcon-
trol of each technological process. In
this paper the former problem, i.e. selection of the operating mode is studied,
which can be based on a hierarchy taking
into consideration the relevant energy
conditions and the demands initiating from
the side of consumer subsystems. A priority principle can be dedicated by the
consumers for a given type of energy
source.

LAYOUT OF A FARM ENERGY/
TECHNOLOGY SYSTEM

A farm energy/technology system includes
the whole available energy sources and
those technological processes which can

be directly connected to the integrated
energy system. The types of energy can
usually be taken into account on a farm
are the following: electrical heating /EH/,
thermogenerator /TG/ with oil or gas burners, solar energy by water collectors /C/
which is a very promising and clean way of
applying the renewable energy even thought
its economical aspects have to be seriously considered.

The bigest energy consumer on a farm are
usually the crop drieres /D/ which need a
preheated air during the drying process.
In spite of this fact, in term of energy
integration, it is more advaisable to use
water flow solar collectors due to their
easy connection to the existing farm arragements, i.e. to link a hot water storage tank, /T/, which can serve simultaneously as an energy storage, too. In such
a way the drier is counected indirectly,
to the energy system throught an air/water
heat exchanger. Making hot water /HWS/ is
possible trought water/water heat exchanger and we can also use heat energy
for different heat consumers /HC/ with the
aid of additional heat exchangers to preheat the biogas containers, for instance.

A simplified layout such a coupled energy/
technology system is shown in Fig.1, where
the necesserry fan, pumps and valves are
also indicated.

Due to water medium is applied the solar
collector can be connected by means of the
pump P1 to the drier, to the hot storage
tank, to the hot water making system and
also to the other technological heat consumer. This gives the advantage of mul-
tipurpose use of solar energy.

The fan F forces the air through drying
bed which can also be preheated by thermogenerator if it is necessary.

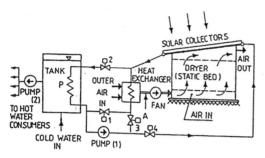

Fig. 1.   Layout of an integrated energy/
          technology system

In the idle run of collector, for instance
at night, the storage tank takes its role.
In this case the drier or other technological heat consumers can be operated from
the storage tank by P1 or P2 pumps. The
service hot water can be produced also in
the tank.

In case of unfavorite weather condition we
need some extra energy source. Such equipments are the thermogenerators and the
electrical heating built into the storage
tank.

DETERMINATION OF OPERATION MODES

On the basis of the scheme shown in Fig.1.
the following questions arrise in conjunction with the operation modes:
- volume of drying claim and its timing,
- is it in operation the solar collector,
- what is it about the loading level in
  the storage in term of energy.
In aswering to this questions we can build
of the available combinations for operation modes.

Table 1.   Categories of operation mode

MAIN OPERATION MODES   SUPPLEMENTARY
                            POSSIBILITIES

  I. Drying by ambient air   V/1

 II. Drying directly from collectors
     1. C→D, →T, →HWS, →HC
     2. C→D, →T, →HWS   V/2
     3. C→D, →T   V/1
     4. C→D   V/1, VI/1, VII/1

III. Charging the storage by collectors
     (no drying)

     1. C→T, →HWS, →HC
     2. C→T, →HWS   V/2
     3. C→T   V/1

 IV. Drying from storage tank
     1. T→D   V/1, VI/1, VII/1

     ADDITIONAL OPERATION MODES

  V. Supply from storage tank
     1. T→HWS⁺, →HC
     2. T→HC

 VI. Charging the storage tank
     1. Electrical heating /EH/

VII. Air preheating by auxiliary energy
     1. Oil/biogas thermogenerator (TG)

The main categories are the following:

  I. Drying with ambient air.
     In that case if the relation between
     the drying capacity of ambient air
     and moisture state of material to
     be dried permits we can continue
     the drying without any preheating
     of air.

 II. Drying from collector.
     The collector works for the drier,
     for the storage, for hot water
     making and for the other technolo-
     gical heat consumers. If the solar
     energy in not enough for all of
     these, then a priority principle is
     valid in the operation of subsys-
     tems according the order listed
     above.

III. Collector works for the storage.
     This case appears when there is
     no drying request. The hot water
     making and the other technological
     processes can be provided as in
     category II.

 IV. Drying from storage.
     If there is no direct solar energy
     then energy demand for drying
     should come from the storage tank.

Additional operation modes are the following:

  V. Other service from storage.
     We can provide hot water from the
     storage tank into the direction
     of HC subsystem or we can pre- or
     overheat the service hot water in
     the tank. These functions are
     served mainly as an additional,
     alternative ways.

 VI. Loading the storage tank.
     In some extra cases, when we need
     the provide hot water or to con-
     tinue the drying from storage then
     the energy loading level of the
     storage should be high enough.
     This can be reached by the built
     in electrical heat source. First
     of all, the electrical heating is
     advised to be applied during the
     night.

VII. Heating by thermogenerator.
     The additional heating by thermo-
     generator is required in the oper-
     ation modes II/4 or IV/1 when the
     required temperature level of dry-
     ing air cannot be secured by the
     solar system.

The most important operation modes are
II/3-4, III/3 and IV/1, respectively.

172

THE OPERATIONAL CONDITION OF SUBSYSTEM

In the following we are going to describe the operational conditions of the subsystems when they are required to put into action. For this, the measuring sensors located in the physical system serve signals. The operational variable $X=0$ means the switch off, while $X=1$ means the switch on states of a given subsystem. Using this assumption the operational conditions of the subsystem are as follows.

Collector.
The operational signal for collector is determinded on the bases of temperature change in flow medium

$$X_C = \begin{cases} 1, & \text{if } t_{C,out} \succeq t_{C,in} + \Delta t_C \\ 0, & \text{if } t_{C,out} \preceq t_{C,in} + \Delta t_C \end{cases} \quad (1)$$

so, in case of $X_C = 0$ there is no water flow in the collector. The value of Δt_C, depending on the flow resistance in the connecting tubes is identified in a range of 1-2 C°,

Drier.
With respect to the construction of drier, i.e. fix bed type in a barn, the first drying request appears at the first packing into the drier and holds on until the total drying of the material at the last packing in. During this period the request of drying can be hanged on only for packing in internal or at an unexpected disturbance in the operation. Let us suppose more drying section, the drying request variebles are

$$X_{Di} = \begin{cases} 1, & \text{if there is drying request in the i-th section} \\ 0, & \text{if there is no drying request in the i-th section} \end{cases} \quad (2)$$

Storage tank.
Supposing a stratified water storage tank three main cases can be considered.

a/ Loading the storage from collector:

$$X_{T,1} = \begin{cases} 1, & \text{if } t_{C,out} \succeq t_{Ti,min} + \Delta t_T \\ 0, & \text{if } T_{C,out} \preceq t_{Ti,min} + \Delta t_T \end{cases} \quad (3)$$

It means, that the storage tank can be loaded if the outlet water temperature from the collector is greater then the lowest temperature in i-th strata in the storage tank by a value of Δt_T.

b/ The storage is capable to send out:

$$X_{T,s} = \begin{cases} 1, & \text{if } C_T \succeq C_{T,cr} \\ 0, & \text{if } C_T \preceq C_{T,cr} \end{cases} \quad (4)$$

So, the storage tank is capable to send out energy if its thermal capacity is greater then a critical value, which can be calculated from the t_{Ti} temperatures of layers. It is important to declare that the $C_{T,cr}$ is selective due to the operational mode

$$C_{T,cr}^{IV/1} \succeq C_{T,cr}^{V/1} \succeq C_{T,cr}^{V/2} \quad . \quad (5)$$

It is obvious, that the loading level in the storage should be highest if beside the additional consumers we need to supply, the drying process.

c/ Charging the storage tank by electrical heating

$$X_{T,EH} = \begin{cases} 1, & \text{if } C_T \preceq C_{T,cr}^{VI/1} \\ 0, & \text{if } C_T \succeq C_{T,cr}^{VI/1} \end{cases} \quad (6)$$

In this condition it is expressed that we need to provide the service hot water at any time.

Service hot water.
The request for making hot water for service purposes appears if its temporary value falls bellow a critical value

$$X_{HWS} = \begin{cases} 1, & \text{if } t_{HWS} \preceq t_{HWS,cr} \\ 0, & \text{if } t_{HWS} \succeq t_{HWS,cr} \end{cases} \quad (7)$$

Other technological heat consumers.
Most of the cases at these technological heat consumers requred a constant temperature level of water

$$X_{HC} = \begin{cases} 1, & \text{if there is request} \\ 0, & \text{if there is no request} \end{cases} \quad (8)$$

Thermogenerator.
It is needed to put into action the thermogenerator if the inlet drying air temperature cannot be rise over the required value $/t_{ri}/$ at any other way, for instance the solar radiation intensity is fairly low or it is not possible to use the storage tank for this purpose. The operational signals in the i-th drying section is as follows

$$X_{TGi} = \begin{cases} 1, & \text{if } t_{D,ini} \preceq t_{ri} \\ 0, & \text{if } t_{D,ini} \succeq t_{ri} \end{cases} \quad (9)$$

The required drying temperatures t_{ri} in each layer are setting up by the local drying controller.

The request for operation of subsystems are shown in Fig.2. including the necessary measuring sensors and switchers, too.

CONTROL STRATEGY OF OPERATION MODES

To set up the appropriate operation mode means a cupled control problem including the optimal collection and distribution of all sorts of energy sources available in the farm taking into consideration, the economical aspects, too. This is a hierarchical control task which can ben derived on the basis of the energy/technology circumstances on the farm. A possible solution for this problem can be described as follows.

Because of the solar system uses the hot water storage tank as an energy storage, therefore it is absolutely important to provide the hot water request at any time. It can be done directly from solar energy, or indirectly from the storage tank. This implies a given loading level in the storage to secure this requrement.

The second absolutely important aim to provide the requested temperature of drying air for not to make any demage in the material to be dried. This can be provided first of all from direct solar energy. If there is not available direct solar energy than we can continue the drying with stored solar energy from the storage tank. If is is also not possible than we can use the oil or gas thermogenerators to preheat the drying air. At the last resort we could only apply the electrical energy, because under the recent price conditions it seems to be the most expences.

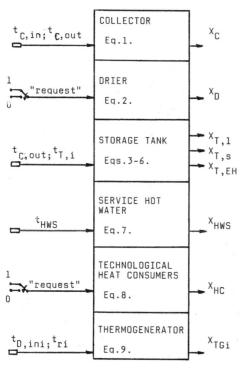

Fig. 2.   Operational requests and conditions of different subsystem

In consuming up the direct solar energy always the bigest energy consumer, the drier has a priority. After that comes the storage tank, because it can be assumed that the tank almost every time can be loaded due to the continous water usage on the farm and due to the physical fact that the temperature in lowest layer in the tank is less than at the highest layer where the water is taken away. After the loading the storage comes the direct hot water making for service purposes and at the end to provide direct solar energy for the other technological heat consumers.

In the operation mode II/1, for example, if the solar energy is already not sufficient to provide the all subsystems than it is needed to take off the subsystems from the direct solar energy providing according to the following order, first the other technological heat consumers, secondly the hot water making and finally the storage tank. Of course, this priority order is valid also in reverse way, if the amount of solar energy increases, we can switch back to the direct solar system first the storage, than the hot water making and finally the other technological heat consumers.

The selection of the appropriate operation mode is based on the hierarchy of energy consumption described above and its flow chart is shown in Fig.3.

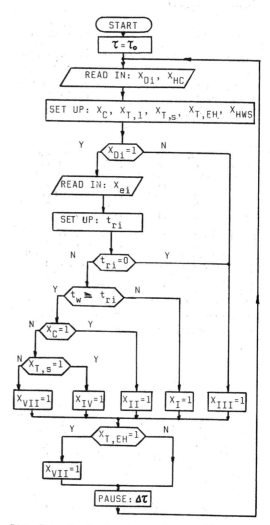

Fig. 3.   Control scheme of operation modes

In the Fig. 3. the X_{ei} signals come from the end switch of control valves of drying heat exchangers. The stage $X_{ei}=1$ means that the control valve is "not totally open" i.e. still we have more solar energy to be used. If all drying valves are open /$X_{ei}=0$/ than direct solar energy cannot be used already for further purposes, as for T, HWS and HC.

To solve the control tasks a block oriented simulation model has been carried out which contains the following units:
- submodel of technological processes,
- organizing submodel which allows to built up an appropriate arrangement from the available subsystems.
- control subsystem which gives also the opportunity to study different control algorithms.

The structure of the simulation model permits flexible extension feasibilities, i.e. including new type of blocks concerning additional technological processes on the farm and/or new control units, too. Additionally, the model was applied to investigated physical behaviour for a cuopled technology system on a middle scale farm.

CONCLUSIONS

The control of the operation modes of an integrated energy/technology system can be solved succesfully by a strategy based on a hierarchical distribution from the technological heat consumers. The priority order of the subsystems are the drier, the storage tank, the hot water making and the other technological consumers.

For the selection of the appropriate operating mode we can use easily measured parameter values in the system, as ambient air temperature, /t_w/, the outlet and inlet water temperature of solar collector $t_{C,in}$; $t_{C,out}$/ the inlet and outlet air temperature of drier /$t_{D,in}$; $t_{D,out}$/ the layer temperatures of storage tank /t_{Ti}/ and the temperature of service not water /t_{HWS}/.

REFERENCES

Farkas, I. (1985.) Control and computer simulation of a complex solar drying system, Ph.D.Thesis, Budapest

Farkas, I. (1990.) Modeling and identification of agricultural driers, 11th IFAC World Congress, Tallin Estonia USSR, Vol.12, 4-8

Sriramulu, V., S.B. Ahmed, M.C. Gupta (1979.) Investigation of thermal storage unit for solar power generation, Solar Energy an Conservation, (ed. by T.N. Veziroglu). Vol.3. Pergamon Press, New Work, 202-226.

Gordon, J.M. A.Rabl (1982.) Design analysis and optimization of solar industrial process heat plants without storage, Solar Energy, Vol. 28, No.6, 519-530

CRACK EVALUATION OF RICE BY COMPUTER VISION SYSTEM

S. Oshita*, H. Shimizu* and H. Ota**

*Faculty of Bioresources, Mie University, Kamihama-cho, Tsu 514, Japan
**Sumitomo Chemical Co., Ltd., Takarazuka Research Center,
2-1. 4-chome, Takatsukasa, Takarazuka, Hyogo 665, Japan

Abstract. A computer vision system was used to evaluate the degree of cracks in brown rice kernel. White spot light applied to a rice kernel parallel to its major diameter in a dark room provided the best viewing conditions. The digital picture has 256 x 256 pixels with 64 gray levels. The principle of the detection of cracks was based on four kinds of indices, two of which were extracted from a gray level profile and the remaining two were extracted from a gray level histogram. Each index was numerized by the membership function. The fuzzy integral was used to evaluate the extent of cracks by putting 4 numerized indices together. With the algorithms developed, damaged kernels were detected with an error rate of 2 %.

Keywords. Agriculture; Rice; Crack detection; Image processing; Fuzzy evaluation.

INTRODUCTION

Crack of rice is one of the most important physical features which affects the postharvest grading of rice. The detection of cracks in rice kernel is done with the naked eyes, however, visual inspection is unreliable since human judgment in identifying the extent of cracks is inconsistent and subjective. This problem is an inevitable accompaniment to visual inspection and there has been interest in using digital image analysis for quality detection of agricultural products.

Sarkar and Wolfe (1985) developed algorithms using digital image analysis and pattern recognition techniques for orientation of tomatoes and their classification. They (1985) reported also on the development of illumination and presentation schemes and the performance of a machine vision tomato sorting system. An image processing algorithms for the detection of defects on the apple surface was proposed by Rehkugler and Throop (1989) and Throop et al. (1989) examined the applicability of computer vision for detecting watercore presence in apple by weight density and by light transmission. A color vision system has also been investigated as an alternative to manual quality evaluation. Miller and Delwiche (1989) demonstrated the feasibility of machine maturity classification of peaches by using color images.

Previous research has shown that an image by transmitted light is useful to detect cracks in the crop. Gunasekaran et al. (1937, 1988) processed the digital image by tansmitted light to detect cracks in corn and soybean kernel and Matsuhisa and Hosokawa (1981) indicated the usefulness of transmitted light image for the crack detection of brown rice.

From image processing applications in agricultural engineering having been expanding as mentioned above, it became clear that there is a limit in accuracy of quality evaluation based on only one index of biological materials.   This paper reports on the application of vision system for extracting features related to the crack in rice kernel and the fuzzy integral based evaluation procedure.

OBJECTIVES

The objectives of this investigation were to
(1) Determine crack indices for brown rice which can be extracted from a processed digital image.
(2) Examine the accuracy of crack classification based on crack indices which are put together by fuzzy integral.

SYSTEM DESCRIPTION

The computer vision system used in this study is shown in Fig.1. The halogen light source having a maximum of 150 W power rating supplied spotlight through light guide to a sample. The sample was held on the black-coated rotary table which can align the rice kernel in the lighting direction. The image produced by transmitted light was acquired by a CCD camera having 378000 pixels. These devices were arranged in a dark room to assure the best illumination condition for acquiring a sample image. The video signal from the camera was digitazed by a image processing boad mounted in a personal computer. The hardware digitazed an image in 0.0167 s. The digital picture had 256 x 256 pixels each of which had a 6-bit value (64 levels) representing the light intensity. Acquired images were stored in a fixed disk for the subsequent processing.

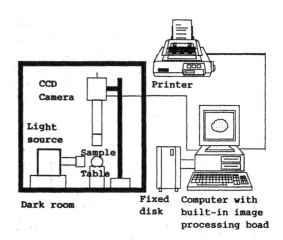

Fig.1 General View of Computer Vision System

PROCEDURE

Sample Preparation

Rice kernels of four varieties harvested in the experimental farm of Mie University, namely, AKIBARE, HATSUBOSHI, KOSHIHIKARI and YAMAHIKARI were used in this investigation. Rough rice kernels were dried by a circulation type grain dryer down to 14 %(w.b.) and hulled by hand so that their surfaces were not injured. Before the computer vision inspection, samples of 402 rice kernels were inspected visually. Table 1 shows the results of visual classification of rice kernels according to crack standards for brown rice proposed by the japanese society of agricultural machinery (JSAE).

Table 1   Classification of Rice Kernels by Visual Inspection

crack	non	half	heavy
AKIBARE	94	5	0
HATSUBOSHI	85	11	1
KOSHIHIKARI	92	10	4
YAMAHIKARI	82	18	0

Illumination

Most of the cracks in rice rise along by the direction perpendicular to the major diameter of kernel as shown in Fig.2. So, spotlight applied to a rice kernel parallel to its major diameter in a dark room contributed to the best condition for acquiring images of rice which had a distinct gray level difference at the cracked portion. The rotary table was black-coated to avoid reflection and it made ease to align a rice kernel in a desired direction.

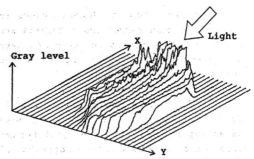

Fig.3 Gray Levles on the Image Plane for Non-Cracked Rice

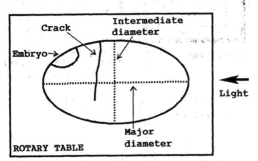

Fig.2   Location of Crack and Lighting Direction

Image Processing

The original image provides with informations on pixel coordinates and their gray level values. Figure 3 shows a three-dimensional representation of the above-mentioned informations for a typical non-cracked brown rice kernel. The similar representation of informations for a typical heavy cracked brown rice is given in Fig.4 in order to see the difference by comparison with Fig.3. The axis expanding perpendicularly upward represents the gray levle value.

Gray levels in a upper right region of a kernel presented very large values. This was because the light reflected from a surface of rice kernel was recieved by the CCD camera. Gray levels of transmitted light decreased monotonously with x-position in case of non-cracked rice kernel (Fig.3), while gray levels along the x-axis changed stepwise at the middle of a heavy cracked kernel where crack was present (Fig.4).

The processing region was limited to a rectangle circumscribed with the rice kernel for the reduction of the amount of digitazed data to be treated.   Given the

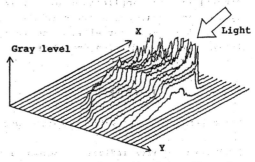

Fig.4 Gray Levels on the Image Plane for Heavy Cracked Rice

limited region of images of individual rice kernels, image analysis consisted of two major phases:

(1) Gray Level Profile -   special crack indices were extracted from the gray level profile on the line which was parallel to the major diameter and passed through the midpoint of the intermediate diameter.

(2) GrayLevel Histogram - crack indices estimating the extent of cracks from the whole information on image were given from the gray level histogram.

Gray Level Profile. The line on which the gray level profile was investigated was determined in an automatic way. Both upper and lower edges of a kernel shown in Fig.2 were detected by searching pixel values of the dark background which were much lower than those within the rice kernel. Then the line passing through the middle point of two edges and parallel to the major diameter of a kernel was identified. The gray level profile on the line was observed on the monitor as it varied in a zig-zag way. This profile was smoothed for

179

the convenience of finding the averaged gray level difference at the cracked portion. Figure 5 shows the smoothed gray level profile extracted from the image of the typical heavy cracked rice kernel. The difference between d2 and d1 increased as the extent of crack augmented.

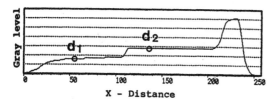

Fig.5 Gray Level Profile along the Major Diameter of Heavy Cracked Rice

The gradient of the gray level profile at the cracked portion changed according to the crack characteristics which couldn't be detected by only the difference d2-d1. Figure 6 shows the differential values of gray levels on the profile. The peak k at the cracked portion increased depending on the extent of crack. The highest peak at the right-side was produced by the reflection from the kernel surface. This was always observed whether the kernel was cracked or not and the algorithm was developed to eliminate the right-side peak and to find the peak k.

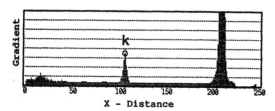

Fig.6 Gradient of Gray Level Values

Gray Level Histogram. After pixel values within the limited region were transformed linearly to sharpen the image, there remained the gray levels where no pixel was found. Linear interpolation was applied to the histogram to fill vacant gray levels with pixels. This modified histogram was smoothed for the sake of subsequent analysis. Figure 7 represents the modified gray level histogram of the same image as used in the gray levle profile processing. Two peaks, P1 and P2,

and the bottom P3 between two peaks were observed in the histogram of cracked rice kernel. The ratio, P2/P1 and P3/P2, took the value corresponding to the extent of crack. These peaks and the bottom value were found automatically by the algorithm developed.

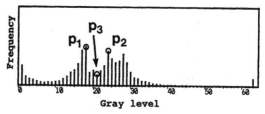

Fig. 7   Modified Gray Level Histogram

Four crack indices stated above were examined and their characteristics were shown in Table 2.

Table 2   Characteristics of Crack Indices

Indices	Extent of Crack	
	Not heavy	Heavy
A1:Gray level difference	small	large
A2:Differential value k	small	large
A3:Ratio P2/P1 (P1>P2)	small	large
A4:Ratio P3/P2	large	small

Fuzzy Evaluation

Digital images involve the fuzziness caused by the projection of 3-dimensional kernels on a 2-dimensional plane. Fuzzy evaluation was used to obtain the integrated value that estimates the extent of cracks in rice kernel after general consideration of four crack indices.

Membership Functions. Four indices, A1 to A4, given in Table 2 were regarded as evaluation items for cracks. Eleven typical kernels of both heavy cracked and non-cracked rice were selected to fix the limits for each item. The limits are shown in Table 3. For each item, membership function was defined by a straight line connecting the limits of both non-cracked and heavy cracked rice given in Table 3. These membership functions represented by μ (Ai) are given by equations (1) to (4).

Table 3 Limit values for each item

Item	Non-cracked	Heavy cracked
A1: d2-d1	0	6
A2: k	12	20
A3: P2/P1	0	0.64
A4: P3/P2	0.76	0.59

$$\mu(A1) = 0.167A1, \quad 0 \leq A1 < 6 \qquad (1)$$
$$\text{where} \quad \mu(A1) = 1 \quad \text{for} \quad A1 \geq 6$$

$$\mu(A2) = 0.125A2 - 1.5, \quad 12 < A2 < 20 \qquad (2)$$
$$\text{where} \quad \mu(A2) = 0 \quad \text{for} \quad A2 \leq 12$$
$$\text{and} \quad \mu(A2) = 1 \quad \text{for} \quad A2 \geq 20$$

$$\mu(A3) = 1.56A3, \quad 0 \leq A3 < 0.64 \qquad (3)$$
$$\text{where} \quad \mu(A3) = 1 \quad \text{for} \quad A3 \geq 0.64$$

$$\mu(A4) = -5.88A4 + 4.47, \quad 0.59 < A4 < 0.76 \quad (4)$$
$$\text{where} \quad \mu(A4) = 0 \quad \text{for} \quad A4 \geq 0.76$$
$$\text{and} \quad \mu(A4) = 1 \quad \text{for} \quad A4 \leq 0.59$$

Fuzzy Integral. The fuzzy integral proposed by Sugeno (1972) was applied to the general evaluation of the extent of cracks. The function h(Xi) was determined as followings. First, h(Ai) was derived by substituting membership functions into the equation

$$h(Ai) = \mu(Ai) / \max\{\mu(Ai)\}, \quad i=1\sim4 \qquad (5)$$

Second, h(Ai) was arranged by magnitude starting with the largest one and assigned by h(Xi). For example, when h(Ai) were arranged as

$$h(A1) > h(A2) > h(A4) > h(A3)$$

Xi should satisfy the condition, namely, X1=A1, X2=A2, X3=A4 and X4=A3.

Fuzzy measure was constructed by means of the λ-fuzzy measure, g_λ, which was proposed also by Sugeno (1973). g_λ was defined by the following equations.

$$g_\lambda \left(\bigcup_{i=1}^{n} Ai \right) = 1/\lambda \left[\prod_{i=1}^{n} \{1+\lambda \, g_\lambda(Ai)\} - 1 \right] \quad (6)$$

$$\text{where} \quad Ai \cap Aj = \phi, \quad i \neq j \quad \text{and} \quad -1 < \lambda < \infty \quad (7)$$

It was easy to give subjectively a weight for each item. Weights 0.3, 0.4, 0.2 and 0.1 were given for A1, A2, A3 and A4 respectively and they were represented by $g_\lambda^{(1)}$, $g_\lambda^{(2)}$, $g_\lambda^{(3)}$ and $g_\lambda^{(4)}$. Substituting the relationships, $g_\lambda(Ai) = c \, g_\lambda^{(i)}$, and the value of λ into eq.(6), the

coefficient, c, can be determined because the right side of eq.(6) should be unity when n=4 is given in this case. The value 0.86 was obtained for the coefficient c when λ was fixed to 0.5. The weights, 0.1 to 0.4, and λ =0.5, were determined by the examination of all combinations of those values to produce the best results.

Fuzzy integral is defined by the equation below.

$$\oint h(Xi) \circ g_\lambda(Hi) = \bigvee_{i=1}^{n} [h(Xi) \wedge g_\lambda(Hi)] \quad (8)$$

$$\text{where} \quad Hi = \{X1, X2, ---Xi\} \qquad (9)$$

Figure 8 shows an example of the fuzzy integral applied to a half-cracked rice kernel. In this investigation, the value of unity was assigned to heavy cracked kernel and the result, 0.73, represented well the extent of crack of rice tested.

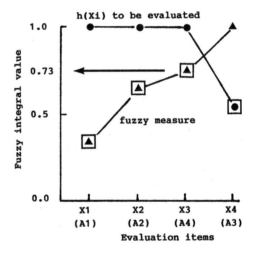

Fig.8 An Example of the Fuzzy Integral Applied to a Half-Cracked Rice

RESULTS AND DISCUSSION

Results of integrated evaluation of the extent of cracks were given in a form of bar graph. Figure 9 shows the results of fuzzy evaluation for 353 kernels of visually sorted non-cracked rice. A large majority of them appeared in the region less than 0.27. This means that the cracks in most kernels were very small. There was only one kernel whose integrated value was unity. This was due to rough surface

which had a bad influence upon the image
produed by transmitted light.

Figure 10 shows the results of fuzzy
evaluation for 44 kernels of visually
sorted half-cracked rice. Rice kernels
were distributed over the region from 0 to
1. Results were in good accordance with
the grade of visually inspected cracks.
Among 44 kernels, 8 kernels were evaluated
to be unity (heavy cracked). The misjudg-
ment was due to the existence of kernels
with rough surface and those almost as
same as heavy cracked rice.

The visually inspected rice kernels
contained only 5 kernels of heavy cracked
rice. The fuzzy integral assigned 1 to
4 kernels and 0.88 to the remaining one.

From the standard defined by the Food
Agency of japan, non-cracked and half-
cracked kernels are classified into non-
damaged rice and only heavy cracked
kernels are recognized as damaged rice. On
the basis of this standard, 10 of 402
kernels in all total were mis-classified
and the error rate was 2 %.

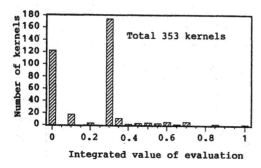

Fig.9 Evaluation of non-cracked rice

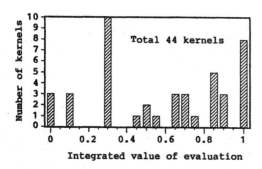

Fig.10 Evaluation of half-cracked rice

CONCLUSION

The computer vision system was used to
obtain images of brown rice kernel. Four
kinds of crack indices were extracted by
the algorithm developed. The extent of
crack was evaluated by the application of
fuzzy integral to four indices. The extent
of cracks in rice kernels were well
evaluated. The classification of damaged
kernels based on the extent of cracks
showed the error rate of 2 %.

REFERENCES

Sarkar, N. and Wolfe, R.R. (1985). Feature
Extraction Techniques for Sorting
Tomatoes by Computer Vision, Trans. of
the ASAE, 28(3), 970-974, 979.

Sarkar, N. and Wolfe, R.R.(1985). Computer
Vision Based System for Quality
Separation of Fresh Market Tomatoes.
Trans. of the ASAE, 28(5), 1714-1718.

Rehkugler, G. and Throop, J.A. (1989).
Image Processing Algorithm for Apple
Defect Detection. Trans. of the ASAE,
32(1), 267-272.

Throop, J.A. et al. (1989). Application of
Computer Vision for Detecting Watercore
in Apples. Trans. of the ASAE, 32(6),
2087-2092.

Miller, B.K. and Delwiche, M.J. (1989). A
Color Vision System for Peach Grading.
Trans. of the ASAE, 32(4), 1484-1490

Gunasekaran, S. et al. (1987). Image
Processing for Stress Cracks in Corn
kernels. Trans. of the ASAE, 30(1),266-
270.

Gunasekaran, S. et al.(1988). Soybean Seed
Coat and Cotyledon Crack Detection by
Image Processing. J. agric. Engng Res.
41, 139-148.

Matsuhisa, T. and Hosokawa,A. (1981).
Possibilities of Checking Cracks of
Brown Rice using Illumination by
Oblique Ray and Image Data Processing
System. J. of the JSAE, 42(4),515-520.

Sugeno, M. (1972). Fuzzy Measure and Fuzzy
Integral. Trans. of SICE, 8(2), 94-102.

Sugeno, M. (1973). Constructing Fuzzy
Measure and Grading Similarity of
Patterns by Fuzzy Integral. Trans. of
SICE, 9(3), 111-118.

MODELLING OF DEEP-BED GRAIN DRYING

S. Morimoto*, K. Toyoda**, R. Takeuchi** and H. Kojima**

*The Graduate School of Science and Technology, Kobe University, Kobe, Japan
**Dept. of Agricultural Engineering, Kobe University, Kobe, Japan

Abstract. Generally grain is dried with a deep-bed after harvest. Its drying characteristics are too complex to investigate directly. Since a deep-bed is assumed to be
composed of a series of thin-layers positioned perpendicular to drying air movement
through a grain dryer, the drying characteristic of a thin-layer of grain needed to
be investigated. The experiment of thin-layer drying was carried out with a
laboratory dryer. First the drying process of wheat was divided into three drying
periods with a view to improving the prediction accuracy of the drying model. Then,
a two-term exponential model could be fitted to those drying periods. The two-term
exponential model could predict the thin-layer drying process. Finally a state equation of the thin-layer drying process was developed from the differential heat and
mass balances equations on the basis of the two-term exponential model. The step
response to inlet drying air was calculated from the state equation. And the exten-
tion of the thin-layer model to the deep-bed drying process was discussed.

Keywords. Drying; Prediction; Models; Thin-layer; Deep-bed.

INTRODUCTION

To predict the drying process of grain by a
mathematical model is very useful for the design
and control of a dryer and to define optimum
drying conditions. In the actual drying operation, grain is dried with a deep-bed, but its
drying characteristic is complex. It is difficult to investigate the drying characteristic
of a deep-bed directly. It is assumed that a
deep-bed is composed of several thin layers.
When deep-bed drying is analyzed using a computer simulation, state values such as drying
air temperature, moisture content etc. are calculated from heat and mass balances for the
drying process on the basis of a thin-layer
drying model. It is important that the thin-
layer drying model should be developed to
analyze the deep-bed drying. Many theoretical
and empirical models were developed in the past.
A thin-layer drying model of wheat is generally
an exponential model. This model is simple and
useful, but prediction errors in an early drying
period are relatively large. To improve prediction accuracy, the diffusion equation is solved
in boundary conditions with the actual drying
process in detail. This method is more accurate;
however, the calculation becomes complex. It
causes the complication of controller or an increase of CPU time etc. in such a model which is
applied to the design of a dryer controller or
to extension to a deep-bed. A thin-layer drying
model should be not only accurate but simple.
One of the simpler and more accurate models is
the two-term exponential model. This model pre-

dicts the drying process of grain during the
falling-rate drying period (*Henderson*,1974; Nellist et al., 1976).

When a model is fitted to drying experimental
data, the following factors must be taken into
account: in drying experimental data in the
early drying period during which the change of
moisture content is large, the data are fewer
than data in the late drying period. Therefore,
when the models are fitted to drying process,
the information in the late drying is likely to
be emphasized and it causes prediction errors in
early drying. The drying operations of wheat in
Japan are finished at 14 to 15%w.b. generally.
The prediction accuracy above moisture content
of 14%w.b. is important. The data under 14%w.b.
may be neglected.

The objectives of this study are to develop a
thin-layer drying model which is simple and accurate and to discuss the modelling of deep-bed
drying. This study consists of two stages. In
the first stage, a thin-layer drying experiment
of wheat was carried out and the drying process
was investigated with a view to improving the
prediction accuracy of the drying model above
14%w.b.. The two-term exponential model was
fitted to the drying process and the relationships between the parameters of the model and
the drying conditions were investigated. In the
second stage, the heat and mass balances equations of the thin-layer were derived on the
basis of the two-term exponential model and a
state equation of thin-layer drying was

developed. Then, the extension of the thin-layer drying model to a deep-bed was discussed.

EQUIPMENT AND PROCEDURE

A schematic diagram of a laboratory dryer is illustrated in Fig.1. A sample holder, with a diameter of 140mm, thickness of 10mm and weight of 31gf, and which contained about 80gf of wheat was hung from the hook of an electric balance (max weight of 300gf). The sample's weight decreases with the evaporation of the wheat moisture. Then the change in the weight of the sample was measured and the grain moisture content is calculated from that. The drying air temperature and velocity were measured with thermo-couples and a hot-wire anemometer. The relative humidity of the inlet and outlet air was measured with electrical humidity sensors. These data were recorded on-line with the data logging system. The sample was Japanease *SHIROGANE* wheat produced in June, 1989. The experiment was carried out at temperatures of 30°C, 40°C, 50°C and 60°C and with the initial wheat moisture content of 20%d.b. to 30%d.b.. The samples of 25%d.b. and 30%d.b. were rewetted. The measurement interval was 30s. in the early drying and then changed 5min. to 10min.. Drying was continued until there was no change in weight.

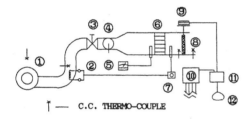

† — C.C. THERMO-COUPLE

① CENTRIFUGAL FAN   ⑦ TEMPERATURE CONTROLLER
② HEATER   ⑧ SAMPLE HOLDER
③ SLUICE VALVE   ⑨ ELECTRIC BALANCE
④ THREE-WAY VALVE   ⑩ DATA LOGGER
⑤ HOT-WIRE ANEMOMETER   ⑪ COMPUTER
⑥ STRAIGHTENING GRID   ⑫ MONITOR

Fig. 1. Schematic Diagram of the Laboratory Dryer.

RESULT AND DISCUSSION OF THIN-LAYER

Figure 2 shows the drying characteristic curve for a drying air temperature of 40°C and initial moisture content of 25%d.b.. This curve could be divided into three periods such as shown in fig.2. This tendency could be observed for all conditions in this experiment. To define the boundary of these periods, $d^2(dM/d\theta)/dM^2$ with a drying rate $dM/d\theta$ differentiated by M was calculated. The $d^2(dM/d\theta)/dM^2$ curve is also illustrated in fig.2. Two peaks were found on this curve. These peaks showed inflection points on the curve, or boundary points. It also indicated

that the drying process could be divided into three periods. These periods were respectively defined as the periods I, II and III. Generally, it is reported that two falling-rate periods can be found in the wheat drying process. As a result of this work, however, three periods, periods I and II, corresponding to the first falling-rate drying period and period III, corresponding to the second falling-rate drying period, were found. Since wheat has no husk such as rice, period I was much shorter than the other periods and finished soon, after about 11 minutes. It is difficult to observe period I with conventional measurement methods. Since moisture content could be measured at a short interval of 30 seconds in the early drying in this work, period I could be observed.

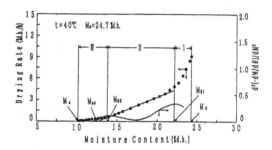

Fig.2 Drying Characteristic Curve

The relationships between the boundary of each period and drying conditions were investigated. Here it is defined that the boundary moisture content between period I and period II is M_{B1} and between period II and period III is M_{B2}. Maekawa(1978) defined that a boundary moisture content between the first falling-rate drying period and the second one was the second critical moisture content. The second critical moisture content was indicated by a first order regression equation for the drying air temperature, but the relationship between the second critical moisture content and the initial moisture content was not reported. The relationship between M_{B1} and M_{B2} and the drying air temperature and the initial moisture content was;

$$M_{B1} = 48.29 - 0.145T + 0.761M_0 \quad (R = 0.989) \tag{1}$$

$$M_{B2} = 58.7 - 0.144T \quad (R = 0.979) \tag{2}$$

M_{B1} and M_{B2} decreased with a increase in the drying air temperature. M_{B1} was dependent on the initial moisture content and increased in proportion to the initial moisture content. M_{B2} was independent on the initial moisture content. M_{B2} was under 15%d.b. in this test.

TWO-TERM EXPONENTIAL MODEL

The wheat drying operation is finished at a

moisture content 14 to 15%w.b.(16.3 to 17.6%d.b.) generally in Japan. So the prediction accuracy for the moisture content above 16.3%d.b. is important. Period III is under 15%d.b. in this experiment and the change in the moisture of wheat is smaller than other periods. Also the number of data is likely to increase in contrast to the other periods. As a result, the accuracy of the prediction in the early drying is adversely affected when the model is fitted to the drying curve by the least square method etc.. Then it is considered that period III can be neglected and the model may be fitted from the start of drying to the end of period II. Since period III is avoided, a new equilibrium moisture content of the model should be required. The linear part of period II (in Fig.2) was extended to the horizontal axial and a point of intersection of the extended line and moisture content line was defined as the new equilibrium moisture content M_{ed}. When the drying air temperature is 30°C, the drying rate is nearly zero for the end of period II. It was defined that M_{ed} was equal to M_{B2}. It was reported that the drying process of rough rice could be divided into three periods and each period could be interpreted by an exponential model (*Motohashi*, 1979; *Kameoka*, 1988). But, since wheat has no husk such as rough rice, it is difficult to consider that drying periods are independent of each other. Aiso the drying characteristic curve is not linear before and after M_{B1}. Moisture content on the surface of grain varies exponentially (*Brooker et al.*, 1981) and wheat drying is predicted by the exponential model. So the two-term exponential model was applied to the wheat drying process. The two-term exponential model is:

$$\frac{M - M_{ed}}{M_0 - M_{ed}} = C_1 \exp(-k_1 \theta) + C_2 \exp(-k_2 \theta) \qquad (3)$$

Procedure of Fitting

First the following exponential model, which is part of Eq.(3), was fitted to period II.

$$\frac{M - M_{ed}}{M_0 - M_{ed}} = C_2 \exp(-k_2 \theta) \qquad (4)$$

The reason this procedure was used was that wheat surface drying might finish for period I, so it was assumed that period II can be interpreted by only a one-term exponential model(Eqn.(4)). Parameters k_1 and k_2 of the two-term exponential model were assumed to be the drying constants of the grain surface and the intra-particle. The surface drying rate is much faster than the intra-particle one. Concerning to property of the two-term exponential model, if k_1 of the first exponential term is much greater than k_2 of the second exponential term, the first exponential term can be neglected in contrast to the second one with an increase in value of θ. Therefore, parameters C_2 and k_2 were calculated in this manner.

Secondly, since period I might consist of two exponential models as mentioned above, the one is a surface drying model and the other is an intra-particle drying model. The two-term exponential model, which added one more exponential model to Eq.(4), was fitted to period I. C_2 and k_2 were required for the first procedure. Eq.(3) was fitted to the period II and the other parameters C_1 and k_1 were calculated. The model could be fitted adequately using the procedures mentioned above.

Fitness of the Two-term Exponential Model

The prediction accuracy for wheat drying by the two-term exponential model was investigated. To compare with the two-term exponential model, an exponential model and a sphere model were fitted to the drying process which was not divided into drying periods. The exponential model and the sphere model are:

Exponential model:

$$\frac{M - M_e}{M_0 - M_e} = A_0 \exp(-k\theta) \qquad (5)$$

Sphere model:

$$\frac{M - M_e}{M_0 - M_e} = \frac{6}{\pi^2} \sum_{i=1}^{\infty} \frac{1}{i^2} \exp(-i^2 k\theta) \qquad (6)$$

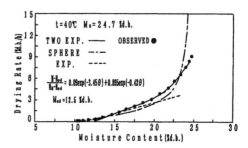

Fig. 3. Comparison of Drying Models.

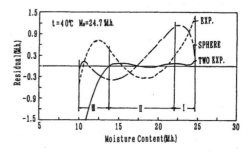

Fig. 4. Residuals vs. Value predicted by Models.

The parameters of the two models were calculated by the non-linear least square method. The results of the calculation are illustrated in Fig.3. It shows the accuracy of the two-term exponential model. In order to indicate the fitness of the model visually, the residual, which is the difference between the observed value and the predicted value by the model, are shown in Fig.4 (*Byler et al.*, 1987). The cross axis in Fig.4 is the observed moisture content. It was shown that the two-term exponential model is superior to the other models.

The Relationships between Drying Condition and Parameters of The Two-term Exponential Model

New equilibrium moisture content M_{ed}. M_{ed} was applied to Henderson's equation. The calculated value is:

Henderson's Eq.: $1 - rh = \exp(aTM_e^n)$

$$M_{ed}: \quad a = -7.563 \times 10^{-9}, \quad n = 4.65 \quad (R = 0.929) \quad (7)$$

M_{ed} was larger than the equilibrium moisture content. M_{ed} is closer to the equilibrium moisture content with a decrease in drying air temperature. Since M_{ed} was defined as a point of intersection of the extended line from period II and the moisture content line, it is larger than the equilibrium moisture content which is the end of period III. The gradient of period II in Fig.2 becomes smaller with a decrease in the drying air temperature. Then M_{ed} becomes closer to the equilibrium moisture content.

Drying constants k_1 and k_2. The parameters k_1 and k_2 were assumed to be drying constants of the grain surface and the intra-particle. In diffusion theory, the relationship between grain drying constant and grain temperature is usually of Arrhenius-type. k_1 and k_2 were applied to the Arrhenius equation;

$$k_1 = 4.98 \times 10^6 \exp(-4.38 \times 10^3/T) \quad (R = 0.922) \quad (8)$$

$$k_2 = 2.53 \times 10^8 \exp(-6.31 \times 10^3/T) \quad (R = 0.992) \quad (9)$$

The drying constant k of the exponential model (Eq.5) for comparison with k_1 and k_2 is:

$$k = 5.51 \times 10^6 \exp(-5.24 \times 10^3/T) \quad (R = 0.945) \quad (10)$$

k_2 is close to k in relation to the order, but k_2 is greater than k.

Parameters C_1 and C_2. It was found that C_1 and C_2 were not dependent upon the initial moisture content and the sum of both was about 1. The result is appropriate. The relationship between C_2 and the drying air temperature is:

$$C_1 = 1 - C_2$$

$$C_2 = 1.336 - 0.0014T \quad (R = 0.834) \quad (11)$$

C_2 could be indicated by the regression form of the drying air temperature. When C_2 is assumed to be the shape factor of the exponential model, it is constant. But in the beginning of period II, the drying characteristic curve has a slight curvature. So, C_2 varies according to the range

of the period to which the model is fitted. Dependence of C_2 upon temperature was relatively small. It could be considered that C_2 was constant. The average C_2, 0.89, is slightly greater than the exponential model's shape factor, 0.86.

THE STATE EQUATION OF THIN-LAYER DRYING PROCESS

Heat and mass balances are required to develop a state equation of the thin-layer drying process (*Spencer*,1969; *Toyoda et al.*,1985). Here, the following assumptions are made to simplify the equation: (1) the volume shrinkage is negligible during the drying process, (2) the particle to particle conduction is negligible, (3) the bin walls are adiabatic, with negligible heat capacity, (4) the heat capacities of moist air and of grain are constant during short time periods.

Heat Balance Equations

Heat balance of the drying air. The heat of the air flowing into thin-layer in time $d\theta$ is:

$$Q_1 = G_a C_a T_{ain} d\theta \quad (12)$$

The change in latent heat of the air in time $d\theta$ is:

$$Q_2 = L_v \rho_s SX(1 - \varepsilon) dm \quad (13)$$

The change in sensible heat of the air in time $d\theta$ is:

$$Q_3 = C_s \rho_s SX(1 - \varepsilon) dT_s \quad (14)$$

The heat value of the air flowing through thin-layer in time $d\theta$ is:

$$Q_4 = G_a C_a T_{aout} d\theta \quad (15)$$

The heat balance equation is:

$$Q_1 = Q_2 + Q_3 + Q_4 \quad (16)$$

The differential heat balance equation of the air is:

$$T_{ain} - T_{aout} = SX(1 - \varepsilon)\frac{L_v \rho_s}{G_a C_a}\frac{\partial m}{\partial \theta} + SX(1 - \varepsilon)\frac{C_s \rho_s}{G_a C_a}\frac{\partial T_s}{\partial \theta} \quad (17)$$
where

$$\frac{\partial m}{\partial \theta} = \frac{1}{100}\frac{\partial M}{\partial \theta}$$

The heat balance of the grain. The change in sensible heat of grain in time $d\theta$ is:

$$Q_{s1} = (1 - \varepsilon)\rho_s C_s SX dT_s \quad (18)$$

The change in sensible heat of moisture evaporation in time $d\theta$ is:

$$Q_{s2} = (1 - \varepsilon)\rho_s L_v SX dm \quad (19)$$

The change in sensible heat of grain that results from heat transfer in time $d\theta$ is:

$$Q_{s3} = hav(T_{ain} - T_s)SX\,d\theta \qquad (20)$$

The heat balance of the grain is:

$$Q_{s1} = Q_{s2} + Q_{s3} \qquad (21)$$

The differential heat balance equation of the grain is:

$$\frac{\partial T_s}{\partial\theta} = \frac{hav(T_{ain} - T_s)}{(1-\varepsilon)\rho_s C_s} + \frac{L_v}{C_s}\frac{\partial m}{\partial\theta} \qquad (22)$$

The mass balance of thin-layer. The amount of evaporated moisture from grain in time $d\theta$ is:

$$(1-\varepsilon)\rho_s SX\,dm \qquad (23)$$

The change in humidity of the air in time $d\theta$ is:

$$(H_{in} - H_{out})G_a\,d\theta \qquad (24)$$

Since the amount of evaporated moisture from grain is equal to the change in the humidity of the air, the differential mass balance of thin-layer is:

$$H_{out} - H_{in} = \frac{S}{G_a}\frac{\partial m}{\partial\theta}(1-\varepsilon)\rho_s X\,dm \qquad (25)$$

The State Equation

The state equation of the thin-layer drying process is helpful in designing a controller for the dryer or when applied to deep-bed drying. Then the state equation is developed from the differential heat and mass balances equations. The two-term exponential model is inadequate to be applied to the state equation because it causes non-linear factor. So the model was transformed into the following equations:

$$M_1 = (M_0 - M_{ed})C_1 \exp(-k_1\theta) \qquad (26)$$

$$M_2 = (M_0 - M_{ed})C_2 \exp(-k_2\theta) + M_{ed} \qquad (27)$$

$$M = M_1 + M_2 \qquad (28)$$

The Eqns.(17),(22) and (25) to (28) are transformed into the following differential equations:

$$T_{ain} - T_{aout} = A\frac{\partial M}{\partial\theta} + B\frac{\partial T_s}{\partial\theta} \qquad (17)'$$

$$\frac{\partial T_s}{\partial\theta} = D\frac{\partial M}{\partial\theta} + E(T_{ain} - T_s) \qquad (22)'$$

$$H_{out} - H_{in} = F\frac{\partial M}{\partial\theta} \qquad (25)'$$

$$\frac{\partial M_1}{\partial\theta} = -k_1 M_1 \qquad (26)'$$

$$\frac{\partial M_2}{\partial\theta} = -k_2(M_2 - M_{ed}) \qquad (27)'$$

$$\frac{\partial M}{\partial\theta} = \frac{\partial M_1}{\partial\theta} + \frac{\partial M_2}{\partial\theta} \qquad (28)'$$

where

$$A = SX\frac{L_v\rho_s}{100G_aC_a}(1-\varepsilon), \quad B = SX\frac{C_s\rho_s}{100G_aC_a}(1-\varepsilon)$$

$$D = \frac{L_v}{100C_s}, \quad E = \frac{hav}{\rho_s C_s(1-\varepsilon)}, \quad F = \frac{SX\rho_s}{100G_aC_a}(1-\varepsilon)$$

It is assumed that the thin-layer drying process can be expressed as the following system.

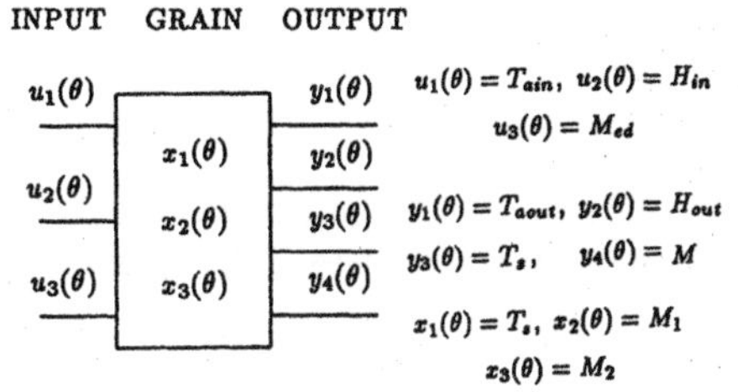

INPUT   GRAIN   OUTPUT

$u_1(\theta) = T_{ain}$, $u_2(\theta) = H_{in}$
$u_3(\theta) = M_{ed}$

$y_1(\theta) = T_{aout}$, $y_2(\theta) = H_{out}$
$y_3(\theta) = T_s$, $y_4(\theta) = M$

$x_1(\theta) = T_s$, $x_2(\theta) = M_1$
$x_3(\theta) = M_2$

Fig. 5.   System of Thin-layer Drying Process

The state equation of the thin-layer drying process was developed from the system in Fig.5 and the differential equations. The state equation and the output equation are also shown and time θ is omitted to simplify the equations.

$$\begin{Bmatrix} \dot{x}_1 \\ \dot{x}_2 \\ \dot{x}_3 \end{Bmatrix} = \begin{bmatrix} -E & -k_1D & -k_2D \\ 0 & -k_1 & 0 \\ 0 & 0 & -k_2 \end{bmatrix}\begin{Bmatrix} x_1 \\ x_2 \\ x_3 \end{Bmatrix} + \begin{bmatrix} E & 0 & k_2D \\ 0 & 0 & 0 \\ 0 & 0 & k_2 \end{bmatrix}\begin{Bmatrix} u_1 \\ u_2 \\ u_3 \end{Bmatrix}$$

$$(29)$$

$$\begin{Bmatrix} y_1 \\ y_2 \\ y_3 \\ y_4 \end{Bmatrix} = \begin{bmatrix} BE & (A+BD)k_1 & (A+BD)k_2 \\ 0 & Fk_1 & Fk_2 \\ 1 & 0 & 0 \\ 0 & 1 & 1 \end{bmatrix}\begin{Bmatrix} x_1 \\ x_2 \\ x_3 \end{Bmatrix}$$

$$+ \begin{bmatrix} 1-BE & 0 & (A+BD)k_2 \\ 0 & 1 & -Fk_2 \\ 0 & 0 & 0 \\ 0 & 0 & 0 \end{bmatrix}\begin{Bmatrix} u_1 \\ u_2 \\ u_3 \end{Bmatrix} \qquad (30)$$

Result of Calculation using the State Equation

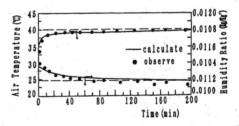

Fig. 6.   The Result of Calculation of Step Response

The change of outlet air in the experiment of this study could be regarded as the step response to the inlet air. The result of the calculation for the step response at the inlet air temperature of 40°C with initial moisture content of 30%d.b. is indicated in Fig.6. Figure 6 indicates that the calculated values demonstrated clearly the tendency of the change in temperature and humidity of the air. It shows the effectiveness of this state equation.

Modelling of deep-bed drying. A transfer function derived from the state equation of a thin-layer is defined as $G_{Ti}(s)$. The deep-bed is assumed to be composed of a series of thin-layers. If the deep-bed is composed of m layers, then the transfer function of deep-bed $G_d(s)$ is:

$$G_d(S) = \prod_{i=1}^{m} G_{Ti}(S) \tag{31}$$

The Eq.(31) is a basic equation for the modelling of deep-bed grain drying in this study.

CONCLUSION

The thin-layer drying process of wheat could be divided into three drying periods with a view to improving the prediction accuracy of the drying model. It was assumed that the drying process of wheat was composed of the surface drying and the intra-particle drying. Then the two-term exponential model could be applied to each drying period except period III. The two-term exponential model could predict the drying characteristic during period I to II accurately. When a drying model is fitted to the drying process, it is an effective method because the drying process is divided into a few periods with a view to improving the prediction accuracy of the model. The state equation of thin-layer grain drying was developed from the differential heat and mass balances equations on the basis of the two-term exponential model. The step response to the inlet drying air was calculated from the state equation and its value predicted well the tendency of change in state of the outlet air. A model of grain deep-bed drying can be derived from the state equation of the thin-layer. However, this state equation is too complex to be extended to the deep-bed. More simplification of the state equation is required.

NOMENCLATURE

A_0 :shape factor (-)
C_n :parameters of the two-term exponential model (-)
C_a :specific heat of air (kJ/kg/°C)
C_s :specific heat of grain (kJ/kg/°C)
G_a :air flow rate (kg/h)
H :humidity ratio of air (kg/kg')
hav :heat conductivity (W/m²/K)
k :drying constant (h⁻¹)
k_n :drying constant of two-term exponential

model (h⁻¹)
L_v :latent heat of evaporation (kJ/kg)
M :moisture content (%d.b.)
M_0 :initial moisture content (%d.b.)
M_{B1}:boundary moisture content between period I and period II (%d.b.)
M_{B2}:boundary moisture content between period II and period III (%d.b.)
M_e :equilibrium moisture content (%d.b.)
M_{ed}:new equilibrium moisture content of two-term exponential model (%d.b.)
R :coefficient of correlation (-)
rh :relative humidity (%)
S :drying area (m²)
T :drying air temperature (K)
T_a, t:drying air temperature (°C)
T_s :grain temperature (°C)
X :thickness of grain thin-layer (m)
ε :void ratio (-)
ρ_s:bulk density of grain (kg/m³)
θ :time (h)

Subscripts
in:inlet
out:outlet

REFERENCES

Brooker,D.B., F.W.Bakker-Arkema and C.W.Hall. (1981). Theory and Simulation of cereal grain drying. Drying cereal grains. The Avi publish., 185-221.

Byler,R.K., C.R.Anderson and R.C.Brook. (1987). Statistical methods in thin layer parboiled rice drying model. Trans. of ASAE, 30(2), 533-538.

Henderson, S.M. (1974). Progress in developing the thin layer drying eqution. Trans. of ASAE, 17(6), 1167-1168.

Kameoka,T. (1988). Thin layer drying characteristics of rough rice(2). Journal of JSAE, 50(4), 57-65.

Maekawa,T. (1978). Studies on convective drying for agricultural products. Memoirs of Tokyo Univ. of Education, No. 24, 48-57.

Motohashi, K. (1979). The application of forced air drying theory to the deep bed drying of rough rice(2). Journal of JSAE, 41(4), 593-601.

Nellist,M.E. (1976). Exposed layer drying of ryegrass seeds. Agric. Engng. Res., Vol. 21,49-66.

Spencer,H.B. (1969). A mathematical simulation of grain drying. J. agric. Engng Res, 14(3),226-235.

Toyoda,K., T.Maekawa and S.Yamazawa. (1985). An investigation of drying conditions of a circulating grain dryer. Sci. Rept. Fac. Agr. Kobe Univ, Vol. 16(2), 499-508.

QUALITY EVALUATION OF RICE AND COFFEE
GRAINS BY USING NEAR-INFRARED RAYS

M. Matuda and K. Kagawa

Dept. of Development and Engineering, Satake Engineering Co., Ltd., Japan

Abstract.   The taste analyzer developed by the Satake Engineering Co., Ltd. is a measuring
apparatus systematically bridging the gap between the sensory perception of rice taste and the
instrumental analysis.   The theory for the measurement is that the major ingredients values
are obtained through light absorption of rice with the near-infrared rays of various
wavelengths, and the taste of rice is then quantitized based on the values.   Since the sensory
values as the reference data for the quantitization has a certain degree of ambiguity, the
application of the Fuzzy Theory has been attempted.   In addition, the neural network theory is
used in the process of deriving taste values from the ingredients compositions, and a good result
has been obtained.

Keyword. SENSORY EVALUATION, NEAR-INFRARED SPECTRO-ANALYSIS,
FUZZY SYSTEMS, NEURAL NETS

INTRODUCTION

As a means to evaluate the tastes of rice or coffee,
there has been used a technique called sensory test, in
which selected persons proficient in the test actually
try the food or beverage.   With this testing method,
some fluctuations are unavoidable due to change in
the physical conditions of the panelists irrespective of
their proficiency, and when the tests involve large
quantity of test samples or are carried out at different
occasions, it is hard to expect the consistent and
reproducible results of the evaluation.   moreover, the
test requires a lot of workload and time, which
hampers the reproducibility of the test operation
itself.

The prices of goods are usually determined by the
demand-supply balance.   In case of food and
beverages, the demands are believed to be dependent
on their taste besides the nutritious values.
Therefore, the prices of food should be in proportional
relations with the taste.

The proportional relationship, however, is extremely
hard to be realized, because, as mentioned earlier, the
sensory tests are costly, time-consuming, and also
inferior with respect to the objectivity and the
reproducibility.   The prices, then, are actually
determined only by the kind and the grade (e.g.
amount of foreign matters mixed in) of the materials.
That cannot be helped because of the lack of simple
and economical methods for determining the taste
quality.

If any simple and economical methods for inspecting
the taste quality are provided in food processing
plants dealing with a lot of products, products of
consistent taste quality can be supplied to the
consumers at reasonable prices to the great benefit of
both processors and consumers.

SYSTEM FOR QUALITY
EVALUATION

In the 1970s, a non-destructive inspection methods
called near-infrared rays analysis began to diffuse in
the United States, and in the 1980s it also began to
spread in Japan.   The frequencies around the near-
infrared region match those of harmonics and
combination tones of frequencies proper to the
molecules comprising organic materials.   By
analyzing the amount of absorbed near-infrared rays
in the material, such information as the amount of
ingredients in the organic materials can be obtained.
In the early stage, the characteristics of near-infrared
rays are utilized to develop simplified methods for
quantitative analysis of ingredients.   After that, by
using the fact that it enables quantitative analysis of
multiple ingredients at the same time and in a very
short time, it was made possible to carry out the
quantitative analysis in a short time with respect to
the overall material properties determined by the
combinatorial effects of the ingredients.

The near-infrared analyzing apparatuses include the
fixed-filter type and the scanning type.   The fixed-
filter type is equipped with several types of
interference filters, each allowing only rays of a

particular frequency pass, and the target values are calculated based on the degrees of absorption with respect to the rays passing through the filters. Since this type of apparatuses only allows the measurement of particular wavelengths, it is not suited for research purposes and is marketed as apparatus for some particular ananlyzing purposes. (See Fig.1 for the basic construction.)

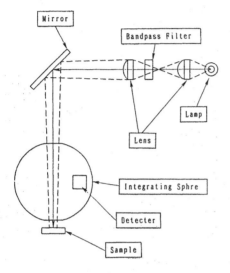

Fig. 1

The scanning type has the capability of measuring the absorbance of near-infrared rays (700 - 2500 nanometer wavelengths) at an interval of approx. 1.0 nanometer at minimum with the use of diffraction gratings. This type is used chiefly for research purposes to determine the effective wavelengths for the analysis of a kind of specimen, and the results are applied to the fixed-filter type units. This type has the drawback of costliness due to the complication of the constructions. (Fig.2 shows the light absorbing characteristics of polished rice and raw coffee beans.)

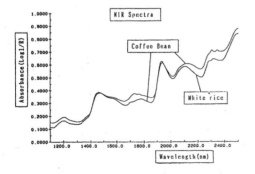

Fig. 2

The light absorbance is calculated by the following equation:

$$absorbance = \log (\text{incident light intensity} / \text{reflected light intensity})$$

The intensity of the incident light is usually substituted by the reflected light flux from the standard reflecting material.

The configuration of an analyzing system is as indicated in Fig.3. The aforementioned fixed-filter type is equipped with the interference filters of wavelengths identified effective as a result of analysis by the scanning type unit. Polished grains are used as specimen for rice, and in the case of coffee, raw beans are processed to a degree of grains by crushers and tempered in thermo-static devices (for approx. 1 hour). The samples should desirably be of consistent-sized grains: In the case of rice, they should of grains size of 200 micrometers or less in diameter with the peak distribution of 140 micrometers; and with respect to coffee, 500 micrometers or less with the peak distribution of 350 micrometers. The crushed and tempered samples are encapsulated into the cells having quartz glass on one side, and the absorbance are measured with the near-infrared rays using the filters. (An example of transmitted light curve is shown in Fig.4.) The values are fed to the analyzing equations stored in the personal computer to work out the amounts of ingredients.

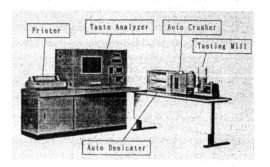

Fig. 3

The measurement of the ingredients by the near-infrared rays analyzers are generally carried out based on the following equation:

$$A = F_0 + F_1 * \log T_1 + \cdots + F_n * \log T_n$$

Where A : amount of ingredient to be obtained (%)
F_0 : intercept
F_n : coefficient for each filter
$\log T_n$: light absorbance for each flter

In the case of rice, the ingredients to be measured are amylose, proteins, water, and fatty acids ; and for coffee, proteins, chlorogenic acid, sucrose, lipid,

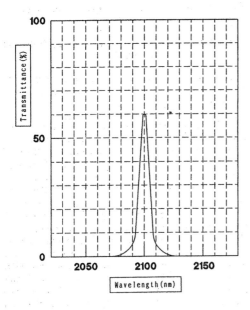

Fig. 4

anhydrous caffeine and water. These substances are major contributor to the taste quality of rice and coffee. Especially with rice, the relationship between the ingredients and the taste quality is fairy understood, and so far the following equation has been used to calculate the gustatory values.

Gustatory value = $P + Q * (amylose)^a * (protein)^b * (water)^c * (fatty\ acids)^d$

Where   P, Q, a, b, c and d are constants.

With respect to coffee, it is known that the aforementioned ingredients are working as premonitory substances to draw out its taste, but the contribution of their amount and interactions is not so far made clear.

The taste analyzer shown in Fig.3 contains in it the near-infrared spectro-analyzer, personal computer, CRT monitor, and voltage stabilizer unit.   In the analysis of coffee, the test rice polisher in the figure is not required.

Table 1 represents the correlation of the results from the wet chemical analysis and the near-infrared analysis of major ingredients of rice.   The same correlation is shown about coffee in Table 2.

Table 1

Ingredient	Wet analysis	Correlation coefficient
Amylose	Light absorbance method	0.850
Protein	Kjeldahl method	0.958
Water	Dry-out method	0.916
Fatty acids	Standard method by Japanese Food Agency	0.896

Table 2

Ingredient	Wet analysis	Correlation coefficient
Protein	Kjeldahl method	0.943
Chlorogenic acids	Light absorbance method	0.881
Sucrose	Liquid chromato-graph method	0.891
Lipid	Petroleum ether method	0.941
Anhydrous caffeine	Liquid chromato-graph method	0.903
Water	Dry-out method	0.923

Fig.5 illustrates the conceptual representation of the quality evaluation by the near-infrared rays spectro-analysis method from the measurement stand point.

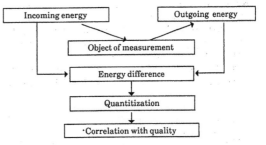

Fig. 5

Despite their high price level, more than 100 units of rice taste analyzers have been so far shipped and are now in use, which exhibits their usefulness and practicality in actual use.

Advantages for using this sort of analyzers are :

1) Sensory values can be expected on a consistent basis under the same standard irrespective of time and place, and only a small amount of sample is required.

2) Results can be obtained at very low costs and in a short time, and can be fed to the next process.

3) No particular proficiency and knowledge are required in the analysis operations.

4) Since no chemicals are used in the analysis, they are safe and have high adaptability to work environment.

Moreover, the analysis of the tastes will facilitate economical simulations of blend effects, and could be used for the applications to quality checkup in the improvement of grain types and culturing technologies. The taste analysis will have extensive fields of applications.

What we are reporting is the result of the application of Fuzzy Theory and neural network to the "Correlation with quality" as shown in Fig.5.

APPLICATION OF FUZZY THEORY

The representation of the object variable or sensory value is the key subject in the process of determining the quality of rice or coffee, or of working out their sensory value of taste, based on the amount of their major ingredients quantitized by the near-infrared rays spectro-analysis. As mentioned earlier, human senses have essentially some vagueness, and it is very hard to represent the perception with an absolute value. With the efforts of various institutions, however, some methods were worked out and the quantitization has been carried out by the methods. In case of rice, about 20 panelists evaluate the samples with respect to each of the glutinosity, toughness, flavor and appearance in comparison with a standard sample to score by the measure of three points both to positive and negative directions with one point of being equivalent, and then they evaluate the overall taste value for the same samples by the same measure. Similar method is adopted for coffee. Major factors of evaluation include flavor, bitterness, acidity, sweetness, and body.

We have attempted to apply the Fuzzy Theory to the process of deriving the overall taste value from the taste factors which are relatively specific sensory evaluated values. The evaluations are carried out by the 20 panelists rather proficient in sensory tests. A total of 105 kinds of rice samples and 30 kinds of coffee samples are selected which are considered deferent in taste for their evaluation with respect to the factors and the overall taste value in comparison with a standard sample. The obtained data are averaged and normalized on [0,1] interval, and the obtained values are adopted as sensory values for each material.

The concepts of Fuzzy measurement and Fuzzy integration are applied to the process of deriving the overall taste value based on the evaluated values of taste factors for each material, with the Fuzzy integration as the overall taste value.

The Fuzzy integration on E (⊂ X) by the Fuzzy measurement (g) is defined as follows :

$$f_E \; h(x) \cdot g \;\; = \sup[a \wedge g(E \cap H a)]$$

Where g : Fuzzy measurement for function
                 h : X → [0, 1]
Ha: {x | h(x)≥a }
E :   area of integration

The function h (x) is fitted to the evaluated values of the aforementioned sensory evaluation factors, and the Fuzzy measurement to the value obtained by AIC (Akaike's Information Criterion) and normalized in [0, 1] interval. The AIC is a value derived by the following formula :

$$AIC = n \cdot (\log_e 2\pi + 1) + n \cdot \log_e (Se/n) + 2(P+2)$$

Where n :   number of date items used for multiple regression annalysis

   P :   number of explanatory variables
   Se: residual sum of squares

Fig's 6 and 7 show the result of the calculation.

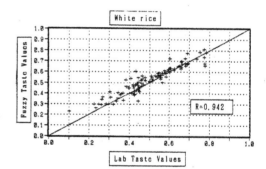

Fig.6

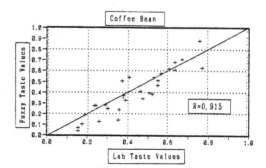

Fig. 7

192

APPLICATION OF
NEURAL NETWORK

We applied the neural network to the analysis, with the amounts of ingredients (see Table 1 and 2) affecting the taste of materials obtained by the near-infrared rays spectro-analysis as input data, and with the overall taste value derived through the Fuzzy integration as previously discussed as the tutorial data. Software used for the analysis is RHINE Ver 2.0 by Century Research Center. Figures 8 and 9 illustrate the interrelation between the overall taste value and the tutorial data calculated by the process.

The correlation chart for coffee is shown in Fig.10 indicating the inter-relation between the evaluation value derived by the neural network method and the original tutorial data, with the composition of each ingredient as input data, and with mean value for each evaluation factor as the tutorial data.

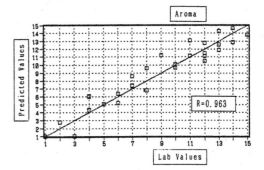

Fig. 10

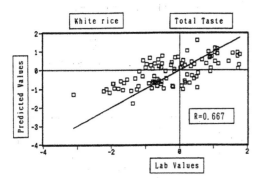

Fig. 8

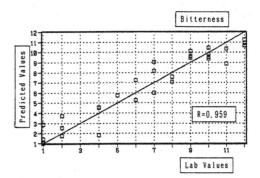

Fig. 11

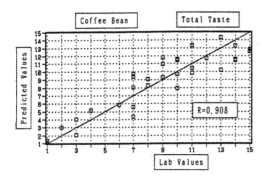

Fig. 9

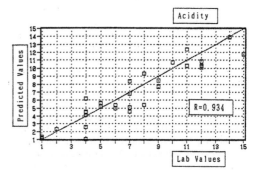

Fig. 12

193

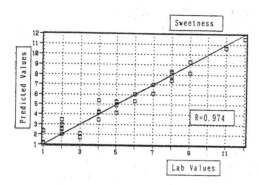

Fig. 13

REFERENCE

○ T.Terano, K.Asai, M.Sugeno, "Fuzzy Systems theory and Its Application," Ohm sha Co., Ltd." 1987

○ Nakano, Iinuma, Kiriya," Neuro Computer," Gijutu Hyoron sha Co., Ltd." 1988

○ Goro Tomoda," Coffee Technology," Hourin Co., Ltd." 1986

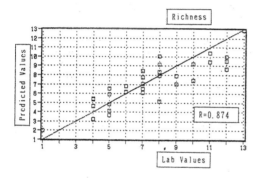

Fig. 14

CONCLUSION

1) By the application of Fuzzy measurement and Fuzzy integration concepts, an objective method has been developed for the evaluation of the taste of rice and coffee effectively overcoming the subjectivity of individuals.

2) The neural network has been used in working out the taste value based on the ingredients compositions affecting the taste of rice and coffee, thus enabling the use of non-linear inference, resulting in a good correlation between actual sensory true values and calculated values.

194

Copyright © IFAC Mathematical and Control Applications in Agriculture and Horticulture, Matsuyama, Japan 1991

QUALITY EVALUATION OF ARTIFICIALLY DRIED RICE BY TASTEMETER

T. Abe, O. E. Chukwudi, Y. Hikida and J. Yamasita

Dept. of Biomechanical Systems, Faculty of Agriculture, Ehime University, Tarumi 3, Matsuyama 790, Ehime, Japan

Abstract. Increasing demand by the general public here in Japan for high quality rice, has led to greater efforts by researchers to develop drying techniques that would enhance the quality of dried rice grains, and new methods of evaluating the palatability of rice. A new method of evaluating the taste of rice is being developed. In order to analyzed the taste of rice by nondestructive method, near infrared (NIR) spectrum analysis is used to measure the quantities of certain key components which make up taste such as amylose or amylopectine, starchs, protein, moisture, and free fatty acid. These measured quantities are then computed by a mathematical formular designed to produce results coinciding with taste appraisals obtained in sensory testing. This paper presents experimental results of the work under taken for the purpose of developing new generation of efficient grain drier and grain drying techniques that would enhance the taste of the dried grain. Newly harvested rice grains were dried under different drying conditions, after which the palatability was evaluated using tastemeter. From the taste evaluation results, it is evident that drying method affects the taste of rice.

Keywords. Agriculture; Computer evaluation; Estimation; Drying; Grain; Postharvest technology.

INTRODUCTION

Japanese rice culture techniques are especially advanced. The introduction of high yielding cereals, well planned and effective irrigation system and efficient weed and pest control mechanism led to increased cereals yield per hectare. Once grown, the grain crops must be harvested, processed, stored, transported and marketed without excessive losses and waste. Any loss or reduction of quality after maturity cause a double waste. First the waste product does not fulfil its primary objective of providing food, and second, the low quality grain provides little returns on investment for its production.

The mechanization of harvesting has rapidly progressed in recent years and more than 55% of the total area of paddy fields is now harvested by head-feeding combines. Paddy harvested by head-feeding combines contains much moisture and, if it is left in bags or in heaps it will deteriorate through putrefaction or fermentation. So it must be dried to a safe moisture content level using a mechanical drier. The high demand and higher price in Japan for high quality and tasty rice has resulted in efforts by both researchers and

designers to develop more efficient method of drying rice grains and also a faster and efficient method of evaluating the taste of rice.

Conventionally, the taste of rice is judged by an overall evaluation of the results of sensory taste in which the five factors of appearance, taste, viscosity, hardness and flavor are evaluated. The physical chemistry method of rice taste analysis, which has a high correlation with the overall evaluation in conventional sensory testing, examines the protein content of milled rice, the amylographic characteristics of powdered milled rice, the maximum and minimum viscosity and brake down in the area of amylographic characteristics of powdered milled rice, the iodine color degree. In the area of cooked rice, the hardness and stickiness, texturometer characteristics are examined. These method are effective, but time consuming.

The near infrared spectrum analyzer is being developed for the purpose of analyzing the taste of rice by non destructive method. It was a general believe in the past that grains dried at a low drying temperature usually has better quality, this assumption was based on visual inspection of the dried grain. However, in 1989 summer, we carried out series of drying experiment using continuous flow tempering drier. It was observed contrary to the general believe when grains were dried at low temperature(40 °C and below) because of the low drying rate, the grains spend considerable length of time in the drier's tempering tank where the temperature is on the avarage 35 °C, because of the high moisture content of the grains and the high ambient temperature in the tempering tank the grains starts to deteriorate. This phenomenon was confirmed when the taste of the dried grains were evaluated using the near infrared spectrum analyzer taste meter.

Drying experiments were carried out under different drying condition such as drying air temperature, tempering condition and grain-airflow ratio. Taste evaluation test was later carried out to determine the effect of drying methods on the taste of rice. Grain dried at 50 °C, grain-airflow rate of 0.3-0.5m³/min/100kgf had a high taste value. In Japan at present there are four organizations working to develop and commercialize the near infrared spectrum analyzer tastemeter. The basic working principles are the same, however differences exist in the measuring procedure and the use of the key components quantities in calculating the taste value.

METHODOLOGY

The study reported here concerns a new method or evaluating the taste of rice and effect of drying parameters on grain crack ratio, viability, free fatty acid content and taste of the artificially dried grain. An experimental dryer shown in Fig. 1 was built and used for drying newly harvest rice grain of Japonica specie(Natsuhikari) in the summer month August and (Nipponbare) in the autumn month of October under each of the following different drying conditions.

TABLE 1 Drying condition

Drying air temp.	40, 45, 50, 55, 60 °C
Grain air ratio	0.1 to 0.5m³/m/kgf
Tempering temp.	25, 35 °C
Tempering time	30, 45, 60min
Exposure time	3, 5, 10min
Weight of grains	1kgf
Initial m.c.	27.1%w.b.
Final m.c.	14.5 ∼ 15.0%w.b.

While using the test drier for the experiment, to replicate the condition that exist in the tempering tank of the commercial tempering drier, after the grains were dried for a specified period, it was removed from the drier and put in a polyethylene bag and placed in an enclosed

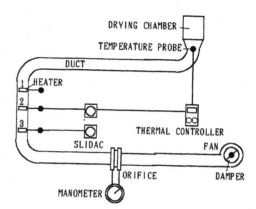

Fig. 1. Drying arrangement

chamber kept at constant temperature of 25 °C and 35 °C respectively. A sample of 10g of the grain was used to determine the initial moisture content in an air oven at 135 °C for 24hrs (JSAM standard). The dry matter thus calculated in the 10g sample and the weight change during the oven drying were used to determine the initial moisture content wet basis.

Effect of drying parameters such as drying temperature, grain air flow ratio, tempering conditions and exposure time on the quality of the grain were investigated. Crack ratio of grain under a particular drying condition was determined by inspecting a sample of the dried grain under fluoresent light. The number of cracked grains per given sample was used to calculate the crack ratio for a particular batch of grain. Free fatty acid content of the grains, which is one of the important indices for the change of rice qualities during storage, were determined. A colorimetric method, which is improved Duncome method, for fatty acid determination, was used.

Taste evaluation procedure. The use of near infrared absorption characteristics of protein, moisture, starch and free fatty acid is a relatively new technique that holds promise for automatic online taste evaluation of rice and control of wide range of rice grade. For the S base evaluation(STU), 2.0g of white rice is grind and sample is put into the sample holder and placed in the measuring chamber. A typical near infrared spectrum analyzer system irradiate the grain sample with near infrared rays, the different components of the grain absorbs the near infrared energy and reflects back some energy based on their respective absorption properties. The radiance flux from the sample is received by infrared sensors. This reflected energy is converted into an electrical signal proportional to the intensity of the radiance recieved, processes the analog electrical signals into a digital data usable by the mini computer incorporated in the system to represent all the quality attributes components and the respective quantities of these components. These measured quantities are then computed by a mathematical formular designed to produce results that coincide with taste appraisal obtained in sensory testing. The measure taste is displayed as a digital number from 0 to 100.

Seed germination. Effort was made to investigate the effect of drying parameter on grain germination, and also to verify the general believe that grains having low germination rate often has low taste value. 10 days after the grains were dried, germination test was carried out to break the seed dormancy during the summer experiment. The seed were soaked in distilled water for 48hours, then kept at 40 ℃ for 8hours and 5 ℃ for 16hours everyday for the duration of the experiment.

RESULTS AND DISCUSSION

Generally rice lower in amylose produces a cooked rice high in viscosity. Rice high in protein produces a cooked rice that is less sticky. Rice of high moisture content 16% wet basis produces a cooked rice that is softer and sticker. The higher the free fatty acid content, the less favorable the taste of the cooked rice becomes. Many other factors contribute to the taste of

rice, such as, rice species, cultural practice, soil type and climatic factors. However the batch of rice evaluated were the same environment and cultural practices.

Fig.2 show the taste evaluation results obtained using the near infrared spectrum analyzer for grains dried with the test drier. High taste values were obtained from grains dried at 40 ℃ exposure time of 10min., grain air ratio of 0.2m³/min/kgf, tempering temperature of 35 ℃ and grains dried at 50 ℃, grain air ratio of 0.2m³/min/kgf, tempering temperature of 25 ℃. Usually most tempering driers has circulation time of 50min, drying time of 5min and tempering time of 45min. From Fig.2 it appears increasing the drying time from 5min. to about 10min. would enhance the taste of the dried grains. Low temperature drying within a particular range gives good results, as can be observed from Fig.2, however it shows clearly that drying temperature of 40 ℃ and below is not adequate for grain drying. High temperature drying 55 ℃ to 60 ℃ gave an interesting result. Grains dried at 55 ℃ had lower taste value when compared with those dried at 60 ℃.

Grains are living matters containing enzymes that facilitate most of the biochemical activities taking place in the grains. At temperature of 60 ℃ most of the enzymes are inactivated resulting in a slow down in biochemical activity. Consequently re- ducing the production of such biochemical by product like free fatty acid which is a major indices for taste evaluation.

Tempering temperarure during summer is on the avarage 35 ℃, however it can be observed from Fig.2 that tempering of grain at 25 ℃ resulted in a high taste

Table 2 Rice taste unit.

Rice composition	40 ℃	45 ℃	55 ℃
Amylose(%)	19.72	19.64	19.77
Protein(%)	6.69	6.63	6.73
Moisture(%)	13.80	14.10	13.50
Fatty acid(KOH%)	0.039	0.023	0.047
STU	68.0	71.0	68.0

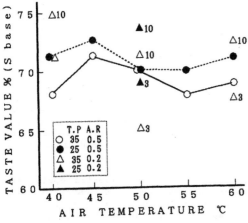

Fig. 2. Taste values for 1kgf dryer

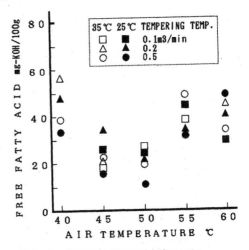

Fig. 3. Free fatty acid for six different drying conditions.

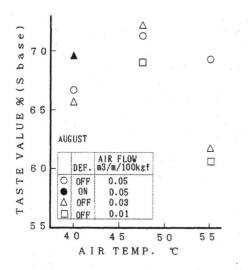

Fig.4(a). Taste values for 100kgf drier.
(S base)

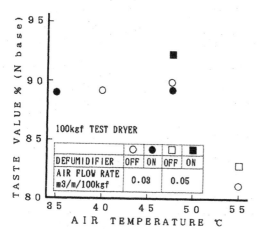

Fig.4(b). Taste values for 100kgf drier.
(N base)

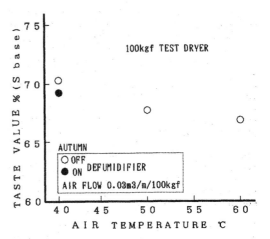

Fig. 5 Taste values for 100kgf drier.
(S base, Autumn)

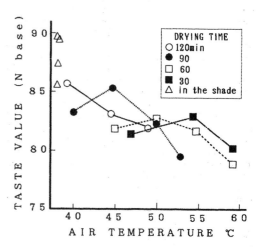

Fig. 6. Taste values for layer type drier

value. It is therefore considered that
lowering the tempering tank temperature
from 35 ℃ to 25 ℃ would improve the taste
value of the dried grain. The taste result
comfirms the believe that the amount of
free fatty acid content is a major factor
that determines the taste of rice. The
free fatty acid content of the batch of
rice which the taste results is shown in
Fig.2, is also shown in Fig.3, it shows
clearly that the batch of grain with the
least amount of free fatty acid content
had the highest taste value. Fig.3 shows
the free fatty acid content for the batch
of grain which taste value is shown in
Fig.2. From Fig.3 it can be observed that
grains tempered at 25 ℃ had a lower free
fatty acid content as compared to those
tempered at 35 ℃. It is also observed that
at a high grain-air ratio the free fatty
acid content of the grain are lower as
compared to when the grains were dried at

a lower grain-air ratio. It is therefore
considered that increasing the grain air
ratio would give a good quality dried
grain.

Fig.4 and Fig.5 shows the taste evaluation
result for the batch of grain dried using
the commercial cross flow tempering drier
(grain capacity 100kgf). The results also
follow the same trend as those reported
for the test driers. The taste evaluation
results of the batch of grains dried
during the summer month of August, comparing the results with the taste evaluation results of the batch of grains dried
during the autumn month of October, the
grains dried in summer had less taste
values.

Generally the grain harvested during the
summer months are the early variety
(Natsuhikari) which is a more tasty rice
as compared to the autumn harvest (Nipponbare). Comparing the taste evaluation

results for the grains dried the summer and those dried during the autumn month of October. It would be observed that at drying temperature of 40 ℃ the taste value for the grains dried in the month of autumn was higher when compared to the grains dried in summer. This result came from the fact that the milder autumn temperature slow down the rate of grain detrioration in the tempering tank where as the higher temperature of summer accelarate the rate of grain detrioration in the tempering tank during the summer. This result may come from the fact that it takes longer time to dry grains during the humid months of summer than during milder months of autumn.

The free fatty acid content of the grains harvested and dried during the month of August was also higher than those harvested and dried during the month of October, even when the drying parameters were the same. The difference in the free fatty acid content of the grains harvested and dried in summer and those of autumn is as a result of the longer time it takes to dry the grains to a safe moisture content level of 15% w.b. during the humid month of August. The longer it takes to reduce the moisture content of the grain to a safe level for storage, the less favorable the taste will become, because biochemical activities taking place in the grains are prolonged resulting in high free fatty acid content. Drying of grains in autumn is faster because the air relative humidity is low and the mild ambient temperature help slow down the rate of grain deterioration.

Fig.6 shows the taste evaluation results of rice grain dried using non tempering drier(continuous drier) and the grain dried under shade with ambient temperature. The grains batch dried for 120min at 40 ℃ had the highest taste value recorded in this particular experiment, followed by the grain batch dried for 90min at 45 ℃. The taste value of the grain decreases as the drying temperature increases, it is not clear if the difference in taste was as a result of the drying temperature, because at the end of the drying operation, the moisture content of the grains were still higher than 15% which is the safe moisture level required for save strage. The grains had to be dried under shade until they attain the safe moisture level of 15%, before taste evaluation test was carried out. Continuous non tempering drying taste evaluation results shows that as the drying temperature increases the taste value decrease. Drying temperature is a primary factor that affects the taste of the grains dried under this condition. The interesting aspect of this experiment was the effectiveness of the near infrared spectrum analyzer technique. The rate of

grain detrioration is function of grain temperature and moisture content of grain. The batch of grain dried at 50 ℃ to 60 ℃ for a short period of time attained high grain temperature, but still contain moisture above the safe level. The increased grain temperature and high moisture content accelerated the rate of biochemical activities in the grain resulting in quality deterioration, which accounts for the lower taste values obtained from the grain batches dried for short period of time at high temperature and then dried to the final moisture content of 15%w.b. under shade.

Fig. 4(a) shows the taste evaluation result of the grain batch dried during the summer month of August using a commercial cross flow tempering dryer assisted during some experimental runs with a dehumidifier. The grain are exposed to the drying air for a period 4.5min after which the grains leave the drying chamber and move into the tempering tank, where it remains for 54min. From the taste evaluation result, it was observed that the batch of grain that spent a total longer period of time in the tempering tank had a low taste value, as compared to the batch that spent a shorter period of time in the tempering tank. It was also observed that the tempering tank air temperature was on the average 35 ℃ when the drying air temperature was 47.5 ℃. High moisture grains kept at 35 ℃ for considerable length of time would start to deteriorate in quantity. The grain batch dried at had a high taste value. The grain batch dried at 55 ℃ had a low taste value when compared with the results obtaind from the batch of grain dried at lower temperatures. It is therefore possible that high temperature

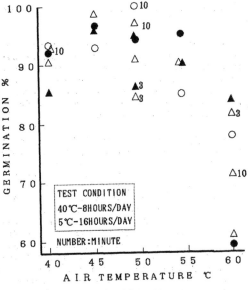

Fig. 7. Germination

affects the organoleptic property of the grains.

Fig.7 show the result of the germination experiment carried out to investigate the effect of drying parameter on grain germination. It can be observed from Fig. 7 that drying grain at temperature range of 40 ℃ to 55 ℃,drying time ranging from 5 to 10min. gave above 90% germination. However at 60 ℃ drying temperature, drying time of 3min. gave a better germination percentage as compared with the germination percentage obtain when drying time was 10min.. Drying time of 10 min. would naturally affect the grain germination because of the heat exposure. Drying at 40 ℃ and drying time of 10min. gave a good result, however grains dried at temperature range of 45 ℃ to 50 ℃ had a better result.

CONCLUSIONS

Drying parameters affects the quality of dried grains. Drying air temperature range of 45 ℃ to 50 ℃, grain air flow ratio ranging between 0.03m³/min/kg to 0.05m³/m /kg,exposure time ranging between 5min to 10min and tempering temperature of 25 ℃ is considered most appropriate for obtainning good quality and tasty rice. Grains must be dried prompty and fast after harvest, grains that took longer period to dry to a safe strage moisture content of 15% wet basis usually turn out to be of less quality and taste incorporating a cooling device in the tempering tank of continuous flow driers to keep the ambient temperature at 25 ℃ is recommended. Near infrared spectrometeric method have been successfully used to evaluate the taste of rice grain. This new method of evaluating the taste of rice holds promise for automatic online evaluation of rice taste. New equipments and new management technology for grain drying and quality evaluation when fully develloped will give agriculture and grain farmers the ability to utilize production inputs more effi-

ciently. The results obtained should serve to demonstrate that a large potion of reserch needed to improve agricultural profits for tommorow can come from the development of technology that will help farmers to dry their grains more efficiently and increase their ability to evaluate the quality of their produce. Until now, the control of drying operation has a tendency to concentrate its attention to phisical phenomena like cracking etc, however chemical phenomena such as taste of grain etc,should also be included in the control processes.

REFERENCES

Kawamura,S., Itoh,K., and Ikeuchi,Y., (1983). Effects of physical properties of brown rice on milling characteristics. Memoirs of the faculty, agriculture, Hokkaido University, 13(4), Japan, 471-474.

Ban,T., (1971), Experimental studies on cracks of rice in artificial drying. Tech. Report, Agric. Machi. Inst., Omiya, Japan, 8, 62-66.

Bekki, E., Kunze, O.R., (1988). Flash drying and milling techniques for high moisture brown rice. Trans. Am. Soc. Agric. Engng., 31, 1828-1833.

Kameoka, T., (1988). Thin layer drying characteristics of rough rice. J. Jap. Soc. Agric. Machinery, 50(4), 57-65.

Toyota, J., (1987) Studies on drying conditions of circulating type dryer. J. of Agric. Machinery, 46 meeting. 157

Driscoll, R.H. and Adamczak T. (1988). Drying systems for the humid tropics. ACIAR proceeding, No.22. 58-68.

Hall, A.W., (1987), Handling and strage of food grains in tropical and subtropical areas. F. A. O. agric. development paper, No.90, 15-22

Rubenstein,I., Phillips,R.L., Green, C.E., and Gengenbach,B.G., (1984), Seeds and germination. The plant seed development preservation and germination. Academic press, New York. pp. 209-245.

IYOKAN (Citrus iyo hort. ex Tanaka) STORAGE HOUSE WITH AIR CIRCULATING UNIT

T. Tsurusaki, F. Tarui and Y. Hikida

Dept. of Biomechanical Systems, Ehime University, Matsuyama, Japan

Abstract. The Iyokans are warehoused after the harvest, which ranges from about the middle of December to the begining of January. For a period about two weeks, the air in the normal temperature storage house should be changed frequently for the preparation of storage, with the natural ventilation or the forced one. In that place, we should like to report on the air circulating unit in the citrus storage room which have designed and found suitable for practical application. The effects of this unit were showing by measuring vertical temperature distribution in the storage house with the simple ventilator. After that, the optimum arrangement of the unit was determined and its mode of use, i.e. whether an upward duct is suitable or a downward one, was discussed with recommendation. Simultaneously, the change of interior conditions were examined in relation to fruit weight loss, and fruits' coloring on storage.

Keywords. Agriculture; air conditioning; quality control ;ventilation; storage house.

INTRODUCTION

It is surely desired that fruit harvested at a proper period is held in a fresh state for a long period of time and can be delivered to consuming areas. Because of this, the development of an epoch-making storage method is required.

The quality control during a storage period is important to Iyo fruit and the after-ripening of fruit has a great meaning in the enhancement of quality from such a viewpoint that the color of the rind becomes deep reddish orange, the balance between sugar content and acid content becomes proper to increase to sweetness and the weight reduction during storage is suppressed low.

A Iyo fruit-producing farmer is taking a measure to meet the quality control of fruit in a normal temperature storage house by natural ventilation using a ventilating opening and a window or forced vantilation using various ventilation fans. In the present situation, as the result of the investigation on the temperature distribution and air flows distribution in a storage house, the stagnation of air is confirmed in several places and both distributions are non-uniform.

Tereupon, an air circulating unit was designed to be made on an experimental basis and arranged in the storage house of a farmer to be subjected to a practical test (FIG.1). The temperature and relative humidity of the objective storage house, the weight of fruit, the color of the rind or the like were measured and the capacity and effect of this unit were investigated from a change of the measured values with the elapse of time.

FIG.1 Air circulating unit in the citurs storage house.

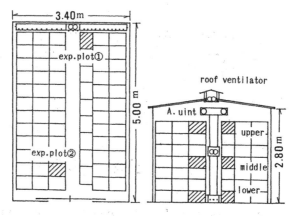

FIG.2 Air circulating uint and positions of storage fruits.

FIG.3 Dry ice method (the air movement direction from a duct was observed according to a dry ice method to be photographed)

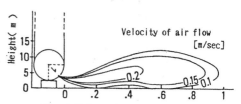

Throw(distance from the outlet to the point where velocity becomes 0.10m·sec⁻¹) (m)

FIG.4 Vertical flows of outlet air of the unit (in case of 45°).

MATERIALS, METHODS AND EQUIPMENT

Iyo fruits (30 fruits) to be tested were distributed to the lower parts (15cm above floor level), middle parts (105cm above floor level) and upper parts (195cm above level) of experimental groups ①, ② by five fruits to be introduced into the stored fruit gruop(FIG.2). The weight of fruit and the color of the rind were respectively measured at 7 ~ 10 day-intervals after the start of storage by an automatic balance and digital color meter.

Dry-bulb temperature and wet-bulb temperature were measured at one point provided outside the storage house and at 6 points provided within the storage house using a thermo-junction type 20-channel recorder. Further, air flow velocities at respective parts in the storage house were measured by a hot-wire anemometer and the air movement direction from a duct was observed according to a dry ice method to be photographed (FIG.3).

The air circulating unit used in the experiment consists of a perforated type small-sized fan (AC 100V, 0.92m³·min⁻¹) and a duct (polyvinyl chloride type, inverted T-sharpe, D=105mm ; air emitting orifice d=10mm , the number of orifices 100) and was arranged on the surface of the floor along side wall of the storage room.

The arrival distance of air flow velocity 0.10m·sec⁻¹ considered to accelerate air circulation is about 1m and the vertical height from the surface of the floor was about 15cm. A proper air emitting angle is considered to be about 30° to the surface of the floor in a downward direction. In this case, it is effective to hold a clearance under a container by planking. This time, a circulation system blowing air in the vicinity of the ceiling to the surface of the floor was used (FIG. 4).

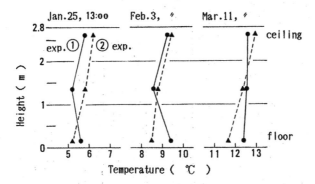

FIG.5 Temperature profiles in the citrus storage house.

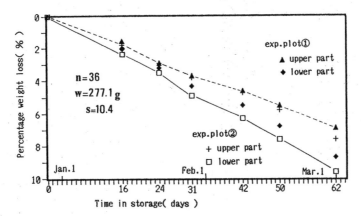

FIG.6 Effect of experimental group①,② (upper,lower parts) on fruit weight loss.

RESULTS AND DISCUSSION

The temperature in the storage house showed a value of 6 ~ 14℃ near to proper storage temperature (8 ~ 10℃) and the effect of the outside air temperature(1 ~ 12℃) is consider-ed to be scarce. Considering a temperature profile, a temperature gradient was lower in the experimental group ① near to the air circulating unit as compared with the experi-mental group ② far from said unit and the temperature said unit and the temperature dis-tribution in the vertical direction was uni-form in the experimental group ① (FIG. 5). On the other hand, the relative humidity in

the storage house was 85 ~ 95% and slightly higher than relative humidity (about 85%) said to be proper.

The reduction of the weight of fruit was 8.2% on an average throughout the total period and the reduction ratio in the lower part was slightly hight in both of the experimental groups ①, ② (FIG. 6). Considerably high positive correlation(r =0.78) was confirmed between the weight reduction of fruit and the saturated deficit (by periods , by measuring points) of the storage house. The saturation deficit was high in the lower part much in the weight reduction of fruit (FIG. 7).

203

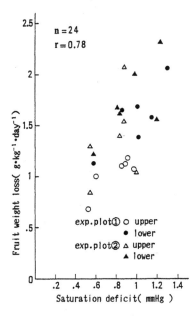

FIG.7 Relationship between the weight reduction of fruit
and the saturation deficit of the storage house.

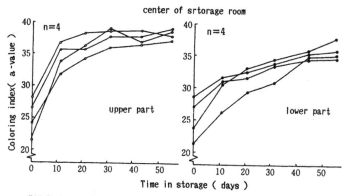

FIG.8 Increasing on coloring index(Hunter a-value)
of citrus Iyo (1985).

A Hunter color specification method was used
in the expression of the color of fruit, and
a hue and chroma were shown as an a-value and
a b-value ,while blightness was shown as an
L-value.   As the rind became reddish orange,
the a-value increased, the b-value and L-
value became decreased. In general, the advance
of the coloring of Iyo fruit during storage has
been said to be largely affected by temperature and it was reported that coloring became
well as storage temperature became high from
the 3℃-group toward the higher temperature
groups of 8℃ and 13℃.

Up to now,the author and others have conducted
the storage experiments of Iyo fruit on the
spot and the following results were obtained.

The coloring of fruits positioned at the upper
part of the storage house always increases
rapidly in the a-value(red degree) as compared with that of fruits at the lower part
and the a-value at the time of the completion
of storage is also high. This phenomenon was
estimated to be generated because the upper
part was a high temperature region and the
lower part was a low temperature region.
Consequently, it was cleared that it was
important not only to raise the temperature of
the lower part but also to average the temperature in the storage house during the so
called coloring period of the initial storage
stage by blowing the air in the high temperature region near to the ceiling toward the
surface of the floor (FIG. 8).

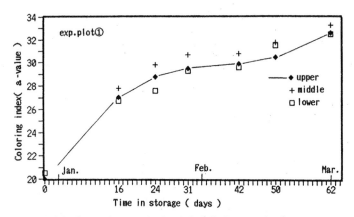

FIG.9 Increasing on coloring index(Hunter a -value)
of citrus Iyo (means of five fruits) (1990).

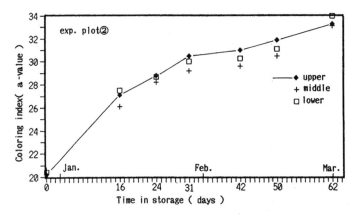

FIG.10 Increasing on coloring index(Hunter a -value)
of citrus Iyo (means of five fruits) (1990).

As the results of this experiment, the a -value of the color of the rind increased from a ≒ 20 to a ≒ 34 with the elapse of storage and the difference between the a -values of the upper, middle and lower parts was low in both of the experimental groups ①, ② and the color of the rind became a state possible to deliver about 30 days after the start of storage in both experimental groups ①, ② (FIG. 9, 10). This effect is considered to be achieved be- cause the air from the air circulating unit reaches the wide range of the stroage house to accelerate a natural convection and the averag-

ing of the temperature in the storage house is advanced.

As the conditions to be possessed of the stor- age house of Iyo fruit , a structure hard to receive a change of the outside air temperatue, the proper arrangement of ventilation equipment or the like are designated.

It is considered that this air circulating unit plays an important role in the improvement of the control technique of the storage house by the application to conventional equipment.

REFERENCES

S.Shiraishi(1972). Study on the degreering of citrus fruit. Research Bulletin of the Fukuoka Horti. Exp.Sta.,2,1-53.

T.Tsurusaki,F.Tarui and D.Sahara(1976). Changes of interior conditiona and fruit's specific gravity on storage of Iyokan. Research Bulletin for Farm Business of the Ehime Univ.,19,83-92.

T.Habu(1977). Effects of handling method on Iyokan fruit during storage. Research Bulletin for Farm Business of Ehime Univ., 20,77-84.

E.Beppu,Y.Ishida,E.Watanabe,A.Oowada and T.Mukai (1979).Studies on the precuring and storage coditions of citrus Iyo.Research Bulletin of the Ehime Fruit Tree Exp.Sta.,7,1-18.

T.Tsurusaki,F.Tarui and Y. Hikida(1979). On the storage house with simple cooling apparatus of Iyokan. Research Bulletin for Farm Business of the Ehime Univ.,22,43-51.

K.Yoshimatsu and H.Tanaka(1984). Studies on the storage condition of Miyauchi Iyokan. Research Bulletin of the Yamaguchi Agri.Exp. Sta.,36,49-52.

J.Waks,E.Chalutz,Mina Schiffmann-Nadel and Ella Lomeniec(1985). Relationship among vantilation of citrus storage room,internal fruit atmosphere,and fruit quality. J.Amer. Soc. Hort.Sci.,110(3),398-402.

T.Tsurusaki,S.Kiyama(1986). On the Iyokan storage house with simple ventilator. Research Bulletin for Farm Business of the Ehime Univ.,28,45-56.

A.Hino,S.Li,S.Kawahara,and K.Kadoya(1990). Effect of seal-packaging with high density polyethylene, waxing and temperature on Iyo fruit during storage. Mem.Coll.Agr., Ehime Univ.,34(2),327-336.

Copyright © IFAC Mathematical and Control Applications
in Agriculture and Horticulture, Matsuyama, Japan 1991

MECHANICAL WEED CONTROL IN SUGAR BEET GROWING: THE DETECTION OF A PLANT IN A ROW

J. Bontsema, T. Grift and K. Pleijsier

Wageningen Agricultural University, Dept. of Agricultural Engineering and Physics, Systems and
Control Group,
Duivendaal 1, NL-6701 AP Wageningen, The Netherlands

Abstract. The detection of the position of an individual plant in a row of sugar beet is discussed. Precise detection of the position is necessary for efficient mechanical weed control in the row in the growing of sugar beet. It is shown that with a simple sensor consisting of a combination of infrared light sources and photoelectric cells, together with some signal procesing, it is possible to reconstruct the individual positions of the sugar beet plants from the measured data, even despite the presence of weed in the row. The signal processing methods used are correlation and fast Fourier transform (FFT) techniques.

Keywords. Mechanical weed control; signal processing; fast fourier transformation; correlation methods.

INTRODUCTION

At present weed control in the growing of sugar beet is mainly done by the use of herbicides, although between the rows also weeders are used. Recently the Dutch government has decided to a reduction of the use of pesticides in agriculture of 50% in the year 2000. In order to achieve this reduction there is among others a need to introduce more mechanical weed control, where in our research in first instance we will focus on the weed control in growing of sugar beet. (see f.i. Kouwenhoven, Wevers and Post, 1990 and Terpstra and Kouwenhoven, 1986).
Since on a field rows can easily be tracked, mechanical weed control between rows is very well possible. However for a human being on a moving vehicle it is impossible to locate individual plants in a row. So in order to do mechanical weed control in the row as well, one needs a fully automated detection system to locate the plants in the row.
Our research started in 1989 with experiments in a laboratory set-up (Pleijsier, 1990) and in the spring of 1990 we started with field experiments.
This paper is organized as follows:
- a simple model of a row of sugar beet is proposed together with a model for the weed.
- a discussion of the possibilities of the use of the autocorrelation function to reconstruct the individual positions of the plants in the row.
- as a second method for this reconstruction the use of the Fourier transform will be considered.
- the field experiments will be discussed and the two above mentioned methods will be applied on real data of this experiments.
- finally some conclusions and remarks for further research are given.

A MODEL OF A ROW OF SUGAR BEET WITH WEED

In practice sugar beet is sowed with a precision drill, so the distance between the plants will be approximately constant. For the model we assume that the distance is approximately 20 cm. In order to model the variations in the in-between distances we assume that this distances have a Gaussian distribution with mean 20 cm and a certain standard deviation. This matches reasonable the practical situation (Gego, 1968). We assume that that the width of a plant is 2 cm. The weed is assumed to cover a certain percentage of the space between the plants. Furthermore we assume that the place where weed appears is random according to the uniform distribution. For the model it is assumed that the row is sampled with a frequency of 3 cm^{-1}.
The actual measured data from such a model is a sequence of zeros and ones, where a one means that there was a (part of) plant or weed. A typical data sequence is given in Fig. 1.

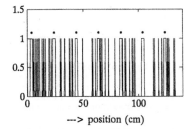

--- > position (cm)

Fig. 1. The data from a model of a row of sugar beet, distance 20 cm, 16% weed, '*' actual position of a beet.

AUTOCORRELATION

A well known method to study the properties of a time series is to determine the socalled autocorrelation function of the series. For a time series $\{x_i\}$, $i=1,..,N$, the autocorrelation function is defined by:

$$r_{xx}(k) = \frac{1}{N} \sum_{i=1}^{N-k} x_i x_{i+k} \qquad (1)$$

For our application we recall the following properties of the autocorrelation function:
1. if $\{x_i\}$ is a completely random series then $r_{xx}(k)=0$ for $k\neq0$.
2. if $\{x_i\}$ and $\{y_i\}$ are mutually independent random series and $\{z_i\}=\{x_i\}+\{y_i\}$ then $r_{zz}(k)=r_{xx}(k)+r_{yy}(k)$.
3. if $\{x_i\}$ is a periodic series with period T then for large N also $r_{xx}(k)$ is periodic with period T.

Since in our model of beet plants and weed we assume that the position of the weed is random and independent of the position of the beets we can expect that if we take the autocorrelation of our measured data, then the influence of the weed will be filtered out. The beets will appear in a repeating way in the data, so we can expect that from the autocorrelation we can determine the position of the individual beets.

In Fig. 2 the autocorrelation is shown of a model of a row of sugar beet, where the in-between distance is 20 cm, with a standard deviation (sd) of 1 cm. We see that the peaks of the autocorrelation function give a good approximation of the individual plant positions. The first beet is positioned at zero, since the autocorrelation only gives the in-between distances and not the actual positions. The fact that the peaks are decreasing for increasing position is due to the fact the we use a finite series (see property 3 of the autocorrelation function).

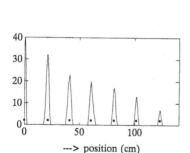

Fig. 2. The autocorrelation of a model of a row of sugar beet, distance 20 cm, sd=1 cm, '*' actual position of individual beet.

In Fig. 3. the autocorrelation is shown for the case that the in-between distance is 20 cm, with a standard deviation of 2 cm. It is clear that the peaks now give much less information about the actual positions.

The autocorrelation function of the model given in Fig. 1 is given in Fig. 4.

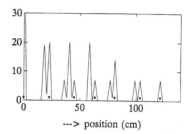

Fig. 3. The autocorrelation of a model of a row of sugar beet, distance 20 cm, sd=2 cm, '*' actual position of individual beet.

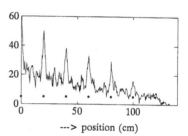

Fig. 4. The autocorrelation of a model of a row of sugar beet, distance 20 cm, 16% weed, '*' actual position of individual beet.

Figure 4. illustrates clearly the feasibility of the autocorrelation to filter out the influence of the weed. Summarizing we can say the contribution of weed to the signal is suppressed, but that the irregularities in the beet distances cause problems. Other calculations also showed that missing beet and different widths cause problems.

FOURIER TRANSFORMATION

Another useful tool to study time series is the Discrete Fourier Transformation (DFT), which is usually implemented by the Fast Fourier Transformation (FFT). If the time series $\{x_i\}$ has length N then the DFT is defined as:

$$X_k = \frac{1}{N} \sum_{n=0}^{N-1} x_n e^{-j2\pi k \frac{n}{N}} \qquad (2)$$

Once we have the transformed series $\{X_k\}$ we can do the inverse transformation (IDFT) to determine the series $\{x_k\}$:

$$x_k = \frac{1}{N} \sum_{n=0}^{N-1} X_n e^{-j2\pi k \frac{n}{N}} \qquad (3)$$

The absolute values $|X_k|$ of the complex numbers X_k give a measure for the importance of the frequency $f_k = k f_s/N$ in the measured data $\{x_i\}$. Here f_s is the frequency by which the series is sampled. The absolute values of the $|X_k|$ considered as a function of the frequencies is called the spectrum of the series $\{x_i\}$.
Analogous to the autocorrelation the DFT

has some useful properties for our purposes:
1. for a purely random signal all the X_k's will be equal.
2. for a periodic signal only one X_k will be unequal to zero.
Intuitively it is then clear for our data that the contribution of our beet plants to the spectrum of the measured data will only be in a restricted frequency band, whereas the contribution of the weed will approximately be the same for all frequencies and it is clear that this last contribution will in general be much smaller that the specific contribution of the beets. Since the frequencies of the beets will be in the lower frequency range (for the above mentioned models around 0.05 cm^{-1}), it is reasonable to assume that X_k is almost zero for the high frequency range. Setting $X_k=0$,for $k>k_b$, in equation (3) will give a filtered version of the original sequence $\{x_i\}$.
In Fig. 5 and Fig. 6. the spectra are given of a model of the row of sugar beet with in-between distance of 20 cm and standard deviation of 2 cm and a model with in-between distance of 20 cm and 16% weed.

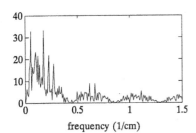

Fig. 5. The spectrum of a model of a row of sugar beet, distance 20 cm, sd=2 cm, sample frequency 3 cm^{-1}.

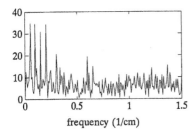

Fig. 6. The spectrum of a model of a row of sugar beet, distance 20 cm, 16% weed, sample frequency 3 cm^{-1}.

In both spectra the first peak appears for the frequency 0.05 cm^{-1}, which corresponds to the in-between distance of 20 cm. For the cut-off frequency needed in the inverse transformation we choose 0.37 cm^{-1}, since for the higher frequencies the contribution in the spectra is negligible, especially in Fig. 6.
The transformed data are shown in Figs. 7 and 8.

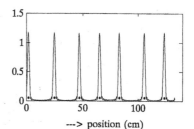

---> position (cm)

Fig. 7. The transformed data of a model of a row of sugar beet, distance 20 cm, sd=2 cm, '*' actual position of individual beet.

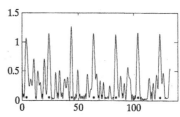

---> position (cm)

Fig. 8. The transformed data of a model of a row of sugar beet, distance 20 cm, 16% weed, '*' actual position of individual beet.

From these figures it is clear that the highest peaks in the transformed data agree well with the actual position of the individual plants. If we compare Figs. 1 and 8 we can conclude that the influence of the weed is sufficiently reduced by the transformation.

EXPERIMENTS

The measurement on a real row of sugar beet were performed in the spring and summer of 1990. The sugar beets were sowed with a precision drill on a parcel of arable land with sandy soil near Wolfheze, the Netherlands. When we started the experiments there was a lot of weed on the land, mainly pigweed and goosefoot. The sensors were placed on a measuring frame which was mounted in front of a tractor (see Fig. 9).

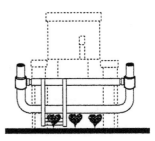

Fig. 9. The measuring frame in front of a tractor.

The sensor consists of 3 pairs of infrared light sources and photoelectric cells, which enables us to measure the row at three different heigths, all at the same time (see Fig. 10). The reason to do this is that we expect that in this way a sugar beet plant will give more signal than weed.

The velocity of the tractor was kept constant during the measurements.

Our idea for the detection system is to transform repeatedly a certain part (f.i. 100 to 150 cm) of the whole row and use this information for actuating the weed control system. For this reason in this paper we only study small parts of the rows.

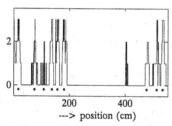

Fig. 10. The detecting sensor.

As a first example we study a part of a row of about 500 cm (see Fig. 11). As can be seen from this figure the individual sugar beet plants give quite different signals and also can be seen that there were a lot of beets missing and that the distances between the plants are not very constant. The signal near about 400 cm is no beet, but is problbably noise due to some vibration.

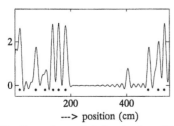

Fig. 11. Measured data from a row of sugar beet, no weed, '*' actual position of a beet.

Taking the Fourier transform and setting the high frequency components to zero and then performing the inverse transformation results in the transformed data as given in Fig. 12.

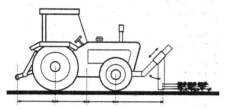

Fig. 12. Transformed data (using FFT) of a row of sugar beet, no weed, '*' actual position of a beet.

As can be seen the peaks in the data give a quite good approximation of the actual position of the beets.

As a second example we take a shorter part of the row before, namely we take only the first 150 cm. In this part there are four beets (see Fig. 13).

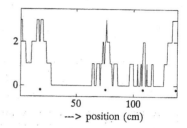

Fig. 13. Measured data of a row of sugar beet, no weed, '*' actual position of a beet.

The spectrum is shown in Fig. 14 and we see that the frequency 0.0111 cm^{-1} gives the largest contribution to the spectrum. This contribution is mainly due to the width of every beet. Furthermore we see at the spectrum that the contributions for frequencies greater than 0.3 cm^{-1} can be neglected.

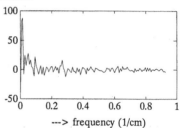

Fig. 14. Spectrum of the measured data of Fig. 13.

The transformed data from the Fourier method is shown in Fig. 15 and we see that the peaks give a good approximation of the actual positions of the plants.

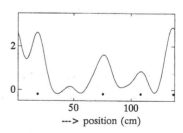

Fig. 15. Transformed data (using FFT) of the row of sugar beet of Fig. 13, '*' actual position of a individual beet.

The autocorrelation of this series is given in Fig. 16 and we see here that this a far worser approximation of the individual positions.

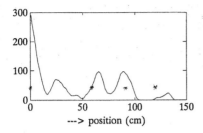

Fig. 16. Autocorrelation of the row of
sugar beet of Fig. 13, '*'
actual position of individual
beet.

As a third example we consider a part of a
row of 150 cm with a lot of weed. We only
consider the lower and the upper part of
the three signals. The measured data and
the transformed data are given in Fig. 17
respectively Fig. 18. The amount of weed
is a little unrealistic but we see that in
Fig. 18 still the actual positions match
reasonable well with the peaks. The weed
however is not sufficiently suppressed in
the transformation.

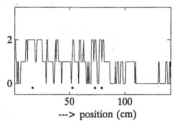

Fig. 17. Measured data of a row of sugar
beet, with weed, '*' actual
position of individual beet.

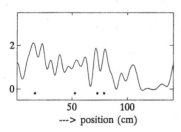

Fig. 18. Transformed data (using FFT) of
the row of sugar beet of Fig.
17, '*' actual position of
individual beet.

CONCLUSIONS AND REMARKS

In this paper we showed that the
combination of a simple sensor and some
transformation techniques gives a
promising method to detect the individual
position of a (sugar beet) plant in a row,
even if there is also weed in-between the
plants. Especially the transformation via
the FFT seems to be powerfull and this
method has the advantage that it is
numerically very fast and it can both be
implemented in software as in hardware.

The autocorrelation method is very useful
for filtering out the influence of the
weed in the measured signal, but is
severely disrupted by irregular distances
and shapes of the plants. The irregularty
in the plant distances was probably due to
the fact the the precision drill did not
worked correctly, with a good precision
drill one can expect that the plants will
be on approximately the same distance. The
fact the we had a lot of missing beets was
probably due to the fact that the
circumstances for our experiments were not
optimal.
For further research we make the following
remarks. In the first place we need more
information about the shape of a sugar
beet plant especially in comparison with
that of the weeds. Also the sensivity of
the sensor needs further study.
In the transformation method with the
Fourier transform we did the truncation of
the higher frequency components in an ad
hoc way, this should be replaced by a more
systematic method. Also we have to
consider if windowing the signals will
improve the results.
Finally in this research on a detection
system we are aiming to end up with a more
mathematical description for a row of
sugar beets. For this description one can
for instance think of a stochastic point
process. Such a model then can be used to
get an even more accurate estimation of
the actual position of individual plants.

ACKNOWLEDGEMENTS

We would like to thank the IRS, Bergen op
Zoom for providing us a testing field and
the IMAG, Wageningen for lending us the
tractor.

REFERENCES
Chatfield, C. (1989). The analysis of time
series, an introduction, Chapman and Hall,
London.
Gego, A. (1968). Ein Beitrag zum Problem
der gestreuerten maschinellen Vereinzelung
von Zuckerruben, Aachen.
Kouwenhoven, J.K., J.D.A. Wevers and B.J.
Post (1990). Possibilities of mechnanical
post-emergence weed control in sugar beet,
Report Tillage Laboratory, Wageningen.
Pleijsier, K. (1990). Mechanische
onkruidbestrijding in de rij bij
suikerbieten, MSc Thesis University of
Wageningen, Wageningen.
Terpstra, R. and J.K. Kouwenhoven (1986).
Inter-row and intra-row weed control with
a hoe-ridger, J. agric. Engng Res., vol.
26, pp. 127-134.

AUTOMATED PLANT HANDLING AND PROCESSING IN A ROBOTIC WORKCELL

W. Simonton

Dept. of Agricultural Engineering, The University of Georgia, Georgia Experiment Station, Griffin, Georgia, 30223-1797, USA

Abstract. Sensing and control techniques and decision strategies for effective robotic handling of greenhouse plant material have been developed. Machine vision performs the critical role of automatic plant part classification. Based on the resulting knowledge of the plant structure, a technique for determining an appropriate robotic processing strategy for each geranium cutting has been implemented. The technique includes local scene analysis for defining a non-damaging grasp tactic. An end effector controllable based upon both position and force has been developed, integrated into the workcell, and shown to be very effective in handling geranium cuttings. In general, results indicate the importance of successful fusion of sensor information, a robust decision structure, and flexible control for assisting robotic systems to accommodate nonuniformity.

Key Words. Sensors; robust control; image processing; end effector; agriculture.

INTRODUCTION

Robotic systems are prevalent in many industries for conducting repetitive and/or precise operations to improve productivity and quality, and reduce hazardous or uncomfortable tasks for workers. Current industrial applications of robotics are most often performed with uniform objects in a defined spatial configuration (Considine, 1986). However, contrary to the uniform conditions in industry, most tasks performed in agriculture involve biological products of variable size, shape, color, position, orientation, and stress/strain relationships. Such nonuniformity presents significant challenges for the development of reliable and cost effective robotic systems (Moncaster, 1985). Fundamental research is underway to establish analytical methods for design of intelligent machines which can autonomously operate in uncertain environments (Saridis, 1989). However, at the present time there is no formal methodology for design of the organization, coordination, and execution levels inherent in intelligent and hierarchical systems.

At the Georgia Station Mechatronics Laboratory research is focused on investigating pragmatic techniques which allow robotic systems to accommodate significant variability in materials handled and processed. The laboratory is based upon three primary systems (Figure 1): an industrial robot arm and controller; an end effector and controller; and a supervisory computer. The robot is a six degree-of-freedom arm with positional accuracy of 0.1 mm. The supervisory computer is a real-time multitasking system which communicates serially with the robot controller and the end effector controller. Command information is unidirectional and data is bidirectional. The robot controller and end effector controller communicate through a parallel interface. The supervisory computer serves as the "master", performs all machine vision processing, and transmits control information to the robot controller. The robot controller performs all position and velocity control of the robot arm along with simple logic operations. The end effector controller performs position and force control of a two-fingered gripping mechanism.

As an application area, the greenhouse industry is a segment of agriculture which is very labor intensive. Many materials handling and processing tasks from propagation to harvest are performed manually. Tasks include processing for propagation, pruning, quality sorting, transplanting, chemical application, and packaging. As a case study for investigation of robotic handling and processing of living plant material, the author has developed a robotic workcell for processing geranium cuttings for propagation (Simonton, 1990a). The robot grasps singulated cuttings from a conveyor and performs unit operations of leaf removal, stem trimming, stem bend measurement, and plug insertion.

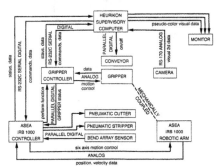

Fig. 1.   Block diagram of hierarchical control and data flow in the Mechatronics Laboratory.

The workcell for processing geranium cuttings (Figure 2) has been designed such that fixtures and sensors mounted within the workspace assist in the processing. A conveyor brings singulated geranium cuttings into the robot work envelope. Using data from an RS-170 camera, the supervisory computer locates a cutting and classifies all primary plant parts (Figure 3). Knowledge of the plant structure is used to grade and determine an appropriate processing strategy. Control information is passed to the robot and end effector controllers. The cutting is grasped by the robot, moved to a pneumatic device for leaf removal if necessary, to a pneumatic cutter for trimming the main stem, to a sensor for measuring the stem bend, and finally to a plug tray for insertion into a propagation medium. At the end of this approximately 6 s cycle, the robot prompts the supervisory computer for data on the next cutting on the conveyor. Mechanical processing of a cutting by the robot occurs concurrently with the visual analysis of the next cutting on the conveyor by the supervisory computer.

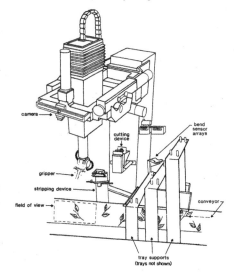

Fig. 2. Workcell layout for geranium cutting processing. Fixtures are used to assist in process unit operations.

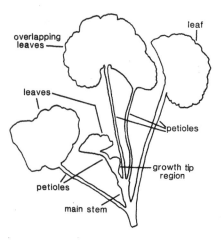

Fig. 3. Typical geranium cutting with major plant parts labelled.

In a 1989 performance evaluation of the workcell with two geranium varieties, 'Crimson Fire' and 'Yours Truly', results showed propagation efficiencies within an average of 5% of manual propagation (Simonton, 1990a). However, the tests revealed limitations in the size of plants which were handled adequately and in reliability of the image analysis. In 1990 a position/velocity and force controlled end effector was developed to investigate the mechanics and control of robotic handling and manipulation of plant material (Simonton, 1990b). Results showed that the end effector could handle a wide range of sizes of cuttings (4-13 mm

caliper) with rare indications of damage to the main stem and growth tip. However, relatively frequent damage occurred to petioles (45% of specimens) and thus significant improvements in control and finger design were indicated.

The primary goal of current research is development of sensing and control techniques and decision strategies for effective, non-damaging robotic handling of greenhouse plant material. Specific objectives of this paper are to:
(1)   describe a machine vision technique for automatic plant part classification of singulated, randomly positioned, and randomly oriented geranium cuttings;
(2)   describe a technique for determining an appropriate robotic processing strategy for each geranium cutting;
(3)   describe the design, control, and performance of an end effector developed specifically for handling cuttings.

MACHINE VISION CLASSIFICATION

The goal of machine vision in the robotic workcell is to locate singulated cuttings on a black conveyor belt, identify their orientation, and classify all plant parts as main stem, growth tip, petiole, or leaf. Identification of the growth tip, the most actively growing portion of the plant, is critical for processing. Image data is binary with 512x484 pixels corresponding to 0.717 mm x 0.557 mm spatial resolution on the conveyor surface. Optics are configured currently for high contrast between plant and background, with small intensity deviations of plant pixels. Previous work (Simonton and Pease, 1990) on automatic plant part classification demonstrated effective segmentation based on the branching structure of the plant; however the need for improved classification and for elimination of orientation constraints were indicated. The following outlines a more robust method of segmenting image data into an undirected graph (Aho et al., 1983) and allows for completely random position and orientation of the plant in the image.

Image Segmentation

Cuttings are generally located by crossing an artificial bar in the image, and the conveyor belt is momentarily stopped. The region of interest is complex and consists of a connected group of pixels called object pixels which define the shape of a single cutting (or possibly another object that is not a cutting). The region defined by the cutting is subdivided into simpler and smaller regions called segments. The image is segmented for the purpose of plant part classification, and in general, each segment represents a portion of a single plant part.

Data representing a cutting is stored in an undirected graph, which is a group of data records linked together by pointers (Figure 4). In the corresponding object, these linked records represent groups of object pixels that are adjacent to one another. Each data record represents one segment. Segment boundaries are determined by the forking, or branching, structure of the image at the pixel level. Segmentation is performed from the top of the object to the bottom tracking row to row, or left to right tracking column to column, so that each segment consists of one or more rows (or columns) of object pixels. For example, when using top to bottom tracking, each row in a segment consists of a single contiguous group of object pixels that overlap at least one pixel on the next row. Segments end when there are no object pixels on the next row that overlap object pixels on the current row, or, when there are two or more distinct groups of object pixels on the next row that overlap those on the current row. In the first case, a segment terminates, and in the second case, a segment spawns a new segment for each distinct group of object pixels on the following row. This technique is very effective in separating the object into small regions amenable to classification.

Data stored in the segment records include segment height, area, average width, and perimeter. A shape factor (height/width), and a compactness measure (perimeter/area²), are utilized in classifying plant parts and determining orientation. Two additional record elements are also available in every segment for intermediate and final classification results. Lastly, an element is available for average pixel intensity of the segment although this information is not utilized for binary image data.

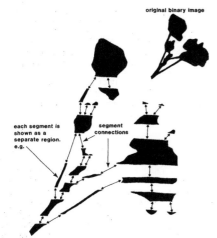

original binary image

each segment is shown as a separate region. e.g.

segment connections

Fig. 4.   Exploded view of a segmented binary geranium cutting image.

Orientation and Preliminary Classification

The angle from the base of the main stem to the growing tip of a cutting defines cutting orientation. To determine orientation, therefore, the main stem must first be identified. Segments are assigned a preliminary class based on the average width of the segment. The shape factor of the segment can also be used to distinguish portions of stems from portions of leaves. A simple sorting rule classifies segments into four groups:

if (segment.height < 5 mm)
   segment.class is 'undecided'
   else if (segment.width < 4 mm)
      segment.class is 'petiole'
      else if (segment.width < 15 mm)
         segment.class is 'main stem'
         else segment.class is 'leaf'.

Using such a simple decision procedure produces a rough estimate of plant parts. This classification is improved upon once main stem orientation has been determined. The main stem is located by searching through the graph for the longest unidirectional path of segments that are classified as undecided or main stem. The direction of the main stem is still unknown, and classification of petioles is used to determine main stem orientation. Petiole angles are measured from the point they branch from the main stem. The main stem direction is chosen so that the relative angles measured between the petioles and the main stem are less than $\pi/2$ rad. Verification is available through examination of segments with a preliminary class of leaf type and their position relative to the start and end of the main stem.

Examining the segmented data as described above yields accurate cutting orientation from $\pm \pi/4$ measured from the vertical and assuming a top/down segmentation. The main stem at other orientations cannot be identified as well because widths of main stem segments do not correspond accurately to true widths. For a cutting at orientation of π,

for example, the segments representing the main stem may appear as wide as the true length of the main stem. In other words, the cutting lies nearly horizontal and main stem segment widths will not likely be measured in the 4 - 15 mm expected range.

To overcome this dependence on an orientation constraint, segmentation is performed twice: from top to bottom, and from left to right in a similar fashion. Tracking the object twice produces two entirely different segmentations. A main stem is located in each representation of the cutting, and the list of segments which best fits a dimensional model of a main stem is chosen as the true main stem. The segmentation that created the true main stem list of segments is used for all further plant part classifications. Thus orientation is determined by location of the main stem, and this likewise fixes whether the top/down or left/right data representation will be used for the complete classification of the image.

Complete Classification

Final classification of segments (Figure 5) is achieved by combining segment record data with global information about the cutting including the location of the base of the main stem and the orientation of the main stem. A scoring function is calculated for each segment whose range is 0.0 to 1.0. Values less than 0.25 strongly indicate that the segment is a portion of a leaf; values greater than 0.75 strongly indicate that the segment is a portion of a stem (petiole or main stem). All values between 0.25 and 0.75 are uncertain values with 0.5 indicating maximum uncertainty.

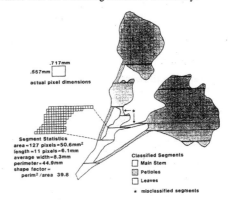

.717mm
.557mm
actual pixel dimensions

Segment Statistics
area = 127 pixels = 50.6mm²
length = 11 pixels = 6.1mm
average width = 8.3mm
perimeter = 44.9mm
shape factor =
perim²/area 39.8

Classified Segments
☐ Main Stem
▦ Petioles
☐ Leaves

* misclassified segments

Fig. 5.   Final classification of an image, with data associated with individual segments also shown.

Starting values for the scoring function are set depending on a segment's preliminary class as described above: 0.25 for a segment identified as a leaf, 0.5 for a segment classified as undecided, and 0.75 for a segment classified as petiole or main stem. The distance from the base of the main stem to each segment divided by the total plant length (d/l) provides a measure of the distance of a segment into the cutting. If P is the current score of a segment,

if (P < 0.25 or P > 0.75)
   P = P
   else if ((d/l) < 0.4)
      P = P + (0.4 - (d/l))
      else if ((d/l) > 0.6)
         P = P + (0.6 - (d/l)).

This adjustment to the scoring function increases the score for those segments close to the base of the cutting, and the score is decreased whenever segments are very distant from the base of the cutting. Other factors that the scoring function incorporates include area and compactness.

215

Grouping of segments into logical objects which represent plant parts is the final step of classification. For example, the list of main stem segments is updated to reflect the reclassification of segments. This helps to ensure that any segments in the main stem list that are classified as leaf type are removed from the list of main stem segments. Segments classified as stem type by the scoring function that are not in the list of main stem segments are considered petioles. Any segments with a score $0.25 < P < 0.75$ must be classified. This is accomplished by using classes of neighboring segments to determine a class for the undecided segment. Current results indicate that orientation of the plant is correctly determined on virtually 100%, with accurate classifications of main stems and other plant parts of approximately 90%.

ROBOTIC PROCESSING STRATEGY

Goals of the Processing Strategy

Once the image data is classified into a description of position, orientation, and interconnection of the primary plant parts, decisions must be made as to how the robot is to process the cutting. The grade of the cutting must be determined to specify a desired post-processing main stem length and to specify the appropriate tray in which to perform the insertion. An ideal grasp location must be located based upon the growth tip position and the desired cut location along the main stem. To avoid striking small leaves and long petioles of larger leaves, allowable end effector finger positions must be found. Once a near-optimal finger position is located along each side of the main stem, a transformation from image coordinates and image angles must be made to robot coordinates and angles. Lastly, it must be determined if leaf removal is required to ensure no stem-to-petiole fork locations will be in the rooting zone of the propagated cutting. The following section briefly explains the structure of the calculations and decisions needed to ascertain all parameters required by the robot for effective handling and processing.

Decision Structure

The grade of the cutting is determined from the average main stem diameter 25 mm below the growth tip. There are three categories of grades, each within a particular diameter range. Each grade has an ideal final main stem length which is used to locate the desired cut, or excision, point along the stem. The ideal grasp position is calculated along the main stem a fixed distance from this cut location. However, to protect against errors in the growth tip classification, the ideal grasp position is checked to ensure the plant part size to be grasped is reasonable for a main stem. If not, a search is made to locate a reasonable position.

Previous results (Simonton, 1990b) indicated that constant end effector finger width and symmetric finger placement about the main stem attributed to damage to the cutting during grasping. To alleviate this, a technique was developed to identify near-optimal finger placement within the branching structure of the cutting. Based on the ideal grasp location, a rectangular region is bounded symmetrically about the main stem. This region of size 25 mm by 40 mm is used to constrain a search for locations that the fingers can be placed which will not damage the stem, small leaves, or petioles of large leaves.

Image data within the bounded region is scanned and searched for possible finger positions which will not strike the cutting either during the end effector approach or during closure until the main stem is grasped. Using an overlay of a size of the finger tips, the region is searched from the top of the stem downwards for acceptable locations of finger placement. Three levels of decision criteria are used to

choose the near-optimal locations, with the following parameters driving each level: (1) proximity to main stem; (2) minimal deviation from the angle of orientation of the main stem; and (3) proximity to the uppermost section of the main stem. The effect of this search is a location on each side of the cutting to which the fingers should be placed during grasping, with resulting data of a grasp center, a grasp width, and an end effector orientation.

Figure 6 shows an example of the search technique for proper finger positioning. Tests were conducted on 25 stored images each of the geranium varieties 'Sincerity' and 'Crimson Fire' to evaluate the capability of the finger positioning algorithm. The software requires 0.3 to 1.0 s, with an average of 0.5 s, depending upon the complexity and size of the cutting image. On 48 of the 50 cuttings, satisfactory locations were found for both fingers to be positioned without putting any of the plant segments at risk of damage. On the remaining two sample images, a satisfactory location was found only on one side of the cutting.

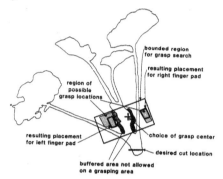

Fig. 6.   Graphical display of local scene analysis for determining proper end effector finger positions to be used to grasp a cutting.

Once the final grasp location and orientation have been determined, the data are transformed into a set of (x,y,z) coordinates in robot workspace and a set of (q1,q2,q3,q4) quaternions for the joint angles. The 'z' coordinate is fixed along the surface of the conveyor belt, as geranium cuttings can be approximated as two-dimensional objects. The quaternions are calculated assuming a vertical approach to the geranium cutting on the conveyor surface once the robot places the end effector above the grasp location. Such a trajectory requires only rotation of the turn disc for proper end effector orientation. Obviously good vision-to-robot coordinate calibration is required for this application, with 1.0 mm being the maximum allowable error.

The final step of the decision structure is to determine the need for leaf removal. Classified image data, in RAM, is checked for petiole-to-main stem branching locations within the stem region between the grasp location and the cut location. Results of the processing strategy are transmitted from the supervisory computer to the robot controller including: grade, grasp position and orientation, and leaf removal requirement. Also, the required end effector finger width for grasping is transmitted from the supervisory computer to the end effector controller.

ROBOTIC HANDLING

Design and Control

The end effector, which links the mechanical portion of a robot to the object being handled or processed, is a critical component in any robotic system. The primary objective of the end effector in the geranium processing workcell is to grasp the plant material from a conveyor in a firm manner without bruising or fracturing the main stem, growth tip, or petioles. The end effector (Figure 7) is a servo-operated, parallel-linkage, two-fingered gripping mechanism with a PC/AT compatible computer as a controller.

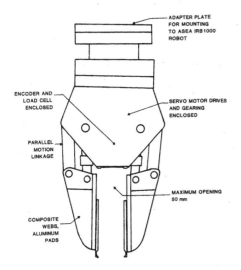

Fig. 7. End effector with primary components indicated.

Fingers (Figure 8) are fabricated from aluminum with curved surfaces to distribute applied forces over the perimeter of the grasped object. A 1.5 mm layer of rubber with a durometer rating of 10 is used for a soft and conforming surface, and the fingers are coated with a liquid rubber to cover all metal surfaces and protect the padding.

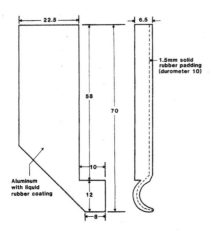

Fig. 8. Geometry and materials of construction for end effector fingers. All dimensions in millimeters.

The gripping mechanism can be controlled based on position and velocity of the fingers or force applied to an object by the fingers. An encoder is used to calculate finger position. A load cell is used to measure applied force. The end effector control system utilizes a two-stage controller based on (1) three-mode feedback control of position and velocity profile and (2) two-mode feedback control of force. In the first stage, the end effector is under closed-loop control of position and velocity using proportional-integral-derivative action. This allows for strict control of the velocity profile

before an object is impacted by the fingers, while also aiding in fast operation. In the second stage, initiated by the applied load to an object exceeding a threshold value, the end effector is under closed-loop control of force using proportional-integral action (Figure 9).

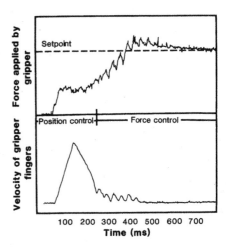

Fig. 9. Load response and velocity profile during grasping a geranium cutting. For cuttings under 18 mm in diameter, the fingers are assured of being within 20% of the minimum velocity before impacting the plant.

The normal sequence of operations is for the robot controller to signal the end effector to *HOME* (zero and initialize sensors and parameters for the upcoming cycle) and to *OPEN* (position fingers to a width determined by image analysis of object to be grasped). The robot controller utilizes position information from the supervisory computer and places the end effector at the desired location. A *FORCE* (perform the two-stage control for grasping) request is then issued from the robot to the end effector. Once the robot controller has completed handling/processing the grasped object, another *OPEN* signal is issued to release the object. The end effector controller signals to the robot controller each time a command or cycle is completed.

Plant Handling Test Procedures and Results

To evaluate grasp performance by the end effector, two tests were conducted. In both tests, singulated geranium cuttings were placed onto a conveyor in the robot workspace. Cuttings were grasped by the robot, moved over to the very end of the belt, held for 6 s, then returned to the belt surface. Data were taken on main stem diameter, proper handling, and the type and extent of damage if damage occurred to a cutting. Damage was categorized into occurrences on the main stem and on any petiole, and as severe (cut, bruise, crushed) or slight (small mark or discoloration). A grasping setpoint force of 11 N was used throughout each test.

The first test was conducted to examine the performance of the finger geometry, the materials of fabrication, and the two-stage controller in comparison to the results in Simonton (1990b). In this test, cuttings were placed onto a fixed position on the conveyor belt and the robot traversed the identical path for each specimen. The second test was conducted to examine the combined benefit of the end effector design and control with the vision software for avoiding any plant parts during the grasp operation. In this test, cuttings were placed randomly in the field of view of the camera on the conveyor belt. The supervisory computer performed the plant part classification as described in Simonton and Pease (1990) and the finger position search as described above, then transmitted position and orientation of a calculated grasp point to the robot controller.

In the first test geranium cutting specimens ranged in size from 3.5 to 8.0 mm in main stem diameter 15 mm from the growth tip, with an average of 5.6 mm. No cuttings were mishandled; all cuttings were located in the fingers of the end effector after the grasp and pickup, and no cuttings were dropped. As can be seen in Table 1, only 18% of petioles and less than 1% of main stems were observed to have signs of damage. All petiole damage occurred either during the movement of the arm just preceding the grasp, or by striking as the fingers were closed. This is a significant improvement over previous results (Simonton, 1990b) which reported 45% of cuttings with damaged petioles. The improvement can be attributed to the following factors: (1) Reduction of finger velocity upon impact by altering velocity profile; (2) Reduction of force level from 17 to 11 N; (3) Alternative finger geometry based on larger curvature and reduced lateral contact area; and (4) Reduction of impact load and improved load distribution due to the 1.5 mm rubber padding.

In the second test cutting specimens ranged in size from 4.5 to 8.0 mm in main stem diameter 15 mm from the growth tip, with an average of 6.0 mm. As in the first test, no cuttings were mishandled. Table 1 shows the significant benefit of using local scene analysis to control the end effector placement and thus avoid striking petioles. Only 1.5% of cuttings were observed to have signs of petiole damage, and each was only a small mark or discoloration. While 2% of main stems showed damage, these were very slight marks which appeared not likely to have an effect on propagation or plant hardiness. By determining a strategy for controlling the placement of the end effector fingers among the branching structure of the cutting, the opportunities for damage were minimized and thus the ability of the robot to handle and process the plant material in an effective manner was greatly enhanced.

SUMMARY

Sensing and control techniques and decision strategies for effective robotic handling of greenhouse plant material have been developed. Machine vision performs a critical role for automatic plant part classification of singulated, randomly positioned, and randomly oriented geranium cuttings. Based on the image classification and resulting knowledge of the plant structure, a technique for determining an appropriate robotic processing strategy for each geranium cutting has been implemented. The technique includes local scene analysis for determining a non-damaging grasp tactic. An end effector controllable based upon both position and force has been developed, integrated into the workcell, and shown to be very effective in handling geranium cuttings in a non-damaging fashion. Testing revealed only 1.5% of petioles and 2% of main stems with slight marks or discolorations from the robotic handling.

In general, results indicate the importance of visual data and the interpretation of that data to assist the robot in accommodating nonuniformity in the plant material. Likewise in importance for effective plant handling are end effector finger geometry, materials of construction, and control method. The successful fusion of sensor information, a robust decision structure, and flexible control has been shown to be significant in assisting the robot in accommodating nonuniformity.

ACKNOWLEDGEMENTS

This research was made possible by the efforts of the following individuals: Jim Pease (algorithm development and vision software for plant part classification); Don Crosby (fabrication of fixtures and end effector fingers); Brian Farmer (electronic interface for end effector controller); Mary Story and Doris Walton (secretarial support). Also acknowledged is the cooperation from Oglevee Products in McDonough, Georgia for donation of plant material, trays, and plugs. Brand names mentioned within do not imply endorsement. The work was supported by a grant from the NASA Marshall Space Flight Center and by the Georgia Agricultural Experiment Station.

REFERENCES

Aho, A.V., J.E. Hopcraft and J.D. Ullman (1983). *Data Structures and Algorithms*. Addison-Wesley, Reading, MA.

Considine, D.M. (1986). *Standard Handbook of Industrial Automation*. Chapman and Hall, New York, NY. pp. 260-320.

Moncaster, M.E. (1985). Applications of robotics to agriculture. Farm Electronics and Computing International Symposium, October 20-25, Stratford-upon-Avon, Great Britain.

Saridis, G.N. (1989). Analytic formulation of the principle of increasing precision with decreasing intelligence for intelligent machines. *Automatica 25(3):461-467*.

Simonton, W. (1990a). Automatic geranium stock processing in a robotic workcell. *Transactions of the ASAE 33(6):2074-2080*.

Simonton, W. (1990b). Robotic end effector for handling greenhouse plant material. ASAE Technical Paper No. 907503. St. Joseph, MI.

Simonton, W. and J. Pease (1990). Automatic plant feature identification of geranium cuttings using machine vision. *Transactions of the ASAE 33(6):2067-2073*.

TABLE 1   Robotic Handling Effectiveness

Test	Variety	Total	No damage	Petiole Damage Severe	Petiole Damage Slight	Main Stem Damage Severe	Main Stem Damage Slight
1	'Ben Franklin	140	118	22	0	0	0
1	'Risque'	100	76	22	0	0	2
		240	194 (81%)	44			2
2	'Risque'	200	193 (97%)	0	3	0	4

Note: Test 1 was conducted without machine vision. Test 2 was conducted with vision including local scene analysis for near-optimal grasp location.

AUTOMATIC GUIDANCE OF FARM VEHICLES

T. G. Nybrant

*Dept. of Agricultural Engineering, Swedish University of Agricultural Sciences,
S-750 07 Uppsala, Sweden*

Abstract. Automatic guidance of two different types of farm vehicles is
described and results from field experiments are presented. The paper deals
mainly with the control theoretical aspects and the sensors and
implementations are only briefly described.

The first vehicle is an experimental 9 m span horticultural gantry running
on crawler tracks. A leader-cable system was used as the reference line.
Field experiments showed that the path of the gantry could be controlled
with an accuracy better than 6 cm.

The second vehicle is a tool carrier with a row-weeder for cereal crops.
The control system sensed mechanically the position of the plant rows. The
field experiments showed that the position of the vehicle could be controlled
with an accuracy better than 2 cm.

Keywords. Guidance systems; control systems; farming; vehicles; modeling.

INTRODUCTION

This paper deals with automatic guidance of
two different types of vehicles. The first
is an experimental horticultural gantry
running on crawler tracks. The second is a
tool carrier with a row-weeder for cereal
crops.

Automatic guidance of farm vehicles may be
desired for different reasons. In some cases
the vehicle, for example a gantry, moves so
slowly that manual steering becomes a very
tedious and unproductive task. When automatic steering is used, the driver is
relieved of this task and allowed to do
something more productive. If the automatic
system is able to steer more accurate, better
land utilization may also be achieved.

In other cases accurate manual steering at
relatively high speeds is a very difficult
and demanding task which a driver can cope
with for just a short period of time. This
is the case for the tool carrier with a
row-weeder, where the steering accuracy
should be better than just a few centimetres.

Different principles can be employed to sense
the reference line. Telle and Perdok (1979)
describe a system which uses a leader cable
system. The system is based on an electric
cable which is buried in the field and
connected to an ac source. Vehicle mounted
coils sense the magnetic field generated by
the cable in such a way that the vehicles
position relative the cable can be calculated
with an analogue electronic circuit. The
system was used to guide agricultural
tractors.

In the previous case the reference lines are
fixed in the field and they are not primarily

related to the crop's position. Another
possibility is instead to sense the crop-rows
or soil-furrows and steer the vehicle with
respect to their position. Such systems were
previously reported by for example Busse and
co-workers (1977) who describe a serial
produced steering system for corn combines
and forage harvesters. Beside the systems
described above, other systems using for
example radio beacons, laser optics, digital
image processing and satellites can be found
in the literature.

Automatic guidance of farm vehicles has been
subject for much research over the years.
The work found in the literature, however,
mostly describe the hardware and the sensors
to detect the reference line. More unusual
are articles dealing with the control
theoretical aspects of the problem, i.e. how
the controller should be designed with
respect to the vehicle dynamics.

In this paper the emphasis is put on the
control theoretical aspects of the two
problems dealt with. The sensors and the
controller implementations are only briefly
described. In both cases it turns out that
fairly simple controllers can be used and
no advanced control theory is really needed.
Instead, the classical continuous time
theory used is meant to give some physical
insights and explanations to the fundamental
dynamic properties of the vehicles.

Work related to this paper is previously
reported in Nybrant (1987), Tillett and
Nybrant (1990) and Wilhelmsson (1990).

PROCESS MODELING

Although the two different processes to be controlled exhibit some similarities they differ with respect to how they are steered. The gantry is steered by varying the relative speeds of the crawler tracks at its ends, while the tool carrier is steered conventionally with its front wheels. Therefore, modeling of the two processes are described separately.

Basic modeling of the gantry.

For control purposes it is suitable to view the gantry as in Fig. 1. It consists of a 9 meter long beam (essentially a tool carrier) with crawler tracks at its ends. The crawler tracks are separately driven by hydraulic motors which are supplied with oil from a valve. This valve divides the main oil flow with an adjustable ratio and steering is achieved by adjustment of this valve.

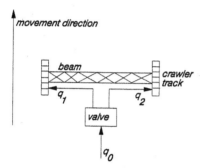

Fig. 1. Schematic view of the gantry. The main oil flow q_0 is divided into q_1 and q_2, which are fed to the crawler track motors.

A mathematical model of the gantry's movement on the field can be derived using the geometries illustrated in Fig. 2. The figure shows what happens during a short time interval dt, when the crawler tracks have velocities v_1 and v_2, respectively. The gantry is assumed to move in the y direction with a heading α defined to be zero when the gantry is parallel to the x-axis.

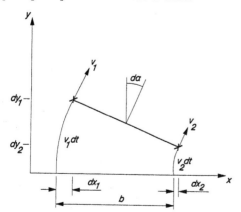

Fig. 2. Basic geometry of gantry movement, with variables used for modeling.

Based on Fig. 2, a simple differential equation for the heading now can be expressed

$$\frac{d\alpha}{dt} = \frac{v_1 - v_2}{b} \qquad (1)$$

where b is the width of the gantry.

From Fig. 2 it also follows that the co-ordinates for the left end of the gantry, where the position sensor is mounted, can be expressed

$$\frac{dx_1}{dt} = v_1 \sin\alpha \qquad (2)$$

$$\frac{dy_1}{dt} = v_1 \cos\alpha \qquad (3)$$

The three equations above form a model that relates the heading and position of the gantry to crawler track velocities at its ends. These velocities, in turn, are determined by the total oil flow supplied into the dividing valve and the setting of the valve. When it is assumed that the action of the valve is linear, it can be included in the model by describing the velocities as

$$v_1 = v(1 + 2C_v\psi) \qquad (4)$$

$$v_2 = v(1 - 2C_v\psi) \qquad (5)$$

where v is nominal gantry speed (the speed at the centre of the gantry), C_v is a constant and ψ is the setting of the valve.

When Eqns. (4) and (5) are substituted into Eq. (1) it gives

$$\frac{d\alpha}{dt} = \frac{4C_v}{b} v\psi \qquad (6)$$

which relates the heading to the valve setting.

In the above discussion it is assumed that the time lag caused by inertia of the gantry is negligible. This was considered to be the case, since during normal operation the accelerations are very small. Also the wheel slip and the influence of actuator dynamics are considered to be negligible.

Linearization of the gantry model.

In the following discussion a linear and speed independent model of the gantry will be derived.

In the controlled situation when the gantry moves along the reference line with only small deviations, the heading α will generally be small and the nominal speed will be close to the speed at the ends. This means that $\sin\alpha = \alpha$, $\cos\alpha = 1$ and $v_1 = v$ are good approximations. Using these approximations, Eqns. (2) and (3) can be rewritten

$$\frac{dx_1}{dt} = v\alpha \qquad (7)$$

$$\frac{dy_1}{dt} = v \qquad (8)$$

Eqns. (6) and (7) now yield

$$\frac{d^2 x_1}{dt^2} = \frac{4 C_v v^2}{b} \psi \qquad (9)$$

This double integrating process would probably be fairly easy to control if it had not included a gain factor v^2. This factor is obviously strongly nonlinear with respect to the forward speed of the gantry, which may vary during operation. This speed dependent gain, however, can be eliminated by substitution of Eq. (8) into Eqns. (6) and (7) which then become

$$\frac{d\alpha}{dy_1} = 4 \frac{C_v}{b} \psi \qquad (10)$$

$$\frac{dx_1}{dy_1} = \alpha \qquad (11)$$

These two equations lead to the model

$$\frac{d^2 x_1}{dy_1^2} = 4 \frac{C_v}{b} \psi \qquad (12)$$

This model is also a double integrator, but by using Eq. (12) instead of Eq. (9) we are now referring to x_1 and ψ with respect to distance y_1 instead of time t, giving a model that is independent of forward speed v. From a practical point of view it means for example that when a discrete time controller is used, sampling should be performed with respect to travelled distance, and not as usual with respect to time.

Modeling of the tool carrier.

The tool carrier is built as a conventional four wheel vehicle which is steered with the front wheels. The actual row-weeder (with hoes running between the rows) is mounted under the vehicle with the position sensor mounted close to it. The basic geometry of the vehicle and its movement while turning is illustrated in Fig. 3.

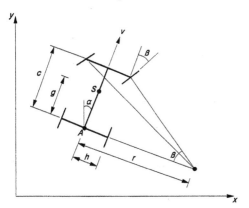

Fig. 3. Geometry of tool carrier.

It is here assumed that the steering mechanism is designed according to the Ackerman principle. This means that the lines perpendicular to the front wheels intersect with the line perpendicular to the rear wheels so that the turning point is uniquely defined. In the figure point A describes the middle of the rear axle and point S is the location of the position sensor.

The vehicle moves with forward speed v and turning radius r. It is then elementary that the rate of change of the heading α can be expressed

$$\frac{d\alpha}{dt} = \frac{v}{r} \qquad (13)$$

The turning radius r is related to the front wheel angle β through

$$\tan\beta = \frac{c}{r-h} \qquad (14)$$

When the vehicle is steered along a reference line (i.e. the crop rows), β will generally be small. The approximation $\tan\beta = \beta$ can then be used and it also means that h (half the rear wheel base) will be negligible compared to r. Using these approximations Eqns. (13) and (14) lead to

$$\frac{d\alpha}{dt} = \frac{v}{c}\beta \qquad (15)$$

In accordance with the derivation of Eqns. (7) and (8), the co-ordinates for point A (middle of the rear axle) are approximated by

$$\frac{dx_A}{dt} = v\alpha \qquad (16)$$

$$\frac{dy_A}{dt} = v \qquad (17)$$

Eqns. (15) and (16) now give the model

$$\frac{d^2 x_A}{dt^2} = \frac{v^2}{c}\beta \qquad (18)$$

which is essentially of the same type as the gantry model, Eq. (9).

The x co-ordinate for the sensor, located at distance g in front of the rear axle, can be expressed

$$x_S = x_A + g\sin\alpha \qquad (19)$$

which together with Eq. (18) and the approximation $\sin\alpha = \alpha$ lead to the model

$$\frac{d^2 x_S}{dt^2} = \frac{v^2}{c}\beta + \frac{gv}{c}\frac{d\beta}{dt} \qquad (20)$$

Like with the gantry model, this model can also be made independent of the speed by substituting v = dy/dt into it, which gives

$$\frac{d^2 x_S}{dy^2} = \frac{1}{c}\beta + \frac{g}{c}\frac{d\beta}{dy} \qquad (21)$$

Again, the speed dependence is eliminated by regarding the input and output signals as functions of y instead of time.

The above model only explains the dynamics caused by the vehicle's idealized movement, shown in Fig. 3. In practice, especially at high speeds, other dynamics caused by for example inertia and wheel slip, may occur. In the actual case, however, it was assumed that the effects of these other dynamics are negligible.

CONTROLLER DESIGN

Gantry controller

When Eq. (12) is Laplace transformed it gives the transfer function

$$X_1(s) = 4\frac{C_v}{b}\frac{1}{s^2}\psi(s) \tag{22}$$

Note that the complex frequency s in this case is related to distance, and not time.

The model (22) is a pure double integrator, which indicates that a PD controller would be a good choice. Such a controller would in this case give opportunity to place the closed loop poles at any desired location. The differentiating term in such a controller is

$$\frac{d(x_r - x_1)}{dy_1} = -\frac{dx_1}{dy_1} \tag{23}$$

when x_r is constant. Since it follows from Eq. (11) that

$$-\frac{dx_1}{dy_1} = -\alpha \tag{24}$$

it means that a measurement of α can alternatively be used, in order to avoid differentiation in the controller. The control law will then consist simply of two feedback loops with static gains.

The position sensing system used consisted of a buried cable as the reference line, and vehicle mounted coils to detect the magnetic field generated by the alternating current in the cable. The coil assemblies were mounted as shown in Fig. 4. The position error is $x_r - x_1 = (e_1 + e_2)/2$ and since α is small it becomes approximately $(e_1 - e_2)/d$. The whole controller was implemented using conventional analogue electronics, and the oil dividing valve used for steering was actuated by a dc motor servo system.

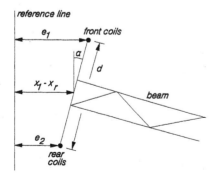

Fig. 4. Sensor arrangement for the gantry

The tool carrier controller

The transfer function for the tool carrier is, cf Eq. (21)

$$X_s(s) = \frac{1 + gs}{cs^2}\beta(s) \tag{25}$$

When this process is controlled by a proportional controller with gain K, the natural frequency and damping of the closed loop system become

$$\omega_o = \sqrt{\frac{K}{c}} \tag{26}$$

$$\zeta = \frac{g}{2}\sqrt{\frac{K}{c}} \tag{27}$$

This shows that the closed loop properties can be arbitrarily chosen, provided the sensor can be mounted at any desired distance, g, from the rear axle. Note that when g<0, i.e. the sensor is mounted behind the rear axle, a nonminimum phase system is obtained, which is undesired from a control point of view.

The controller was implemented using an Intel 8031 one-chip microprocessor, which also included the controller for the dc motor driven steering servo. The position sensor consisted of a light metal profile running between the crop rows. The angle of this profile, and thereby the position, was sensed by a potentiometer, cf Fig. 5. The distance between the rows was 25 cm.

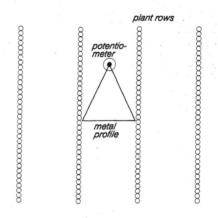

Fig. 5. Sensor for the tool carrier (view from above).

Remark

The process models presented above relates the input and output variables to distance. This means that the closed loop properties also are related to distance. When, for example, the controlled vehicle is started beside the reference line, it will follow the same track back to the line regardless of forward speed.

FIELD EXPERIMENTS

A field experiment with the gantry is illustrated in Fig. 6. The controller used was the one previously described, i.e. static feedback from the position and the heading. The position was measured through a stream of emulsion paint discharged from the centre of the gantry. The position is shown with an offset of half the gantry width, so it can be easily compared to the reference line (the buried cable). If the initial response is neglected, the trace of the gantry centre was consistent to ± 6 cm over four experiments.

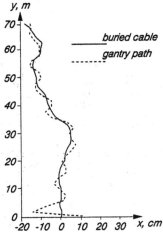

Fig. 6. Control experiment with the gantry.

In Fig. 7, an experiment with proportional control of the tool carrier is illustrated. The position of the sensor was chosen to give good damping characteristics. The results show that the path was consistent within ±2 cm from the reference line. This was the case also in the other experiments performed.

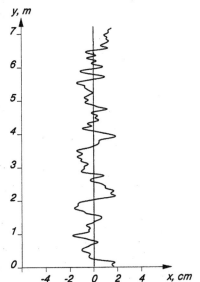

Fig. 7. Control experiment with the tool carrier.

CONCLUSIONS

Despite the simple controllers used, the experimental results obtained showed that accurate control, meeting the practical requirements, was maintained for both vehicles. An explanation might be that the vehicle speeds used in the experiments were relatively low (a few km/h). This means that higher order dynamics, caused by for example inertia of the vehicles and wheel slip, become negligible. It is most likely that there exist upper speed limits, above which the steering accuracy should be too low to meet the requirements. These limits, however, have not yet been investigated, since they probably are well above the upper practical limits determined by the application and the implements.

ACKNOWLEDGEMENTS

The work with the gantry was carried out during a stay as a visiting researcher at AFRC Institute of Engineering Research, Silsoe, U.K. The author wishes to acknowledge the contribution made by N. Tillett, who was responsible for the controller implementation as well as the field experiments.

The author also wishes to acknowledge the contributions made by P. Wilhelmsson and S. Klensmeden, who implemented the control system for the tool carrier and sorted out a lot of important practical problems.

REFERENCES

Busse, W., H. Coenenberg, F. Feldman and T.F. Crusinberry. (1977). The first serial produced automatic steering system for corn combines and forage harvesters. Proc., First Int. Grain and Harvesting Conf., Am.Soc. of Agr. Eng. Publ. 1-78. pp. 43-47.

Nybrant, T.G. (1987). Automatic Guidance of an Experimental Field Gantry, Divisional Note, DN. 1428. AFRC Institute of Engineering Research, Silsoe, U.K..

Telle, M.G., and U.D. Perdok (1979). Field experiments with a leader cable tractor guidance. Am. Soc. of Agr. Eng., Paper No. 79-1069, 1979 Summer meeting of ASAE and CSAE.

Tillett, N.D. and T.G. Nybrant. (1990). Leader cable guidance of an experimental field gantry. J. Agric. Engng. Res., Vol.45, pp. 253-267.

Wilhelmsson, P. (1990). Automatic Guidance of a Row-weeder. Report 147. Dept. of Agr. Eng., Swedish University of Agricultural Sciences, Uppsala, Sweden. (In Swedish)

ROBOT FOR MASSPROPAGATION

A. Kinase* and H. Watake**

*Manufacturing Engineering Research Center, TOSHIBA,
8, Shinsugita-cho, Isogo-ku, Yokohama, 235, Japan
**Control and Instrumentation Division, TOSHIBA, 1-1, Shibaura 1-chome, Minato-ku,
Tokyo, 105-01, Japan

Abstract. TOSHIBA developed a robot system for masspropagation of plantlets. This system consists of two robots, a conveyer and tray stockers. The sensing robot measures the plantlet shapes and decides the cutting positions. Then, the cutting robot grips the stem of the plantlet and cuts it between the nodes, and transplants the cut-out segment into the other tray. It repeats until all the nodes of the plantlet are cut out.

The main features of this system are as follows: Firstly, this system can recognize plantlet profiles by using laser beam and can decide the cutting positions and angles. Secondly, it has a gripper which can grasp the plantlet without damage. Thirdly, this system can carry out automatically all the procedures from loading of plantlets to unloading of the transplanted trays. The system was demonstrated at The International Garden and Greenery Exposition (EXPO'90), and transplanted one segment in fifteen seconds.

Keywords. Robots; biotechnology; cameras; cutting; force control; sensors; signal processing; strain gauges.

INTRODUCTION

Industrial production of fine plantlets for delivery to farms is now a flourishing trade. It is often said that the quality of a plantlet decides half of a flower quality or vegetable quantities in producing a harvest. Most plants are affected by viral or bacterial infections, in other words, they are virtually sick. The demand for virus-free plantlets is therefore high.

Plant improvement by biotechnology is producing new types of plants. Many of these new plants have difficulty in making seeds and therefore are not propagated by planting seeds. Instead, tissue culture is used to produce virus-free plantlets cloned replicas of the parent plant without sowing seeds. An apical meristem trimmed from a superior parent plant is cultivated in a clean room and the resulting shoots are transplanted and grown separately. These are cut and transplanted as cuttages and grown. Plantlets are thus multiplied by repeating these procedures. Nowadays some flowers are propagated through the use of such methods. The problem of this type of production is the requirement of a clean room and the high labor costs which is reflected in the cost of the plantlets. Labor costs account for about 70% of plantlet costs. Automation can bring down the cost of plantlets developed through tissue culture to make them available on a larger scale and cause far-reaching future agricultural changes.

Fig.1. The developed robot system

Toshiba has developed a robot system (see fig.1) to automate the cutting and transplanting procedures which are the most time-consuming and where most of the labor is concentrated. This robot system can recognize undefined profiles of plantlets and handle delicate plantlets without damaging them. The following is a description of the new robot system.

PLANTLET PRODUCTION BY ROBOTS

Plantlet Production Procedures

Fig.2 shows the production procedures of a typical system producing carnations and potatoes. First, an apical meristem cut from a parent plant is grown to produce plantlets. Manual labor is used at this stage since there are only a few plantlets.

In the next stage Robots handle the plantlets. The robots have three-dimensional sensing capabilities to recognize the nodes of a plantlet on a tray. Using the result of recognition, the robots grasp the plantlet gently with a soft-gripper, cut it and transplant it to another tray. When one plantlet has been processed in this way, the robots works on the next plantlet and so on. The tray filled with the

transplanted segments of plantlets is automatically transported to a nursery room. These procedures are performed by robots. The number of plants are thus increased by repeating to cut, transplant and nurse. This stage is the most labor dependent stage, so it is better to use robots for this stage.

In the final stage the plantlets are passed through the procedure of acclimatization in which the plants acclimatize to the outside environment. They are delivered to the farms after this process.

Robot System

The robot system makes it possible to automate all procedures from loading plantlet trays to unloading transplanted trays, which include recognition, cutting and transplanting. The plantlets are cultivated and transported with plastic trays.

The structure of the system is shown in fig.3. Two tray stockers store four types of trays which are

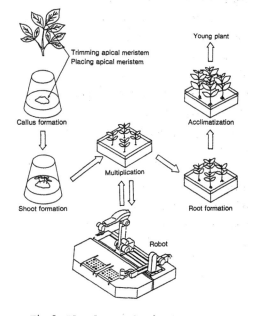

Fig.2. Plantlet production by Robots

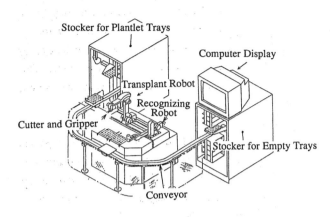

Fig.3. System Configuration

with plantlets, empty, cut-out, and with cut-out segments. These trays can be stored or ejected as required. A monorail type conveyor moves two carriers on which trays transport between the stockers and the robots. The system is highly efficient and compact as the two carriers can move independently each other on the monorail.

The conveyor transports two trays which are with and without plantlets to the front of the robots. And the two robots perform the procedure of a multiplication. The recognition robot, a Cartesian robot, is equipped with a head by which the robot can recognize three dimensional profiles of plantlets. The transplantation robot is a 6-axis articulated robot of which first axis is a traverse at the base, and has a hand with a soft-gripper and a cutter.

Two controllers manage the robot system, each of them has several 32-bits microprocessors which are hierarchys for controlling the robot and peripheral equipments.

HAND

One of the most important requirements is that the hand should be able to handle the plantlets without damage. Plantlets cultivated through tissue culture are frail. The hand has gripper and cutter. The gripper must be able to pick up and hold plantlets with just the right pressure. If the pressure is too much, the plantlet is damaged, and if it is too little, the plantlet slips out of the gripper. The hand is compact and contains the cutter since the plantlets it handles are small and much of the work done is of a delicate nature.

The researches using shape memory alloys(Miwa,1986) and plate springs(Miles,1987) had investigated. However, these alloys and springs are not rigid or accurate enough to correctly transplant the plantlets or to apply the pressure for the various size and weight of the plantlets.

The hand developed for the system is shown in fig. 4. As shown the gripper and the cutter are integrated and the gripping pressure is measured by strain gauges in the fingers of the gripper. The

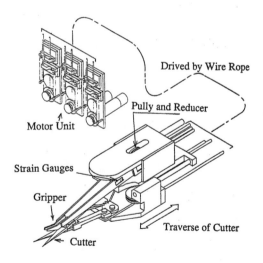

Fig.4. Configuration of the Hand

227

gripper feeds back pressure data to open or close the fingers. Usually this type of hand is used for small size objects, and plantlets are closely together in a environment of plant factorys. To minimize the size of the hand unit, it is split into a main unit and a drive unit connected together with a wire rope for transmitting power.

The gripper consists of tweezer-like fingers with a pulley and reducer at their base; a scissor-like cutter with a pulley and a reducer at its base; guides which move the cutter back and forth; a pulley and reducer required to control back and forward cutter movement; a wire rope and outer tube used for each drive axis and a motor unit to power them.

Wire Drive Mechanism

The gripping force cannot easily be controlled via a wire rope.

The reason is due to the nonlinearity caused by friction and backlash in the transmission system from the motor to the fingers, and the low rigidity of the wire rope. This means that the control gains cannot be sufficiently increased to maintain accuracy. And that the control parameters considerably differ between when a plantlet is held

and when not (i.e. the parameters which are valid when the fingers are open produce limit cycle when they hold a plantlet).

The following three measures were taken in the design of this system.

1. Wire rope and outer tube are coated with Teflon.
2. The speed reduction ratio of the fingers to the wire rope is increased.
3. Finger rigidity is reduced (not greater than plantlet rigidity)

These measures have the following effects.

1. Friction is reduced.
2. Less driving power is required whereby friction is reduced.
3. The equivalent rigidity of the wire rope for the fingers increases to the square of the reduction ratio.
4. The fingers are less affected by the backlash of the wire rope, which decreases to one by reduction ratio.

The third countermeasure means that the stiffness of the plantlet has no effect on the control system. This makes it possible to use the same servo gain in holding the plantlet.

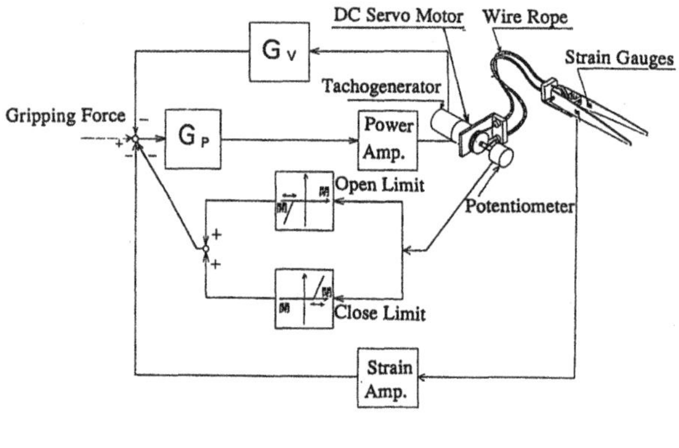

Hybrid Control of Gripping Force and Position

A force controlled gripper must stay at a desired position, that is open or closed position, when it is not gripping an object. To hold finger at a desired position, one solution is to install a mechanical stopper. But the stopper cannot easily be installed and the stop position cannot be changed. Instead, hybrid control of the force and position of the fingers was adopted. As shown in fig.5, strain gauges, a potentiometer and

Fig.5 Control Block

tachogenerator are used to feedback data. The feedback of the potentiometer goes through two nonlinear functions. When the fingers are between open and closed position, the output is zero, and when they go beyond the closed or open limit position, the output corresponds to actual finger position. The position of the breakpoint of the function corresponds to the limit position and this position can be changed easily. In this system, when the fingers are between open or closed position, the potentiometer feedback is zero which permits true force control. If the fingers should exceed the open or close position and there is no feedback from the strain gauges, the feedback from the potentiometers will cause the fingers to stop in open or closed position.

RECOGNIZING PLANTLET PROFILE

Let us assume that the plantlets are placed in a lattice pattern. The stems of the plantlets have nodes and they should be cut between these nodes. Usually, a CCD camera is used to measure the profiles of objects. The problem is that since visual data is 2-dimensional, it is difficult to distinguish the plants from the side view as outlines overlap.

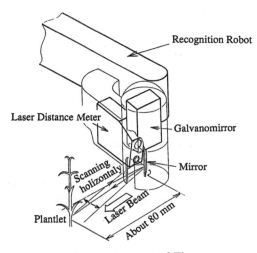

Fig.6. The structure of Three Dimensional Vision

Instead, laser triangulation is used to pick up the profile of a single plantlet nearest the camera. And the robot recognizes the profile and determine where to cut.

Measuring Plantlet Profile

To measure the plantlet, laser beam is scaned on it from the side. The measuring unit is the head part of the recognition robot whose configuration is shown in fig.6. The laser distance meter uses the triangular measurement method to measure the distance from the point where the laser beam hits the plantlet. The laser beam is scanned horizontally with a galvanomirror to measure the horizontal sections of a plantlet. The three-dimensional profile of the plantlet is measured by repeating these measurements at different heights. Thus the profile of an objective plantlet can be sensed by excluding depth data outside a specified range.

Actual Measurements and Determining Cutting Positions

The position of the cut is determined through the procedures described below. Since the time of the laser scan depend on the number of points to be measured within the range, two types of scanning are performed to speed up the process: rough scanning and fine scanning.

In a rough detection, laser beam scans the area which is likely to be occupied by the plantlet to sense the presence of a plantlet. The height of a plantlet is scanned at wider intervals and its depth is scanned within a specified range. A part which has a thickness within a certain range of value is assumed to be the stem. After this the position and height of the plantlet is checked. This is followed by a fine detection process. This process uses the rough detection data to measure the stem at closer range and makes a more detailed vertical inspection of it.

Now the robot uses the measured three dimensional profile to check the diameter of the stem and presence of a branch, and recognizes the positions of the nodes. The robot then cuts the plantlet between two nodes. The positions of the cut are determined according to the following formulas.

1) when $1 > 6$

   then $X = 3.5 + 0.4 * (1 - 6)$

2) when $1 < 6$

   then don't cut.

1 : The length of the stem between two nodes

X : Cut position (from the lower end of the stem)

The stem is cut when 1 is large to make the cutout segment long enough. When it is short, the robot can not cut the stem since there is not enough space for the cutter and gripper operate in. The result of the plantlet recognition is shown on the display (see fig.7). The profile of the measured plantlet is shown on the left. The lines indicate the cutting positions and the angles of the cut. The coordinates of the cutting positions and the attitudes of the transplanting robot are shown on the left.

CONCLUSIONS

The above is a description of the robot system used to automate the cutting and transplanting of plantlets. The robot system consists of a gripper that can gently handle the frail plantlets without damage and a sensor that can pick out the profile of a single plantlet among several plantlets and recognize nodes from the acquired data to automatically determine the position of the required cuts. The robot system performs practically all the tasks involved in plant propagation quickly and efficiently.

It performed one transplant operation including the sensing operation within 15 seconds. Future objectives include faster speed and lower cost systems.

This robot system was demonstrated at the International Garden and Greenery Exposition (EXPO '90 OSAKA JAPAN) and was on display in its Japanese Government Garden from April to September, 1990.

REFERENCE

Miwa, Y., et.al. (1986). Research into automation of plant tissue culture. Proceedings of the 4th symposium of the Robotics Society of Japan, pp. 465-466.

Miles, G.E., et.al. (1987). Robotic transplanting of bedding plants. trans. of the ASAE, vol.30-3, pp. 586-590.

Kinase, A., and H. Watake (1988). Robot for plant tissue culture. Robot, 64, Japan Industrial Robot Association, Tokyo, pp.74-79.

Kinase, A., et.al (1989) Development of a plantlet propagation system -- Methods for detecting plant nodes --. Proceedings of the 7th symposium of the Robotics Society of Japan, pp.275-276.

Kinase, A., et al. (1990) Development of the Robot System for plant factorys. Proceedings of the 8th symposium of the Robotics Society of Japan, pp1033-1036.

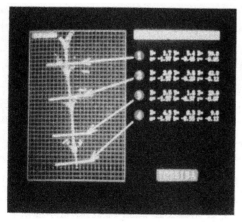

Fig.7. Result of Detection

DISCRIMINATING ROBOT SYSTEM FOR CARNATION SEEDLING WITH FUZZY LOGIC

H. Fujiwara

Japan Tobacco Inc., Engineering Research Laboratory, 1-31, Kurobeoka, Hiratsuka, Kanagawa 254, Japan

Abstract. A carnation seedling is an agricultural product and its shape
is very various. Therefore, it is difficult to discriminate this object
by using the industrial method of some fixed pattern matching. We proposed
the evaluation and the judgement of this seedling based on the subjective
knowledge of a connoisseur, by using the image processing of the top and
the side view and the fuzzy calculation such as fuzzy matching and fuzzy
integral. In an online test with 500 samples, a good judgement rate of 97
% was achieved for a carnation seedling. This technique, mentioned here,
will make it possible to evaluate and judge other agricultural product.

Keywords. Fuzzy logic; image processing; carnation seedling;
discriminating robot; features extraction.

INTRODUCTION

The connoisseur of a carnation seedling is
quickly and correctly judging whether the
seedling is good or not, without comparing
the model pattern or measuring the details.
From this point of view, it follows that the
connoisseur has evaluated the seedling by
using the subjective knowledge through the
repeated experience. This discrimination
method is different from the one of inspecting
an industial product, which has many standards
and can be checked with some fixed methods.
Agricaltural products, like a carnation
seedling, don't have any standard.

This paper presents a discriminating system
of a carnation seedling that uses the image
processing and the fuzzy logic. This system
was demonstrated at the Japanese Government
Garden, ” EXPO ’90 ”, in Osaka city.

First of all, some characters of the seedling
are extracted with the real time image
processing, which is carried out at three
stages (i.e., an image data input, a

preprocessing and a features extraction)
using the top and the side views from two CCD
cameras. As a matter of course, these
extracted characters are based on the
connoisseur's knowledge of a carnation
seedling represented by natural language.

Secondly, the fuzzy integral is performed with
the subjective measure of the connoisseur of
each character and each degree of fuzzy-
matching between the extracted data and the
stored qualitative knowledge. Therefore, this
system can subjectively evaluate and judge a
carnation seedling like human senses.

MASS PRODUCTION SYSTEM FOR A MICRO-PROPAGATED PLANT

Fig.1 is the scheme of the mass production
system for a micro-propagated plant. The
cutting robot cuts the virus-free plant at
every joint and transplants them to plugs.
The cut-plant is brought up in a nursery
chamber, and then cut and transplanted again.
In this way, the plants continually propagated

themselves. The purpose of the next chamber is to acclimate the plant to the outside. The carnation seedling, taken out from this chamber, is then transplanted to a rockwool cube. This transplanting system was developed by JapanTobacco Inc..

In this transplanting system, there were two problems we had to solve. One was to define the qualitative characteristics of a carnation seedling. The second was to develop an evaluation method to determine whether the seedling is of good quality. By using image processing technique and fuzzy logic, we were able to solve these problems.

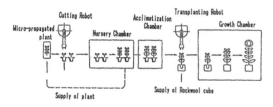

Fig.1 Mass production system for Micro-propagated plant

SYSTEM CONFIGURATION OF A DISCRIMINATING ROBOT

Fig.2 is a scheme of the discriminating robot system. This system consists of a robot, an image processing unit, two CCD cameras and an air duct. The movement of this system is as follows. First, a plug tray comes out from the nursery. Next, this robot grasps and sets the carnation seedling at a center position in order to obtain the image data from two CCD cameras. Based on evaluation by the image processing and fuzzy calculation, using the fuzzy image processing unit, the robot either transplants the seedling to a rockwool cube if it is good, or throws it away. A rockwool cube is automatically transported by pneumatic force.

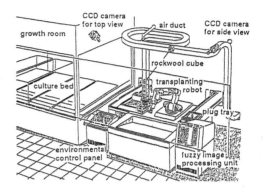

Fig.2 System Configuration of a Discriminating Robot

A SCHEME OF DISCRIMINATING METHOD

Fig.3 is a scheme of the discriminating method. The connoisseur of carnation seedlings quickly and correctly judges whether the seedling is good or not. The method of this discrimination is different from those used in industrial inspection, employing fixed pattern recognition.

We developed a discriminating method for carnation seedlings that uses an image processing and a fuzzy calculation for evaluation. There are four stages of discrimination: image data input, pre-processing, features extraction and fuzzy calculation. I want to explain in detail the features extraction, based on a linguistic representation of a good seedling, and fuzzy calculation including fuzzy matching and fuzzy integral.

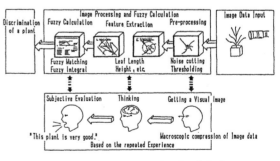

Fig.3 A Scheme of Discriminating Method

pipe-line image processor and so on. The robot controller is also employed to control optimizingly the vertical articulated robot with six joints and six degrees of freedom.

METHOD OF FEATURES EXTRACTION

As mentioned before, there are three stages at the image processing, that is, an image data input, a preprocessing and a features extraction. At the stage of an image data input, the image data taken with the two CCD cameras are stored in the image memory of 512 × 480 pixels with 8 bit gray levels when the robot grasps and sets the seedling at the picture-taking position.

Over the above processing, these digital image data are smoothed by the 3 × 3 matrix digital filter and then sliced in binary image data. This threshhold selection is carried out automatically with the method of the discriminant analysis.

The features extraction is based on these binary image data. The extraction from the top view leads to the detection of four features such as a top leaf-area, a leaf-length, a leaf-width and a leaf-number. In the same way, the extraction from the side view leads to the detection of three features such as a side leaf-area, a seedling-height and a seedling-direction. The outline of the extraction algorithm is as follows:

A top view shown in Fig.4 has some characters, that is,
(1) Detect the value of area by counting the pixels of the binary image data with 255 level point in brightness.
(2) Calculate the center of gravity (CG) in the seedling pixels.
(3) Extract the contour of this image data in the counter clock wise manner and, at the same time, obtain the coordinates.
(4) Calculate the distances from CG to each edge points and recognize the part, which is continuously longer than a fixed level for some period, as a leaf.
(5) Recognize the length from CG to the tip of a leaf as a leaf-length and the width at a middle position of a leaf as a leaf-width.
(6) Calculate the average length and the

average width based on the recognized leaf number.

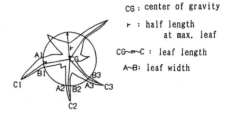

Fig.4 Image Processing from Top View

A side view shown in Fig.5 has some characters, that is,
(1)~(3) In the same way as the top view processing.
(4) Calculate the least square fit line of the contour points and detect the seedling's height on the new axis taken by using the affine transformation.
(5) Detect the seedling's direction by calculating the ratio of the leaf pixels to all pixels within a circular arc of 30 degree.

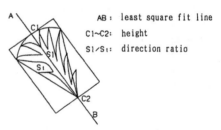

Fig.5 Image Processing from Side View

FUZZY MATCHING AND REAL-TIME EVALUATION WITH FUZZY INTEGRAL

According to the qualitative knowledge of the connoisseur, we make the support set of membership functions about the good seedling. They are represented in Fig.6 from the offline extracting test. Table 1 is the details of all the functions, described as the parameters A, B. The parameter A or B stands for the value of 'M - 3σ' or 'M - σ'. (where M is the average and σ is the standard deviation in offline test.)

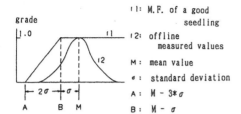

Fig.6 Decision of Membership Function

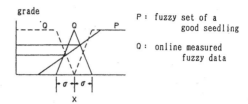

Fig.7 Fuzzy Matching

TABLE 1 Parameter A,B for Membership Function

item	unit	A	B
① top area	pixel	4000	12000
② number	count	2	4
③ width	pixel	5	15
④ length	pixel	100	160
⑤ side area	pixel	4500	6000
⑥ height	pixel	260	340
⑦ direction	%	5	15

In the next step, we construct the fuzzy data base with these parameters and load the image processing unit with the above mentioned image processing and fuzzy calculation programs. At this moment, the preparation of the system has been finished. The discriminating algorithm of a carnation seedling with the fuzzy logic is as follows:

(1) Fuzzy matching

The fuzzy matching tecnique is shown in Fig.7. Each character of the seedling is extracted without vagueness by carrying out the real time image processing. However, it is doubtless that a carnation seedling has a linguistic vagueness in the description of these characters. In order to represent this vagueness, we allocate to each character a fuzzy number Q in Fig.7, based on the online extracted value and the standard deviation found in the offline test.

The degree of matching, h(xi), between two fuzzy set P and Q are calculated by the below equation.

$$h(xi) = \{MAX(P \cap Q) + MIN(P \cup 'Q)\} / 2$$

where P stands for the fuzzy set of the good seedling.
    'Q stands for the negation of Q.

The values, from h(x1) to h(x7), rearrange themselves in order of largeness for the calculation of a fuzzy integral.

(2) Real-time evaluation and judgement with fuzzy integral

In the evaluation and judgement of the seedling, we must consider the subjective measure of the connoisseur as well as the above fuzzy-matching degrees. Therefore we adopt a fuzzy integral to perform the synthetic evaluation like human senses.

The calculation of a fuzzy integral with a λ-fuzzy measure is as follows :

$$e = \int h(x) \bigcirc g\lambda = \overset{7}{\underset{i=1}{V}} [h(xi) \wedge g\lambda(Hi)]$$
$$g\lambda(Hi) = gi + g\lambda(Hi-1) + \lambda \cdot gi \cdot g\lambda(Hi-1)$$

where h(xi) : the degree of fuzzy matching
              at xi, $h(xi) \geqq h(xi-1)$
    $g\lambda(Hi)$: λ-fuzzy measure
    gi   : g({xi})

Solving the equation, $g\lambda(H7) = 1$, we see that λ equals -0.5.
As a criterion of the judgement, we choose the value of this fuzzy integral at 0.6 from the offline test.

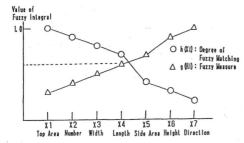

Fig. 8 Example of a Synthetic Evaluation
by Fuzzy Integral

CONCLUSION

A carnation seedling is an agricultural
product and its shape is very various.
Therefore, it is difficult to discriminate
this object by using the industrial method of
some fixed pattern matching.

We proposed the evaluation and the judgement
of this seedling based on the subjective
knowledge of a connoisseur, by using the image
processing of the top and the side view and
the fuzzy calculation such as fuzzy matching
and fuzzy integral.

In an online test with 500 samples, a good
judgement rate of 97 % was achieved for a
carnation seedling. This technique, mentioned
here, will make it possible to evaluate and
judge other agricultural product.

ACKOWLEDGEMENT

The author is indebted to Dr. T. Terano and
Dr. S. Masui, Hosei University, for valuable
suggestions during the course of this work.

REFERENCE

1. T. Terano, S. Masui, and et al; Recognition
of crops by fuzzy logic, Preprints of second
IFSA Congress, Tokyo(1987)

2. Hideyuki Fujiwara, Masaki Shouji;
Evaluation of a carnation seedling by fuzzy
integral, Proceedings of the International
Conference on Fuzzy Logic & Neural Networks,
Iizuka(1990)

Copyright © IFAC Mathematical and Control Applications
in Agriculture and Horticulture, Matsuyama, Japan 1991

DEVELOPMENT OF THE ELECTRIC DRIVEN AUTOMATIC GUIDED VEHICLE FOR USE IN GREENHOUSES AND ITS TRAVELLING PERFORMANCE

J. Yamashita*, K. Satou**, M. Hikita***, T. Imoto****
and T. Abe*

*Faculty of Agriculture, Ehime University, Tarumi, Matsuyama, 790, Ehime, Japan
**Tokushima Technical Junior College, Itano, Tokushima, 779-01, Tokushima, Japan
***Takamatsu National College of Technology, Chokushi, Takamatsu, 761, Kagawa, Japan
****Shikoku Research Institute Incorporated, Yashima, Takamatsu, 761-01, Kagawa, Japan

Abstract. We designed and manufactured the prototype of microcomputer installed railless type unmanded vehicle that travels along ridges furrows and in between water culture bed rows in greenhouses, for the purpose of labor saving in hard transport work. To prevent the vehicle from running on the ridges, the vehicle are equipped with front axle supported by a couple of ball joints, spin turn mechanisum, which allows turning with minimum space, and microcomputer control unit for autonomous traveling. It is a four-wheel vehicle driven independently by 12 V, 150 W DC geared motors, and is capable of traveling at 40 cm/s speed with a maximum of 80 kg carring capacity. The route and direction of movement of the vehicle is controlled by optical sensors mounted on the body of the vehicle, and signs placed at desired locations in the greenhouse. Bed rows traveling vehicle is equipped with 4 wheel reverse phase steering mechanism which measure the distance between vehicle and bed wall face, and straighten itself by two ultrasonic wave sensors attached at the side of vehicle. The results of traveling test showed that continuous five hours operation was possible with standard load. The practical application of the prototype vehicle is now envisioned.

keyword. Computer control; automatic control; microprocessors; ultrasonic transducers; transport vehicles; greenhouses.

INTRODUCTION

In Japan, the area of greenhouse cultivation has been increasing year after year, and official statistics show that it reached about 31,700 ha in 1986. Thus Japan is now one of the few great greenhouse countries in the world. However, vegetable cultivation in the greenhouse is very much labor intensive, and the mechanization of seeding, control of insect pests, harvesting and transportation is not so simple. This means that severe working conditions are not improved noticeably. Of these, the rates for harvesting and processing work are high. Atmospheric conditions in the greenhouse are characterized by high temperature and high humidity, and ridges are thickly wooded by leaves and canes thereby restricting the mobility of workers and generating discomfort. Much harvest is transported by a manual cart or a manual barrow which can carry only small amounts at a time. This eventually results in long walking distances, and much fatigue, and poses a health problem. At the same time, people who can afford their labor to agriculture are aged, and many people are not of course available at present. Such being the case, it is of utmost importance to develop an unmanned vehicle for the purpose of labor saving in transportation work.

As a means to overcome these circumstances, we designed and manufactured the prototype microcomputer installed railless type unmanned vehicle. This paper discribe outline of the electric driven transport vehicles which travels automatically along ridges furrows and in between water culture bed rows in greenhouses, and the results of investigation of traveling performances.

OUTLINE OF THE PROTOTYPE VEHICLE WHICH TRAVELS ALONG RIDGES FURROWS IN GREENHOUSES

Fig. 1 shows the prototype unmanned transport vehicle which travels automatically along ridges furrows in greenhouses. This is a four-wheel vehicle, and its rear wheels are driven independently by 150 w geared motors. Principal particulars of this vehicle are shown in Table 1. This vehicle is equipped with the following control functions.

Mechanical Control Unit

So far various kinds of unmanned vehicles utilizing automatic guiding technology have been studied (Tsumura), and these are primarily used on production sites in the factory. The majority of these vehicles are guided by optical, magnetic, electric wave or ultrasonic wave means. If these are used in the greenhouse where rough ground is heavily covered by leaves and canes, travelling performance and ease of guiding may be lowered significantly. It is therefore considered that the automatic direction correction method by simple and mechanical composition is more adaptable. Thus we

237

developed a steering system equipped with the automatic direction correction function.

Fig. 2 shows in simplified form, the principle of operation of direction self-correction mechanism. This mechanism is composed of a front axle, and a steering shaft which is normal to the axle and is composed of two ball joints (imaginary shaft a-b in Fig. 1). The front axle is mounted to the body frame via these two ball joints. Correction of the direction is automatically achieved by setting the mounting position of these ball joints to provide an inclination angle (caster angle) to the steering shaft.

In case the vehicle deviates from desired traveling path and one front wheel run on the ridge as shown in Fig. 2, the front axle is rotated about the steering shaft by as much as required according to the height of wheel riding. In other words, the design is such that the front axle is turned around shaft a-b within the plane normal to shaft a-b. Thus as a front wheel is raised a distance H, it is at the same time rotated about an axis c-d perpendicular to a-b. As a result, steering angle β occurs independently for the right and left front axle, and the vehicle descends the ridge side while the direction of motion is changed. Namely, the direction is corrected by utilizing the torsion angle generated between the front axle and the rear axle. In case both front wheels and rear wheels run on the ridge, this torsion angle becomes small resulting in reduced correction effect. To prevent this difficulty, the rear tread is made 35 mm shorter than the front tread so that running of the rear wheel onto ridge side may be delayed. We considered that this mechanism might also contribute to prevention of tilting of the machine body.

TABLE 1 Specifications of an Automatic Transport Vehicle

The body of a vehicle

Vehicle weight	75 kg
Maximum carring capacity	85 kg
Overall length	800 mm
Overall width	590 mm
Overall hight	440 mm
Wheel base	540 mm
Front tread	375(+20)mm
Rear tread	340(+20)mm

Traveling device

Battery	12V(36Ah/5h)
Driving motor	150W × 2
Steering motor	100W × 2
Traveling speed	40 cm/s

Microcomputer unit

CPU	Z80A
Interface	8255
ROM	256 K bit
RAM	64 K bit
Appoint one's destination switch	$D_0 \cdots D_5(6)$
Function switch	GO, STOP, STEP, HOME, BACK

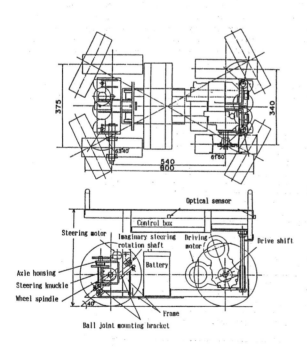

Fig. 1. An automatic transport vehicle for use in greenhouses.

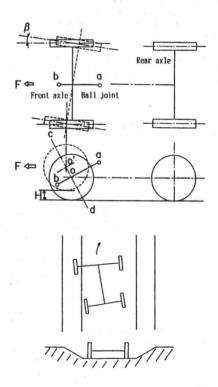

Fig. 2. Schematic diagram of the self carriage correct mechanism.

In addition, this vehicle is equipped with the spin turn mechanism which allows turning within minimum space (Fig. 1). This mechanism is primarily composed of two steering motors and pinions gear, one each for the front and rear axles, and reduction gearing for each of the four wheels. If the steering motors are operated, as shown in Fig. 1 the gears to the right and left of the pinion gears turn in opposite directions. Thus the steering knuckles are also turned in opposite directions for the left and right wheels of each axle, and each wheels axis passes through the center of the vehicle body p. At the same time the directions of the left and right rear wheels drive motors (reduction ratio 1/88) are controlled so as to achieve a spin turn. Two limit switches are used to control the turning angle (open/close angle) of the steering knuckle. The maximum turning angles for the front and rear wheels are 63° 40' and 61° 30' respectively. The turning radius is 440 mm.

Microcomputer Control Unit

Table 1 gives details of the microcomputer control unit mounted on this vehicle. This unit is composed of the minimum necessary system that is primarily composed of Z80A microprocessor in order to reduce the cost. The sensor unit is composed of two mirror reflection type optical sensors mounted at the rear, left and right side of the vehicle body. Reflected infrared rays coming from the reflection plate are detected by these sensors. The driving unit is composed of a 150 w traveling motor (geared motor) used for driving each of the right and left rear wheels and a 100 w steering motor used for steering each of the front and rear wheels. The former is controlled by normal / reverse speed control (PWM method), and the latter is controlled by ON / OFF control by relays. The operation panel is composed of data entry switches and function switches. Control software for this vehicle was prepared using the Z80 macro-assembler.

Example of Traveling

A map guided traveling method was adopted. Marking (200 mm x 200 mm reflector plates pasted by Scotch tape) was provided at strategic locations in the greenhouse. Their position were stored in the ROM of the microcomputer as map information. The vehicle moves towards the prescribed position by checking present location using information given by the reflection plate and the map.

Fig. 3 shows the traveling path used for experiments. This is made of decomposed granite soil. The slope of the ridge side was 40° and ridge width was 50 cm. The following description shows experimental examples of unmanned traveling.

After the data such as designation of destination (D_2) in the figure is input at HOME (starting point), the vehicle moves straight forward. When the reflected infrared ray coming from the reflector plate is detected by the left receiving sensor provided at the left of the vehicle, that position is detected by the controller, and the vehicle slowly reduces its speed and eventually stops. Then the turning mechanism described previously actuates a spin turn clockwise. However, reflected light is detected by the rear sensor on the way. When the vehicle is turned by 90°, the rear wheel driving motor is then stopped and the vehicle moves towards the row spacing. Then next traveling at row spacing is started. If vehicle direction is deviated or the vehicle runs onto the ridge, it is mechanically corrected by direction correction mechanism. Normally picking work is being done between ridges.

After it is fully loaded with the harvest and the HOME switch is pressed, the vehicle starts traveling in arrow direction along with the path between ridges since present location and traveling path are stored in the microcomputer. A timer is provided to the controller so that the vehicle may stop after passed through the reflection plate at the end of ridge (D_{2-a} in the figure). Further, it is detected by the left sensor to execute 90° turning. After that, the vehicle starts traveling toward the marking D_{0-a} on the road running north and south along with west side wall.

At markings D_{0-a} and D_{0-b}, the right sensor and the rear sensor are actuated respectively. The vehicle

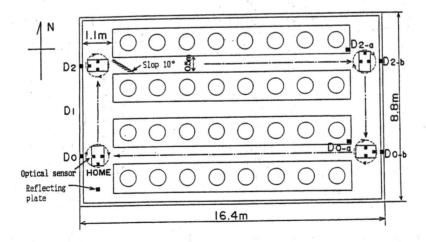

Fig. 3. A case of traveling rout in the greenhouse.

moves to the home while traveling automatically between ridges. After turned 90° again at the end of redge, the vehicle stops and the container is unloaded.

Traveling Performances

The effectiveness of direction self-correction function was confirmed by actual traveling on the test course. In this experiment, deviation of the front axle center from the reference line and convergence distance were measured when the vehicle ran onto the ridge with a ridge side angle of 40°. A maximum of 18 mm deviation and 1.8 m convergence distance were measured respectively at an approach angle of 12°. It is therefore revealed that the functions of this vehicle do not pose any problem for practical application.

Power consumption of the two driving motors was

TABLE 2 Specifications

The body of a vehicle	
Vehicle weight	87 kg
Maximum carring capacity	100 kg
Overall length	1100 mm
Overall width	400 mm
Overall hight	450 mm
Wheel base	500 mm
Tread	318 mm
Rear tread	340(+20)mm

Traveling device	
Battery	12 V(36Ah/5h)
Driving motor	120 W × 2
Steering motor	78 W × 2
Traveling speed	50 cm/s

measured while traveling and turning. Values obtained were 87.6 W and 63.6 W respectively at a traveling speed of 40 cm/sec and maximum carrying capacity of 80 kg. It is then revealed that there is enough balance of power.

Battery durability was checked and it was found that six hours continuous operation is possible with a standard load of 52 kg. Thus practicability of the battery was confirmed.

Power consumption while ascending with a road inclination angle of 14° and maximum carrying capacity of 80 kg was 184 W which does not allow traveling. It is revealed from this that the inclination angle of wooden plates and the like laid over heating pipes should be less than this angle.

OUTLINE OF THE PROTOTYPE VEHICLE WHICH TRAVELS ALONG WATER CULTURE BED ROWS IN GREENHOUSES

This is the automatic transport vehicle which utilized the wall of culture bed as the delection medium used in the water culture in the greenhouse. The vehicle travels automatically by means of ultrasonic wave guided method.

Steering Mechanism

This vehicle (Fig. 4) is electric driven four-wheel type. The rear wheels are driven, and the front wheel turns in the opposite direction of that of the rear wheel with the same angle by four-wheel steering mechanism. Specifications for this vehicle is shown in Table 2. The steering motor (78 W) to steer the front wheels, reduction gear, and steering angle sensor (encoder) are provided in the steering housing provided at the front wheel. The steering is transmited to the rear wheels via link rod. The rear wheels are driven by two geared motors (120 W) via intermediate reduction gear.

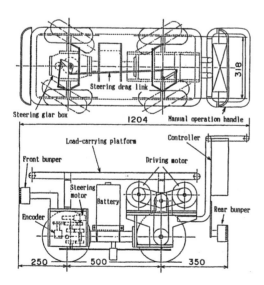

Fig. 4. An automatic transport vehicle for use in water culture.

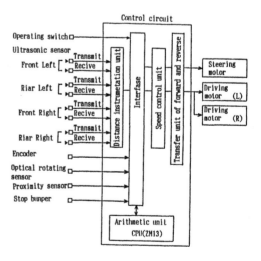

Fig. 5. Control system

Control System

Fig. 5 shows the control system mounted on this vehicle. This system consists of an ultrasonic wave sensor, encoder, optical detection type rotary pulse counter for turning control, and proximity sensor for detection of the marking. The speed control (electronic chopper control) of the driving motor, start, stop, and emergency stop are done by this system.

Steering Control

The following controls were provided to ensure that the vehicle posture would be parallel with the wall. The first, the distance with regard to the wall is measured by the ultrasonic sensors mounted on the fore and aft of the vehicle side. The steering direction (straight forward, right turn or left turn) was determined by comparison of these two. The amount of the steering is calculated by the equation shown below, the steering motor is revolved normally or reversely as much as the amount of encoder indication. The following description shows a concrete example.

In Fig. 6, when the vehicle body is approaching to the target line set beforehand ($|L_1-d| \leq 10$ mm) or it is separeted from the target line by more than 5 cm ($|L_2 - L_1| \geq 50$ mm), the vehicle will move straight forward. In other cases, the steering is done to close the vehicle to the target line according to the encoder indication obtained from the amount of steering as shown below (Fig. 7).

> where
> Steering amount = $\alpha \times |L_1-d| \leq 60$
> Encoder indication = $127 \pm$ steering amount
> (α : constant, right turn for $L_2 - L_1 > 0$,
> left turn for $L_2 - L_1 < 0$)

Detection of Marking and Turning Control

A marking (steel plate with 5 cm (W) × 30 cm (L)) was provided on the road where the vehicle stops or executes turning. This is detected by the proximity sensor mounted at the vehicle bottom to identify present location of own vehicle. Electromagnetic energy is generated at this instance. Then the electric circuit is disconnected thereby stopping the vehicle. Immediately after this, the vehicle turns 90° in the set direction. The turning is done up to 60° by reading pulse counting of the optical detection type rotary sensor (number of slits 30) mounted to the rear wheel. After that, posture control is done by means of the ultrasonic wave sensor. Therefore, judgment of the bed where the vehicle should stop is made by counting the number of marking detected.

Evaluation of the Posture Control

The following experiments were carried out to clarify the follow-up performance of the vehicle against the wall faces.

Measurements of turning track and yaw angular velocity of the vehicle body. In oder to check the running track at the center of the vehicle body, a bending wall face making of plywood boards was manufactured (Fig. 8). A water tank with 50 ml capacity was mounted onto the vehicle body. Water was dropped during vehicle travel from the bottom center position of the vehicle body via vinyle tube. Deviation of the center position of vehicle from the target line was measured from the trace of water left on the road by the distance of 10 cm from the bending point. To measure the yaw angular velocity of the vehicle body, a rate gyro was mounted at the center of the platform. The data was saved in a 4-channel data recorder mounted on the vehicle and then recorded into an analyzing recorder with the time axis expanded.

Results and discussions. As shown in Fig. 9, the follow-up performance of the vehicle against the bending wall was checked in several modifications

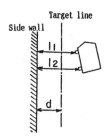

Fig. 6. The relation between transport vehicle and side wall.

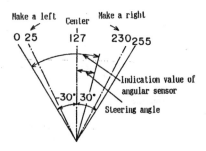

Fig. 7. The relation between encoder value and steering angle.

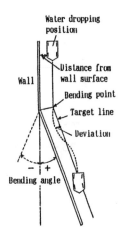

Fig. 8. Experiment of measurement of running track.

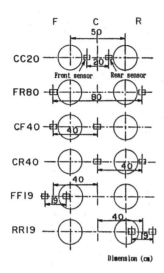

F   C   R

50

CC20

Front sensor   Rear sensor

FR80   80

CF40   40

CR40   40

40

FF19   19

40

RR19   19

Dimension (cm)

Fig. 9. Mounting position of ultrasonic wave sensor.

of the position at which the ultrasonic wave sensors were attached at the side of vehicle. The parameters F, C and R in the figure indicate the front, center and rear positions of the vehicle at the distances shown by numerals in cm, respectively.

Fig. 10 shows the experimental result when the vehicle ran at the velocity of 34 cm/s with the distance from the wall surface having the bending angle of 20˚ set to 30 cm (which was provided by software). This indicates that, for FC40, FR80, and FF19, the running tracks were apart from their target tracks by 7 to 10 cm after passing a bending point by 0.5 m, meandered after the passing, and converged to the target values after about 3 m run.

However, CC20 and CR40 could not follow the bending of the wall to cause their tracks to be diverged in the middle of the way. In particular, the tendency appeared remarkably for RR19 and thus RR19 could not be used for the experiment. Based on the results, the sensor was mounted at the position of FC40, at which the sensor could be easily mounted on the vehicle.

CONCLUSIONS

Two automatic transport vehicles were trial manufactured for the purpose of reduction of transport labor in the soil culture and the water culture in greenhouses, and their traveling performances were investigated.

With the automatic transport vehicle which travels along with the ridges furrows under severe conditions such as rough ground, mechanical posture control is superior to electronic control with regard to high reliability traveling and stability.

High practicability was revealed in such that the vehicle can carry as long as five hours after the battery was charged once. In the meantime, with the ultrasonic wave sensors, posture control was possible even by a simple proportional control. If the bed-to-bed distance is made slightly large, follow-up traveling was possible. In addition, the vehicle is of four-wheel sterring type, turning at the end of culture bed was easy. The practical application of the prototype vehicle is now envisioned.

REFERENCES

Tsumura, T. (1987). Control system of automatic guided carriers. Journal of the Society of Instrument and Control Engineers, 26, 593-598.
Yamashita, J. (1989). Automatic carrier for greenhouses. Journal of the Society of Instrument and Control Engineers, 28, 142-143.

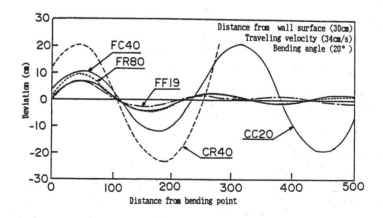

Fig. 10. Test traveling track based on position change of ultrasonic wave sensor.

STUDY ON GRAPE HARVESTING ROBOT

N. Kondo

Faculty of Agriculture, Okayama University, 1-1-1, Tsushima-Naka, Okayama, Japan

Abstract. Grapevine is planted in trellis training in Japan. It is considered that this training has much adaptability of robot harvesting than other training. A robot to harvest individual bunches of grapes was made as a trial and the experiment was done. This robot consisted of a manipulator with 5 degrees of freedom, a hand which could hold and cut rachis and push bunch, a visual sensor in which TV camera and optical filters were used, and a crawler type travelling device.

From the experimental result, it was observed that the grapevine in trellis training in Japan was easier to detect and to harvest bunches of grapes than that in other training and that the training of the other fruit trees was necessary to be reconsidered for agricultural robots development.

Key Words. Robots; Sensors; Feedback control; Grape Harvesting; Trellis Training.

INTRODUCTION

In Japan, robots which harvested tomatoes, cucumbers, and citrus fruits in the greenhouse or the field were reported[1-3]. The training of these objects was not easy to detect and to harvest the fruits because they grew almost in vertical plane or sphere so that the leaves and the stems might be obstacles for the robots. It was reported that the feasibility of a robot to harvest grape was also studied in France and that the robot seemed unfeasible in the field conditions that the plantation were on rows 1.2m high and 0.5m width[4].

As against the French training, grape, Chinese gooseberry, and pear are planted in the trellis training in Japan, so that only the fruits hang down from the trellis. It is considered that this training has much adaptability of robot harvesting, but it is hard work for the manual harvest.

In this report, basic studies were done in order to develop a grape harvesting robot. At first, three-dimensional positions of grape bunches were measured, and a polar coordinate manipulator with 5 degrees of freedom was made as a trial. Secondly, physical properties of rachis were researched, and a hand which could hold and cut rachis and push bunch was

made as a trial. In visual sensor, after spectral reflectance of grape plant was measured, suitable wavelengths were selected, and the experiments to discriminate the object using optical filters and TV camera and to detect the position were done. Finally, the experiment to harvest the grape bunch by the robot which was mounted on a crawler type travelling device was done.

GRAPE HARVESTING ROBOT AND EXPERIMENTAL METHOD

Fig.1 shows grape harvesting robot made as a trial. This robot consisted of a polar coordinate manipulator, a hand for harvesting grape, a visual sensor, and a travelling device.

Fig.1 Grape Harvesting Robot.

Manipulator

Fig.2 shows the positions of the grape bunches on the horizontal plane in trellis training in Okayama University. The height from the ground to the bunches was 170 cm, and the variety was Muscat of Alexandria. It was considered that the robot could work efficiently when it traveled along the scaffold.

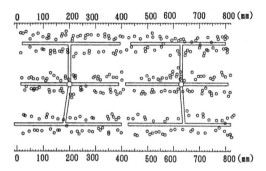

Fig.2 Positions of Grape Bunches.

From this result, a manipulator as shown in Fig.3 was made. This manipulator had 5 degrees of freedom, and the joints of waist, shoulder, arm, and wrist were able to rotate at a various speed. The length of the arm was 1.6 m, and the stroke was 1 m.

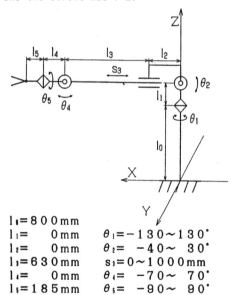

$l_0 = 800$ mm
$l_1 = 0$ mm $\theta_1 = -130 \sim 130°$
$l_2 = 0$ mm $\theta_2 = -40 \sim 30°$
$l_3 = 630$ mm $s_3 = 0 \sim 1000$ mm
$l_4 = 0$ mm $\theta_4 = -70 \sim 70°$
$l_5 = 185$ mm $\theta_5 = -90 \sim 90°$

Fig.3 5 Degrees of Freedom Manipulator.

The hand was controlled to move horizontally below the trellis at a constant speed, because there was fewer obstacles in this training than the other training.

Hand for Harvesting Grape

A hand was made to hold and cut a rachis, since the grape bunch might shatter when it was grasped hardly at harvest time and it had white powder which enhanced its marketing value. The grasp force and the cutting force were obtained from the relationship between grasp force and frictional resistance and between cross-sectional area of rachis and twig and cutting resistance in order to make the hand. The grasp force was assigned 10 N, and the cutting force was 100 N.

The hand made as a trial from these results is shown in Fig.4. A function which pushed bunch was added to this hand, because it could hold also very short rachis at harvest time, reduce bunch swing at carrying time, and oriente the bunch at placing time. The cutter and finger were droven by one DC motor and two springs, and pushing device was moved straight by changing rotating direction of DC motor using rack and pinion.

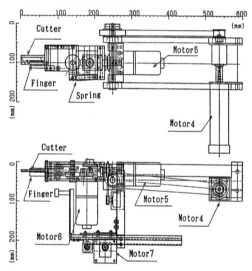

Fig.4 Hand for Harvesting Grape.

Visual Sensor

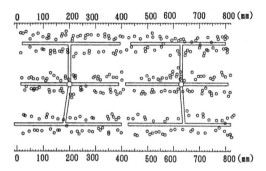

1,2: Half Mirror
3: Mirror
4-6: Optical Filter
7-9: Image Sensor

Fig.5 Optical System of Visual Sensor.

A visual sensor which consisted of some image sensors and optical filters was assumed as shown in fig.5 in order to discriminate grape plant. Light entering through lens is divided by half mirror, and makes images on the image sensors after it passes the optical filters. The images of the parts of the grape plant are obtained by calculation of the outputs from the image sensors.

Fig.6 shows spectral reflectance of Muscat of Alexandria whose fruit color is white-green. It was observed that the object color was realized from the reflectance in visible region and that the difference of the parts of the plant induced the difference of their reflectance in near infrared region.

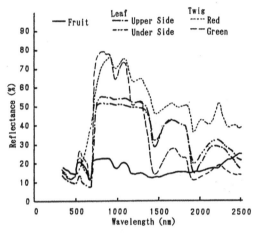

Fig.6 Spectral Reflectance of Grape Plant.

Interference filters of 500, 550, 670, 850, and 970 nm were selected in order to discriminate the parts of the grape plant from this figure. The discriminating experiment was done by using the filters and a visual sensor which had sensitivity to the wavelength of 400-1200 nm.

In this study, the visual sensor was attached near to the hand. This position was effective to get higher accuracy of detecting fruits, because picture elements of fruits increased with coming near to fruits. An experiments to detect a fruit with the movement of manipulator toward a fruit were done by method of average diameter of fruit, and by method of visual sensor moving toward fruit. If fruit diameters of same variety are nearly constant, the picture element number of fruit gives an approximate distance from the visual sensor to the fruit as equation (1) in fig.7. When the whole fruit is in the image, the distance from the visual sensor to the fruit is calculated from the picture element numbers of fruit as equations (2) and (3).

Travelling Device

As a travelling device of the robot, crawlers were used, because the field was

non-tillage. The width of the crawlers were 360 mm, and the ground contact length were 1010 mm. The width of the travelling device was 1400 mm, the length was 2300 mm, and the height from ground to the plate on which robot was mounted was 420 mm. The travelling speed of the device was changed to 2 m/s. In the experiment, this device was steered by manual, and the engine was stopped when the robot worked.

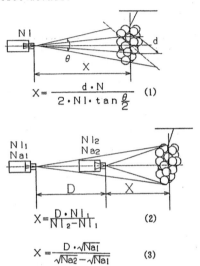

$$X = \frac{d \cdot N}{2 \cdot Nl \cdot \tan \frac{\theta}{2}} \quad (1)$$

$$X = \frac{D \cdot Nl_1}{Nl_2 - Nl_1} \quad (2)$$

$$X = \frac{D \cdot \sqrt{Na_1}}{\sqrt{Na_2} - \sqrt{Na_1}} \quad (3)$$

X:distance from visual sensor to fruit
d:average diameter of fruit
N:picture element number of visual sensor on a line
Nl:picture element number recognizing fruit on a line
θ :visual angle
D:visual sensor moving distance
Na:picture element number recognizing fruit on a area

Fig.7 Methods of Detecting Position.

Method of Harvesting Experiment

The harvesting experiment was done by the robot in room and the field. The robot approached to a fruit by visual feedback control based on the detecting result from the visual sensor, assuming that the plant parts were discriminated by their spectral reflectance and shapes.

EXPERIMENTAL RESULTS

Discriminating Results

Fig.8 shows the experimental result when Muscat of Alexandria was discriminated using R,G filters. The vertical axis indicates the percentage of picture elements and the horizontal axis indicates the character corresponding to ratio of (G output)/(R output + G output).

It was observed that discrimination of the fruit was difficult because the fruit color was similar to the leaves and the green twig.

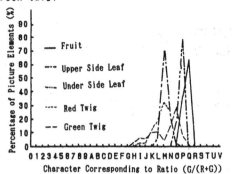

Fig.8 Discrimination Result (1).
(R,G Filters)

Fig.9 shows the result when 550 nm and 850 nm interference filters were used. From this figure, it was easy to discriminate when the near infrared wavelength was used. As for Delaware and Campbell Early whose fruit colors were different from their leaves and twigs, it was easy to discriminate them even using RGB filters.

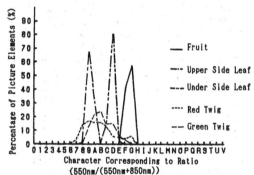

Fig.9 Discrimination Result (2).
(Interference Filters)

Detecting Results

Fig.10 shows the experimental result of detecting positions of the fruit and the rachis. From this figure, it was observed that the error depended on the difference between fruit diameter and average diameter remained when the method of average diameter was used. When the method of moving visual sensor toward fruit was used, the error was about -20 mm. When the rachis was detected by this method, the error was bigger, because the picture element number of the rachis was too small to detect.

Experimental Result to Harvest

It was observed that the hand was able to harvest, since the structure of the hand permitted the reasonable error, although the detecting error of visual sensor was about 20 mm. The pushing

device also worked effectively except for too short rachis. Fig.11 shows the example of the experimental result to harvest.

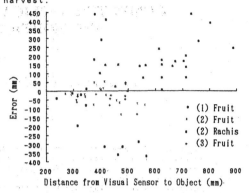

Fig.10 Detecting Result by Visual Sensor.

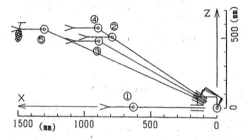

Fig.11 Experimental Result to Harvest.

CONCLUSION

It was observed that the robot could harvest the fruit which was in an ideal state from the experimental result, but that it took long time for this visual sensor to detect obstacles, and that the visual sensor needed automatic exposure in the field, and so on. Therefore, it seemed adequate that the visual sensor was used with another sensor which could measure three-dimensional shape of the object.

It was considered that the grapevine in trellis training in Japan was easier to detect and to harvest bunches of grapes than that in other training and that the training of the other fruit trees was necessary to be reconsidered for agricultural robots development.

REFERENCES

1)KAWAMURA,N.,NAMIKAWA,K,FUJIURA,T,URA,M.:Study on
   Agricultural Robot (Part2), Journal of the Japanese
   Society of Agricultural Machinery,47-2,177-182,1985
2)KAWAMURA,N.,FUJIURA,T,URA,M.,KONDO,N.:Fruit Harvesting
   Robot, Journal of the Japanese Society of Agricultural
   Machinery, 47-2,237-241,1985
3)AMAHA,K.,TAKAKURA,T.:Development of a Robotic Hand for
   Harvesting Cucumber Fruits, Journal of Agricultural
   Meteorology,45-2,93-97,1989
4)SITTICHAREONCHAI,A.,SEVILA,F.,FATOU,J-M.,CONSTANS,A.,
   BRONS,A.,DAVENEL,A.:A Robot to Harvest Grape, ASAE
   Paper, No.897084,1989

Copyright © IFAC Mathematical and Control Applications
in Agriculture and Horticulture, Matsuyama, Japan 1991

COMPUTER APPLICATIONS IN AGRICULTURE
AND HORTICULTURE: A VIEW

W. Day

Process Engineering Division, Silsoe Research Institute, Wrest Park, Silsoe, UK

Abstract The opportunity for computer applications in agriculture and horticulture is
presented in three example areas: image analysis, crop models, and information
technology. Image analysis techniques that are capable of coping with the variability
typical of biological targets have great scope for use in robot control. Crop models
have been a continuing target for research, and in the greenhouse sector are being
actively considered as the basis of environment control strategies. Inormation
technology opens the door on the use of computer technology throughout farming,
from selective control of field operations to expert systems for crop and farm process
management.

Keywords Agriculture, computer applications, image processing, models, expert systems,
process control.

INTRODUCTION

As computer technology continues to advance in
processing power and speed, compactness and
acceptably low cost, new applications in
agriculture become feasible. These applications
will be as varied as agriculture itself, and will
bring with them specific challenges to the
technology. Biological variability between items
of produce, the complex interactions of
physiological processes that underlie crop growth
and the uncertainties associated with the
prediction of processes that will be influenced by
future weather are issues in many agricultural
problem areas. Developments in hardware systems
and software capability will allow practical
implementation of computer techniques in produce
grading, crop yield prediction, and crop and
animal management.

In this paper, a number of example areas will be
presented in which computer applications can be
expected to develop rapidly in the next ten years.
In each the relationship between scientific
research, computing systems and the application
area will be discussed, and future opportunities
commented on.

IMAGE ANALYSIS

Horticultural processes frequently involve human
operators making decisions on the basis of visual
information. This is particularly apparent for
operations such as produce harvesting and grading.
It also applies more widely in agriculture, with
stockmen assessing health and well being of their
animals in part from visual information on activity
and gait, and farmers identifying plant diseases
through their visible symptoms. The technique of
computer-based image analysis can be applied to
these problems, and the application raises a number
of interesting research challenges. In particular
the objects of interest are generally biological
and this brings considerable variability from
object to object within a batch. The objects are
also often viewed in an unstructured environment,
again complicating the decision making task.

Two key attributes of the revolution in computing
hardware have been crucial to the advance of
practical applications of image analysis in
agriculture and horticulture - system cost and
processing speed. Standard TV camera technology
can produce images resolved into 512 x 512 pixels,
with each pixel giving intensity as an 8-bit
number, representing a grey level from 0 (black) to
255 (white). Such information is quite adequate to
represent many of the scenes of interest, and
standard colour cameras can expand this capability
to more complex discrimination. Because this
capability is available in TV based technology that
has a mass market, unit costs are low and the
agricultural applications become feasible. But
just creating the information is not sufficient,
and decisions in agricultural operations must be
made rapidly if process automation is to be
successful. To deal with a quarter of a million
items of information rapidly has required the
availability of fast processors. It is estimated
that processing speeds are currently doubling every
three years, so that speeds have now reached
around ten million instructions per second with
memory access times down to tens of nanoseconds.
Even more complex image processing tasks may be
possible as these speeds rise and access times
fall.

In the short term, the practical examples of the
application of computer-based image analysis in
agriculture and horticulture are likely to be
limited to straightforward repetitive tasks like
those associated with produce grading. Research
can be expected to make progress with more complex
tasks, and specific applications will continue to
be vitally important to establish the scale and
form of the real-world problems. However progress
on a broad front will depend upon the establishment
of generic approaches to issues like 3D structural
interpretation and training of computer programs.
The development of more rapid processing systems,
and particularly transputers which can provide
parallel processing and sophisticated
intercommunication between processors, will help.

Advanced image interpretation

The use of artificial intelligence to advance the sophistication of image interpretation is likely to be a major route in the coming years. The biological variability inherent in agricultural objects has already been referred to, and it is accompanied by geometrical complexity and considerable subtlety in the differences in form or colours that are important in judging quality. Intelligent knowledge-based systems can provide a way of utilising past knowledge whilst making decisions about the current task. The knowledge is entered by the operator, whilst the capability to reason is present in the computer in the form of an 'inference engine.'

An alternative approach to achieving intelligence is to allow the computer to learn from examples. The 'neural network' concept is based in principle on the learning and memory abilities of the human brain. The training period is used to teach the program from examples of 'correct' and 'incorrect' objects. The neural network can then apply this knowledge to classify other objects. The approach is particularly useful where the quality attributes are fairly easy for human decision but are difficult to quantify as rules or numbers.

An example of this approach has been to determine the position of nodes in an image of a chrysanthemum plant for micropropagation. The nodes are characterised by rapid changes in direction of the image outline. Classifying the whole outline in this way will specify many positions which could be nodes, whereas only a few are. A training routine can then be used in conjunction with a number of images for which nodes have been separately identified. After this training, the program is capable of identifying nodes with a high degree of success (Davis, 1991). The use of intelligent systems can be expected to grow in importance, and it will probably have particular significance in agriculture where the variability in size and geometry will make definition of precise rules and formulae difficult.

Image analysis as a sensor for robot control

Advances in image processing methods will allow many more interpretation tasks to be attempted - and this will lead to opportunities for use of the information in real-time control of automation. Robotics -the use of robots to undertake mechanical tasks that entail a high degree of 'mental' and physical agility - is very dependent on sensors, and automatic guidance of robots using image analysis will be a major feature of developments in this area.

The robotic handling task will set major requirements on image analysis and computation if accurate control is to be achieved. An example can be taken from initial approaches to robotic mushroom harvesting (Tillett and Reed, 1990). The image analysis task is crucial to the potential success of the operation. Identification of mushrooms from the growing medium is straightforward given the high contrast between dark peat and light mushrooms. Locating the position of the mushroom and sizing it is helped by the regular shape of the mushroom. When viewed and lit from above the mushrooms appear brightest at the centre with the brightness decreasing as the mushroom surface becomes more vertical towards the edges.

The final stage is to choose which mushroom to pick, by analysing sizes against a target grade and assessing the best order in which harvest closely packed mushrooms if damage to adjacent ones is to be minimised.

CROP MODELS AND OPTIMISATION

Computers have been used in research to analyse and predict crop growth and response to environment for many years. Crop simulation models have been constructed from mathematical descriptions of component processes, like radiation interception by the leaf canopy, photosynthesis and dry matter growth. For some time it has been suggested that such models can have direct application to the farming industry.

Computer models of field crop performance

Though field crop models have been developed by many scientists, few have been used yet for real-time prediction of crop performance. This reflects the complex interaction between biological processes, and variables defining soil and atmospheric environment of the crop. Some component processes, like crop development, have been described in models developed for prediction (e.g. Travis et al, 1988), and such models are now being used as part of crop consultants' advice to farmers on the best time to apply fertilisers, for example.

More complete models of crops have proved difficult to construct with sufficient stability in relation to soil or weather factors to provide convincing predictions to farmers. The GOSSYM model of cotton growth has been taken further than most. It has been interfaced with an expert system so that questions of importance to the farmer, for example on fertiliser application strategy, can be evaluated through simulation under the control of the expert system and clear answers given to the farmer (McKinion et al, 1989).

It is likely that this sort of approach will eventually be of direct importance in farm management. At present, knowledge of the detailed interactions between growth processes is incomplete and methods to allow for the uncertainty in the definition of conditions and in the models themselves are also lacking.

In the meantime, a major computer application in relation to crop performance will be in predictions of the effect of changes in environmental conditions or government policies on crops. Concern about the effects of climate change on agricultural systems is widespread, and experimental studies are both difficult and costly if they are to be of direct relevance to field performance. Other environmental issues, like the desire to reduce nitrate emissions from farming systems and to decrease use of pesticides, are also being reviewed through the use of crop models. These evaluations can be expected to increase, though with the proviso that there is still only very limited validation of the capability of such models to represent accurately the responses to variations in field conditions. It will be vitally important to continue to focus on this issue of validation, as ever more extensive policy issues, like the review of national agricultural capability (Maracchi et al, 1988), are tackled with crop models.

Models for optimal control of the crop environment

In protected crop production, many of the environmental variables that complicate prediction of responses for field crops are held constant. The potential simplification in the models required to specify key environmental responses suggests that it is in this environment that crop models will first be involved in real-time control of crop management.

Computers have been in use to interpret and control the physical environment within greenhouses for many years (Bailey, 1991). The next step was to utilise information from the plant that is indicative of its physiological state - the 'speaking plant' approach (Hashimoto et al, 1981). For example, changes in leaf temperature can be monitored and interpreted in terms of the changes in stomatal aperture caused by the onset of water stress (Hashimoto, 1982). Leaf temperature can then be used to control irrigation so that stomata are kept open and CO_2 uptake maximised.

More complex processes can be considered as part of the control strategy, now that computing power has increased to the point where models of crop response to environment can be evaluated in real time. Van Henten and Day et al at this meeting have presented approaches to dynamic CO_2 optimisation. The benefit to be gained from enhanced CO_2 concentrations can be determined by evaluating models of photosynthetic response to light and CO_2, conversion of photosynthate to growth and yield, and ultimately to additional crop value. The cost of additional yield can be evaluated in terms of the price of CO_2 gas and the rate of supply necessary to maintain an enhanced concentration, which depends directly on the greenhouse ventilation rate. Real time optimisation of these models may require some elements to be simplified, but the finite response time of the greenhouse system to changes in CO_2 enrichment is of the order of 5-10 minutes, providing a significant time window for processing. It is likely that currently available personal computers will be capable of providing optimal CO_2 set points to greenhouse climate control systems, at least in the context of experimental systems, in the near future.

INFORMATION TECHNOLOGY ON FARMS

The greatest global impact of the computer revolution has been in the development of Information Technology (IT) - the use of computer-based systems to acquire, store, process, transform and disseminate data and information. IT has made some inroads into farming, but the major computer application in this area is just beginning.

The concept of the IT farm has been described fully by Moncaster (1988), as a comprehensive and integrated management information system. Components already exist, in the form of packages for farm business management, bureaux services for market price information, weather and disease forecasts, and some automatic data collection equipment, eg in dairy farming. The longer term view is that these components will be supplemented by others, and integrated into a system that can provide information for both strategic and tactical decision-making on the farm.

Within the IT farm, some components will operate on quite short timescales and have close control over processes. The protected crops industry provides a current example of IT systems in direct control over a farming process. The greenhouse control computer, gathering information from its sensor array, makes decisions on operational control of

heating and ventilation systems to maintain a preset environment plan. In the previous section, the next stage in the development of this implementation of IT was outlined - with real time optimisation of system models defining the optimal environmental setpoints for current weather and market conditions.

In field agriculture, the opportunities for direct control are not so well formed as yet. An example of current interest is the concept of selective or spatially variable field operations - tailoring operations to meet the requirements of soils or crops in parts of the field, rather than applying the same uniform operation over the whole field.

On the fringe of the IT farm itself, information technology systems will be important to the adviser. They can provide him with rapid access to management information, forecasts and the results of research and development work, so that his decisions are up to date and based on the knowledge of many experts. To this end, expert systems will play an important role.

Selective field operations

The basic unit of most farming operations is the field - whose size can vary from a fraction of a hectare to many hundreds of hectares. Large fields will often contain a range of soil types; there may be considerable variation in slope and in water table height; long term weed problems may exist in some parts of the field but not in others. With appropriate technology, these variations could be taken into account to ensure that operations are closely tailored to the requirements of each part of the field. Benefits would be both economic and environmental. Matching soil conditions and fertiliser supply more closely to the crop's needs will give more cost effective production, often by reducing input costs, and reduced use of agrochemical inputs will be beneficial in meeting environmental demands on agriculture, now increasingly forming part of the legislative framework in developed countries.

What makes selective field operations a possibility now? As a concept it depends upon making decisions about field management on a local basis, and controlling farm machinery in order to implement these decisions. This involves similar issues of information processing and control that we have seen in the use of image analysis to control grading. The prime components in the execution of selective operations in the field will be quantifying the key parameters that vary across the field, interpreting their influence on the optimal form of the field operation and implementing this operation at the appropriate positions in the field. Interpretation may well involve the integration of a number of pieces of information, reflecting both current conditions and past performance for specific parts of the field, so the use of computer packages for coordinating spatial data is crucial. These Geographical Information Systems (GIS) provide a database that can be called on both in the execution of selective operations but also in reviewing their effectiveness by assessing historical changes.

One operation of current interest is spatially selective application of herbicides, or "patch" spraying. The feasibility of spatially selective spraying depends on whether weeds grow in patches or are uniformly spread throughout fields. For cereal fields in the UK, a number of studies (Marshall, 1988; Wilson and Brain, 1990) have shown that some grass weeds grow in persistent patches. Early in the growing season, a widespread weed coverage is common and a field-wide uniform spray may be necessary. However subsequent sprays to tackle persistent weeds could be applied to limited areas. Such an approach could permit significant reductions in total herbicide use, and also result in cost savings as the chemicals to tackle persistent weeds are often expensive. Techniques are now available to control the rate of application of agrochemicals dynamically during sprayer operation. Injection metering systems (e.g. Frost, 1988) allow the concentration of active ingredient to be adjusted, either in relation to vehicle forward speed (to maintain a uniform dose rate) or in response to other information, for example on the need for agrochemical in that part of the field.

Providing the information to control the sprayer is the major challenge, and will call for application of many features of advanced computer technology. Image analysis techniques might be able to be used to identify weeds and quantify weed numbers in real-time as part of a sprayer control system. However the demands on resolution and interpretation are enormous, particularly given the low population level of some weeds that can significantly decrease yields (e.g. cleavers at 1 to 2 plants m^{-2}), and for the present alternative methods are required (Thompson et al, 1990). A possible method is to build a historical database from information gathered at times when weeds are particularly distinct. The contrast between weeds and crop immediately prior to or after harvest could permit rapid interpretation of aerial photographs, and this could be supplemented by direct observation by farmers. The database must then be available in the form of a map, and there must be a means of locating the sprayer vehicle in the field, so that the information on the map can be turned into a decision on spraying (Stafford et al, 1991).

This type of system can provide a mechanism to respond to increasing demands for lower applications of agrochemicals, by targeting them more carefully. It will be demanding of many features in computer technology, and will form a challenging application in coming years.

The expert system link from research to practice

Information technology provides the means to bring knowledge across to the user efficiently. One of the most important areas for this is for strategic decision making, particularly in areas where research is bringing rapid advances. The adviser or consultant is likely to have a continuing role at the interface, but needs to be able to call rapidly on quantitative information, and tailor it to the user's particular requirements. Expert systems provide the essential elements for this knowledge transfer, and will be a major area for application of computer systems.

The expert system provides a heuristic approach to problem solving, utilising judgemental reasoning in conjunction with formal reasoning. The expert system can also be transparent, explaining and justifying the line of reasoning. By separating the knowledge base from the inference procedures, updating the expert system to encompass new knowledge can be particularly straightforward. The expert system can also respond directly to the available information, using it to define the most appropriate path or to pinpoint missing information that will have a major influence on the final decision. The qualitative routes of the expert system can also be coupled to quantitative calculations, of considerable importance to defining the cost effectiveness of the solutions available.

An example is the need to make decisions on the handling of animal wastes, particularly to meet problems of pollution and nuisance. From the farmer's perspective, there is a wide range of constraints on his operation yet he may still be required to adapt as potential pollution or odour nuisance problems arise, or legislation changes. Research has provided quantitative information on the capabilities of different engineering approaches for the handling and treatment of wastes, and current strategies of advice can elicit the options available to the farmer to react to particular problems. The expert system can provide the effective means of harnessing computer technology to link these information sources together, and provide front line expertise to the planning process for the farmer (Beaulah et al, 1990).

CONCLUSION

Agriculture and horticulture are diverse industries, and the scope for computer applications is equally diverse. The problems associated with accurate description of the form or performance of biological systems is a continuing challenge to the practical implementation of some of the new concepts, and continuing advances in computing technology will be needed to ensure many of these applications are realised.

REFERENCES

Bailey, B.J. (1991) Climate modelling and control in greenhouses. In J. Matthews (Ed) "Progress in Agricultural Physics and Engineering", CAB International, Wallingford.

Beaulah, S.A., Brewer, A.J., Hall, C.A., Cumby, T.R. and Phillips, V.R. (1990). The Waste Engineering Expert System (WEES): An expert system applied to pollution from animal wastes. Part I: Initial developments and programming aspects of the odour control and aeration modules. Divisional Note DN 1575, AFRC Institute of Engineering Research, Silsoe.

Davis, P.F. (1991) Orientation independent recognition of chrysanthemums nodes by an artificial neural network. "Computers and Electronics in Agriculture, 5, 305-314.

Frost, A.R. (1988). Methods of controlling the application rate of chemical from a crop spraying machine. In S.W.R. Cox (Ed.) Engineering Advances for Agriculture and Food Butterworths, London. pp61-62.

Hashimoto, Y (1982) Dynamic behaviour of leaf temperatures: a review. Biological Science (Tokyo),34, 68-75.

Hashimoto, Y., T. Morimoto and S. Funada (1981). Computer processing of speaking plant for climate control and computer aided plantation (computer aided cultivation). Acta Horticulturae, 115, 317-325.

Maracchi, G., C. Conese, F. Miglietta, L. Bacci, and J.K.Parikh (1988) An information system for agricultural productivity. In J.K. Parikh (Ed.). Sustainable Development in Agriculture Kluwer Academic, Dordrecht. pp 59-97.

Marshall, E.J. (1988) Field scale estimates of grass weed populations in arable land. Weed Research. 28, 191-198.

McKinion, J.M., D.N. Baker, F.D. Whisler, and J.R. Lambert, (1989). Application of the GOSSYM/COMAX system to cotton crop management. <u>Agricultural Systems</u>, <u>31</u>, 55-65.

Moncaster, M.E. (1988). Introduction to the application of information technology to agriculture. In S.W.R. Cox (Ed.) <u>Engineering Advances for Agriculture and Food.</u> Butterworths, London. pp321-326.

Stafford, J.V., P.C.H. Miller, and B. Ambler, (1991). Mapping weedy selective herbicide application. Proceedings of conference <u>"Locating   systems for agricultural machines."</u> Godollo, Hungary.

Thompson, J.F., J.V. Stafford, and P.C.H. Miller, (1990). Selective application of herbicides to UK cereal crops. American Society of Agricultural Engineering paper 90-1629.

Tillett, R.D. and J.N. Reed, (1990). Initial development of a mechatronic mushroom harvester. In Proceedings of International Conference, <u>Mechantronics: Designing Intelligent Machines.</u> Institution of Mechanical Engineers, London pp109-114.

Travis, K.Z, W. Day, and J.R. Porter, (1988). Modelling the timing of early development of winter wheat. <u>Agricultural and Forest Meteorology</u>, 44, 67-79

Wilson, B.J. and P Brain (1990). Weed monitoring on a whole farm - patchiness and the stability of distribution of Alopecurus mysosuroides over a ten year period. Proceedings of the European Weed Research Society Symposium, Helsinki.

KNOWLEDGE ACQUISITION AND LEARNING
FOR EXPERT SYSTEMS: RESEARCH ACTIVITIES
IN JAPAN

S. Kobayashi* R. Mizoguchi**

*Tokyo Institute of Technology, Japan
**Osaka University, Japan

1 Introduction

The number of projects for research and development of
knowledge-based systems in Japan has exceeded 3,000, with
applications extending to all areas of industry. Despite this,
even liberal estimates put the number of systems which have
entered practical use under 500 and many systems still remain
in the prototype or field test stages. This is caused by several
factors. Among them, the knowledge acquisition bottleneck
has been closed up more and more. As reasons of knowledge
acquisition bottleneck, the following have been pointed out
[1], [2] :

1. human experts rarely recognize or organize their expertise as an explicit model for problem solving,

2. they have not necessarily perfect knowledge which covers
   their domains, and

3. they use to make effort to acquire new knowledge from
   cases of success or failure continually, especially in design
   domains.

Therefore, it is quite important and urgent to establish a
methodlogy for supporting or automating knowledge acquisition. In Japan, sice a few years ago, research activities of
knowledge acquisition and machine learning have become active remarkably.

The purpose of this paper is to review the present of research activities related to knowledge acquisition in Japan.
This paper consists of eight chapters. Chapter 1 is an introduction. Chapter 2 is concerned with knowledge acquisition
systems which support knowledge acquisition from domain experts. And several research activities about knowledge compilation from deep knowledge into operational knowledge are
described in chapter 3. Chapter 4 reviews several activities
about machine learning, focusing on similarity-based learning
and explanation-based learning. Finally, topics of case-based
reasoning are reviewed in chapter 5.

2 Knowledge acquisition support systems

The main role of knowledge acquisition support systems is to
support knowledge extraction or knowledge transformation
through interactions between domain experts and systems
without knowledge engineers. Here, knowledge acquisition
support systems are classified into interview-based systems
and task-oriented systems. The former lays stress on interview strategies and the latter does stress on use of generic
tasks.

2.1 Knowledge acquisition by interview

Interviewing is an intellectual activity and very important for
eliciting new knowledge from domain experts.

---

*Graduate School of Science and Engineering 4259, Nagatsuta,
Midori-ku, Yokohama 227, Japan. kobayasi@sys.titech.ac.jp
**.I.S.I.R., 8-1, Mihogaoka, Ibaraki, Osaka 567, Japan.
miz%ei.sanken.osaka-u.ac.jp@relay.cs.net

(1) SIS

An interviewer has two kinds of knowledge. One is about a
domain under consideration and the other is knowledge about
the interview process itself, which makes the interviewer an
interview expert. The latter seems to be independent of the
domain and consists of several interview knowledge primitives. A shell for an interview system, called SIS , has been
developed on the basis of this idea [3]. It has seven domain-
independent question strategy primitives. Task-specific interview systems can be generated automatically by describing
the interview knowledge in term of these primitives.

To start an interview, SIS first requests initial information
and then constructs an incomplete domain model which takes
the form of a network. The sentences given by an expert
are parsed and attentions are generated based on the primitives. These attentions are analyzed and converted to questions. The domain model is refined through answer-attention-
question cycles.

The generality and effectiveness of SIS have been demonstrated through implementation of the two interview systems,
MORE[4], a well-known knowledge acquisition system for diagnostic expert systems and I^2S-LD [5], an interview system
for logic design of data bases.

The SIS-made interview system only supports knowledge
acquisition based on static analysis. More efficient knowledge
acquisition can be done by solving problems using the acquired knowledge and refine it during the course of problem
solving. I^2S-LD has been extended to include this dynamic
analysis capability [6], based on which a general architecture
of knowledge acquisition by interview has been proposed [7].

(2) ISAK

Knowledge acquisition can be coupled with machine learning
techniques. Along this line, a new system called ISAK, an interview system for acquiring design knowledge of oil-hydraulic
circuits [8][9], is now under development. ISAK is composed
of two major modules: UBL and UBI (Understanding-Based
Learning and Interview). UBL is a module that compiles
design knowledge through understanding some examples of
design drawings. UBI is a module activated when UBL fails
to understand an example and it acquires design know-how
or missing domain theory which contributes to improving the
performance of UBL.

ISAK has some interview strategies as well as initial domain
knowledge. Given a design example, ISAK tries to explain
why the example satisfies the design specification associated
with it according to the domain theory and the topological
structure of the circuit and identifies which component is most
necessary for the specification to be satisfied. Identification
is done on the basis of a generate-and-test paradigm, and an
explanation is generated from which a piece of compiled design
knowledge that maps function to attribute is extracted.

If the proof fails, the explanation process is analyzed and an
interview is initiated according to the strategies. If the failure
is due to a deficiency of the domain knowledge, the knowledge
is refined or added by the expert, and if it is due to something which cannot be deduced theoretically, new heuristic
knowledge is also acquired from the expert. Thus this system acquires not only heuristics but also domain knowledge
and compiled design knowledge. Having the domain theory
in advance helps reduce the number of questions raised by the
interview system. The method used here is a combination of

interviews and an explanation-based learning method (EBL). EBL-obtained design knowledge can be regarded as a specialization of domain theory rather than a generalization of an example, which is a key idea of ISAK.

(3) EPSILON/One

EPSILON/One is a knowledge acquisition system based on expert model consisting of operations which represent small tasks in human expert's problem solving process[10]. Seven types of operations are identified by analyzing real knowledge bases built for diagnosis problems in production rules. An interview method referred to as pre-post method is also developed. It consists of two interview strategies. It stimulates domain experts to remind pre or post operations by asking what operations are performed before and after a specific operation. It enables to elicit knowledge about operations efficiently according to operation types. EPSILON/One is implemented in ESP on PSI machine and is composed of interface supporting graphic multi-windows and mouse, knowledge elicitation module, knowledge refinement module and knowledge transfer module. The knowledge acquisition research at ICOT has been conducted extensively and several results are obtained [10] [11][12].

(4) CONSIST

Most of the existing knowledge acquisition systems are based on top-down elicitation, by which we mean an interview for obtaining task-specific knowledge conducted assuming that the task type is known in advance. On the other hand, there exists another type of elicitation, i.e., bottom-up interview. CONSIST which stands for CONstruction Support system of information STructure) is a system of this type [13]. It is a frame-based system aiming at eliciting information structure underlying the domain of interest using KJ method [14]. CONSIST provides us with a useful assistance when the problem solving knowledge is vague even for domain experts.

(5) GRAPE

When more than one domain expert are available, we have to acquire expertise from all of them. GRAPE, GRoup ware for Acquiring, Processing and Evaluating knowledge, is a system for such situations [15]. It exploits an WHYSIWIS interface and elicit expertise for selection problems using AHP (Analytic Hierarchy Process).

2.2 Task-oriented knowledge acquisition

Building an expert system is still an art. Problem solving based on human information processing tasks has brought great insight into expert system architecture. A generic task is an abstract description of a particular task, and a problem solving process is simulated by the proper combination of generic tasks.

(1) MULTIS

Inspired by pioneering works on the generic task method [16][17] and the half weak method [18], new research is being conducted which seeks to establish a generic method for building an expert system [19][20][21].

In MULTIS, each generic task has its own problem solving method, is implemented as a building block or combination of building blocks, and requires task specific knowledge. A problem solver compiler generates a problem solving engine by structuring these building blocks in accordance with the generic tasks identified for the problem to be solved. MULTIS, MULti-Task Interview System, is now being developed based on this idea.

MULTIS acquires the necessary knowledge for each generic task. Therefore, two kinds of knowledge must be acquired, one being problem solving methods(task identification) and the other being the knowledge used in the task. The latter is acquired by an interview method similar to the one described above. The former is based on an interview of another type,

i.e. task analysis interview aimed at identifying the nature of the task. Since each task is considered as being domain-independent, it is possible to identify the task by dividing the given problem into a set of known tasks.

Collection of vocabularies specific to the task is important in this identification. MULTIS has, as its library, generic vocabularies, and case data of representative expert systems expressed by the generic vocabularies. A rough identification is first made under the framework of the classification problem and then the detailed identification is made based on case-based reasoning (CBR).

After the generic tasks have been identified, it is then necessary to synthesize the building blocks for these tasks into a single problem-solving engine. This part has not been completed yet.

(2) Automated generation of task-oriented problem solver

Another interesting work is an automatic generation of task-oriented expert system [22]. This system takes an approach which is slightly different from the above. It is based on the assumption that each information processing task can be regarded as a search problem with two kinds of control strategies: task-specific (domain-independent) and domain-specific. By identifying and specifying these strategies as searches, it is possible to automate generation of a task-specific expert system embedded with domain-specific knowledge. It is not necessary to prepare a set of generic tasks as construct primitives in advance. A task-specific problem-solving program can be constructed by specifying six element functions of a general search program. Search is defined here, based on a concept of abstract data type, as data structures and functions which operate on the data. These data structures consist of initial nodes, target nodes, intermediate nodes, relation-to-children nodes, node constraints, and node state. A total of 21 primitive functions have been predefined, which represent the node state. They operate on the data to alter the state, and are used to specify the six element functions above.

The user can build an expert system by providing search strategies of both kinds and domain-specific knowledge by answering the instructions given by using the data structure specified. The system is implemented by ES/KERNEL. Three different kinds of existing expert systems on planning problems in nuclear power plants, all of which are also implemented by ES/ KERNEL, have been reproduced using this system.

The generated expert systems exhibit exactly the same performance, but the number of programming steps has been reduced to one third of the original for each case. It is hoped that this approach can be applied to a wide class of tasks, although this still remains to be confirmed.

3 Knowledge compilation

Knowledge compilation is a very effective method for acquiring heuristics from a model describing a working mechanism for a system where deep functional knowledge is available. The need for this approach has been recognized since the early phase of expert system technology [23] and significant work has already been carried out on compiling diagnostic knowledge from deep models [24].

Deep knowledge is objective knowledge which basically depends on domain, not on task. Task knowledge is embedded in the compiler. The underlying concept is that many useful heuristics can be justified by deep objective knowledge. It would be desirable if deep knowledge could be prepared in a task-independent manner and if various task-dependent knowledge compilers worked on this knowledge to produce task-specific shallow knowledge.

3.1 Diagnostic knowledge compilation

(1) KC-II

One major research in this field is the development of a knowledge compiler for a diagnostic expert system which uses qualitative reasoning [25]. It has been determined that there are five kinds of knowledge required for diagnosis: knowledge of 1) device world (DW), i.e. structure and function of the system, 2) physical world (PW), i.e. principles governing the domain, 3) control world (CW), i.e. failure frequency and ease of measurement used to control reasoning, 4) interpretation world (IW), i.e. interpretation of internal state to make correspondence with failure, and 5) failure mechanism world (FMW), i.e. heuristics about how failure takes place. DW and PW are domain-dependent but task-independent, and can be called deep knowledge. CW and IW are both domain and task-dependent. These are forms of knowledge that the knowledge compiler makes reference to during compilation rather than being deep knowledge. FMW is heuristics about failure, but some of them are described in a domain-independent manner, e.g., existence of an obstacle is caused either by production or by movement.

Using these five kinds of knowledge, retrospective qualitative reasoning is first made in a backward direction from a symptom towards possible hypotheses. Then prospective qualitative reasoning is performed to find possible symptoms, and diagnostic rules are generated. The method seems very simple but it is very useful and it has been demonstrated that when applied to diagnosis of air-conditioners, it can generate rules which even domain experts could not think of easily.

The idea of knowledge compilation is also applied to a scheduling expert system of electric power network maintenance [26]. They identified several kinds of deep knowledge by investigating how every shallow knowledge in the expertise could be justified. A unified framework of knowledge compilation is designed based on these two results [27].

(2) KC-II/DST

The latest results related to the above system include a domain- and task-specific shell referred to as KCII/DST [28]. It is designed based on the above mentioned knowledge compiler. Its major characteristics are:

1) Diagnostic rules are compiled from deep knowledge, 2) The rules generated are hierarchically organized and infer intermediate hypotheses according to the hierarchical structure of the device of interest, 3) It can deal with devices with feedback loops and failure mode dependent diagnosis, and 4) Sophisticated iconic interface is available for describing DW.

KCII/DST is implemented in ESP (Extended Self-contained Prolog) developed by ICOT on PSI-II(Personal Sequential Inference machine) also developed by ICOT, and in Common ESP (CESP) on SUN3/260. Although it is a prototype system, a novice user can obtain a diagnostic expert system for the device by describing its structure through the iconic interface.

(3) Compilation using EBL

Compilation of diagnostic knowledge from deep models can be performed based on the explanation-based learning method[29]. This study uses adjustment cases of domain experts as training examples, the causal model as domain theory and observability and strategic importance as operationality criteria respectively. This method consists of generation of explanation structure, reduction of explanation structure, and transformation to rules. First, explanation structure is generated by explaining adjustment case using the causal model under machine adjustment goal. Second, this explanation structure is generalized by structural reduction using the operationality criteria. Finally it is transformed into operational rules. A knowledge compilation system, which has been developed based on this framework, acquired 79 diagnostic rules when appied to 50 actual adjustment cases in cigarette making machines. Comparing these rules with the knowledge base in the practical expert system, the effectiveness of the proposed method has been proven in both knowledge acquisition and knowledge refinement.

(4) QR/P

Qualitative simulation can be applied to knowledge compilation. QR/P is a system for diagnostic knowledge compilation using qualitative simulation, in which diagnosis knowledge rules are generated from a structural model built in terms of qualitative parameters and constraints among them[30]. The difference between the normal behavior and abnormal one activates the rule generation procedure. It has been implemented in ESP on a PSI machine and was applied to glaucoma diagnosis.

(5) BERG

Problem solving and knowledge acquisition is intricately connected. A hypothetical reasoning method called BERG[31] is developed to achieve a function of compiling compact rules for diagnosis. This method first performs a diagnosis of a specific problem based on the RESIDUE algorithm developed in the DART project, and the result is generalized (i.e. compiled) using EBL. Application to electric circuits indicates that the generated rules run about 10 times faster.

3.2 Design knowledge compilation

(1) MECHANICOT

MECHANICOT[32][33] is a tool for design expert system's based on knowledge compilation. Design process is formulated as a constraint satisfaction problem in MECHANICOT, where knowledge about the domain such as constraints, facts and theories is described in a declarative form. Knowledge compiler in MECHANICOT generates existing paths of constraint processing more efficient by translating the input design specifications into design plans assuming that the structure of the design objects has been determined. Its main mechanism is based on the idea that constraints described declaratively can be transformed into procedural representations using the model of design process. MECHANICOT has been implemented in ESP on PSI-II developed by ICOT and been applied to parametric design of gear units.

(2) Compilation using EBL

An EBL-based method for acquiring operational design knowledge in physical systems was proposed[34]. In this study, any object can be interpreted via various design rationalities such as technological, causal and economical ones and it can be modeled as a hierarchy which consists of design goals, subgoals, structures and substructures toward attaining those goals. Through organizing domain-specific knowledge of the EBL system according to the above model, this system yields a systematic explanation of how structures attain design goals. This results in a generalized version of Functional Diagram used in Value Engineering. From this diagram, general design knowledge of various levels can be generated by a single instance of designed objects.

(3) Hierarchical compilation

Another study now in progress is the use of hierarchical qualitative reasoning based on approximations [35][36][37]. This method enables acquisition of heuristics and simultaneous utilization of knowledge expressed on multiple levels with different abstractions. It makes use of equations which express physical principles as deep knowledge, and compiles them into shallow knowledge expressing representative behavior of a particular element or composite element. This shallow knowledge expresses the way a human expert understands a complex composite element. The method serves as a framework for formalizing various heuristics which have so far been difficult to handle without an experiential method. Special care was taken to suppress over-generalization caused by approximation. The method is now being tested for its understanding of electric circuits.

4 Machine learning

Learning is defined as a process to acquire knowledge necessary for problem solving and to refine its performance. The former process is called knowledge acquisition and the latter is called knowledge refinement. Machine learning can be classified into inductive leraning, deductive learning, and others. The inductive learning is often called similarity-based learning, because of its investigation of similarities among examples. The deductive learning is often called explanation-based learning, because of its usage of explanations by domain theory. In the following, research activities of similarity-based learning and explanation-based learning are reviewed.

4.1 Similarity-Based Learning(SBL)

(1) ARIS

ARIS (Adaptive Rule Induction System), an extended version of ID3, is developed [40]. Identification of useful attributes and generalization of the resulting trees play crucial rolls among several issues in decision tree induction. ARIS is designed based on a decision tree induction method but is augmented by introducing an attribute generation mechanism and generalization operators. New attributes are generated in the same way as the original induction method using only samples that are difficult to discriminate. ARIS has an another function which deals with attributes that take numerical values. It is implemented in Common Lisp, Flavors and OPS5 on Symbolics 3620. The latest version of ARIS is extended to coordinate with production rules written by human experts. An integrated environment enables to amalgamate human knowledge and one obtained by machine learning. The performance of ARIS was evaluated through the phoneme recognition in continuous speech. The result shows the decision tree obtained by ARIS is superior to production rules written by human experts.

(2) LS/1

LS/1 is a domain independent inductive learning system, which repeats Question-Response-Answer interactions with a teacher, and learns the structure of the relation between questions and answers[41]. The knowledge representation system of LS/1 is called a label net which represents relational data base, conceptual hierarchy and certainty factors of rules. The learning algorithm of LS/1 includes generalization and merging of rules, generalization by identification of two predicates, and changing ceratinty factors in rules. The effectiveness of LS/1 has been shown by application to the task of learning English-Japanese translation.

(3) CIGOL

CIGOL[42] is one of the approaches that weaken the bias, and it can invent new terms by using inverse resolution operators while it is learning new concepts. Inductive learning methods are important for intelligent systems to acquire knowledge. Since most inductive learning systems are strongly biased by their representation language of concepts, auxiliary concepts must be provided before learning. For this reason, the bias is generally too strong. However, its teacher must answer difficult questions. In order to alleviate the teacher's responsibility, CIGOL adopts a learning method which improves the interaction between the teacher and the learner. That is, any regular set can be learned with only restricted inverse resolution operators and simple queries.

4.2 Explanation-Based Learning(EBL)

The mainstream of machine learning for the last few years seems to have been EBL and its extension. EBL can learn justified generalizations from only one example when the domain theory is perfect, but it does not work when the domain theory is imperfect. Imperfectness of the domain theory can be classified into four levels, i.e. incomplete, intractable, inconsistent and non-operational theories. Recently, most researchers have been concerned with an attempt to solve these imperfect theory problems.

(1) Augmented EBL

An augmented EBL can be formulated to handle plural examples simultaneously to unify EBL and SBL and a concept of the least EBG is introduced to extract similarities from multiple examples[43].

For the non-operational theory problem, it is shown that the least EBG is more operational than the usual EBGs. Theorems about operationality of the least EBG show the utility of the least EBG. For example, well known programs, (e.g., simple list processing predicates such as *member* or *append*) are already operational because they have no backtrackings. In much of existing EBLs, its performance often becomes wrong because it generates the same number of macros as examples, but a learner using least EBGs can learn the most operational EBG.

(2) EBG version space method

For the inconsistent theory problem, a concept of an EBG version space is introduced, and the MSV (the Most Special Version) is shown to be the least EBG[44]. We assume that training examples are given with judgements whether they are positive or negative, and define a domain theory is inconsistent iff some negative examples are provable as positive examples. It is further assumed that a learner swaps rules which conclude the goal concept with EBGs. A learning task is to find EBGs s.t. prove positive examples but never prove negative examples. In SBL, such generalizations make a range called a version space. Its least upper bound and its greatest lower bound are called MGV (the Most General Version) and MSV, respectively. It is a powerful method for incremental learning. It is shown that a generalization space of EBG has the same structure as a version space.

The generalization hierarchy of common explanation structures, which are generated from multiple primitive explanation structures, can be effectively organized based on a concept of maximal covering.[45] The maximal covering reflects similarities among the common explanation structures. By top-down search over the generalization hierarchy, the best set of macro rules, which includes all positive examples and excludes all negatives, can be found. A learning system based on this method has been implemented in Prolog. Some experiments show that it can easily generate valid macro rules under the inconsistent domain theory. Its incremental version has been developed as an extension[46].

(3) EBL with background theory

The intractable theory problem is distinctive in synthetic problems such as planning and design. A learning task for intractable theory problem is to find strategies to solve any problems without search. The least EBG can also be used. However, it often falls into a difficulty of SBL that a learner requires numerous negative examples before convergence. It means that the computational load shifts from EBL to SBL to resolve the imperfectness of the domain theory. Therefore, to solve the intractable theory problem, it is suggested to introduce a meta level domain theory. In practice, it is shown that the use of meta-level domain theory may reduce computational loads[47].

For the incomplete theory problem in classification, an approach using a background theory in addition to the usual domain theory is proposed[48]. A background theory consists of a commonsense knowledge such as text books or dictionaries. A learning task is to extend a domain theory by transferring knowledge from the background theory based on similarities of examples. The least EBG can also be used. Thus, to solve the incomplete theory problem, it is necessary to consider knowledge transfer from other domains. Analogical reasoning or case-based reasoning seems useful methods realizing knowledge transfer.

(4) PiL

Paradigm-based inference learner, PiL for short, is a learning system for acquiring strategic knowledge of problem solving. It first tries to solve a given problem[49][50][51]. When it fails, it asks a solution from which it acquires some strategic knowledge for basic operator applications. The strategic knowledge is represented as macro-operators which are subsequences of operator applications. PiL was applied to generation of strategic knowledge for linear equations. The evaluation function used for selection of macro-operators is called the perfect causality, which is a heuristics for selecting valid macro-operators. Macro-operators are generalized according to EBG. PiL2 is a direct successor of PiL and has the augmented perfect causality. It is applied to robot planning and obtained a successful results.

(5) Learning from partial explanation structures

An integrated learning method with an extended generator of macro operators is proposed[52]. In this study, it is shown that a set of rules can be specialized and generalized by rule composition and decomposition. Explanation structures of examples are generated using domain theory and they are generalized by decomposition. The usefulness of this method was demonstrated by application to synthesis of logical circuits and English-Japanese translation.

5 Case-Based Reasoning

The knowledge-based reasoning(KBR) requires complete and consistent knowledge base. This brings about the difficult problem of knowledge acquisition and refinement. For supplementing KBR, it is useful to introduce analogical reasoning or case-based reasoning that explicitly uses past cases which are experts' experiences with success or failure results.

(1) Activity of ICOT KSA/KAR WG

The KSA/KAR working group, established in ICOT last year, has surveyed the needs and seeds of CBR technologies from a variety points of views[53]. The main results of this group's activities are summarized as follows:

1. practical needs on planning and design problem domains were extracted and the availability and usefulness of the CBR approach was clarified,

2. effective ideas to overcome issues related to indexing of cases, organization of case base, retrieval of analogical cases, modification and repair of cases, and so on were presented,

3. a theoretical connection between CBR and EBL was investigated,

4. the CBR was located at the center of a framework of integrating problem solving paradigms and learning paradigms.

(2) Analogical Reasoning

A new formalism of analogical reasoning is presented[54]. According to the formalism, a single domain for the reasoning is given to a reasoning system. The domain has some rules, constraints, and negative facts. The negative facts are used to reject in appropriate analogies that derive contradictions. Based on the formalism, a possibility of inferring incomplete information by analogy is discussed. Moreover a procudure to find a sound analogy is presented.

(3) Interactive CBR model

To automate the CBR process, domain knowledge for problem indexing and case modification or repair must be prepared in advance. However, it is difficult in practice. To overcome such a difficulty, an interactive CBR model is proposed[55]. This model makes interactions between the user and the system flexible under imcomplete domain knowledge. This model has been applied to a machine adjustment problem. An adjustment case is defined as an instance that goal was set under condition such that some faults were inspected, possible cases were identified, method was performed, and result and its evaluation were recorded. A prototype CBR system based on this model has been developed. Application of this CBR system to machine adjustment problems has shown that it can generate appropriate method plans in cases of failure by the existing expert system.

Roles of domain knowledge and user in interactive CBR systems should be investigated from the viewpoints of costs and benefits[56]. The domain knowledge needed to realize CBR consists of conceptual knowledge hierarchy, causal knowledge, heuristic knowledge and meta-knowledge. Heuristic knowledge consists of case modification knowledge and measurement and testing knowledge. Conceptual knowledge hierarchy is often used in building/managing the case base, problem indexing and case modification/repair. Causal knowledge is often used in case modification/repair. As case modification/repair is a central task in CBR, these three kinds of knowledge are expected to contribute to automating the whole CBR process. Control knowledge for reasoning is important especially in automating case modification process. Costs of knowleadge acquisition and benefits of knowledge utilization at each step of CBR can be estimated qualitatively. The results of such a cost-benefit analysis suggest a form of cooperation between domain knowledge and user.

(4) Case-based process diagnosis

A framework for case-based diagnosis is prensented[57]. In this framework, a mechanism to drive the dynamic organization of the memory is introduced. In the memory, knowledge is represented by a frame which is structured to maintain the number of slots small by sharing the common attribute-values among cases. An architecture for case-based diagnosis is presented, using the memory organization where diagnostic cases are indexed both by the symptoms and by the causal relations among symptoms. The program CAOS(CAse Operation System) is implemented for the prosess diagnosis domain.

(5) Case-based information retrieval

Information retrieval requires a lot of time for users to generate appropriate retrieval conditions. CBR is effective in reducing such an unwilling loss of effort. By introducing the framework of Carbonell's derivational analogy and organizing retrieval cases systematically, a case-based reasoning system can be constructed[58]. In this study, a case is defined as a tuple of intial and final retrieval conditions, transformation rules and trial sequence. The case base is organized as a lattice structure on the basis of similarities among cases. For retrieval of analogical cases, conceptual hierarchical knowledge are necessary. A prototype CBR system has been implemented on Prolog. As the framework of this system is domain-independent, it is applicable not only to information retrieval but also to other problems with analogical structure.

6 Concluding remarks

Before finishing this review, one of new projects in progress in Japan will be mentioned. ASTEM (Advanced Software Technology & Mechatronics Research Institute of Kyoto) is going to set up a project on a large-scale engineering knowledge base under collaboration with academic and industrial organizations. The project aims at building knowledge bases playing as an infrastructure of expert systems which contains a huge amount of knowledge such as domain principles , general knowledge and common sense in engineering. The resulting knowledge base will be used as deep knowledge from which several kinds of shallow knowledge is generated, as domain theory in EBL and as stuff of basic knowledge in interview system, which helps knowledge acquisition to a greate extent. This project will make a useful contribution to the progress of knowledge-based systems research.

Research activities related to expert system building technology are becoming very active not only at universities, but also at industries in Japan. The ones reviewed in this paper cover only a small portion of such an activity, but highlight efforts being made to advance expert system development from an art to an engineering discipline. Research activities of knowledge acquisition and machine learning will make a true contribution to realization of intelligent systems. We have confidence that fruitful results will be expected in the near future.

References

[1] Kobayashi S., and Terano T.,(eds.) *Knowledge Systems Handbook(in Japanese)*, Ohm Co., Tokyo (in Press) .

[2] Terano T., and Kobayashi S., *Problem Analyses, Tool Evaluation, and Verification and Validation Study: Three Steps for Knowledge-Based Systems Development Methodology*, Proc. of IJCAI'89 Workshop on Verification, Validation and Testing of Knowledge-Based Systems, 1989.

[3] Kawaguchi A., Mizoguchi R., Yamaguchi T., and Kakusho O., *SIS: A Shell for Interview Systems*, Proc. of International Joint Conference on Artificial Intelligence, pp.359 -361, 1987

[4] Kahn G., Nowlan S., and McDermott J., *MORE: An Intelligent Knowledge Acquisition Tool*, Proc. of International Joint Conference on Artificial Intelligence, pp. 581-584, 1985

[5] Kawaguchi A., Taoka N., Mizoguchi R., Yamaguchi T., and Kakusho O., *An Intelligent Interview System for Conceptual Design of Database*, Proc. of the 7th European Conference on Artificial Intelligence, pp.39-47, 1986

[6] Kawaguchi A., Matsuyama E., Numata K., Mizoguchi R., Nomura Y., and Kakusho O., *A Database Construction/Manipulation Support System based on Interviewing (in Japanese)*, Journal of Japanese Society of Artificial Intelligence, Vol. 4, No. 7, pp.421-430, 1989

[7] Kawaguchi, A., Motoda, H., and Mizoguchi, R., *An Architecture of Knowledge Acquisition by Interview based on Dynamic Analysis*, Proc. of JKAW90, 1990

[8] Matsuda K., Nomura Y., Baba T., Nakashima Y., Kawaguchi A.,and Mizoguchi R., *An Interview System for Acquiring Design Knowledge based on Understanding of Examples (in Japanese)*, Proc. of Japanese Society of Artificial Intelligence, pp.739-742, 1989

[9] Matsuda, K., and Mizoguchi R., et al., *ISAK: Interview system for acquiring design knowledge*, Proc. of JKAW90, 1990.

[10] Taki H., and Tsubaki K., *Expert model: A knowledge representation for knowledge acquisition(in Japanese)*, Journal of Japanese Society for Artificial Intelligence, Vol.5, No.2, pp.203-212, 1990.

[11] Taki H., *Knowledge acquisition by observation*, Proc. of FGCS, pp.1250-1258, 1988.

[12] Taki H., et al., *Operation presumption: Knowledge acquisition by induction*, Proc. of EKAW89, pp.34-48, 1989.

[13] Shinohara Y. and Terano T., *Support system for construction information structure using frame-based hierarchical relations(in Japanese)*, SIG-AI-86-32, Information processing Society of Japan, pp.31-36, 1986.

[14] Kawakita J., *The method of abductive thinking (in Japanese)*, Chuoukoron Pub. Co. Ltd., 1967.

[15] Ueda H., and Kunifuji S., et al., *An implementation of knowledge acquisition support groupware GRAPE (in Japanese)*, Proc. of the 4th National Conference of Japanese Society for Artificial Intelligence, 17-8, pp.673-676, 1990.

[16] Chandrasekaran B., *Generic tasks in knowledge-based reasoning: high-level building blocks for expert system design*, IEEE Expert, Vol. 1, No. 3, pp.23-30, 1986

[17] Breuker J., and Wielinga B., *Models of Expertise in Knowledge Acquisition*, Topics in Expert System Design, Guida G., and Tasso C. (Editors), Elsevier Science Publishers, pp.265-295, 1989

[18] McDermott J., *Using problem-solving methods to impose structure on knowledge*, Proc. of Intl. Workshop on Artificial Intelligence for Industrial Applications, pp.7-11, 1988

[19] Tijerino Y., Kawaguchi A., Mizoguchi R., and Kakusho O., *A Generic Vocabulary and Generic Task Library for Use in MULTIS*, SIG-KBS-8805-3, Japanese Society for Artificial Intelligence, pp.21-30, 1988

[20] Mizoguchi R., *Towards Establishment of a Methodology for Building Knowledge-based Systems (in Japanese)*, SIG-KBS-8901-4, Japanese Society of Artificial Intelligence, pp.21-30., 1989

[21] Tijerino Y., and Mizoguchi R., et al. *A task analysis interview system that uses a problem solving model*, Proc. of JKAW90, 1990.

[22] Kasahara T., Yamada N., and Kobayashi Y., *Automated Generation Method of Task-oriented Problem Solver for Expert System Architecture (in Japanese)*, SIG-KBS-8905-6, Japanese Society of Artificial Intelligence, pp.41-48, 1990

[23] Chandrasekaran B., and Mittal S., *Deep Versus Compiled Knowledge Approaches to Diagnostic Problem-solving*, Proc. of the National Conference of Artificial Intelligence, pp.349-354, 1982

[24] Sembugamoorthy V., and Chandrasekaran B., *Functional Representation of Devices and Compilation of Diagnostic Problem- Solving Systems, Learning, Experience and Memory*, edited by Kolodner and Riesbeck, Laurence Erlbaum and Associates, pp.47-73., 1986

[25] Yamaguchi T., Mizoguchi R., Taoka N., Kodaka H., Nomura Y., and Kakusho O., *Basic Design of Knowledge Compiler Based on Deep Knowledge (in Japanese)*, Journal of Japanese Society of Artificial Intelligence, Vol. 2, No. 3., pp.333-340, 1987

[26] Adachi K., Yamaguchi T, and Mizoguchi R. et al., *Deep knowledge in scheduling type of expert systems (in Japanese)*, SIG-KEAI 59-18, Information Processing Society of Japan, pp.1988

[27] Yamaguchi, T., and Mizoguchi R. et al., *A unified framework for deep reasoning in expert systems (in Japanese)*, SIG-KBS-8801-3, Japanese Society for Artificial Intelligence, pp.23-31, 1988

[28] Ozawa T., and Mizoguchi R. et al., *A DST based on a knowledge compiler(KCII)*, SIG-KBS-8905-5, Japanese Society for Artificial Intelligence, 1990

[29] Nakamura K., and Kobayashi S., *Knowledge Acquisition from Machine Adjustment Cases by Causal Model and Operational Criteria*, Proc. of JKAW90, 1990

[30] Owada H., Mizoguchi F. and Kitazawa Y., *A method for developing diagnosis systems based on qualitative simulation (in Japanese)*, J. of Japanese Soc. for AI, Vol.3, No.5, pp.617-626, 1988

[31] Koseki Y., *Experience Learning in Model-Based Diagnostic Systems*, Proc. of International Joint Conference on Artificial Intelligence, pp.1356-1362, 1989

[32] Nagai Y., Terasaki S., Yokoyama T. and Taki H., *Expert system architecture for design tasks*, Proc. of FGCS'88, pp.298-317, 1988

[33] Nagai Y. and Terasaki S., *Towards constraint analysis and plan generation of constraint compiler (in Japanese)*, Proc.of the Third National Conference of Japanese Soc. for AI, 11-43, pp.693-696, 1989

[34] Katai O., Kawakami H., Sawaragi, T., Iwai S. and Konishi T. *Acquisition of operational design knowledge from designed objects using explanation-based learning method (in Japanese)*, Trans. of Soc. of Instrument and Control Engineer, Vol.26, No.8, pp.916-923, 1990

[35] Yoshida K., and Motoda H., *An Approach to Hierarchical Qualitative Reasoning - Constructing Shallow Knowledge from Deep Knowledge - (in Japanese)*, Journal of Japanese Society for Artificial Intelligence, Vol. 4, No. 4., pp.447-455, 1989

[36] Yoshida, K., and Motoda, H., *Towards Automatic Generation of Hierarchical Knowledge Bases*, Working Notes of the AGAA-90 Workshop, Boston, pp.98-109, 1990

[37] Yoshida, K., and Motoda, H., *Hierarchical Knowledge Representation based on Approximation*, Proc. of JKAW90, 1990

[38] Nishida T., Kosaka A. and Doshita S., *Towards knowledge acquisition from natural language documents –Automatic model construction from hardware manual–*, Proc. of the 6th IJCAI, Kalsruhe, pp.482-486, 1983

[39] Handa K. and Ishizaki S., *Acquiring knowledge about relation between concepts*, Proc. of EKAW89, Paris, pp.380-390, 1989

[40] Tsujino K., and Mizoguchi R., et al., *Adaptive rule induction system: ARIS (in Japanese)*, Trans. of IEICE of Japan, Vol.J72-D-II, No.1, pp.121-131, 1989

[41] Akama K. *Learning of translation by the inductive learning system LS/1 (in Japanese)*, J. of Japanese Soc. of Artificial Intelligence, Vol.2, No.3, pp.341-349, 1987

[42] Sakurai S. and Haraguchi M. *Improving man-machine interaction in learning regular sets*, Proc. of PRICAI'90, 1990

[43] Yamamura M., and Kobayashi S., *An Augmentation of EBL on Plural Examples (in Japanese)*, Journal of Japanese Society for Artificial Intelligence, Vol. 4 , No. 4., pp.389-397, 1989

[44] Yamamura M., and Kobayashi S., *Towards Unifying EBL and SBL to Imperfect Domain Theory Problems*, Proc. of PRICAI'90, 1990

[45] Kobayashi S., Shirai Y., and Yamamura M., *Acquiring Valid Macro Rules under Imperfect Domain Theory by Topdown Search on Generalization Hierarchy of Common Explanation Structures (in Japanese)*, Journal of Japanese Society for Artificial Intelligence, Vol. 6, No. 3 (to appear)

[46] Ono, T, Yamamura M., and Kobayashi S., *Incremental Refinement of Domain Theory based on Structural and Parametric Generalizations (in Japanese)*, Proc. of 11-th Knowledge and Intelligent Systems Symposium, pp.81-86,1990

[47] Yamamura M., and Kobayashi S., *An Augmented EBL Learner for Acquiring Problem Solving Macrotable (in Japanese)*, Journal of Japanese Society for Artificial Intelligence, Vol. 6, No. 1 (to appear).

[48] Araki, Y., Yamamura M., and Kobayashi S., *An Incremental Learning of Domain Theory with a Background Theory (in Japanese)*, Proc. of 11-th Knowledge and Intelligent Systems Symposium, pp.87-92,1990

[49] Yamada S., Abe N. and Tsuji S. *PiL: A system to learn strategy knowledge in problem-solving (A case study in an equation & inequality of the first degree) (in Japanese)*, J. of Japanese Soc. for AI, Vol.3, No.2, pp.206-215, 1988

[50] Yamada S., Tsuji S. and Abe N. *DSBG: Direct solvability-based generalization –Defining operational SOLVABILITY concept– (in Japanese)*, J. of Japanese Soc. for AI, Vol.3, No.6, pp.783-791, 1988

[51] Yamada S. and Tsuji S. *Selective learning of macro-operators with perfect causality (in Japanese)*, J. of Japanese Soc. for AI, Vol.4, No.3, pp.321-329, 1989

[52] Numao M. and Shimura M. *A learning method based on partial structures of explanations (in Japanese)*, Trans. of Inst. of Electronics, Information and Communication Engineers, Vol. J72-D-2, No.2, pp.263-270, 1989

[53] Kobayashi S. et al, *A Survey of Case-Based Reasoning and Its Applications (in Japanese)*, ICOT Technical Mem., Institute for New Generation Computer Technology, 1990

[54] Haraguchi M. *On inferring incomplete information by analogy*, Proc. of Japan-Czechoslovak Joint Symposium on Theoretical Foundation of Knowledge Information Processing, pp.31-40, Inorga Pub., 1990

[55] Nakamura K., and Kobayashi S., *An Interactive Case-Based Reasoning Model and Its Application to Supporting Machine Adjustment (in Japanese)*, Journal of Japanese Society for Artificial Intelligence, Vol. 4 , No. 6., pp.704-712, 1989

[56] Nakamura K., and Kobayashi S., *Roles of Domain Knowledge and User in Case-Based Machine Adjustment Systems*, Proc. of PRICAI'90, 1990

[57] Ishida T., and Tokumaru T., *A framework for Case-Based Diagnosis - Dynamic Frame and Reasoning on the Representation– (in Japanese)*, Trans. of Soc. of Instrument and Control Engineer, Vol.26, No.6, pp.706-713, 1990

[58] Ase H. and Kobayashi S., *A Case-Based Reasoning System for Intelligent Information Retrieval*, Proc. of PRICAI'90, 1990

INFORMATION TECHNOLOGY AND INDUSTRIAL AUTOMATION TRENDS IN AGRICULTURE

A. J. Udink ten Cate

*Department of Computer Science, Wageningen Agricultural University,
Wageningen, The Netherlands*

Abstract. An analysis is presented of the impact of information
technology (IT) and industrial automation in agriculture. The analysis
is based on the identification of the actors in the market, their
respective roles and the impact of technology in terms of trends to the
future. It is stated that especially in primary agricultural production,
industrial automation, measurement and control will be of most
importance. For the distribution and intermediation of agricultural
products IT is more crucial.

Keywords. Technological forecasting; agriculture; knowledge engineering;
industrial control; economics

INTRODUCTION

Approach

In order to discuss the impact of industrial
automation in agriculture and the related
methodology like measurement and control,
the following approach was selected.

Instead of industrial automation, the impact
of information technology (IT) and
telecommunication on agriculture has been
discussed. This means that the discussion is
not limited to industrial automation as
such, but covers the whole range of IT
applications.

Research has been performed using semantic
models of IT in agriculture (Udink ten Cate
and Van Lopik, 1990). Advances in system
simulation methodology (Elzas et al., 1989)
and related software (Campus Software, 1989)
have led to the concept of semantic models
(Udink ten Cate, 1991).

Most of the findings of this analysis
conform with a recent study on economic
growth and IT carried out for the Dutch
Ministry of Economic Affairs (Butler Cox,
1991). Here Porter's methodology on the
competition of nations was applied.

The semantic models on agriculture contain
statements of potential changes, which have
to be translated to the actual field of
interest. In the following only the
significant results are given.

The approach followed is:

1. Which actors play a part on the relevant
market, which roles do they perform and what
is the relevance of IT in this respect?
2. What are in broad terms the trends in IT,
assuming a technology push situation?
3. Which trends are there in the
agricultural production, and what is the
influence of IT on these trends?
4. Which obstacles are there?
5. Which conclusions are to be derived from
this?

Summarizing it can be stated that

{IT applications} = f {IT technology trends,
    agricultural production trends,
obstacles}

The mapping f{...} on {IT applications} is
determined by the actors in the market.

Actors in the market

In the market various functions are
performed, such as (physical) production,
intermediation, distribution. The related
actors are: supply companies, primary
production (farmers), auction, wholesale
companies, retail companies. In table 1 a
summary is presented of the various actors
and their functions.

TABLE 1. Functions of the various
actors.

ACTOR	FUNCTION
supply company	distribution
farmer	physical production
	distribution
auction	intermediation
	distribution
wholesale	intermediation
	distribution
retail	distribution

For the various functions of the actors an
overview can be presented of the influence
of IT (table 2).

TABLE 2 Influence of IT on various
actors/functions. Scale from 0
(no influence/relevance) to ++
(much influence).

	phys. prod	inter- mediation	distribution
supply company	0	0	++
farmer	+	+	+
auction	0	++	++
wholesale	0	++	+
retail	0	0	+

259

TRENDS IN IT

Computers

- In the field of IT there is a trend to be
seen towards "commoditation", meaning that
IT products can be viewed upon as a
commodity rather than a specialty. As a
result, the production column will be
segmented horizontally. The production of
computer systems, for example, does not take
place within one company but components from
all types of supply companies are assembled
into a product. These supply companies also
deliver to the competition. Well known
trends are the PC's and the minicomputers.
Within this last category especially UNIX
and the multi-processor computers have
contributed to this trend.
  Discussion:
  In the multi-processor architecture 2-6
  processors (of the Intel or Motorola type)
  are operating in parallel. Via the
  operating system or a special processor
  the jobs are processed. By this
  architecture relative low-cost (no R&D
  investments) and slow processors can be
  employed leading to a relatively high
  throughput.

- Also in the field of operating systems a
similar commoditation takes place. The
preference for UNIX is not so much a result
of a drive towards "open computing" in
favour of the end user, but in essence a
drive from industry towards the application
of an already existing operating system with
the related low capital investment
properties.

- In the user perpective, commoditation
leads to lower prizes (and margins) and
increased flexibility in product offerings.
It is a property of a mature industry. An
essential boundary condition is
standardization.

- In the field of hardware (chips)
components are packed upon a chip in
steadily increasing densities. Because of
this the heat dissipation reduces, which in
turn leads to improved reliability. Also the
processing speed of processors increases
considerably. memory chip fall in price.
Because of this the price/prestation of
microcomputer systems decreases strongly.

- Computer peripherals like storage memory
(disks) increase rapidly in terms of
capacity. Introduction of CD-ROM deliver
much capacity on the user's desk. By the
introduction of HDTV computer colour
monitors decrease in price.
- In the computer software development
process the advance using CASE tools remains
slow. Computer supply companies rather seek
to improve their existing software platforms
(software engineering).

Networks

- Physical cabling over longer distances
will increasingly be done by optical fibre,
especially the PTT's will increasinly employ
monomode fibre. This facilitates very high
communication speeds.

- In local networks optical fibre remains
too expensive. For small distances the
communication speeds over copper will be
drastically improved (10 Mb over twisted
pair telephone cable).

- To the private customers, the public
network cannot deliver a higher
communication speed than 64 kb/sec, and in a
few cases 2 Mb/sec -despite the application
of optical fibres. The pricing for this
facility will be relatively high, so that
for the time being a difference remains to
exist beween private cabling and public
cabling.

- Public networks will be provided with ISDN
telephone exchange switches, which means
that (only) 64 kb/sec can be achieved. By
using the digital ISDN synchronization
channel the reliability will improve
significantly.

- Over public ISDN lines all sorts of
services can be offered by all sorts of
companies, as the delivery of (electronic,
voice, transactional) messages (with
guaranteed reception).

- On the local level networks will emerge
for the connection between various types of
measurement and control equipment. These
"field buses" are not standardized as yet.

Industrial automation

- By standardization it will become possible
to connect all types of technical equipment
over a special local network. The chips
required to implement these connections drop
in price.

- Chips for the direct measurement of
physical and chemical variables will not
become available widely. It remains a
problem to achieve the required accuracy.
Throw-away sensors offer a greater
potential.

- Because image processing chips become
widely available as do low cost processors,
the field of robotics will advance.

- Advances in the field of real-time
programming and testing will be few, slow
and costly.

TRENDS IN AGRICULTURAL PRODUCTION

In this section some general trends in
agricultural production will be treated in
more detail, whereby trends with a specific
IT nature will be highlighted. Where it is
possible IT trends will be referred to.

Market trends

- The market is developing from a supply
side market to a demand driven market. On a
mature consumer's market especially impulse
buying is of importance. By quick responses
the market share can be increased.
[IT trends: Networks play a major part in
the relation between producer and consumer.
Relevant consumer needs can be directly
transmitted from a distribution center to
the producer. This will lead to long term
agreements with more risks for the
distributor and guaranteed albeit lower
pricing for the farmer. With a better
predictability of quality and quantity of
the production this will contribute to
intrinsic lower production costs.]

- In the European Common Market the market
potential will strongly grow after 1992.
[IT trends: Networks, the financial services

in the form of payment guarantees will develop quickly; the general demand for business information will increase because of the growing anonymity of the market.]

- Growing wealth will lead to more potential customers, especially in nonfood sectors (cutflowers and potted plants). The life style will become more individualistic –which will be mirrored in a broader range of products both in quantity and in quality: 1000 brands of cheese and biodynamical tomatoes. [IT trends: networking, especially in the marketingsector; for quality control and certifying technical automation will contribute -e.g. chipcards.]

- Competition of other EC countries will increase, whilst the political pressure will increase to produce against world market prices.

Trends in crop production

- Biotechnology will provide opportunities to lower production costs, increase the number of varieties and reduce the variance within one crop. [IT trends: supportive of R&D and company wide data processing; e.g. computers and networks.]

- The development of low cost IC type sensors will increase the potential of measurement of environmental variables (temperature, humidity, radiation). IC sensors for the measurement of chemicals and gases remain problematic. Video based technology for telemetrics will decrease significantly in price, however the reliability will not increase to that extent.

- The widespread use of fertilizers and minerals is understood to be more and more harmful for product and environment.

- Remote visual product information can be delivered by video monitors with HDTV ans the associated image processing.

- Improved knowledge of the relation between crop growth and development and the associated crop treatment, will lead to a higher degree of prediction of crop growth and consequently crop quality. [IT trends: mainly related to R&D, slow development.]

- Integration of process control (technical automation) and business automation at the farmer's holding leads to the generation of more management information. This in turn will lead to a more optimal production. [IT trends: technical automation, local area networks and standards are of importance here.]

- Management information systems (MIS) for crop production, crop protection, yield and labour management become more reliable. Classical investment and labour rationalization methods can be applied per individual holding. Basic data for comparing holdings can be retrieved from centralized data bases. The present information models in use in the Netherlands (formulated for each sector of primary production, as well as the general information models for farm accounting) can be of use here, largely by standardizing basic data via a meta-model of data demand.

- The agricultural extension service kan be coupled directly to the farmer's computer.
Discussion:
Essential for an improved service are:
- standardization of basic data;
- improved monitoring of the performance of the farm;
- on-line CD-ROM type of data banks for estabilshing the pay off of alternatives;
- improved software for optimization (multiple goal programming versus LP algorithms).

- Robots and computer vision are applied in yield, sorting and local transport. With the application of well-insulated and conditioned spaces where handling of material is carried out without human intervention, quality can be improved.

- Artificial intelligence can be employed in trouble-shooting of equipment on the farm.
Discussion:
The increasing complexity of the computer controlled equipment, combined with the tendency to couple various types of equipment makes trouble shooting more difficult. AI can be used in diagnosis systems, which might also be connected via public telephone lines.

Trends in crop protection

- There is a growing tendency to reduce the application of pesticides. Biological crop protection is advocated as a substitute.

- Video sensors and the use of artificial intelligence increase the effectiveness of biological pest control.

- Via the monitoring of environmental factors, exact prediction can be made on the occurrence of pests and diseases.

- Improved information dissemination on new diseases will facilitate a quicker response from Health Authorities.

Trends in storage technology

- Airthight storage by encapsulation foils is a presently viable technology. In future robots can be emplyed for handling products without human contact or intervention. [IT trends: technical automation: robots, chip cards for monitoring of the storage and distribution process.]

- Cooling processes become better controllable. By on-line maesurement and control, and by new measurement equipment like sonic lasers or infra-red measurement of tracer gases or spurious gases, the quality can be improved.

Trends in transport

- Transport will be possible in better conditioned environments, which potentially can improve the quality. Using robots the handling of fresh products can also be performed in conditioned areas. Since handling and transport can take place without physical human intervention, quality degradation in the distribution chain is reduced.

- In the transport sector computer guided routing is possible. This reduces the need for large scale distribution centers. [IT trends: networks, especially EDI=electronic data interchange networks (or: electronic forms); for integration of EDI the computers of the various market actors must be interconnected.]

- Systems for container tracking (based on mobile communication) can be used to optimize transport and reduce delay times in the distribution chains. As a result the buffer supplies will deminish.

OBSTACLES

User-friendliness

- One of the main problems of the application of PC's and larger computers is the lack of user-friendliness and the related flexibility.
    Discussion:
    A core problem is the interaction of the user and the operating software. Command driven operating systems like MS-DOS for the PC are replaced by object oriented operating systems (like Apple's MacIntosh or PC Windows), which are much easier to operate (because data related to a file is coupled directly to the file itself, the connection of new application files is much more easy and reliable). In the minicomputer arena Windows like technology will also become popular.

- Because of increased man-computer interaction the user can operate more complex classes of software. Software developed for business processes can be tuned to the needs and working attitudes of the individual user.

- The user friendliness of (semi-) public databases is low. The database interconnection as well as the human interface can be improved significantly. The related price/prestation ratio is not favourable.

Standards

- Standards related to commoditation of the computer suppliers market will emerge on a supra-national level (EC).

- Standards related to commoditation of agricultural technical automation will not emerge automatically as an industry standard.

- Standards related to computer applications for management information will not emerge as an industry standard autonomously. The information models of farming can significantly contribute here.

- Standards related to the national use of networks (PTT's), e.g. for a distribution chain, will be developed by the agribusiness (EDIflower). A problem here is the coordination of the interests of the various interest groups.

Networks

- In the area of public networks important developments are at hand. In the 90 ties the introduction of ISDN will take place which will increase the capacity for the user of the public telephone network. A bottleneck remains the facility to connect computers at the farm in a reliable manner (requiring true multi-tasking in the present generation of PC's). The public network will mainly deliver store and forward functions.

- At the farms themselves the building of local area networks remains costly in an industrial environment (e.g. in glasshouses or stables).

Lack of knowledge

- For several processes to be automated the avilable knowledge of the properties of these processes is clearly insufficient. Examples are the climate control of glasshouses and stables (especially natural ventilation). For the accurate control of temperature, humidity, ammonium, carbon dioxide, no rigourous dynamical models are available which are the basis of any control.

- Much knowledge on agricultural production is available on the level of low-dimensional time-independent causal relationships, which are a result from the traditional approach of labour organization and rationalization. This knowledge domain can be easily transformed in planning related data (bookkeeping and budgeting), and can be automated for these types of business processes. Also in the earlrier referred to information models, much attention is devoted to data and less for the processes which generate these (time-dependent) data.

CONCLUSIONS

In the following some conclusions are drawn with respect to IT and industrial automation in agriculture. In the introduction statement was made concerning the importance of IT for the various actors in the agricultural marketplace. For primary agricultural production, the relevance of IT is not very high since it plays only a supportive role. In industrial automation and related fields like measurement and control technology IT is of a higher importance, which also holds for management related processes (management information systems). For distribution and intermediation of agricultural products IT is of a high importance.
Following the concept of a technology push which is described under IT trends, some potential developments in agricultural production are described. In the obstacles section some of the boundary conditions for the implementation of the potential technology are described. This leads to the following conclusions.

1. For primary agricultural production IT is a support function and can be understood as a result from mechanization and automation. For management support IT is more essential.
2. In primary production industrial automation is of key importance. Here the technology push in the field of sensors, local area networks and hardware is dominant. For advances in automation the lack of relevant knowledge of the processes is apparent.
3. Commiditation will not take place in industrial automation on the farm. The cost level will remain as it is.
4. Suppliers of farming material will use IT as a management tool: purchasing of lasting inventory, transaction processing (which party guarantees payment?), planning of service delivery.
5. For distribution and intermediation the role of IT is significant. Actors in this market will face changes in the distribution

chain. These changes are non technological by nature, but are enabled by IT.
6. Developments in IT can play a key role for adjusting the food market from a supplier's market to a buyer's market. For example by supplying specific products (biodynamical or organic products, products from integrated agriculture) with the corresponding certification and packaging.

REFERENCES

Butler Cox (1991). Economic survival and Information Technology. Min. of Econ. Affairs, The Hague, the Netherlands. Campus Software (1989).

CM100 User's manual. Campus Software, Wageningen.

Elzas, M.S., Zeigler, B.P. and Oren, T.I. (1989). Modelling and simulation methodology. North Holland, Amsterdam.

Groeneveld, H. (1991). A semantic model of the farming-2000 project. Thesis, Dept. of Computer Sci., Agric. Univ. (in Dutch).

Udink ten Cate, A.J. and Van Lopik, R.A.M. (1990). Technology impact model of the distribution chain of flower production 1990-2000. In: Alleblas, J.T.W. et al. (eds). Telecommunication networks in the flower industry. Research Report 73, LEI (Instit. of Agricultural Economics, The Hague, the Netherlands, pp. 59-74.

Udink ten Cate, A.J. (1991). Semantic modelling for improved cooperation in policy-making procedures. In: Proc. Int. Conf. on "Problems of support, survival and culture", Amsterdam April 2-5, 1991.

ANALYSIS OF KNOWLEDGE INVOLVED IN GREENHOUSE CLIMATE MANAGEMENT - APPLICATION TO THE DETERMINATION OF DAILY SETPOINTS FOR A TOMATO CROP

T. Boulard*, B. Jeannequin** and R. Martin-Clouaire***

*INRA, Station de Bioclimatologie, 84140 Montfavet, France
**INRA, Domaine Horticole d'Alenya, Le Mas blanc, 66200 Elne, France
***INRA, Laboratoire d'Intelligence Artificielle, B.P.27, 31326 Castanet-Tolosan, France

Abstract . In the framework of a project aiming at the design and implementation of a decision making system for daily management of a greenhouse tomato production, the fundamental heuristic knowledge used in the determination of climatic setpoints is identified and articulated. This analysis pertains to modern greenhouses located in Mediterranean France and to winter productions of tomatoes. Since no comprehensive quantitative models of the crop and greenhouse systems are yet sufficiently developed for automated decision making purpose the underlying motivation behind this work is to come out with an adequate body of heuristic knowledge that can then be represented and used within an artificial intelligence approach. A key feature and novel aspect of the presented decision model stems from its integration of short term (three days) weather predictions that are worth considering in the southern France context in order to reduce the stress caused to plants by abrupt and important changes of weather. The exploitation for climate management purpose of the property that tomato plants act as efficient temperature integrators over a several day period (i.e. plants have "reservoir" capabilities) is considered. Chiefly, by enabling anticipatory decisions the intelligent use of predictions together with "reservoir" capabilities can potentially contribute in a significant manner to energy saving.

Keywords . Automatic control ; artificial intelligence ; agriculture ; greenhouse climate control ; humidity control ; temperature control

INTRODUCTION

Greenhouse environment control and production management is a complex task. Automating or at least supporting the decision making process underlying it is becoming an increasingly important issue in order to master this complexity and make a more rational use of resources. Rough estimations concerning the French Mediterranean context in which we are interested showed that for the same agronomic objectives (in terms of quantity, quality and timing) up to 25% of fuel energy could be saved by a more careful and wiser management of the greenhouse climate. Motivated by this established finding some laboratories of our institute have set up an internal research project that aims at constructing a computer-based decision making system designed to tackle the decision problems that are beyond the scope of currently available greenhouse computer systems. The work reported in this conference is still in a preliminary stage. In order to state more precisely the kind of problem we want to address and how, let us briefly recall the greenhouse decision and the computer architecture frameworks in which our work takes place.

Greenhouse decision and computer framework

In agreement with the widely spread decomposition (Udink ten Cate and Challa, 1984; Baille, Boulard and Gary, 1990) of the the decisions that lie between strategy formulation and its imple-

mentation, we consider a decision process structured in a tree level hierarchy going at the bottom (level 1) from on-line climate control through (level 2) the tactical decision level concerning the determination of daily setpoints and to the higher (level 3) seasonal planning. Each level is associated with a particular goal and some decisions must be taken so that these objectives are satisfied. The decision making process at a particular level determines the objectives assigned to the next lower level except at the bottom one where decisions are directly transformed into actions. The first level encompasses a set of regulation algorithms that aims at controlling the important (and controllable) climate parameters. The automation of the short term decision process at this level is fairly well mastered (although some improvements are still possible) in modern, well-equipped greenhouses. The second level deals with deciding the appropriate daily settings depending on the actual status of the crop (growth stage, vigour), weather conditions and timing situation with respect to the specified overall planning. The involved decisions concern the environment setpoints of a particular day but the reasoning needed to reach these decisions must span a several day time scale. Only little research (Challa, 1989 ; Tantau, 1989) has been devoted to the issue of automating middle level decision making. Finally, given the general goal of maximizing the profit of the grower, the upper level, that is still to be explored, deals with the appropriate decomposition of the crop production season in growth stages and the determination of the corresponding mean inside temperatures according to constraints related to mean outside climate, crop growth and development, equipment, cost of production, tomatoes pricing conditions, etc...

Again hitherto only the first level is fully and satisfactorily automated. In practice this is realized thanks to the now classical distributed processing architecture in which real-time constraints on particular greenhouse compartments are handled by dedicated control computers that are connected to a central machine. The latter is typically of the PC type and is essentially used as an information system for collecting, storing and displaying environmental data. Beside its utilization for monitoring the important parameters of the different compartments, it can be used as the media to modify the settings of the control devices that are regulated by the dedicated control computers.

The higher the level in the hierarchy the less often the decisions must be taken and the less structured the decision types. Thus it is not surprising that the only level really automated is the first one. The lack of structure of a decision type means that either all steps of the decision cannot be specified before the decision is taken or the decision depends on qualitative (i.e. incomplete) knowledge or imprecise/uncertain data (e.g. predicted weather). Consequently, such decision types do not yield easily to formal analytical treatment by quantitative decision models.

Moreover no comprehensive dynamic simulation model of the behavior of the tomato crop system in a greenhouse environment has been sufficiently developed for decision making purpose. In the future, when adequate knowledge will be available for building such a reliable model we might be able to use simulation techniques to predict crop growth, development and yield from input data and weather (Challa H. and colleagues, 1988 ; Challa H., 1989). These simulation capabilities might then become the main tool within a decision support system for helping in the management of a greenhouse production. See Jones P., Jones J. and Hwang (1991) for a first attempt in this direction.

The addressed problem

The work reported here is an attempt to address the middle level decision process (daily setpoint determination) outside of the realm of purely quantitative approaches (e.g. optimization algorithms, simulation models) that, we feel, are not yet applicable. Since quantitative modeling of the whole crop and greenhouse systems is hitherto out of reach, we must rely on a decision model of an ideal expert grower that would be very knowledgeable and efficient in coping with the complexity of the reasoning task. Simply stated the decision problem consists in finding the values enabling the requirement of the plants to be satisfied at the lower possible cost (of energy) given the environmental conditions outside the greenhouse. This paper presents an analysis of the basic body of practical knowledge to take into account and how to use it wisely in the determination of daily climatic setpoints. The empirical knowledge we want to articulate pertains to modern greenhouses located in southern France and to winter crops (typically, tomatoes are planted in late November and are grown through June with harvest begin-

ning as early as end of January). Such information is hard to find (very disseminated and incomplete) in the specialized literature because much of it is context dependent (i.e. specific to the climate in the area) or is still informal due to uncertainties pervading many aspects of the subject.

The main motivation behind this work is to come out with an adequate body of heuristic knowledge that can then be represented and used within an artificial intelligence approach. A prototype of this approach, called SERRISTE (which means grower in French), has been implemented and is described in a companion paper (Cros and Martin-Clouaire, 1991). This software is designed to be run on the central PC machine of the greenhouse computing system.

A key feature and novel aspect of the underlying decision model stems from the flexibility and powerfulness provided by an intelligent treatment of historical data (the degree-days and radiation effectively received by the crop) and fairly reliable information pertaining to the future (two or three day weather predictions). Local area meteorological predictions are worth considering in the southern France context in order to reduce the stress caused to plants by abrupt and important changes of weather. In addition taking into account the property that tomato plants act as efficient temperature integrators over a several day period enables anticipatory decisions which can potentially contribute in a significant manner to energy saving. The rationale behind these "reservoir" capabilities of plants and the basic principles behind its profitable exploitation are addressed in the last part of the paper.

BASIC KNOWLEDGE INVOLVED IN THE DETERMINATION OF CLIMATIC SETPOINTS

As a preliminary step this paper focuses only the climate issue. Other important components of the production management problem such as carbon dioxide concentration and nutrition are not considered here although we agree that preferably they should not be treated independently of the others.

In the problem of climatic setpoint determination one has to take care of and integrate the following two classes of parameters describing :
- the situation outside the greenhouse, expressed through air temperature, humidity, solar radiation, direction and intensity of wind ;
- the inside situation that we want to manage and which may be evaluated through both quantitative measures of air temperature, soil temperature and the saturation deficit and a qualitative appraisal of some physical aspects of plants including especially the stage indicators and symptoms of diseases, wilting or too strong vigour.

The setpoints are another class of variables whose values specify the domains outside of which the control computers must command the use of a device such as the heating system (start or stop) or the roof windows (close or open). The setpoint class includes the air heating, soil heating and aeration setpoints. Some parameters of the above-mentioned second class are directly associated with setpoints (e.g. air temperature and the air heating setpoint or air temperature and the aeration setpoint) and some are not (e.g. saturation deficit, vigour).

<u>Main factors to care of</u>

The choice of setpoints must be done by a reasoning process integrating the above-mentioned parameters in an advantageous manner ensuring a profitable, though safe, combination of growth and development factors while keeping the energy spending within acceptable bounds and as low as possible.

Managing the production aims first at controlling the basic physiological functions such as photosynthesis, respiration, assimilation and transpiration that underlie the growth and development of plants. How the above functions individually depend on climatic conditions is rather well known though not sufficiently precisely for a global quantitative modeling (Jones and colleagues, 1989). Roughly one can consider that growth is essentially affected by the intensity and duration of solar radiation that provides the energy needed in the photosynthesis process whereas development is directly linked to the amount of heat (temperature) received over a period. Temperature influences also the rate of photosynthesis and thus the rate of growth.

The main parameters describing the inside climate (light, temperature, saturation deficit) interact with each other ; few interventions (heating, ventilating, shading) act preferentially on one particular parameter but modify also several others. For examples heating affects temperature but also the saturation deficit, ventilating affects both temperature and the saturation deficit and modify the carbon dioxide concentration (though we shall not consider explicitly this parameter in this paper). Each decision has complex repercussions that are sometime opposed in their more or less delayed effects on the crop.

In order to reach the most rational decision in the setpoint determination problem one must rely on the currently available understanding of the phenomena, their interactions and above all the adequate (from a theoretical or empirical point of view) decision attitudes in front of the possible classes of problems. Actually the main rules that must be fulfilled by the parameters in order to be part of an acceptable (i.e. rational according to the decision maker) solution are known, though imprecisely, by expert growers. For instance, the results of experimental works give us guidance on the most beneficial and forbidden light/temperature combinations in a particular period of the year for suitable growth and development of plants. Solving the determination problem is not easy however because several decision alternatives may have to be explored before a solution is reached. The generation and evaluation of these alternatives may be the source of a significant combinatorial complexity that growers may have hard time to cope with contrary to computers.

Besides growth and development factors care must be taken beforehand to prevent undesirable situations. This concerns in particular the incidence and development of diseases (mainly grey mold caused by *Botrytis Cinerea*) or infestations by parasites. Essentially the management decisions must ensure that conditions of high humidity (low saturation deficit to be more exact) are avoided. Again the rules governing such decisions are empirically available.

Another kind of undesirable situation is the lost of balance between reproductive and generative functions, characterized by particular aspects (size, shape, color) of different organs (leafs, stems, flowers, inflorescence) developing in the upper part of plants. Note that the interpretations of these aspects may be different depending on the development stage, climatic data of the close past and the time of their observation. The common symptom that the growers are able to perceive is a too strong vigour where an acceleration of vegetative growth is occurring while at the same time the flower production is decreasing. This vigour symptom may have been caused by fertilization or climate-related problems but can be controlled by proper climate settings. Simple rules, induced from the observed practices of experienced growers, tell that the appropriate reaction in case of a too strong vigour is to increase the mean (over the day) air temperature, decrease the soil temperature and lower the humidity (i.e. increase the saturation deficit). A too weak vigour is also unsuitable and can be corrected by the converse actions.

A specificity of the weather in the French Mediterranean area of France is that periods of beautiful days with intense solar radiation may alternate abruptly with dark and windy days. A sharp and sudden change of weather is harmful to the plants. It stresses them due to the inertia of some physiological functions such as water absorption. The stress phenomenon occurs when a long period of cloudy and damp days (during which plants have gotten used to a low transpiration activity) is followed by a period of dry sunny days. A wilting of the plants ensues and is noticeably visible at the moment when the solar radiation and saturation deficit are important. It may appear during several consecutive days while the plants are not used to the new climatic conditions. From a physiological point of view this wilting phenomenon corresponds to a demand of water from the environment (strong solar radiation and high saturation deficit) that exceeds the amount absorbed through the roots. The stomatic regulation causes then the stomata-closure which in turn impedes the photosynthesis process. A natural solution to this problem is to prepare the plants to the change of weather that can be foreseen using the commonly available local area forecasts. The preparation consists in increasing the soil temperature (to increase the water absorption through the roots) and increasing the saturation deficit (to increasing the water demand of the environment) the last day of the cloudy weather period. One could also reduce the detrimental effect of such a sudden change by increasing the soil temperature and decreasing the saturation deficit the first day of the sunny period.

Finally in deciding the control regime, especially the temperature regime, to be maintained in the greenhouse the decision maker must take into account not only the requirements of the plant but also the cost of providing heat. This economical factor depends on timing, yield of the production, expected level of market prices of tomatoes and price of energy. So far we have not consider the problem in its full generality where the above considerations fluctuate from one year to the other. Implicitly a standard situation is assumed and the solution corresponding to minimal spending in energy is preferred. We shall explain later on how this quest for optimization can benefit from an intelligent use of the local area weather predictions of the next three days.

Stages of growth and periods in a day

It is clear that the above factors are more or less relevant or important depending on the stage of growth of the plants. To each stage correspond specific rules used in the decision process. We are considering four main stages that practically correspond to uniform morphological objectives and monitoring management recommendations. Each stage is characterized by a sum of degree-days that the plants must receive within it. Table 1. describes the four stages in the case of the Capello variety.

Tomato plants are subject to life-cycles that demand special considerations. Therefore, we have decomposed the 24 hours of a day in four periods :
- from sunset to sundown during which light time physiological objectives (concerning photosynthesis, transpiration and respiration functions) must be satisfied ;
- the first part of the night where respiration and assimilation partitionship is taking place as a continuation of the photosynthesis process of the previous diurnal period ;
- the main part of dark night that requires heating although the plant activity is lessen (this is where energy saving can potentially be made by adequate management);
- the dawn where the plants have to be prepared to the next daylight period and where humidity has to be taken care of.

The starting point and duration of the above periods are changing through the season.

A table of suitable ranges of the inside parameters has to be given for each stage of growth in the interval of time that we consider (i.e. from planting time until late in the fourth stage when control is no longer possible due to high temperature and intense solar radiation) and for each of the four considered periods of a day.

STRUCTURE OF THE DECISION PROCESS AND USE OF RESERVOIR CAPABILITIES

Decision process

The management of the greenhouse climate by the growers comes down to finding the best compromise between the aforementioned factors underlying the different stages of growth. The decision process must absolutely ensure a correct handling of aspects related to important or irreversible damages that may be caused to the crop and to the formation and development of fruits: avoid the forbidden zones of the climatic parameters, take care of risks of diseases and crop infestations. Then, outside the possibility of occurrence of dramatic situations, the reasoning process must consider choices of setpoints that can maintain a good growth/development combi-

nation, an adequate photosynthesis/respiration balance, a normal vigour of plants and a fair balance of the auxiliary fauna. The choices must also take into account a possible sharp change of weather. Finally among the possible alternatives the decision process must find out the cheapest one in terms of cost of energy to be provided artificially.

The decision process starts from an overall objective expressed as a desired sum of degree-days over several days that chiefly depends on the solar radiation already received and forecasted for the next days. It goes through successive refinement of this objective by producing sub-objectives associated with smaller intervals of time. At the end of this time-based iteration the setpoints which are nothing but sub-objectives concerning each of the four elementary periods composing a day are reached.

The knowledge to take into account in the course of the decision process is directly represented as or transformed into a set of constraints on the involved inside parameters and setpoint variables. For instance the suitable growth/development combinations are expressed by mathematical functions that associates a solar radiation intensity (which is an imposed datum) with an interval of acceptable air temperatures. This association depends on the time in the season. The way of setting heating and aeration in order to obtain a desired mean air temperature and humidity during day time is another example of an empirical and approximate know-how that ultimately is expressed by a numerical constraint on the difference between aeration and heating setpoints. Actually the constraints can be seen as simple physiological or physical sub-models. Nothing prevents a priori the use of more sophisticated ones if needed.

The prevailing crop and weather conditions of the current day are translated into numerical constraints on the possible values of some of the variables to consider (including setpoints). For example, if the plants are too vigourous increase the mean daylight air temperature of about one or two degrees, decrease the soil temperature of about two degrees and increase the saturation deficit. Another example of such a transformation is given by the following rule : if damp weather is forecasted enforce a smaller difference between aeration and heating setpoints (that is the way of ensuring that aeration will occur frequently thus preventing high humidity that may contribute to disease incidence).

Once all the constraints corresponding and pertinent to the situation at hand have been gathered a reasoning process can try to find the values of the variables that are compatible with all the constraints. In the SERRISTE system (Cros and Martin-Clouaire, 1991) we are using specific constraint satisfaction techniques coming from the field of artificial intelligence. Within this system the determination of setpoints amounts to an intelligent search in a set of alternatives, which is conceptually at the very heart of artificial intelligence.

The reservoir phenomenon and its potential exploitation

Several researchers, including Cockshull, Hand and Langton, (1981) and de Koning (1988), have pointed out that growth and development of many crops respond to the mean air temperature received over a given period rather than to particular day and night temperatures. Thus all what is important is the integral of temperature over the considered period or, in other words, temperature can be compensated within an interval of time. The limits of temperature and time interval within which the compensation works are not completely known yet. This property of temperature integration has been investigated in detail on a 24 hour basis for a tomato crop (de Koning, 1988). Other experiments in which week long periods where considered have shown that kohlrabi (Liebig, 1988) and tomatoes (Hurd and Graves, 1984) still verify the compensation property. Even longer periods seem to be feasible (Heuvelink, 1988) for young tomatoes. De Koning (1989) has conducted experiments that lasted two month and a half, in which several days (up to 12) of low temperature were alternated with equal period of high temperature (with up to 6°C of amplitude). He concluded that these conditions where not extreme enough in the sense that integration was still observed.

A physiological explanation of the temperature integrating capability of tomato plants has been investigated by Gary (1988). He suggested that the temperature integration phenomenon is directly linked to the capacity of storage of assimilates in the plant. The produced assimilates can be stored for a while in the plant before they are transformed into fresh weight growth. Thus plants are acting as reservoirs (buffers) of assimilates, the use of which can be delayed. More specifically, the net amount of assimilates produced by photosynthesis (i.e. amount of carbohydrate synthesized minus what is consumed by the maintenance respiration) is not instantaneously balanced by the amount of assimilates that is translocated to different parts of the plant and becomes fresh weight growth. The conversion process may neither be concomitant nor operate at the same speed than the production. Indeed the conversion process depends on temperature whereas the photosynthesis process is predominantly determined by light and carbon dioxide. Thus a good balance between the demand (requirement of fresh weight growth) and the supply (produce of photosynthesis) is obtained only with particular combinations of temperature and light conditions. Outside of these special conditions when the supply exceeds the demand the assimilates may be stored somewhere in the plant and used later (when temperature is sufficiently high). Conversely a strong demand can be satisfied only up to the level of supply except if some reserve is available (from an anterior production).

So far it is not clear how long the delay between the production of carbohydrates and their use to contribute to growth can be. Gary (1988) has shown that it can be of at least two days for young tomato plants but did not carry out experiments with longer periods and older plants. Moreover nothing is known about the limits of the amount of assimilates that can be stored in a plant. Although it is tempting to draw a parallel between the temperature reservoir and the assimilate reservoir little is available about their relationship at the quantitative level. The possibility of extreme cases raises some other important questions. What happen when the supply largely exceeds the demand ? Could the photosynthesis process be slowed down ? If the demand widely exceeds the supply, could the plant try to take what it needs somewhere in its existing structure in order to feed the growth demand ? The above issues are in need of

further research in order to use safely and confidently the reservoir capabilities in the management reasoning process.

Potentially, the reservoir capabilities provides flexibility because the satisfaction of some of the requirements that are directly linked to temperature can be obtained reactively (afterwards), postponed or anticipated. A mistake in choosing temperature setpoints and the occurrence of unexpected events disturbing the management plans (e.g. breakdown of the heating system) or merely causing the effective (measured) temperature to deviate from the expected one, may have no serious impact on the crop since a compensation of their effects (if not too disastrous) is feasible in the not too far future. The possibility of correcting an undesired situation can be used in a less conservative approach (in terms of risk taking) of the climate management problem. However due to insufficient knowledge about the capacity and time of conservation that reservoirs might have one cannot use it optimally (disasters are still not excluded).

Postponing or anticipating the temperature demand can result in significant energy saving. For instance, if the current weather conditions are unfavorable and a nice nice day is forecasted tomorrow, it is possible (and wise) to let the mean temperature of the present day go below the average of the considered growth stage and take advantage of the free compensation that the solar energy can provide the next day. Conversely, beautiful days may be exploited to constitute some kind of reserve to be used in the coming days if bad weather is announced. This use of the reservoir capabilities relies on weather forecasts which are inevitably uncertain. So its proper exploitation is a matter of rational decision making under uncertainty. We have not addressed this issue yet.

CONCLUDING REMARKS

We have reported on part of the knowledge engineering work behind the problem of determining daily climatic stepoints for a tomato greenhouse production. The main aspects of the underlying cognitive approach are the following:
- we model the decision making process of an expert grower rather than try to use simulation models of the tomato crop and greenhouse systems since the latter are not available and do not directly provide a solution to the problem (an supervisor has to decide which simulation to do and what to conclude from it);
- it is flexible enough to cope with an evolving body of technology and know-how thanks to artificial intelligence representation techniques;
- we take into account qualitatively assessed parameters such as symptoms of diseases and vigour of the plants which relate directly to practical know-how;
- short term (three days) weather predictions are used, enabling to prepare plants to dramatic change of weather and giving opportunities to save energy;
- reservoir capabilities for energy saving and for performing a kind of feedback control on a slow response phenomenon are exploited.

The structuring of knowledge that we have done is a required step before transferring it to a decision making computer system. Moreover it is also worthy for another reason: it helps experts in identifying gaps in knowledge where further investigations are needed.

The work reported in this paper is still in a preliminary stage. In the future we plan to :
- carry on the implementation of the SERRISTE system;
- investigate the decision-making issues related to the use of reservoir capabilities with uncertain predictions;
- incorporate the management of carbon dioxide enrichment and nutrition;
- supplement the knowledge base and constraint satifaction algorithm of SERRISTE with a greenhouse model enabling the computation of the cheapest (in terms of energy) solutions among those proposed by the constraint solver;
- identify and incorporate the important knowledge to be used in problematic situations (such as when a particular devices used to modify the greenhouse climate has not worked properly);
- validate the approach by using it in a real context (real greenhouse);
- evaluate the robusteness of the approach and consider its applicability to other geographic contexts.

Acknowledgements
This work has been partially supported by the AFME agency (Agence Française pour la Maîtrise de l'Energie). We would like to thank our colleagues M.-J. Cros, M. Mermier and P. Reich that are also involved in this INRA project.

REFERENCES

Baille A., T. Boulard and C. Gary (1990). Les critères d'optimisation dans la gestion du climat et de la production sous serre. Compte Rendus de l'Académie d'Agriculture, 30 Mai, 25-36.

Challa H. (1989). Modelling for crop growth control. Acta Horticulturae: 248, 209-216.

Challa H., E.M. Nederhoff, G.P.A Bot and N.J. van de Brack (1988). Greenhouse climate control in the nineties. Acta Horticulturae: 230, 459-470.

Cockshull K.E., Hand and Langton (1981). The effect of day and night temperature of flower initiation and development in Chrysanthemum. Acta Horticulturae: 125, 101-110.

Cros M.-J. and R. Martin-Clouaire (1991). Determination of climatic setpoints by constraint satisfaction. IFAC/ISHS Workshop on Mathematical and Control Applications in Agriculture and Horticulture, Sept. 30 - Oct. 3, Matsuyama, Japan.

Gary C. (1988). Relation entre temperature, teneur en glucides et respiration de la plante entière chez la tomate en phase végétative. Agronomie : 8, 49-54.

Heuvelink E. (1988). Temperature integration and the growth of young tomato plants. CEC Workshop on Temperature Integration by Plants, May 25-27, Hannover.

Hurd R.G. and C.J. Graves (1984). The influence of different temperature patterns having the same integral on the earliness and yield of tomatoes. Acta Horticulturae : 148, 547-554.

Jones J.W., E. Dayan, H. van Keulen and H. Challa (1989). Modeling tomato growth for optimizing greenhouse temperatures and CO2 concentration. Acta Horticulturae : 248, 285-293.

Jones P., J.W. Jones and Y. Hwang (1991).Simulation for determining greenhouse setpoints. Transactions of the ASAE, Baltimore, 33(5), 1722-1728.

de Koning A.N.M. (1988). The effect of different day/night temperature regimes on growth development and yield of glasshouse tomatoes. Journal of Horticultural Science : 63(3), 465-471.

de Koning A.N.M. (1989). Long term temperature integration of tomato. Growth and development under alternating temperature regimes. Accepted in Sciencia Hortic.

Liebig H.P. (1988). Temperature integration by kohlrabi growth. Acta Horticulturae : 230, 373-380.

Tantau (1989). On line climate control. Acta Horticulturae : 248, 217-222.

Udink ten Cate A., H. Challa (1984). An optimal computer control of the crop growth system. Acta Horticulturae : 148, 267-276.

TABLE 1 Stages of Growth and corresponding objectives

Stage	Initial State	Final State	Duration	Morphological Objectives and Recommendations	Climate Management Objectives and Recommendations
1	planting	3rd bud flowering	3 weeks at most	. develop root system . maintain vegetative balance . help fruit setting	. maintain air/sol temperature balance . maintain solar radiation / air temperature balance . avoid low humidity in the first part of the stage
2	3rd bud flowering	6th bud flowering (2 weeks before first harvest)	4 to 5 weeks	. help fruit growth . maintain vegetative balance . prevent incidence of botrytis	. maintain solar radiation / air and soil temperature balance . be careful with high humidity at end of the stage
3	6th bud flowering (2 weeks before first harvest)	harvest of 2nd bud	4 weeks	. maintain vegetative balance . prevent development of botrytis . manage development with respect to photosynthesis performance	. maintain solar radiation / air and soil temperature balance . optimize the choice of mean air temperature with respect to available solar radiation . perform more systematic aeration and drying . prepare plants to abrupt changes of weather
4	harvest of 2nd bud	end of harvest	16 weeks or more	. help fruit growth . maintain vegetative balance . make sure water and nutrition are correctly supplied	. prevent too high air temperature . prevent diurnal low humidity . manage water and nutrition supply with respect to climate conditions

DETERMINATION OF CLIMATIC SETPOINTS BY CONSTRAINT SATISFACTION

M.-J. Cros and R. Martin-Clouaire

Laboratoire d'Intelligence Artificielle, Institut National de la Recherche Agronomique (INRA),
B.P.27, 31326 Castanet-Tolosan Cedex, France

Abstract . The problem of determining automatically daily climatic setpoints to be maintained in a greenhouse producing tomatoes is addressed with an artificial intelligence approach. This approach, named 'SERRISTE', was developed and tested through a preliminary implementation. Its main characteristic is that knowledge is represented by constraints relying on numerical variables and that the solutions are found by a constraint resolution mechanism. This mechanism takes advantage of a tree structure built by analyzing the set of constraints so that interrelations between variables can be dealt with efficiently. The algorithm that searches the solutions (sets of possible values for the variables) combines a technique for filtering candidate values with a process that controls assignments and backtrackings (needed in case of dead end and to seek all the solutions). Besides its generality, this approach has been designed to provide a resolution process understandable by users.

Keywords . Artificial intelligence; constraint satisfaction; humidity control; temperature control

1 INTRODUCTION

In the framework of computer-aided management of agricultural production systems (Cros and Martin-Clouaire, 1990), we address the problem of determining automatically daily climatic setpoints to be maintained in a well-equipped greenhouse. For any day, four different periods associated with important physiological phenomena to be controlled are considered. We want to obtain, for each period, the "best" values of climatic parameters such as temperature and humidity for producing early tomatoes (i.e. tomatoes are planted in late November and grown through June with harvest beginning as early as late January). Finding appropriate setpoints can only be done by integrating several interdependent factors, the environmental status (e.g. outside temperature and solar radiation) and several pre-specified objectives (Boulard and others, 1991). Therefore, the problem cannot be solved by merely using a database containing situation-setpoint relations; one has to resort to a reasoning process which must be in agreement with some kind of "logic" of the grower (i.e. his current practices) and that relies on a body of agronomic and thermodynamic knowledge (physiology of the producing plant and behavior of the greenhouse depending on its physical characteristics). A successful production in terms of yield, quality and cost requires both reactive and anticipatory attitudes in the management process. For this reason, an essential characteristic of our approach is that it operates on data relative to the past (e.g. measured solar radiation) and the future (e.g. weather forecasts).

The complexity and diversity of the involved phenomena make crop management difficult. Moreover, the fast paces at which greenhouse technology, practical know-how and the range of plant variety evolve do not let the grower acquire a robust know-how learned by experience. Thus automating the choices of climatic setpoints (or providing a decision support for these choices) by relying on expert knowledge may result in lower production costs, higher yields, better or earlier produce. For example, rather than maintaining a standard temperature during the nocturnal periods (dawn, dark and dusk) as most growers do, it is possible, without affecting the production process, to reduce heating costs significantly by adapting the setpoints to the solar radiation of the previous diurnal period. By allowing a better approach to situations and their consequences, expert knowledge should take advantage of predictions reducing energy costs. Moreover, a rigorous management of interactive parameters such as humidity and temperature greatly contributes to better prevent diseases (which otherwise cause expensive interventions). At last, automation will avoid harmful effects of management discontinuities that are encountered when several decision makers are involved in the choices of setpoints.

From a computer science point of view, solving such a problem requires a representation and a logical organization of knowledge appropriate to the reasoning task at hand. We have translated the problem of the determination of setpoints into what is known in artificial intelligence as a constraint satisfaction problem (CSP). In this framework, knowledge is expressed through algebraic constraints restricting the possible values of numerical variables. The reasoning process is carried out by a mechanism that searches all the values (of the variables) that are compatible with all the constraints.

The paper describes the conceptual scheme, called SERRISTE (which means greenhouse grower in French), and the main features of its preliminary implementation. Section 2 outlines the artificial intelligence approach underlying SERRISTE. Section 3 shows briefly the kind of knowledge involved and how it is represented. Section 4 presents the main techniques that are used in the resolution mechanism. Section 5 gives some concluding remarks concerning the advantages of our approach and our plans for future developments of the project.

2 AN ARTIFICIAL INTELLIGENCE APPROACH

The problem of determining climatic setpoints is seen as an allocation task in which a set of variables must be assigned values satisfying simultaneously a set of constraints that represent pieces of knowledge to be taken into account. The constraints specify directly or indirectly the allowed (or forbidden) values of the concerned variables. The constraint satisfaction techniques developed in artificial intelligence (Kanal and Kumar, 1988; Mackworth, 1987) address problem-solving by building a constraint network in which nodes are variables having discrete domains and convergent arcs express n-ary constraints (i.e. any constraint involving n variables).

The resolution process consists in choosing a not yet instantiated variable and assigning it a value taken in its domain such that all constraints bearing on this variable are satisfied. In doing so, one may reach a dead end in which no value can be given to the variable under consideration. A backtracking is then necessary; one must withdraw the instantiation of a previously considered variable and assign a new value to it. When all variables have been instantiated: a solution is then found. However, there may be zero or several solutions. In the latter case, resolution may be carried on until all solutions are found. This basic principle of search in a set of alternatives is conceptually at the very heart of artificial intelligence. An original use of it is made in our software SERRISTE.

The essential goal of a resolution mechanism is to make the best possible choices in the following two situations:
- selection of the next variable to be instantiated,
- determination of the value to be assigned to the selected variable.

In the SERRISTE system, we propose two specific techniques to improve these choices taking into account the next two criteria:
- closeness to the grower behavior because this helps validate the reasoning process,
- efficiency of the resolution.

The selection of variables relies on the use of a specific tree structure that is constructed beforehand and that tells anytime which variables to consider given the last instantiated. This tree is produced by a careful analysis of constraint-induced dependencies between variables. Each node of this tree is a cluster of strongly interdependent variables.

Usually, before achieving an instantiation, nothing tells how to identify the values that will lead to a global solution. At most one can use a filtering technique to reduce, for each variable, the number of alternatives to be explored. Such a technique aims at eliminating impossible alternatives (i.e. those leading to inconsistencies). The immediate effect of applying the filtering technique on the variables of a cluster is to reduce the domains (i.e. the set of candidate values) of these, or some of these variables. Reducing the domains of a particular set of variables may result in that some of the previously possible values (i.e. that are assumed to be so) of other variables become impossible. The filtering technique propagates automatically this effect through all the concerned variables in the constraint network.

Finally when several sets of possible values are found by the resolution mechanism (usually around ten, but up to one hundred in some cases), SERRISTE has to perform a sort of tradeoff reasoning by integrating preferences (e.g. attitude with respect to disease occurrence, energy saving strategy) that are not expressed by constraints of the above-mentioned type. This ultimate phase of problem solving that aims at returning the "best" solution is not addressed in this article. We shall only deal with the issue of determining all the collections of setpoints that are compatibles with the constraints.

3 KNOWLEDGE REPRESENTATION

The first step in applying the constraint satisfaction approach to setpoint determination consists in identifying the variables involved, their associated domains and the relations (i.e. constraints) between these variables.

So far, all the variables involved in the SERRISTE system are numerical and most of them concern temperatures. For illustrative purposes, only temperature variables are used. The set of variables includes those associated with setpoints plus some additional ones, called intermediary variables, that are needed for the reasoning process and for constraint expression. All these variables have continuous domains that are nevertheless treated as if they were discrete for two practical reasons, ensuring efficiency and reasoning at the appropriate level of precision. It is assumed that the domain of each variable is known and given in the form of an interval which is interpreted by the resolution mechanism as a discrete and finite set of values. These domains depend on variety of tomato and environmental context. For instance, the domain of ITP_3 (average night temperature) for a Rondello variety in the phase preceding the flowering of the first bunch is the discrete interval [11, 18] such that two consecutive values differ by 0.5°C.

Formally, constraints are relations between variables. Since the variables are numerical, it is natural that most constraints are algebraic. Thus, the types of constraint include:
- unary predicates (e.g. in the early stage of development, the average diurnal temperature ITP_1 must belong to [14, 28]),
- linear equations (e.g. the average temperatures at dawn, night and dusk (ITP_2, ITP_3, ITP_4 respectively) must satisfy

the relation $3.\text{ITP}_2 + 8.\text{ITP}_3 + 2.\text{ITP}_4 = 14.\text{ITN}$; where ITN is the average temperature of nocturnal time),
- inequalities (e.g. if the foreseen solar radiation is high, one must have $2 \leq \text{ITD} - \text{ITN} \leq 6$ where ITD is the average temperature of diurnal time).

A convenient way to visualize the dependencies between the variables is to build a constraint network incorporating all the variables and constraints involved. Figure 1 is a simplified example of such a network. An edge between two variables means that a binary constraint connects these variables. Several edges converging in one point represent an n-ary constraint on the variables located at the extremities of the edges.

4 RESOLUTION PROCESS

As pointed out in Section 2, the problem of setpoint determination is an allocation task in which numerical values must be assigned to a set of variables that are connected to each other through constraints. For the efficiency and the intelligibility (i.e. understandability) of the resolution process a preliminary processing consists in building a structure that will provide an appropriate clustering of the variables and an order for exploring the constructed clusters. The clustering techniques are presented in the subsection below. The overall resolution process combines filtering technique (that aims at eliminating impossible values) with a search algorithm that keeps trying instantiations and withdrawing (backtracking) them when a failure is detected. The filtering and search algorithms are described in the last two subsections (for more details, see (Cros and Martin-Clouaire, 1991).

<u>Clustering of variables</u>
The idea underlying this preprocessing phase is to exploit the structural properties of the constraint network in order to get together variables strongly interdependent. Intuitively, it is wiser to try a simultaneous instantiation of such a group of variables because it is an effective strategy that avoids many backtrackings, and it brings a focussing effect that eases the task of a user who wants to follow the resolution process. The clusters of variables are organized in a tree that can be explored in a systematic way, providing an appropriate order for the instantiation of the variables. Clustering ideas have been investigated by others in the artificial intelligence community. See, for instance, (Dechter and Pearl, 1989) for a different approach.

Let us first define some basic notions used in the following. Two constraints are connected if a particular variable is involved in both (e.g. $C_1(V_1, V_2)$ and $C_2(V_2, V_3, V_4)$ are connected because they share the variable V_2). Two variables are directly connected if they are involved in the same constraint. More generally, two variables X and Y are indirectly connected if there exits a chain of connected constraints such that the first one concerns X and the last one concerns Y (e.g. with $C_1(V_1, V_2)$, $C_2(V_2, V_3, V_4)$ and $C_3(V_3, V_5)$, V_1 is indirectly connected to V_5).

Clusters of variables are achieved and structured into a tree by the iteratively running a procedure that examines exhaustively

the constraint network starting from a given set of variables defined as the root cluster. This procedure, when applied to a cluster of variables, produces the clusters that immediately descend from it (i.e. its sons). Initially, a set of variables playing a particular role is defined deliberately as the root cluster. The above-mentioned procedure is applied to this root cluster and is then used repeatedly for each newly created cluster. Without going into details (see (Cros and Martin-Clouaire, 1991) for the full algorithm), let us describe the basic principles behind the construction of sibling clusters descending from a given cluster G.

Let $\text{Cp}_1,...,\text{Cp}_r$ be the constraints in which at least one variable of G is involved and that have not already been considered by the clustering procedure. Let us suppose that C_j is one of these constraints. $\sigma_j(G, C_j)$ denotes the set of variables concerned by C_j but not contained in G. This is called the development of G according to C_j. The construction of new clusters is obtained by merging the developments associated with $\text{Cp}_1,...,\text{Cp}_r$, if one of the three criteria given below is satisfied, until merging between developments or newly merged developments is no longer possible. Two developments or merged developments σ_1 and σ_2 are merged if they satisfy one of the three criteria:

- σ_1 and σ_2 have a non-empty intersection,
- a variable of σ_1 is directly connected to a variable of σ_2 through a constraint that has not been considered yet (i.e. that has never been involved in the development of a group),
- a variable of σ_1 is indirectly connected to a variable of σ_2 through constraints that have not been considered yet .

In our simple example illustrated in Figure 1, the root cluster which is the starting point of the instantiation process includes only the variable IT_1 that stands for the average temperature desired over a 24 hour period and has only one constraint involving IT_1. The development of the root cluster according to this constraint is the set {ITD ITN}. Since no other development has to be taken into account, this set is the son cluster of the root. Two constraints involving ITD and ITN have to be considered. The developments of {ITD ITN} according to these constraints are {ITP_1} and {ITP_2 ITP_3 ITP_4}. As these two development do not satisfy any of the merging criteria, they give birth to two clusters that in turn are considered in a similar fashion. In this simple example however, no merging actually takes place. Ultimately, this construction process provides a tree of clusters.

Let us assume that a particular systematic procedure for exhaustive exploration of a tree structure is given. Then, the tree of clusters may be used as it stands for the purpose of guiding the instantiation process. However, efficiency and user friendliness of the resolution process may be improved by arranging the order of the sons of each cluster. As far as efficiency is concerned, it is preferable to consider first the clusters containing variables that are difficult to instantiate. In addition, a particular arrangement may turn out to be closer to the grower's pattern of reasoning. These late arrangements are not automated because they are user-dependent). The network in

Figure 1 enables the construction of the tree structure shown in Figure 2. The numbers near the groups of variables indicate the order in which the clusters are explored. Technically, this order corresponds to a left to right depth first search. In other words, given an instantiation of some intermediary variables, this strategy is intended to find as soon as possible if there exists a possible instantiation for the setpoint variables which are contained in terminal or leaf clusters.

Filtering

This process aims at reducing the set of candidate values. In SERRISTE, a so-called Waltz (1975) filtering is used. A general study of this general technique which has been employed in several artificial intelligence applications was made by Davis (1987). Given a set of variables subjected to some constraints, the filtering detects and eliminates the values that cannot take part in a solution. It relies on an intelligent search process that is iterated through the constraints and the variables involved in these constraints.

Let V_1, ..., V_n be the variables involved in constraint C and A_i the possible values for variable V_i. The Waltz filtering is based on an essential operation, embedded in the procedure Reduce_domain which when applied to the constraint $C(V_1, ..., V_i, ..., V_n)$ and the variable V_i, returns the subset of A_i obtained by removing all values incompatible with all conceivable sets of values that might be assigned to the other variables. A formal definition of this operation is:

Reduce_domain $(C, V_i) = \{a_i \in A_i \ / \exists \ a_j \in A_j, \ j=1,...,n \ ,$
$j \neq i \ C(a_1,...,a_i,...,a_n)\}$

where $C(a_1,...,a_i,...,a_n)$ means that constraint C is satisfied by the value a_j for the variable V_j, $j = 1, ..., n$.

For continuous variables, the programming of this procedure depends on the nature of the constraint concerned. If C allows to be expressed each variable as a function of the other variables (this is the case for linear constraints) and if the A_is are intervals, the Reduce_domain operation can be performed by a simple computation on intervals (calculus on the bounds). For example, given the constraint $V_1 + V_2 = 10$, the initial domains $A_1 = [0 , 2]$ and $A_2 = [5 , 9]$. The new domains A'_1, A'_2 are obtained as follows: $A'_1 = A_1 \cap [10 - 9, 10 - 5] = [1, 2]$ and $A'_2 = A_2 \cap [10 - 2, 10 - 0] = [8 , 9]$. Note that the result would not be improved by reusing Reduce_domain with A'_1 and A'_2 instead of A_1 and A_2 (i.e. the returned sets would not be smaller).

Such a filtering takes advantage of the linearity of the constraints and amounts to treat the variables as continuous. We shall see that the general search algorithm presented in the next subsection works with discrete variables. The domain discretization that is necessary to carry on the resolution is done after the filtering has been performed. It is a particularity of our approach to use the algebraic nature of constraints and variables to switch conveniently from continuous to discrete domains and vice versa. So, rather than an expensive filtering on discrete universes, SERRISTE performs a much simpler filtering using interval calculus on continuous universes.

In the Waltz algorithm, the Reduce_domain operation is used in the procedure Revise which, for a given constraint C, detects among its associated variables those with a domain that can be reduced and eliminates the corresponding values. The Waltz algorithm applies to a set of constraints that one is willing to take into account. Each of these constraints are considered in turn by the Revise procedure. Each time some values can logically be removed from the domain of a particular variable involved in the constraint under consideration, the algorithm removes these values. The other constraints involving variables with a modified domain are then pushed in the queue of constraints to be considered by the Waltz filtering. In this way, the consequences of reducing a domain are propagated and induce modification in the domains (i.e. of other variables) through the appropriate constraints. The propagation of domain reduction is carried on until there is no more constraint to be considered (natural quiescence) or until a termination condition applies.

Termination criteria (such as the number of times any constraint may be considered) are needed because, in some cases, the Waltz filtering algorithm may go into infinite loops or reach quiescence only after a long time. Even in these cases however, filtering is worth doing since it reduces the number of failures in the instantiation process. The Waltz algorithm is sound, that is, it cannot eliminate good values (it cannot cause a solution to be missed), but incomplete (except with simple constraints) since the Cartesian product of reduced domains still contains elements that are not solutions.

Search algorithm

The resolution process considers one cluster after the other and, for each of them, determines the possible instantiations of its variables. This process uses a search algorithm which continuously aims at completing a partial solution (i.e. the current assignment of the variables contained in the clusters already considered) by trying to instantiate the variables of the next cluster. The clusters are considered in the order specified by the tree structure built for this purpose.

More formally, let p be the number of clusters, $V_{g,j}$ the j-th variable of the g-th cluster and $a_{g,j}$ a feasible instantiation of variable $V_{g,j}$. The addressed problem can be expressed in an abstract way as the task of finding all the possible sequences $(a_{1,1},...,a_{1,j1} ...., a_{p,1},...,a_{p,jp})$ of assignments such that the property $P_p(a_{1,1},...,a_{1,j1} ...., a_{p,1},...,a_{p,jp})$ is verified (j_g denotes the number of variables in group g). The property P_p holds if all resolution mechanism expands incrementally (following the order specified in the tree structure) a series of instantiation in such a way that, at each step, a property P_k bearing on the k first groups holds. More specifically, $P_k(a_{1,1},...,a_{1,j1} ..... a_{k,1},...,a_{k,jk})$ is verified if all the constraints concerning exclusively variables taken in the k first groups are satisfied with the sequence $(a_{1,1},...,a_{1,j1} ...., a_{k,1},...,a_{k,jk})$ of instantiation. In other words, for the cluster under consideration at a given point in time, the algorithm aims at instantiating its variables so preserve compatibility with the other (already made) instantiations. When this is impossible (dead end), the algorithm backtracks and reconsiders previously made instantiations. This algorithm is given below.

Search algorithm for all the solutions

1. INITIALIZATION

   $k := 1$

2. COMPUTING OF THE FEASIBLE INSTANTIATIONS I_k OF THE VARIABLES OF CLUSTER k

   $I_k := A_{k,1} \times ... \times A_{k,jk}$ where $A_{k,j}$ is the set of possible instantiations *(after using Waltz filtering)* for the j-th variable of the k-th cluster taking into account the already made instantiations *(i.e. the ones of groups 1 to k-1)*

3. ARE THERE OTHER POSSIBLE INSTANTIATIONS FOR THE VARIABLES OF GROUP k ?

   if $I_k = \emptyset$, then go to step 7

4. CHOOSING OF INSTANTIATIONS FOR THE VARIABLES OF GROUP k

   Choose in and remove from I_k one element.

   Let us denote it by $(a_{k,1},...,a_{k,jk})$

5. CONSIDER THE NEXT CLUSTER IF THE LAST CHOICE IS OK

   if $Pk(a_{1,1},...,a_{1,j1},...,a_{k,1},...,a_{k,jk})$ holds, then $k := k+1$

6. IS A SOLUTION FOUND ?

   if $k > p$, then store $(a_{1,1},...,a_{1,j1},...,a_{p,1},...,a_{p,jp})$

   (this is a solution to the problem)

   else go to step 2

7. BACKTRACKING

   $k := k-1$

8. END OR CONTINUATION ?

   if $k < 1$, then stop *(no other solution)*

   else *(other possible instantiations)* go to step 3

In step 2, the I_k set of possible instantiations is a result of the Waltz filtering that returns a Cartesian product of intervals. These intervals are used in a discrete form. To save computer memory, no element of the Cartesian product is generated before it is needed. In some cases where there are as many variables as there are constraints, the Waltz filtering is replaced by an exact Kramer resolution.

The backtracking step occurs in the case of a dead end and also after a solution has been found, in order to carry on the search for other possible solutions.

From an artificial intelligence point of view, the salient features of our approach are the processing of variables by clusters rather than individually and the incorporation of a filtering operation before any instantiation of the variables contained in a cluster. We also used advantageously the fact that the domains of the variables may considered as continuous for filtering purposes and discrete in the final stage of search.

5 CONCLUDING REMARKS

Basically, SERRISTE solves an assignment problem through a general artificial intelligence approach. Although so far the constraints have chiefly been linear, we did not use mathematical programming techniques for the following reasons:
- lack of a clear objective function,
- desire to produce an understandable resolution for users,
- wish to keep open the possibility of incorporating heuristic choices if necessary.

Beyond the computer sciences aspects, the formalization of knowledge and know-how has brought valuable benefits. Experts make progress in the field because they have to think about their practices, clarify what is implicit and uncover gaps in knowledge. Another significant by-product of this work is that it will contribute to facilitate knowledge transfer.

Presently, the first version of SERRISTE has been implemented on a PC computer with GURU which is a commercial software including an expert system shell, a procedural command language, a database management and spreadsheet software. However, the tree clustering of variables has been implemented separately in Lisp.

So far only temperature setpoints have been determined by SERRISTE. The system returns all the solutions (if asked to do so) and provides a step by step explanation of the resolution process; thus making the output easily understandable by the user.

Although not fully satisfactory in its present implementation (due to some limitations of the underlying software), this prototype of SERRISTE has proved the suitability of this approach due in particular to its intelligible resolution: solutions as well as the paths toward them can be evaluated.

Further research will aim to:
- expand the knowledge base incorporating new variables and constraints and validate it,
- rewrite the program in a more appropriate representation language,
- design and implement a module for selecting among the collection of setpoints the one best satisfying the user preferences,
- experiment the system under real conditions in order to evaluate the significance and robustness of our automated management approach to greenhouse climate.

Acknowledgements

This work has been partially supported by the AFME agency (Agence Française pour la Maîtrise de l'Energie). It has been achieved in the framework of an INRA project involving colleagues from Bioclimatology (T. Boulard, M. Mermier, P. Reich) and Agrarian Systems and Development (B. Jeannequin, J. Lagier) Departments.

REFERENCES

Boulard, T., B. Jeannequin and R. Martin-Clouaire (1991). Analysis of knowledge involved in greenhouse climate management. IFAC Workshop on Mathematical and Control Applications in Agriculture and Horticulture, 30 Sep. - 3 Oct., Matsuyama, Japan .

Cros, M.-J. and R. Martin-Clouaire (1990). Management of agricultural production systems. Proc. of the Third International Congress for Computer Technology: Integrated Decision Support Systems in Agriculture, May 27-30, Frankfurt a. M., Germany.

Cros, M.-J. and R. Martin-Clouaire (1991). Détermination de consignes par satisfaction de contraintes. <u>Conférence générale apllications</u>, EC2, Avignon, France.

Davis E. (1987). Constraint propagation with interval labels. <u>Artificial Intelligence</u> : 34(1), 1-38.

Dechter R. and J. Pearl (1989). Tree clustering for constraint networks. <u>Artificial Intelligence</u> : 38(3), 353-366.

Kanal, L. and V. Kumar (editors) (1988). Search in Artificial Intelligence. <u>Springer-Verlag</u>, New-York.

Mackworth A.K. (1987). Constraint satisfaction. In the <u>Encyclopedia of Artificial Intelligence</u>, Wiley-Interscience Publication (Shapiro S.C. Ed.), pp. 205-211.

Waltz D. (1975). Understanding line drawing of scenes with shadows. In the <u>Psychology of Computer Vision</u> (P.H. Winston Ed.), McGraw-Hill, New-York, pp. 19-91.

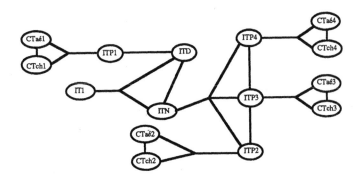

Fig 1 . Network of constraints

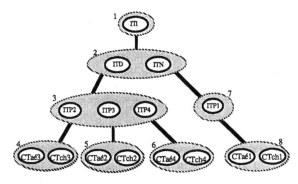

Fig 2 . Tree structure of arranged clusters

AN OBJECT-ORIENTED ENVIRONMENTAL CONTROL MODEL IN PROTECTED CULTIVATION

T. Hoshi

*Dept. of Biological Science and Technology, School of High-Technology for Human Welfare,
Tokai University, Numazu City, Shizuoka, Japan*

Abstract. A prototype computer model based on object-oriented approaches for
environmental control in protected cultivation was discussed. The model
handled to convert the control setpoints having generic form into the real
manipulations of specific control instruments for measurement and control.
The requests for environmental control were realized to send a message to
specific object, and the object started to manipulate the control. The model
did not need to take notice of any instruments specification, because each
instrument specification was controlled by the specific object. A prototype
system based on the concept was developed and tested. From the result of
testing, the object-oriented approaches provided the prototype system with
having flexibility, having reliability and saving development costs.

Keywords. Artificial intelligence; Computer control; Hierarchically
intelligent control; Object-oriented model; Plant production.

INTRODUCTION

According to rapidly evolution of protected
cultivation technology, various environmental
control instruments, for example, heat pumps,
heat storage systems with phase changing
materials, plant moving robots, hydroponics,
intelligent sensors and so forth, have been
developed and introduced. Also, conventional
environmental control instruments, for
example, roof ventilators, oil heaters,
irrigators, thermistors and so force, have
been still used. Computers for controlling
plant growth environment in protected
cultivation must manipulate the various
control instruments to use universally. In
addition, Growers require superior and
exclusive control ability of the computers.
That is mainly why software of the computers
becomes complicated and huge, and programming
costs of the software become high.

Udink ten Cate and co-workers(1978) indicated
that the computer control of plant growth
environment was divided among three levels:
1) the realization of desired environment
factor; 2) the determination of the desired
values of these environment factor; and 3) the
strategy for an optimal yield. Hoshi and
Kozai(1984) proposed to use the knowledge-
based and hierarchically distributed on-line
control system correspond with the three
levels, namely, 1) subprocessors, 2)
hostprocessors, and 3) coreprocessors.
Environmental control setpoints represent
every the desired environment factor in the
computers. The final aim at subprocessors is a
realization of the setpoints by manipulating
the control instruments. The setpoints have
generic form, but the real manipulations
change remarkably dependent on the assortment
of the control instruments at each protected
cultivation equipment.

It is important to develop a universal
conversion model, which handles to convert the
control setpoints having generic form into the
real manipulation of the specific control
instruments, because the model is a basic
element in development of the environmental
control software development of the
environmental controller. The objective of
this research is to propose a prototype
conversion model based on object-oriented
approaches.

OBJECT-ORIENTED MODEL AND PROTOTYPE SYSTEM

Features of Object-oriented Approaches

Procedures and functions are core of software
in the almost all traditional programming
environment. We apply data to the procedures
and/or functions for getting results. Object-
oriented approaches are fundamentally
different from traditional one. Objects
encapsulate both private data and methods of
processing the data. Only messages are a
request for an object to perform one of its
methods and to obtain results of the private
data processing.

Object-oriented approaches has following
features: 1) It allows to handle as if objects
are parts for universal use, because it has
high reusability with overloading and
genericity (Meyer,1987); 2) Differential
programming is possible, because a new object
inherits data type and its methods from the
other objects; 3) parallel processing with
every objects is possible. Moreover, Booch
(1986) notes, "an object-oriented approach
leads to improved maintainability and
understandability of systems whose complexity
exceeds the intellectual capacity of a single
developer or a team of developers."

Model Design Criteria

Object-oriented approaches has been already
used for control software development.
Gauthier and Guay(1990) developed a prototype

system for a greenhouse climate control. They concluded that the object-oriented model had shown great potential for the type of application considered in greenhouse climate control. On top of their results, the primary goal in this research is to obtain the generic and reliable conversion model through using object-oriented approaches.

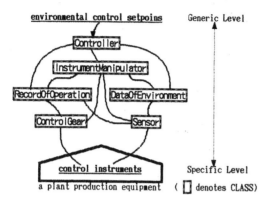

Fig. 1. The Concept of Object-oriented Environmental Control Model in Protected Cultivation.

Figure 1 shows the message network from generic control setpoints to specific control instruments using the object-oriented approaches in protected cultivation. A classified objects is defined as the class. The class ControlGear represents properties, abilities, conditions, real manipulations and running costs of the control instruments. The class Sensor represents properties, abilities, conditions, measurement compensations and substitution plans of the detective control instruments. The classes RecordOfOperation and DataOfEnvironment are database of the classes ControlGear and Sensor. The class of InstrumentManipulator represents available control instruments, and alternative control plans. The class Controller represents control patterns.

Implementation of Prototype System

A prototype System was done in the Smalltalk/V (Digitalk Inc.,1986), that was an object-oriented programming environment, and MS-DOS ver.3.10 disk operating system(Microsoft Corp.). The hardware environment was PC-9801 desktop micro-computer system(NEC Corp.) using i80286 micro-processing unit and c80287XL mathematic co-processor(Intel Corp.).

The real object were defined as an instance in the world of Smalltalk. In the prototype system, the objects for environmental control were represented as sub-classes or instances in the seven classes of Fig. 1.

RESULTS AND DISCUSSION

The prototype system was tested in a simple greenhouse heating and cooling simulation model. Desired environmental control setpoints were given the system from keyboard as the smalltalk message. Test of the prototype system could be done successfully by the inside air temperature controlled.

The objects of all the control instruments which were installed in the plant production equipment were made in the system. When a new objects installed, the system adapted to new assortment of the instruments only making 2 or 3 new objects. In the case of making a new object, the bulk of input was little, because the new object of the object-oriented approach inherited the common specifications from belonging to the object (i.e. its super-classes). Hence, the system was adapted to many types of the plant production equipments with little modification. Therefore, the object-oriented approaches provided the prototype system with having flexibility and saving development costs.

The specification of each instrument was managed by the each object. The system did not need to take notice of any instruments specification. The object of the class InstrumentManipulator searched the most effective object of the class ControlGear, and sent the message of control or measurement request. The object of the class ControlGear or Sensor changed the received message into the specific manipulation of the instrument.

If any instruments were broken, the object of the class Sensor made up a phantom using the other objects, or the object of the class InstrumentManipulator searched the next best object of the class ControlGear.

In the test, when the inside air temperature sensor was removed, the system was able to maintain inside air temperature. The system predicted the inside air temperature changes using the just before operation of air conditioning and outside air temperature, and the system executed rough control the next step operation of air conditioning using the predicted inside air temperature. The result indicated that the prototype system had high reliability.

A trial for applying a full system to control real greenhouse environment has just been started. Unfortunately, Smalltalk spends much working memory and is hard to use a peripheral I/O port, because Smalltalk is a programming environment suitable for experimental prototyping (Diederich and Milton, 1987). Additionally, the multiple inheritance is not available. At the trial, I need to consider the utilization of a new programming environment, which is adding object-oriented concepts on top of conventional languages, for example, Flavors, Objective-C, C++ and Ada (Cox, 1984; Gorlen, 1987).

CONCLUSION

An environmental control model was discussed, and a prototype system was developed and tested. From the testing described above, we may conclude that the object-oriented approaches provided the prototype system using the model with having flexibility, having reliability and saving development costs.

REFERENCES

Booch, G. (1986). Object-Oriented Development. IEEE Trans. Software and Eng., SE-12, 211-221.
Cox, B.J. (1984). Message/Object Programming: An Evolutionary Change in Programming Technology. IEEE Software, Jan., 50-61.
Diederich, J. and J. Milton (1987). Experimen-

tal Prototyping in Smalltalk. _IEEE Software_, _May_, 50-64.

Digitalk Inc. (1986). _Smalltalk/V Tutorial and Programming Handbook._ California.

Gauthier, L. and R. Guay (1990). An Object-Oriented Design for a Greenhouse Climate Control System. _Trans. ASAE_, _33_, 999-1004.

Gorlen, K.E. (1987). An Object-Oriented Class Library for C++ Program, _Software-Practice and Experience_, _17_, 899-922.

Hoshi, T. and T. Kozai (1984). Knowledge-Based and Hierarchically Distributed Online Control System for Greenhouse Management. _Acta Hort._, _148_, 301-308.

Meyer, B. (1988). _Object-Oriented Software Construction._ Japanese Edition, ASCII Corp., Tokyo.

Udink ten Cate, A.J., G.P.A. Bot, and J.J. van Dixhoorn (1978). Computer Control of Greenhouse Climates. _Acta Hort._, _87_, 265-272.

DEVELOPMENT OF A REAL COMPUTER INTEGRATED CULTIVATION SUPPORT SYSTEM

Y. Nakanishi, T. Uchida and T. Sakano

Shikoku Research Institute Inc., Takamatsu 761-01, Japan

Abstract. It is no easy matter to use computer systems in agriculture. To develop a really useful computer system requires to design it with a deep insight as to what computer system should be ideal in agriculture. The system discussed in this article is what we think is an ideal environmental control computer system for horticulture.

Keywords. Environment control; Microcomputer based control; Expert systems; Neural nets; Man-machine systems.

INTRODUCTION

The recent advance in computer technology is surprising enough and everybody may think of applying it to agriculture in general and horticulture in particular. But the fact that the use of computer-based environmental control systems developed so far is extremely limited (in Japan) shows that such application is no easy matter.

Growing a plant requires not only deep knowledge of the plant concerned but the ability to observe its growth continuously and adjust its environment dynamically based on the results of the observation.

Without consideration to these things, computer-based automated environmental control would seemingly reduce the burden of the cultivator but would in practice make it difficult for the cultivator to control the environment based on his or her own judgement.

If a computer is to be used in agriculture, it shuld be incorporated in the system comprising the cultivator, cultivation facility and plants in a well balanced manner such that the cultivator is not required to be conscious about its use. This is also an objective for our horticulture computer system development project now under way.

This article discusses the total design of our environmental control computer system for horticulture, ESPRIT (Expert System for Planting, utilizing Rational Interface Technology), together with the required technologies and the current status of our work.

THE TOTAL SYSTEM DESIGN

The hardware of the ESPRIT system comprises a work unit to be installed in the greenhouse, a sensor system to be connected to it, and a communication unit carried by the cultivator (Fig.1 and 2). The work unit is a computer provided with measurement and control interfaces. It measures environmental data in the greenhouse and controls various environmental control machines. The communication unit is a notebook-type personal computers with a graphic display and keyboard. It serves as a window through which the cultivator communicates with the ESPRIT system. The work unit and communication unit are linked by radio (when the communication unit is carried by the cultivator) or via a telephone line (when he or she is at home) and function as a system.

Most ESPRIT functions are realized by means of software. The following is an explanation of the functions of each unit based on the software block diagram shown in Fig. 3.

The upper part of the software block diagram is mounted on the work unit and the lower part on the communication unit.

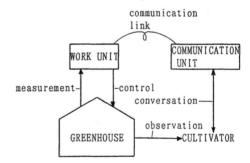

Fig. 1. The system comprising greenhouse, work unit, communication unit and cultivator.

Fig. 2. A trial unit of ESPRIT system.

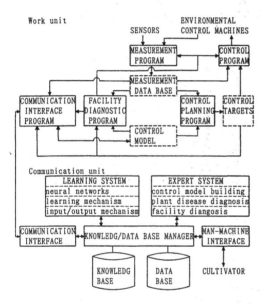

Fig. 3. Software block diagram of ESPRIT.

Work unit

The role of the work unit is to control the operations of various environmental control machines installed in the greenhouse based on the environmental control model for a short-term environmental control schedule. The reason why the environmental control model is designed for short-period use is that the environment in the greenhouse depends on the cost of energy required for environmental adjustment, the target crop and quality as well as the status of plant growth. These need to be scheduled strategically by taking the unit price of fuel, the market price of the produce and other factors into account. An environmental control model cannot therefore be built on a work unit alone. The use of a communication unit that provides higher-level functions and the involvement of the cultivator are also required.

The functions of the work unit are realized by five programs which are executed concurrently. The basic function of each of these programs is detailed below:

- Measurement program

The measurement program reads signals from various sensors connected to the system, convert their units for practical use, and stores the data in the system memory as a measurement data base. The measurement data base is accessible from all other programs freely.

- Facility diagnostic program

The diagnostic program always checks the sensors and environmental control machines for any abnormal condition. If an abnormal condition is detected, the program diagnoses the trouble. The result of the diagnosis is converted into codes and stored in the system memory.

- Control planning program

The control planning program gives the current environmental control targets based on the environmental control model.

- Control program

The control program controls various environmental control machines to bring the actual environmental conditions to the target levels. To do this, the facility diagnostic codes are used to select the sensors to be referenced, the machines to be controlled and the method of control properly.

- Communication interface program

The communication interface program provides a logical linkage between the work unit and communication unit.

Communication unit

The role of the communication unit is to allow the cultivator to use the sophisticated functions of the work unit, such as the building of an environmental control model, culture facility diagnosis, plant disease diagnosis and accumulation of cultivation techniques, with as simple machine operations as possible.
What follows is an outline of the basic design of the communication unit (the lower part of the block diagram) to realize this. The basic design is still in the research phase and its field applicability is yet to be proven.
The system integrates an expert system, neural networks, knowledge base, data base and the cultivator by using very flexible interfaces.

- Expert system

This system, based on rule base inference mechanism, is intended to solve logical problems. It comprises an environmental control model building system. a plant disease diagnosis system, and a facility diagnosis system (this provides more complex diagnosis than the facility diagnosis built in the work unit), among others.

- Learning (neural network) system

This system applies the mechanism of learning to solve problems which are difficult to solve by the expert system because of the difficulty in defining inference rules (knowledge). It comprises a neural networks, input/output mechanism and learning mechanism.

- Knowledge/data base manager

This is an internal interface feature to ensure efficient use of knowledge and data.

- Communication interface

This interface is used to connect the work unit and the communication unit. The interface permits each unit to use the other as its part.

- Man-machine interface

This realizes communications between the ESPRIT system and the cultivator.

The total design of ESPRIT is characterized by the fact that it is provided with knowledge-based problem solving and learning capabilities and that the interface between each unit is designed to ensure efficient operation of the system as a whole with the work unit, communication unit and cultivator treated as an integrated system.

CURRENT STATUS OF OUR WORK

Devolopment of work unit

The development of the hardware and software of the work unit has almost been completed. At present, a trial unit is under field testing for further improvement in performance. The work unit is required to provide adequate computing ability to realize the total design, be reliable enough for long-term stable operation in the greenhouse environment, and have fault tolerance capacity, or environmental control ability to maintain the life of plants even in the event of a system trouble. The sensor system also needs to have long-term reliability and stability. To meet these requirements, we designed and

made all the electronic circuit modules built in the work unit (a computer, measurement control interfaces, sensor signal conditioners) and their software (operating system and five programs stated above).
The work unit has been in use for one year for greenhouse environment control in a severe temperature condition (max. 50℃ in summer; min. -3℃ in winter). No trouble or deterioration in measurement accuracy in excess of the designed tolerance (±0.5℃ in the case of temperature measurement) has arisen so far. The following is an outline of the electronic circuit modules we developed:

- Computer module (Fig.4)

The computer module is a single electronic circuit module using a state-of-the-art 16-bit microprocessor with 1.5MIPS in operating speed, 1MB of RAM and 1MB of ROM.

Fig. 4. Computer module.

- Control circuit module

The control circuit module is provided with 24 digital input channels, 24 output channels, 3 pulse count input channels and 8 analog voltage ouput channels to control various environmental control machines.

- Measurement circuit module

This is an A/D conversion module with 64 analog voltage input channels. It is capable of setting the gain of the built-in amplifiers in the range from 1 to 256 based on the instruction given from the computer module.

Development of communication unit

The hardware of the communication unit is a notebook-type personal computer available on the market. There are available on the market a number of such machines with adequate performance to realize our total design.
Each software component (an expert system, neural networks, and man-machine interface) is now

under development based on the total design. Specifically, we have built an environment that simulates the communication unit on a high-performance work station (SUN4/1+) and are developing and evaluating prototype components. In the total design, the work unit functions as the environment measurement data base of the communication unit. In the present stage of things, however, since an interface to link the work unit and the work station has not been developed yet, we transfer to the work station the environmental conditions and the operating status of each environmental control machine measured on the work unit as a data base on an off-line mode.

- Development of expert systems

At present, we are doing work on expert systems based on the rule base inference mechanism. Specifically, we are developing a facility diagnosis system, plant disease diagnosis system and environmental control model building system.
We use OPS83 as the language to describe our expert systems. OPS83 can be used for both rule base programming and procedure base programming. The language resembles the C language and is therefore most suitable for the development of expert systems to be incorporated in a total system such as ESPRIT.

- Development of learing(neural network)system

At present, basic research is being done to determine whether or not it is possible to have a neural network learn the relations between greenhouse environmental conditions and the growth of plants for purposes to eventually develop a system capable of retrieving the neural network as a function from any expert system. We use Neural Works, a neural network simulator developed by NeuralWare Inc., as our research tool. The neural network developed through simulation will be restructured by using the C language and made a system.

- Development of knowledge/data base manager

The knowledge/data base manager is an internal interface mechanism that controls the knowledge base and data base and provides the necessary knowledge and data at the request of the expert system and/or learning system. At present, we are doing basic research on the building and maintenance of the knowledge base and on the effective use of the data base.

- Development of man-machine interface

What we intend to develop is a flexible interface to allow the cultivator to obtain all necessary functions through conversation with the system. At present, our research is on how to provide the interface with some inference ability for improved flexibility in conversations between the cultivator and the system. C and Ops83 are used as system description languages.

CONCLUSION

When we started the development of ESPRIT, our original concept (initial total design) was to have the communication unit function simply as the operating console of the work unit. When the original concept almost took shape in April, 1990, it was found that the system could be expanded to provide more sophisticated functions. We have since been making efforts to design and realize the current total design. Our purpose is to realize ESPRIT as a total system, not to be complete individual component technologies. We do not hesitate to modify the current total design and incorporate new technologies if necessary to do so.
Our thanks go to Professor Y. Hashimoto, College of Agriculture, Ehime University, who gave us valuable advise.

REFERENCES

Forgy, C.L. (1985). ops83 User's Manual. Department of Computer Science, Carnegie Mellon University.
Hoshi, T., and T. Kozai (1984). Knowledge-based and hierarchically distributed online control system for greenhouse management. Acta Horticulture, 148, 301-308.
Kozai, T. (1985). Ideas of greenhouse climate control based on knowledge engineering techniques. Acta Horticulture, 174, 365-373.
Nakanishi, Y., and T.Uchida (1989). Development of a Stand-Alone Computer Integrated Cultivation Support system. Environ. Control in Biol., 27(4), 145-147.

IDENTIFICATION OF WATER AND NUTRIENT SUPPLY TO HYDROPONIC TOMATO PLANTS BY USING NEURAL NETS

T. Honjo and T. Takakura

University of Tokyo, Dept., of Agricultural Engineering, Yayoi 1-1-1, Bunkyo-ku, Tokyo 113, Japan

Abstract. Water and nutrient supply to hydroponic tomato plants in a greenhouse is identified by using neural nets. Learning procedure of the nets is back-propagation. Inputs are time series data of 5 environmental conditions (solar radiation, inside temperature, inside humidity, nutrient water temperature and CO_2 concentration) and outputs are water and nutrient supply to hydroponic tomato. All data are measured in a commercial greenhouse. The factors of the neural nets structure, such as span of input layer, number of hidden layers and number of units in hidden layer are investigated. As a result, by using simple nets (input data of 8 hours and single hidden layer which consists of 4 units), the amount of water and nutrient supply is calculated well.

Keywords. Neural nets; water supply; nutrient supply; identification.

INTRODUCTION

Recently, many researchers focus on the automation of complicated and intellectual control in the greenhouse. The AI approach such as expert systems or machine learning have received attention (Takakura, Shono and Honjo, 1983; Kurata, 1988; Hoshi, Hirafuji and Honjo, 1990). System identification of plant behavior is an approach to this problem (Hashimoto, Strain and Ino, 1984) . Neural net is also one of the promising techniques and is being applied to this area (Hirafuji, 1988; Ohara, 1989).

In the present study back-propagation algorithm (Rumelhart, Hinton and Williams, 1986) is used for the identification of water and nutrient supply to hydroponic tomato plants in a greenhouse. The factors of the neural nets structure, such as span of input layer, number of hidden layers and number of units in hidden layer are investigated.

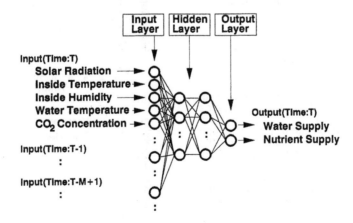

Fig. 1 Neural nets for the identification of water and nutrient supply.

METHOD

Neural Nets for the Identification

The structure of the nets used in this study is shown in Fig. 1. Inputs are time series data of 5 environmental conditions (solar radiation, inside temperature, inside humidity, nutrient water temperature and CO_2 concentration). Outputs are the amount of water and nutrient supply to hydroponic tomato plants. The input/output data are obtained every hour. To calculate the water and nutrient supply at the time T o'clock, M hours of data (from T o'clock to T-M+1 o'clock) are used, where M is a span of input data. Therefore, the number of input units is 5xM and the number of output units is two. Several cases of constant M is tested and the optimal number is determined empirically.

Between input and output layer, at least one hidden layer exists. The number of hidden layers and the number of units in hidden layer is also determined by trials.

The program of back-propagation used in this study is written in Fortran and is available as a public domain software (Honjo, 1989).

Data

Data used in this study are measured in one of the commercial greenhouse of Yokogawa Green Farm Co. (Itsukaichi, Tokyo) from Dec. 9, 1990 to Jan. 17 1991. The area of the greenhouse is 806 m^2 and hydroponic tomatoes are grown in the greenhouse.

Three typical days (fine day:Dec. 20, cloudy day:Jan. 2, rainy day:Dec. 21) are chosen from the data and are used in the calculation below. The data of each day consist of 24 data vectors and each data vector includes 5xM input data and two output data.

The set of 24 data vectors of each day is learned by the nets and the nets are valid only for the calculation of the day.

Structure of the Nets

The optimal structure of the nets is determined by the following procedure.

1. Using the nets which have 1 hidden layer with 2 units, the span of input data M is tested in the cases of 2, 4, 8, 16 and 24.
2. Using the nets which have 1 to 3 hidden layers where each layer consists of 2 units, the number of layers are tested. In this case, M is set 8.
3. Using the nets which have 1 hidden layer, the unit number of the hidden layer is tested in the case of 2, 4 and 8. In this case, M is also set 8.

The result is evaluated by comparing measured (desired) data d_{ij} and calculated data c_{ij}, where i means category of output layer and j means cases of data vector. The error, E which is defined by the following equation.

$$E = \sum_i \sum_j (d_{ij} - c_{ij})^2 / 24 \tag{1}$$

After 2000 sweeps of learning through the data of each day, the case of smaller error is considered to be better.

RESULTS AND DISCUSSION

Structure of the Nets and Identification

One of the result of the procedure 1 described in the previous section is shown in Fig. 2. Fig. 2 is the example of the cloudy day. From Fig. 2, 8 hours of span of input layer is considered enough. Results of the other days also show the same tendency.

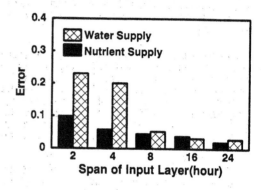

Fig. 2 Relation between span of input layer and error (Jan. 2 1991).

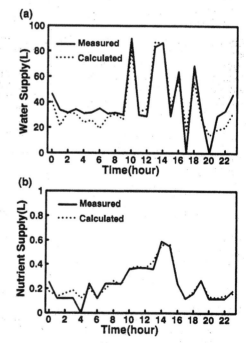

Fig. 3 Measured and calculated values of water and nutrient supply on Jan. 2 1991.

286

From the results of the procedures 2 and 3, we choose the nets of single hidden layer which consists of 4 units as the optimal nets. Therefore, the number of units in input layer, hidden layer and output layer of the nets are 40, 4 and 2, respectively. This nets are used in the following analysis. Calculated values by using this nets are compared to measured values in Fig. 3. Both agree very well.

Prediction and the Number of Learning Sweeps
In each typical day, the error which is defined by the equation (1) decreases as the number of learning sweeps increases. In Fig. 4, the iteration number of the sweeps is prolonged to 10000 times and the relation between the error and the number of learning times is shown in the case of water supply. Only the data of the rainy day is very fluctuating because of the small amount of water supply.

It can be said that the more nets learn, the more accurate the identification is. But this is not true in the case of prediction. Fig. 5 shows the prediction of water supply on Jan. 3 by using the nets which learn the data of the previous day. The nets after 10000 learning sweeps and the one after 2000 sweeps are compared. Predictions by both nets are not so fitted as the calculated data shown in Fig. 3.

In Fig. 4, the error of the nets after 10000 sweeps is almost half of the error after 2000 sweeps as for the case of Jan. 2. Meanwhile, the error is bigger in the case of 10000 sweeps of learning in Fig. 5.

CONCLUSION

Back-propagation algorithm is applied to the identification of water and nutrient supply to hydroponic tomato plants and the factors of the neural nets structure are investigated. By using simple nets (input data of 8 hours and single hidden layer which consists of 4 units), the amount of water and nutrient supply is calculated from the time series of 5 environmental factors. As a practical tool for the identification of greenhouse system, neural nets show powerful performance in the examples of this study.

ACKNOWLEGEMENTS

We would like to express our appreciation to Mr. Tsuyoshi Okuya and Mrs. Akiko Okuya (Yokogawa Electric Co., Ltd) for their great assistance with this study. We also thanks Mr. Kiyoshi Nakajima for his cooperation.

REFERENCES

Hirafuji, M. , Ono, Y. and Kobayashi, K. (1988). A knowledgebase system and a neural-computer system for agricultural information processing. *Acta Hort.*, **230**, 253-259.

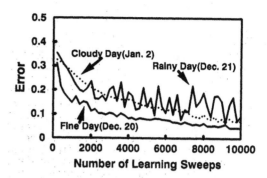

Fig. 4 Relation between the number of learning sweeps and error.

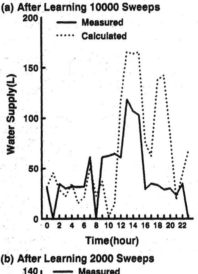

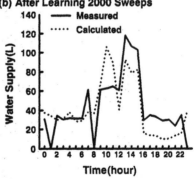

Fig. 5 Prediction of water supply on Jan. 3. The nets are calculated by 10000 sweeps of the data (a) of Jan. 2 and by 2000 sweeps (b).

Honjo, T. (1989). Program of back-propagation. Agr. info. systems in 90's, Commission for the information systems, Soc. Agr. Met. Japan (in Japanese).

Hoshi, T., Hirafuji, M. and Honjo, T. (1990). Expertsystems for biotechnology and agriculture. Corona Publishing Co., 208pp (in Japanese).

Hashimoto, Y., Strain, B. R. and Ino, T. (1984). Dynamic behavior of CO_2 uptake as affected by light: system identification based on spectral analysis. *Oecologia*, 63, 159-165.

Kurata, K. (1988). Greenhouse control by machine learning. *Acta Hort.*, **230**, 195-200.

Ohara, G. (1989). Proposition of new crop management method with neural-network system. Agr. info. systems in 90's, Commission for the information systems, Soc. Agr. Met. Japan (in Japanese).

Rumelhart, D. E., Hinton G. E. and Williams R. J. (1986). Learning representations by back-propagating errors. *Nature*, **323**, 533-536.

Takakura, T., Shono, H. and Honjo, T. (1983). Application of knowledge engineering to crop cultivation, *J. Agr. Met.*, **39**, 113-116 (in Japanese).

Copyright © IFAC Mathematical and Control Applications
in Agriculture and Horticulture, Matsuyama, Japan 1991

A TIME-CONTINUOUS QUANTITATIVE PLANT GROWTH MODEL BASED ON NEURAL NETWORKS AND LOTKA-VOLTERRA EQUATIONS

M. Hirafuji

*National Agriculture Research Center, Ministry of Agriculture, Forestry and Fisheries,
Tsukuba 305, Japan*

Abstract.
A continuous-time quantitative plant growth model is proposed. The model
is implemented as a neural network, which is a three layers neural network
and has mutual connections on the middle layer. The neural network is derived from Lotka-Volterra equations which describe mechanism of physiological interference between organs. The plant growth model includes the
specific organ growth models which are represented by units which have
same functions as neurons. It is shown with computer simulation that solutions of the plant growth model implemented as a neural network are approximated to that of Lotka-Volterra equations. The relation between the
Lotka-Volterra system and neural networks is discussed and an algorithm to
estimate unknown parameters in the models is proposed.

Keywords. Biology; computer simulation; ecology; environment control;
neural nets.

INTRODUCTION

A mathematical model is needed to control
growing plants in greenhouses and fields.
Especially for optimum control and strategic decision making with computers, plant
growth model is indispensable. In addition
to these engineering needs, developing
mathematical models which simulate on
computers as real plants as possible is
useful for investigating the fundamental
theory of growth and life.
There are many types of plant growth
models such as dynamic simulation, multivariate analysis, logistic and polinominal
exponential regression(Hsu,F.H., C.J.Nelson
and W.S.Chow, 1984), Markov mod-
els(Richard L.Olson, Jr., Peter J.H.Sharpe
and Hsin-I Wu,1985) and so on.
Dynamic simulation model is a mechanistic
model, which can describe plant growth in
terms of physiological mechanisms. However we can not let the model include all
physiological process because of the excessive complexity of the growth process and
our insufficient knowledges. Even if we
succeed to develop a dynamic model, it
would be difficult to identify the model
parameters.
Multivariate analysis models and regression models are statistical models, which

can be identified based on least square error method. Because its methods have robustness for noise, measured noisy data of
plant growth can be used for developing
models. However it is difficult to let the
models include mechanistic (metabolic and
physiological) process.
Markov models have a intermediate character of mechanistic and statistic models,
but they lose the advantages of statistical
models.
A modeling method of the plant growth
model on computers should be provided
the above both advantages. Plant growth
has two considerable important elements
for modeling, increase of size (and weight)
and geometric form, which depend on
mutual interference between organs such
as heterotrophic relation on source-sink
translocation and competition for limited
resource. Consequently plant growth models must include the competitive mechanism between organs and specific organ
growth models respectively.
In this paper, a continuous-time quantitative model which describes increase of
weight (or size) of specific organs is proposed. The model is derived from Lotka-
Volterra equations which describe the interference between organs. In this model,
a sigmoid function which is used empiri-

cally as activation functions in neural networks is derived naturally from Lotka-Volterra equations. And It is shown that the model base on Lotka-Volterra equations is same as neural networks mathematically. The relation between the Lotka-Volterra system and neural networks is discussed and an algorithm to estimate unknown parameters in the models is proposed.

DERIVIATION FROM LOTKA-VOLTERRA EQUATIONS

Organs and tissues grow competitively under restriction of limited resources. Growth of vegetative organs encourages growth of other organs, but increase of its consumption. The mechanism similar to competition and symbiosis in ecosystems. Lotka-Volterra equations have been used as general model of population growth, and are thought to be the simplest model of plant growth models. Here with Lotka-Volterra equations a plant growth model is described as following.

$$\frac{dg_i}{dt} = g_i(r_i - \sum_{j=1}^{m}\mu_{ij}g_j) \qquad (1)$$

where g_i is amount of growth of i-th organ, r_i is its intrinsic growth rate and μ_{ij} are the interference coefficients of competition and symbiosis. Although the equations are simplest, it is difficult to solve analytically the equations without the case of m=1. The numerical integration by a computer is used to solve (1) in the case of m≥2. For lack of analytical solutions of Lotka-Volterra equations, it is difficult to estimate the unknown parameters from measured data with least square error methods.
Here we introduce the assumption that the magnitude of competitive interactions between organs (μ_{ij}, i≠j) are very weak relatively to magnitude of competition in individual organs ($=\mu_{ii}$).

$$|\mu_{ii}| >> |\mu_{ij}| \qquad (2)$$

On this condition, the Lotka-Volterra system has a nonnegative equilibrium point because of existence of Lyapnov function(N.Shigesada, K.Kawasaki and E.Teramoto, 1984). The path to the equilibrium point means a pattern of plant growth.
We further assume that the value of amount of growth are regularized from 0 to 1.0.

$$0<g_i<1.0 \ (i=1,2,...,m) \qquad (3)$$

Namely g_i is interpreted as relative characteristic value of growth in this model. Let

$$z_i = \sum_{i\neq j} \mu_{ij} g_j \qquad (4)$$

be sum of interference from other organs. Under assumption (2), the effect of z_i is so small that z_i can be treated as stationary while a short time span. Then (1) can be integrated for time,t, and is solved under the restriction (3) as:

$$g_i = \frac{1}{1+\exp(-(r_i-z_i)t+\theta_i)} \qquad (5)$$

where θ_i is:

$$\theta_i = \ln\left(1 - \frac{r_i-z_i}{g_{i(t=0)}\,\mu_{ii}}\right) \qquad (6)$$

where $g_{i(t=0)}$ are initial value of g_i. According to above assumptions, θ_i are treated as constant. And θ_i are a parameter which concern to self-consumption in a organ. For example, if μ_{ii} increases, θ_i increases too.
Structure of the model represented by (4) and (5) can be arranged into a network (Fig.1). This network is similar in the structure to multi-layers neural networks. Here we can equate this model with a neural network by following interpretation. In Fig.1, the open circle means a unit whose transfer function is (5). Where input value of the unit is r_i-z_i and output value is amount of growth, g_i. These units are interpreted like as neurons in neural networks, and simultaneous equations of (5) are interpreted as a neural network. All the units has a internal status for t (time), r_i (growth rate) and θ_i. The value of t is common in all the units. Growth rate, r_i, and initial constant, θ_i, is specific on each organs. For correspondence to a neural network, t is equivalent to maximum gradient on the sigmoid function which represent transfer function of neurons and r_i is input value from the input layer. If the output value of the unit on the input layer is constant (=1.0), r_i can be treat as synaptic weight connecting from the input layer to the middle layer. θ_i is equivalent to threshold value of neurons.
The solutions of equation (5) are amount of organ growth on competitive condition in a plant body. Where amount of whole plant growth, Ψ, is equal to sum of amount of organ growth, gi. However we can not develop and not identify the organ growth models, because we can get only a little information (measurable data) for the parts of organs. Here it is assumed that

representative organs such as stems, leaves, roots and fruits has a strong correlation to amount of whole plant growth. Ψ is described with regression model using linear or sigmoid function as following.

$$\Psi = \sum_{i=1}^{m} w_i g_i \qquad (7)$$

$$\Psi = \frac{1}{1+\exp(\sum_{i=1}^{m} w_i g_i)} \qquad (7)'$$

The whole plant growth model based on (5) and (7) or (7)' is implemented as a three layers neural network (Fig.2). The neural network model has one unit on input layer whose value keeps 1.0 and one unit on output layer whose value means amount of whole plant growth. Number of units on hidden (middle) layer is equal to number of representative organs. The whole plant growth model consists of parameters, t, $\{r_i\}$, $\{w_i\}$, $\{\theta_i\}$ and $\{\mu_{ij}\}$.

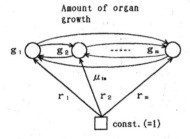

Amount of organ growth

Fig.1 An organ growth model implemented as a neural network.

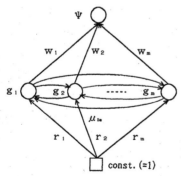

Amount of whole plant growth

Fig.2 A whole plant growth model implemented as a three layers neural network.

A PLANT GROWTH MODEL EXTENED FOR ENVIRONMENTAL FACTORS

Although r_i is assumed to be constant on above plant growth model, actually r_i is variable depending on environmental conditions. There are many environmental factors such as temperature, humidity, CO_2 concentration, PAR(photosynthetically active radiation), water and nutrients that affect ri. These variables can be added to above plant growth model by the means of making the input unit be a function of the environmental factors. The simplest model of r_i and environmental factors is a linear model:

$$r_i = \sum_{k=1}^{n} v_{ik} x_k \qquad (8)$$

where x_k is characteristic value of k-th environmental factor, v_{ik} is weight of x_k and n is number of considering environmental factors.

This extended whole plant growth model based on (5)-(8) is implemented as a three layers neural network (Fig.3). This neural network model has n units on the input layer whose value are $\{x_k\}$. Number of units on the middle layer is equal to number of representative organs. i-th organ has specific weight vector $\{v_{ik}\}$ according to environmental factors. This extended whole plant growth model consists of parameters, t, $\{r_i\}$, $\{w_i\}$, $\{\theta_i\}$, $\{\mu_{ij}\}$ and $\{v_{ik}\}$.

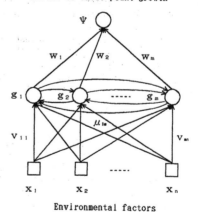

Amount of whole plant growth

Environmental factors

Fig. 3 Extended whole plant growth model including effects of environmental factors.

COMPARIOSN OF NEURAL NETWORK MODEL AND LOTKA-VOLTERRA MODEL WITH AN EXAMPLE

Because proposed plant growth models have assumptions such as equation (2), deference between Lotka-Volterra equations and a corresponding neural network model is investigated using following simple example.

$$g_1 = \frac{1}{1+\exp(-(20-0.5g_2)t+10)}$$

$$g_2 = \frac{1}{1+\exp(-(10-g_1)t+3)}$$

(9)

These equations can be solved with algorithm shown in Fig.4. The solutions are shown in Fig.5 by dashed lines. Equations (9) correspond to following differential equations.

$$\frac{dg_1}{dt} = g_1(20-20g_1-g_2)$$

$$\frac{dg_2}{dt} = g_2(10-2g_1-10g_2)$$

(10)

Namely, $\{r_i\}$ and $\{\mu_{ij}\}$ in (1) are:

$$\{r_i\} = \begin{pmatrix} 20 \\ 10 \end{pmatrix}$$ (11)

$$\{\mu_{ij}\} = \begin{pmatrix} 20 & 0.5 \\ 1 & 10 \end{pmatrix}$$ (12)

Initial constants of (10) are determined from (9) as followings.

$$g_{1(t=0)} = \frac{1}{1+\exp(10)}$$

$$g_{2(t=0)} = \frac{1}{1+\exp(3)}$$

(13)

Equations (10) are solved with numerical integration. The solutions are shown in Fig.5 by solid lines.
The solutions of (9) agree with sufficiently the solutions of (10). This result means that we can solve Lotka-Volterra equations with a neural networks.

<Step 1> Let t in equation (5) equal to specific time on all the units.

<Step 2> Initialize $\{g_i\}$ with constant or random value.

<Step 3> Calculate new $\{g_i\}$ according to equation (4) and (5) using initial value of $\{g_i\}$.

<Step 4> Calculate next $\{g_i\}$ using new $\{g_i\}$.

<Step 5> If $\{g_i\}$ become stationary then end, else go to <Step 4>.

Fig.4 An algorithm to calculate $\{g_i\}$ with a neural network.

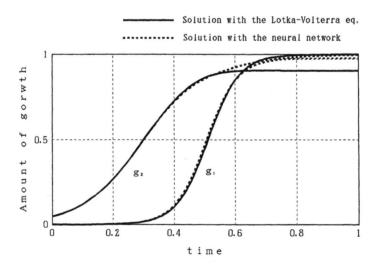

Fig.5 Comparing solutions of the neural network model and solutions of the Lotka-Volterra equations.

292

AN ALGORITHM TO ESTIMATE UNKNOWN PARAMETERS

Neural network models and Lotka-Volterra models can be transformed into each other. According to above relations, parameters in a Lotka-Volterra system can be estimated with learning algorithms for neural networks. For example, backpropagation algorithm which was developed for multi-layers neural networks (Rumelhart, D.E., G.E.Hinton and R.J.Williams, 1986) is the typical algorithms. However conventional backpropagation algorithm cannot be applied to neural networks which have feed back connection (i.e. mutual connections). Neural networks for plant growth models have mutual connections $\{\mu_{ij}\}$ between units on a same layer. To estimate unknown parameters in plant growth model, I propose a learning algorithm as follows.

Suppose there are measured data set, whose p-th data consist of time t_p, environmental vector $\{x_{pi}\}$ and amount of organ growth $\{g_{pi}\}$ or whole plant growth Ψ_p. In the case that we measure amount of organ growth $\{g_{pi}\}$, sum-squared error E_p is:

$$E_p = \frac{1}{2}\sum_{i=1}^{m}(g_{pi}-g_i)^2 \qquad (14)$$

In another case that when we measure Ψ_p, sum-squared error E_p is:

$$E_p = \frac{1}{2}(\Psi_p-\Psi)^2 \qquad (15)$$

Synaptic weight $\{w_i\}$, $\{v_{ik}\}$ and $\{\mu_{ij}\}$, and threshold $\{\theta_i\}$ are estimated by following equations with extending backpropagation algorithm.

$$\frac{dE_p}{dw_i} = 0$$
$$\frac{dE_p}{d\theta_i} = 0 \qquad (16)$$
$$\frac{dE_p}{dv_{ik}} = 0$$

$$\frac{dE_p}{\mu_{ij}} = 0 \qquad (17)$$

E_p is a function of all parameters in the neural network. With conventional backpropagation algorithm, (16) can be solved, but added (17) cannot be solved. In backpropagation algorithm, error signal propagates from an output layer to an input layer. This procedure in error propagation is same calculation in the expanded equations of (16). On the neural network for plant growth model, error signal propagates through mutual connection μ_{ij}. This error propagation is thought as calculation in the expanded equation in (17). Then propagated error signal δ_i of i-th unit is:

$$\delta_i = \delta_{up} + \sum_{j=1(i\neq j)}^{m} \delta_j\mu_{ij} \qquad (18)$$

where δ_{up} is error signal propagating from the upper layer (output layer), which is calculated with conventional backpropagation algorithm.

By using this error signal, μ_{ij} is corrected with following learning rule (delta rule).

$$\mu_{ij} := \mu_{ij} + \varepsilon\,\delta_i\,g_i \qquad (19)$$

By the same manner with the error propagation and updating for $\{r_i\}$, $\{w_i\}$, $\{\theta_i\}$ and $\{v_i\}$, every parameters can be estimated.

Using sample data from earlier example equations (9), this algorithm was tested. In about 10,000 iterations on learning, average E_p becomes less than 10^{-4}. Then parameters of (9) were estimated as sufficiently accurate values. However the estimation was sensible for magnitude of E_p, because there exist local minima of E_p.

DISCUSSION

A continuous-time plant growth model including mutual interference between organs is developed with a neural network which is derived from Lotka-volterra equations. In this model, Lotka-volterra equations can be solved without numerical integration. And parameters in Lotka-Volterra equations have been identified with the corresponding neural network. This method is effective to find the solutions of unknown parameters. Using the characteristic parameters of the model, amount of plant growth can be predicted quantitatively. The plant growth model will be used for optimum control. And estimated the parameters will be useful for characterization of plant growth.

We may be able to maximize (minimize) size of specific organs by the means of changing environmental factors or interference between organs (for example, removing other organs) into the optimum conditions predicted by this plant growth model. However, in the neural network, strong interference between organs is inhibited because of the assumed restriction for stability of its solutions. In the case that interference between organs is strong, the parameters of the plant growth must be identified in Lotka-Volterra models.

By appending another differential equation which minimize deviation error, E_p, the parameters may be estimated. For example, following equation may be considered.

$$\frac{\mu_{ij}}{dt} = -\epsilon \frac{E_p}{dt} \qquad (20)$$

By (20) $\{\mu_{ij}\}$ will change for the direction to decrease E_p. Though this procedure is not practical algorithm for model identification, this idea shows us interesting view points. The effect with equation (20) is thought as a kind of adaptation or evolution, and then $d\mu_{ij}/dt$ is interpreted as adaptation rate or selection pressure. Namely adaptation and selection in ecosystems correspond to learning in neural networks. We may find new learning algorithms in ecosystems and inside plants, because actual plants have remarkable ability for such adaptation.

REFERRENCE

Hsu,F.H.,C.J. Nelson and W.S.Chow (1984). A Mathematical Model to Utilize the Logistic Function in Germination and Seedling Growth, *Journal of Experimental Botany*, Vol.35, No.160, 1629-1640.

Richard L.Olson,Jr., Peter J.H.Sharpe and Hsin-I Wu(1985). Whole-plant Modelling:A Continuous-time Markov (CTM)Approach, *Ecological modelling*, 29,171-189.

Rumelhart,D.E., G.E.Hinton and R.J.Williams (1986). Learning Internal Representations by Error Propagation. In D.E.Rumelhart, J.L.MaClelland and the PDP Research Group(Ed.). *Parallel Distributed Processing*, Volume 1. MIT Press, pp.354-361.

Shigesada,N., K.Kawasaki and E.Teramoto (1984). The effects of interference competition on stability, structure and invasion of a multi-species system, *Journal of Mathematical Biology*, 21, pp.97-113.

AGRICULTURE INFORMATION NETWORK
TREND FOR THE 21ST CENTURY

N. Izumi, S. Fujita and H. Ikemoto

*Nippon Telegraph and Telephone Corp., Shikoku Telecommunications Service Region,
Ichibancho 4-4-3, Matsuyama, Ehime 790, Japan*

Abstract. This article introduces an example of a greenhouse control system and explores the present condition and the development direction of agriculture and horticulture information systems. This system is a remote control type greenhouse cultivation system used in Japan and it is a joint-use type system.

Keywords. Automatic alarm transmission; greenhouse controller; greenhouse cultivation system; high-tech system; ISDN; joint-use type system; mandarin orange; remote control; telenetwork; weather robot.

PRESENT CONDITION OF HORTICULTURE SYSTEMS OF
MANDARIN ORANGE GREENHOUSE

The recent rapid progress in microelectronics technology
has contributed to computerizing all areas of society.
This includes the field of agriculture, where a great
deal of research and experiments using information
communication systems is being carried out.
These systems have been rapidly put to practical use.
Many of the experiments and much of the research for
agriculture and horticulture in Japan are performed in
Ehime Prefecture. Ehime Prefecture is located in the
southwest part of Japan and its climate is mild.
Nippon Telegraph and Telephone Corporation(NTT) and
Shikoku Telecommunications Service Region Office have
developed a new greenhouse control system at the
request of the town of Ikata(Fig.1).
This system is the first remote control type system
used in Japan and it is a joint-use type system.
This system has the following functions.

i) Joint-use of the control center
ii) Automatic control and remote control
function of the greenhouses
iii) Support functions to save energy and
human labor
iv) Database of environmental conditions
in growing stage
v) Graphic display and supply management
data including daily and monthly reports
vi) Alarm transmission function in case of
emergencies

The next section will describe this system
in greater detail.

OUTLINE OF THE HIGH-TECH MANDARIN
ORANGE CULTIVATION

Basic system concept

A joint-use system utilizing personal
computers has been developed to support
greenhouse control in mandarin orange
cultivation.
We call this system the "high-tech mandarin
orange cultivation support system", or the
"high-tech system", for short.

Fig.1 The geographic features of
Ehime Prefecture

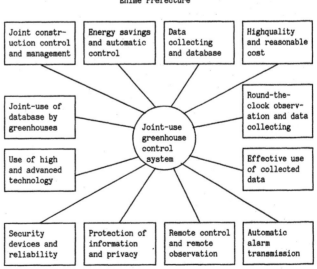

Fig.2 Basic system concept

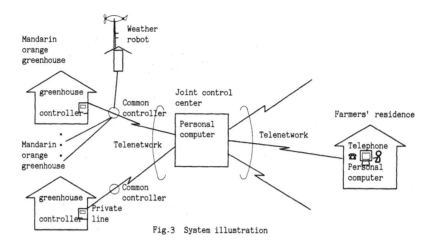

Fig.3 System illustration

This system was developed to save energy and human labor, to allow automatic and remote control, and to enable the growing amount of environment control data to be scientifically controled. The basic concept of these purposes is shown in Fig.2.

System Illustration

The hardware of the mandarin orange greenhouse cultivation support system is composed of a greenhouse controller, control center, sensors, common controller, weather robot, personal computers in farmers' residence and telephone lines. The system's communication network is illustrated in Fig.3.
GreenHouse controllers are installed near each greenhouse, controlling the temperatures and conditions inside. The greenhouse controller collects information on the inside of the greenhouse at fixed intervals and sends that information in addition to weather information to the control center. This information is placed into the database at the control center.

The concept of the high-tech system was born out of the problem of inefficient mandarin orange cultivation the town of Ikata, where clusters of mandarin orange-cultivation greenhouses are located in mountainous areas far removed from thier farmers' domiciles.
With the high-tech system, a farmer can contact the system control center merely by flicking on the switch of his personal computer. He can then utilize the control center to adjust and regulate the greenhouse controller from his home.

The farmer can observe the inside environment and outside weather conditions of distant greenhouses by viewing the personal computer display at his house. He can set the most moderate environmental conditions for each stage in mandarin orange growth by controlling the temperature in the greenhouses, using the remote control function. In case of trouble with the temperature control system, the sensors detect the problem with the temperature and automatic alarm transmitting function, which sends

TABLE 1 Functions for growth control in mandarin orange greenhouses

no	Required function
①	A system which can be jointly used by many farmers.
②	A remote control, power-efficient system which operates 24 hours a day.
③	A temperature control by the total amount of sunshine for the purpose of saving energy.
④	Sorting, accumulation and processing of data at the control center.
⑤	A system best suited to farmers, which will provide them with useful and appropriate information.
⑥	Remote observation of greenhouse equipment and environment and alarm transmission in the event of abnormal conditions.
⑦	Precise measurement of rainfall, humidity, wind, soil temperature, etc.

alarm messages to the farmer immediately. This allows farmers to be informed of trouble in the greenhouse quickly and they can prevent any serious problems from occurring as a result of the high temperatures.

Characteristics of High-Tech Support System for Growing Mandarin Oranges

The functions farmers need to realize using high-technology are shown in TABLE 1.
We must consider that farmers have developed very professional cultivating skills through years of experiments, but need new high-technology cultivating techniques to support them.
The problem lies in how to apply both the farmers' skills and the high-technology cultivating techniques to the greenhouse. The following are the main functions needed for highly-organized systems to support growth control in mandarin orange greenhouses, in reference to TABLE 1.

Joint-use of the control center. To effectively use the functions of the joint-use system, farmers need to make greater use of the source data.
Eventually, it will be possible to coordinate research to analyze the gathered data. In a joint-use system, all greenhouse data is housed in the control center. Data files have individual security devices to protect each farmers' privacy.

Complete automation of the greenhouse and the remote control function. In the growing season, doing greenhouse inspections is apt to be a very laborious job, but using this automatic, remote-controlled system makes doing this work much easier, since a number of tasks which required human labor previously can now be done automatically.
If the system should break down, so that it cannot be remotely controlled, an off-line partial environmental control system is available as a support unit.

Saving energy and human labor. In winter, a hot air heater maintains the temperature of the greenhouse during the day. Temperature is fixed to minimize the operating time of heaters. If we use the difference between the minimum and maximum temperature during the day, or the total amount of sunshine, we can effectively control the temperature by using heaters and fans, and save in night power or energy.
This allows flexible automatic greenhouse control.
The method of temperature control is shown in Fig.4.

Collecting data and saving it in a database.
Once the heating process begins, data on temperatur changes is accumulated for the blossoming and ripening time in each greenhouse for each stage of growth, and stored so as to form a database for growing conditions. Using a weather robot, weather data is periodically gathered and stored it in the database at the control center as well.
An outline of the weather robot is shown in Fig.5.

Using the graphics function to provide greenhouse management information for daily and monthly reports.
The system must enable us to obtain greenhouse management information from the control center merely by calling it up on a personal computer. Analyzing this information enables us to grow mandarin oranges effectively on the basis of scientific data rather than relying completely on experience. By using the system's graphics function we can display various graphics, charts and so on, on the screens and call up greenhouse management information for daily or monthly reports.
An example of the types of graphics that can be called up on the screen is shown in Fig.6.

Transmitting warning of abnomal conditions. It is necessary to protect greenhouses from natural phenomena such as thunder or snow.
However, greenhouses equipment and sensors are not sufficiently protected against the sort of conditions which occur in mountainous areas. If an unusual condition occurs in a greenhouse,(for example, if the temperature rises to an uncommon degree), the greenhouse controller detects it immediately, and automatically sends a synthetic-voice message to the farmer that the greenhouse hot functioning properly. This enables the farmer to take immediate actions to correct the problem.

Automatic alarm transmission. If the farmer is working outside the house, he can make use of a pocket-bell alarm transmission function.
This is illustrated in Fig.7.

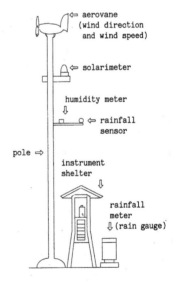

Fig.5 View of weather robot

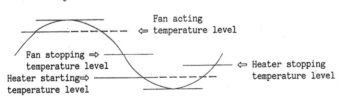

Fig.4 Method of greenhouse temperature control

date (4.25.1989)

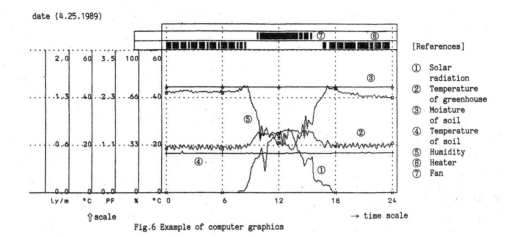

[References]

① Solar radiation
② Temperature of greenhouse
③ Moisture of soil
④ Temperature of soil
⑤ Humidity
⑥ Heater
⑦ Fan

⇑scale   → time scale

Fig.6 Example of computer graphics

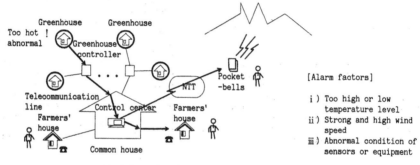

[Alarm factors]

i) Too high or low temperature level
ii) Strong and high wind speed
iii) Abnormal condition of sensors or equipment

Fig.7 Automatic alarm transmission

FIELD TEST RESULTS

The results obtained in field tests conducted on the system over the last forth years are desribed below.

Database efficiency. Because there were insufficient database applications, improved method of data analysis and apllication were needed.

Problems of developing equipment and sensors.
Because of high temperatures in summer, and snowstorms and low temperatures in winter, it was essential to develop ways of protecting sensor and equipment. Therefore, protective devices such as a noise-cutfilter were developed. We pulled up the sensor code in summer for preventing from destruction such as thunder. However, the cost of the system increased as a result.

Protection of equipment(controllers and so on) under unfavorable conditions. Noises caused by turning heaters and fans on and off frequently resulted in abnomal controller operation.
These noises had a very serious effect on the system and offen caused it to shut down. Therefore, we needed to develop adapters to silence these noises, which resulted in increased costs.

Feedback on experimental results to the farmer.
Because farmers are interested in the data obtained in

experimental results, a device to provide such results instantly is needed.
As we cannot expect farmers to mutually exchange information we must consider how to provide them with data pertaining to year-to-year changes in climate and so on, which can be used for mandarin orange cultivation during the following year. Since data can now be analyzed by AI(Artificial Intelligence), full-scale incorporation of AI methodolosy may bring about many changes in cultivation techniques.

This system has not yet flourished because of the fact that many farmers rejected the use of personal computers, and because of problems with differences among system costs as compared to the farmers' incomes. At present, simple wireless communication systems are selling well in areas of Shikoku (an island of Japan (cf.Fig.1)), because of their low construction cost. However, because wireless communication systems do not have the two-way transmission capability of this system, their main function is limited to alarm transmission in the event of abnomal conditions.

As a result, these systems only need a receiver, and do not require key operation with a personal computer.
However, they cannot scientifically manage of data.

Therefore, they are not suitable for farmers.
This system eliminates the faults of wireless systems, and makes it possible to realize new discoveries in cultivation techniques, and to save energy and human labor. It has the following functions.

ⅰ) Automatic temperature control in the greenhouse
ⅱ) Remote control of greenhouse by a farmers' personal computer
ⅲ) Collecting and accumulating data at the control center
ⅳ) Real-time collecting function of greenhouse environmental data by personal computer
ⅴ) Abnormal condition alarm transmission function

These functions cannot wholly satisfy farmers' needs, but adding more function could increase the cost of the system. Therefore, it is necessary to realize a lower-cost system which any farmer can afford.
We must develop systems which are as economical as wireless communication systems and which have the minimum necessary functions according to field test results.

GREENHOUSE NETWORKS

High-technology networks using computers for mandarin orange cultivation have only recently begun to be used. The circumstances surrounding mandarin orange cultivation in Ehime are difficult today, with the move to internationalization bringing about a more open market for imported oranges. Under such circumstances, it is more important than ever to make the younger generation aware of the importance of agriculture and to get them to stick with farming, so that the region can continue to produce agricultural products of high quality and added value.

The testing system reported above has already had problems, including a decline in the initial cost, problems with the system gaining acceptance, difficulties in obtaining precise analysis of database information, actually providing the functions offered to farmers, performing maintenance on equipment, and realizing systematic analysis and use of data by many different types of organizations and research institutions. The system needs to be developed further to make it more functional and economical.

Our aim is to solve these problems through the formation of an information network. We feel that it will be necessary for greenhouses, universities, and other research bodies to be able to freely exchange data amongst themselves towards this end. The basic framework of the high-tech system(fig.3) should remain unchanged. However, it is felt that we will need to develop and expand the system into a large network to accommodate an increased number of users. We will also use remote controlling TV-cameras to observe the blossoming stage and transmit the visual data to farmers' homes and research organizations. Since the demmand for this visual data will be high, the system will need to be able to transmit a large volume of data, maintaining high picture quality at low cost.

In the future, it is possible that this system will be used to transmit visual data to laboratory and colleges as well. Forming an integrated information network will

require us to utlize NTTs' Integrated Services Digital Network(ISDN). ISDN is now being expanded into NTT Shikoku Telecommunications Service Region office. This will provide the basis of the telephone communication and data communication network for the advanced information society, of the next century, and enable us to meet user needs througout the entire region. With ISDN being expanded to provide service on a nationwide basis, it will be possible for us to transmit a large volume of highquality data to points throughout the country at high speed and low cost.

The system can also be expanded to handle multimedia communication rather than simply remain an agriculture information system. Thus it will easily be able to handle the requirments of the fully integrated agriculture information system of the future.
The difference between ISDN and present telephone communication networks is shown in Fig.8.
ISDN can be connected with networks in other countries for the purpose of international standardization. Hence, we will be able to use larger and more flexible international systems in the future.

In Japan, the government is pushing a new trend called "Green-topia" within the country as part of its agricultual policy. Ohzu city in Ehime Prefecture was designated to implement this concept which we call "Green-topia of the Hizi-kawa basin". Under the "Green-topia" concept, for example, one user can send information to another user via facsimile by using his personal computers to search for database information in the electronic filing system. So long as we attempt to establish the same agricultural and horticultural center in the future, we hope that agricultural networks can be connected to each other. The development of the integrated information agricultural network is shown in Fig.9.

Because there are economic problems to consider regarding this network, industrial, educational and governmental organizations must cooperate with each other from a global perspective to alleviate these problems. Engineers and organizations need to actively pursue the development of integrated information systems by cooperating with young managers of agriculture and other people who are tackling the problem of modern agriculture.
These are problems yet to be solved. It will be necessary to seek and listen to the opinions of a large number of engineers, specialists and researchers, and to hold timely conferences to discuss these opinions afterwords in order to eliminate these problems.

During the next few years, we will need to develop the basis of support for agriculture in Ehime Prefecture in the 21st century.
The cooperation of everyone involved-farmers, researchers, network developers and engeneers and others —— will be necessary, and we expect that the day in which this cooperation will be forthcoming is not far off.
I hope the results of the field tests of this system will be helpful to those people who are concerned with and hope to improve modern agriculture.

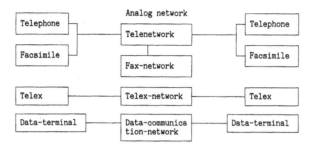

i) Style of present telenetwork

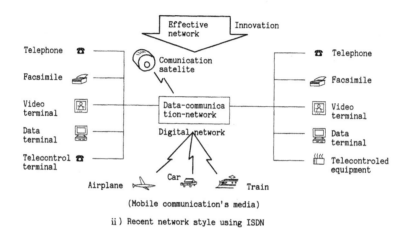

(Mobile communication's media)

ii) Recent network style using ISDN

Fig.8 The style to use ISDN which is different with telenetwork

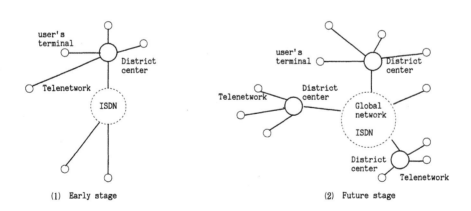

(1) Early stage   (2) Future stage

Fig.9 Extended style of integrated information network for
       agriculture and horticulture

300

COMPUTER SUPPORT SYSTEM FOR TOMATO
CULTIVATION IN PLANT GROWTH FACTORY

K. Hatou*, H. Nonami*, M. Itoh**, I. Tanaka**
and Y. Hashimoto*

*Dept. of Biomechanical Systems, Ehime University, Tarumi, Matsuyama, 790, Japan
**Corporate Planning Office Idemitsu Kosan Co., Ltd., Anegasaki, Chiba, 299-01, Japan

Abstract. A support system of tomato cultivation in plant growth factory was designed and examined. The purpose of the system lies mainly in the support of horticultural operation in the special plant growth factory developed by Idemitsu Kosan Co., LTD. The system is found to be effective in the horticultural operation and management. It seems evident that operation and management for tomato cultivation in such a plant growth factory could not be carried out without a computer–aided cultivation support system. Furthermore, diagnosis of physiological disorder and disease is examined by introducing the artificial intelligence into the cultivation support system. Initial disorder caused by environmental stress and nutrient deficit could be diagnosed in the system. The computer support system as shown in this paper, may be expected to be helpful for any horticultural cultivation in plant growth factories.

Keywords. Computer applications; artificial intelligence; control applications; support system; agriculture

INTRODUCTION

In greenhouses and plant growth factories, many computers have been introduced for controlling environmental conditions and mineral components and concentrations of nutrient solution, supporting the decision of horticultural operation and management and operating the robots and mechanization. In these applications, knowledge processing as well as usual numerical data processing is being introduced. In the previous papers (Hatou, 1990a, 1990b), we examined computer integrated system including an AI computer for a plant growth factory. In this paper, a cultivation supporting system was examined, based on horticultural operation and diagnosis of biological evaluation, while the diagnosis of system without biological elements and expert system for control engineering are investigated in another paper (Hatou, 1991a).

MATERIALS AND METHODS

Plant Growth Factory

A plant growth factory was developed by using a glass greenhouse whose floor area was 120 m². Figure 1 shows the plant growth factory and its air conditioning systems. The air conditioning systems were combinations of a pat-and-fan system, a balloon curtain system, a mist system, a top window system and a heat pump system. The balloon curtain system was developed particularly for this plant growth factory in order to control air temperature effectively. The balloon curtain was made of transparent plastic materials which contained infrared shielding materials, and did not inhibit photosynthetically active radiation. The balloon curtain was used when the surrounding temperature became excessively cold in order to prevent radiation cooling during winter time and when solar radiation was excessively strong during summer time in order to prevent heating caused by the solar radiation. When the balloon curtain was not in use, the air in the balloon was removed and the curtain was set aside compactly. When amount of photosynthetically active radiation decreased due to weather condition, supplementary

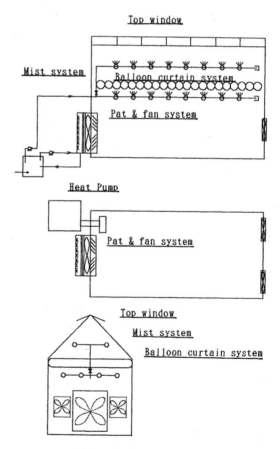

Fig. 1. Plant growth factory and air conditioning systems.

light illumination was given by using high-pressure-sodium lamps in order to promote photosynthesis of plants. Additionally, CO_2 concentration in the atmosphere in the plant growth factory was elevated until noon to promote photosynthesis.

Plant Materials

Tomato plants (_Lycopersicum esculrntum_ Mill.) were grown hydroponically in the plant growth factory. Figure 2 shows a 3-dimensional planting system for tomato plants used in the plant factory. This system allows to grow about 2-fold density of plants compared with a ordinal flat-bed planting system used in hydroponic culture. In order to maximize the efficiency of tomato cultivation in terms of labor cost and stable marketing, three different stages of tomato plants were grown simultaneously in this system. When a harvest target was set to be six clusters of fruits for each harvesting time, more than 3 harvests were possible in a year.

Computers and Program Languages

A tomato cultivation support system used in the plant factory was operated by using two types of computers; two Intel 80386-based computers and one AI computer. The Intel 80386 computers were PC-98XL[2] (NEC Co., Japan) and the AI computer was PSI II (Mitsubishi Electric Co., Japan). Languages used in the Intel 80386 computers were C and BASIC, and a language used in the AI computer was Extended Self-contained Prolog. Because C and BASIC are convenient to use for calculation and picture drawing, C and BASIC were used to program screen manipulation on the

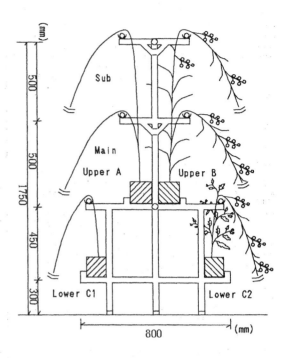

Fig. 2. Diagram of a 3-dimensional planting system used for tomato cultivation in the plant growth factory.

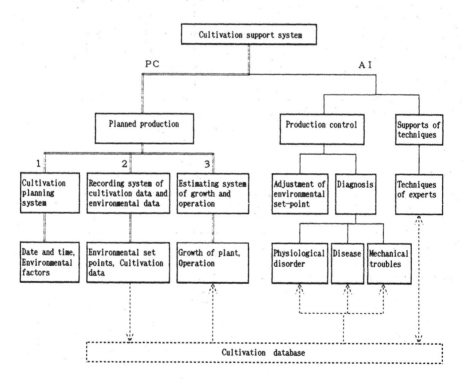

Fig. 3. Schematic diagram of hierarchical operation of cultivation support programs and data acquisition operation for tomato cultivation support system used in the plant growth factory. Solid double lines indicate operational flows of software programs used the 80386 computer (PC), solid single lines indicate operational flows of programmes used in the AI computer (AI) and dotted lines indicate data exchange pathway between programs and data storage area in computer memories.

302

computer monitors and calculation for data analysis. Thus, one 80386—computer was used as a host computer of the tomato cultivation support system, and the other 80386 computer was used for data acquisition and environmental control operation. Extended Self—contained Prolog is suited for set operation and inference of logic, and hence, the AI computer was used to conduct diagnosis of tomato plant diseases.

RESULTS

Tomato Cultivation Support System

A tomato cultivation support system consisted of programs which were operating the 80386—computers and programs which were operating the AI computer. Figure 3 shows hierarchical program structure of the tomato cultivation support system. Because management of the standard cultivation method was memorized in the host computer prior to operation of the plant growth factory, when the support system was initially started, PC—operated production system was automatically chosen in the program operation flow shown in Fig. 3. If operators specified to choose the AI computer operation for diagnosis of plant disease or knowledge support for cultivation set—points, programs of "Production control" and "Supports of techniques" shown in Fig. 3 were activated.

Planned Production System

This system consisted of 3 parts; 1: preset cultivation procedures, 2: data acquisition by measurements of environmental conditions and tomato plant conditions, and 3: control and estimation of tomato production. When tomato cultivation was started, at first the preset cultivation procedures were operated as feedforward regulation (i.e., "Cultivation planning system" in Fig. 3). Subsequently, environmental conditions and growth conditions of tomato plants were monitored (i.e., "Recording system of cultivation data and environmental data" in Fig. 3). Then, the monitored data were fed back to the computer, and estimation of growth and fruit production of tomato plants was made (i.e., "Estimating system of growth and operation" in Fig. 3). Simultaneously, environmental conditions were adjusted to the desired conditions by feedback operation.

Cultivation Planning System

Preset cultivation plans were programed in this system ("Cultivation planning system" in Fig. 3). Figure 4 shows a tomato cultivation plan used in the plant growth factory. This plan was constructed from the standard database of meteorological conditions where the plant factory was located and physiological characteristics of tomato plants used in the present study (TABLE 1). Physiological information regarding transplanting, flowing and harvest time was expressed as functions of time period of days since seeds were sown, the accumulative temperature, and the accumulative solar radiation (TABLE 1). According to this plan, tomato fruits could be harvested 3.5 times a year.

Recording System of Cultivation Data and Environmental Data

This data acquisition system received environmental condition data such as air temperature, air humidity, CO_2 concentration, light intensity, nutrient solution temperature, electric conductivity and pH of the nutrient solution, and physiological data such as stem length, differentiation of flower buds, hormone treatment, removal of lateral buds and harvest of fruits. Environmental condition data were averaged in a period of one hour, and

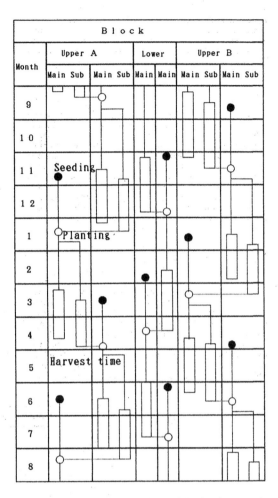

Fig. 4. Tomato cultivation plan used in the plant growth factory.

TABLE 1 Standard Database of Cultivation System

		Day (day)	Accumulative Temperature (°C)	Accumulative solar radiation (ly)
Seeding		1	0	
Trans— planting		20	400	
Planting		40	800	0
Flowering	Start	44	880	400
	End	100	2000	9000
Harvest	Main 1st start	110	2200	10000
	Sub 1st start	120	2350	10800
	Main 2nd start	122	2400	11000
	Sub 2nd start	132	2650	12300
	Main 3rd start	135	2700	12500
	Sub 3rd start	145	2900	13500
	Main end	165	3500	16500
	Sub end	180	3800	18000

303

stored in the computer every hour. Also, the maximum and minimum values of data in a day were stored.

Estimating System of Growth and Operation

Growth conditions of tomato plants could be estimated from data acquired with the "Recording system of cultivation data and environmental data". Estimated growth conditions of tomato plants could be visually displayed on the computer monitor (Fig. 5).

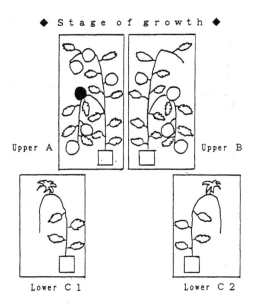

◆ S t a g e   o f   g r o w t h ◆

Upper A   Upper B

Lower C 1   Lower C 2

Fig. 5. An example of display of three different growth stages of tomato plant estimated by using the cultivation planning system on a computer monitor. ● fruit on harvest time, ○ growing fruit, ❀ flower, ⚘ flower bud.

Furthermore, operation plans for tomato cultivation could be listed on the computer monitor so that workers in the plant growth factory could effectively manage to work on schedule efficiently. An example is shown in Fig. 6.

◆ O p e r a t i o n   on   a p p o i n t i n g   t h e   d a y ◆

A	B

C1	C2

* Check point of machine *
Today: Light
Tomorrow : Fan

* Operations to plant *

Today	Tomorrow
A: Hormone application Harvest	Hormone application Harvest
B: Hormone application	Hormone application
C1, C2: Prepare seeding	Seeding

Fig. 6. An example of display of suggested operations by the support system.

Diagnosis System in Production Control

If growth condition and fruit production of tomato plants deviate from the scheduled condition set by the cultivation support system, operators can choose AI-operated diagnosis programs as shown in Fig. 3. The diagnosis program contained judgement programs for physiological disorder of plants, diagnosis of plant disease and mechanical troubles in the plant growth factory.

A schematic diagram of operation of the program was shown in Fig. 7. It is known that plants have protection mechanisms against there parasites such as fungi and bacteria. Thus, when plants suffer from diseases, plants usually exhibit physiological disorders prior to the occurrence of plant disease. Diagnosis was conducted by the AI computer, when workers in the factory noticed abnormal symptoms of plant growth. In the present study, the plant growth factory was isolated, and thus, noticeable intrusions of insects into the factory did not occur and plant diseases caused by pest parasites were not observed. Therefore, diagnosis of physiological disorder was studied.

Diagnosis System for Physiological Disorder of Tomato Plants Based on AI

In this diagnosis system, we make production rules based on following symptoms. When workers input the following information;

1) Leaf rolls.
2) Growth inhibition of leaves.
3) Growth inhibition of stems.

the computer was programed to choose the most light symptom for plant disorder. Such an order of choices were programed in the knowledge database. In the above input, the computer chose information related leaf rolls, and then, was programed to ask further questions as follows;

Where are the leaf rolls in plants?
How much did the leaf rolls occur?

When the workers input a statement that the leaf rolls occurred at lower nodes, the computer was programed to diagnose as follows;

Such leaf rolls occur in the process of maturation of plants. If there are no other symptoms, plants can be considered healthy.

then, the computer terminated the diagnosis program. If workers input a statement that the leaf rolls occurred at the position close to the apex, then the computer was programed to diagnose as follows;

Physiological disorders related with cell expansion are occurring.

The computer proceeded to ask questions related with more severe physiological disorders, and was programed to ask the following question;

Was there any symptoms in flower bud formation?

If the workers answered "NO", the computer recommended to check environmental conditions related to water stress or salinity stress such as irrigation frequency or concentrations of the nutrient solution.

If the workers answered "Yes", the computer was programed to ask questions related to higher orders of physiological damages. This diagnostic program was guided by using physiological knowledge of tomato plants, which was stored in knowledge database in computer memories. Therefore, this

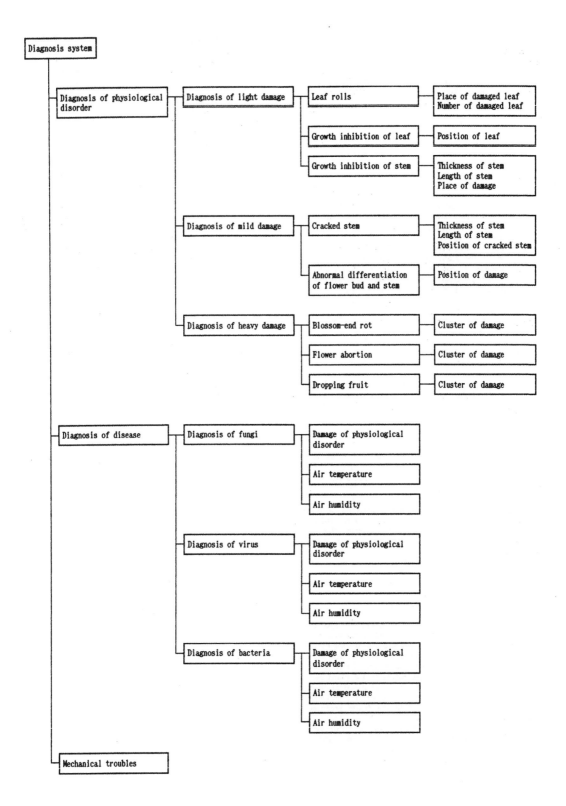

Fig. 7 Algorithm of knowledge on diagnosis system in the plant growth factory.

process cannot be processed in simple mathematical calculations used in PC computers. The AI computer was programed to memorize when the workers asked the questions and what kind of questions was asked. When the computer was asked a similar question in a short interval, the computer could skip some steps of questions and could reduce time required for inference.

DISCUSSION

We developed the computer—aided tomato cultivation support system for the plant growth factory. Yield of tomato fruits produced by using the cultivation support system in the plant growth factory was compared with yield of tomato fruits produced in conventional cultivation methods (TABLE 2). The plant growth factory operated by this support system produced 3.3 times higher yield than the conventional methods, although we could not attain the target yield (TABLE 2). Quality of fruits was compared by using sugar content measured with a Brix meter. Fruits produced in the plant growth factory were sweeter than those produced in conventional methods (TABLE 2). It is evident that the computer—aided tomato cultivation system improved both yield and quality of tomato fruits.

TABLE 2 Result of Tomato Cultivation in Plant Growth Factories in 1990

Yield (t · 10000plant/1000m²)			Brix (%)		
Former	Target	Result	Former	Target	Result
24	100	80	5	6	5.7

The diagnosis system used here also contributed higher production of fruits in the plant growth factory, because significant physiological disorders were not observed when the plant factory was operated.

Additionally, strawberry (Fraguria glandiflora Ehrh.) and lettuce (Lactuca sativa L.) plants were cultivated hydroponically in similar plant factories by using computer—aided cultivation support system for those plants. Therefore, this computer—aided cultivation support system can be applied to any plants grown in plant growth factories.

CONCLUSION

The tomato cultivation support system was developed in the present study. The cultivation support system was operated by combinations of PC computers and the AI computer. The PC computers were used as the host computer, and managed cultivation plans including measurements of environmental conditions and controls of set—points in the plant factory. The AI computer was used to diagnose physiological disorders of tomato plants and to support technical knowledge for tomato cultivation. Yields of tomato fruits produced by using this support system was significantly higher than those produced with conventional methods. Furthermore, quality of tomato fruit was improved by the cultivation support system. Similar cultivation support systems could be used in plant growth factories in order to produce high yield and better quality products.

REFERENCES

Hatou, K., H. Nishina, and Y. Hashimoto. (1990a). Tomato Cultivating Support System Based on Computer Integrated Agricultural Production. Proc. 3rd International DLG Congress for Computer Technology Integrated Decision Support System in Agriculture: 100—107.

Hatou, K., H. Nishina, and Y. Hashimoto. (1990b). Computer Integrated Agricultural Production. Proc. 11th IFAC World Congress, Vol. 11: 306—310, Pergamon Press, Oxford.

Hatou, K., Y. Kamio, and Y. Hashimoto. (1991). Computer Integrated Plant Factory based on Artificial Intelligence. 9th IFAC/IFORS Symposium on Identification and System Parameter Estimation, (in press).

Copyright © IFAC Mathematical and Control Applications
in Agriculture and Horticulture, Matsuyama, Japan 1991

IDENITIFICATION, ESTIMATION AND CONTROL
OF GLASSHOUSE SYSTEMS

P. C. Young, W. Tych and A. Chotai

Centre for Research on Environmental Systems, University of Lancaster, Lancaster LA1 4YQ, UK

Abstract   The paper outlines the *True Digital Control (TDC)* design philosophy and describes the major aspects of the *Proportional-Integral-Plus (PIP)* design procedure for systems described by transfer functions in either the backward shift (z^{-1}) or the delta (δ) operator. This procedure has been implemented in the *TDC* computer aided control system design package, and the use of this package is illustrated by an example concerned with the *LQ*-optimal *PIP* control of an *Nutrient Film Technique* system used in glasshouse horticulture.

Keywords   True Digital Control (*TDC*), Identification, Parameter estimation, Proportional-integral-plus (*PIP*) control, Pole assignment, Optimal *LQG* control, Self-adaptive control, environmental control, Nutrient Film Technique (*NFT*)

INTRODUCTION

Although the rapid development of the digital computer during the 1970's stimulated the move towards *Direct Digital Control (DDC)*, it failed to make too many inroads into the underlying faith of most traditional control systems designers in continuous-time system design methods. And even today, despite the tremendous developments in microcomputers over the 1980's and the now common use of microprocessors for the implementation of control systems, the majority of practical automatic digital controllers used in industry are still based on the mechanical digitisation of continuous-time designs, such as the ubiquitous *Proportional-Integral-Derivative (PID)* controller. In this paper, we try to promote an alternative philosophy; one based on the idea of *True Digital Control (TDC)*, as propounded in a number of our previous publications (Young et al, 1987b, 1988, 1991a; Young, 1989, 1991; Chotai et al, 1990a, 1991b).

The *TDC* philosophy rejects the idea that a digital control system should be initially designed in continuous-time terms. Rather, it demands that the control systems analyst considers the design from a digital, sampled-data standpoint, even when fast sampling, near continuous-time operation is required. The *TDC* design procedure consists of the following three major steps:-

(1) **Recursive identification and estimation** of discrete-time models based on the analysis of either planned experimental data, or via model reduction from data generated by (usually high order) continuous or discrete-time simulation models.
(2) **Off-line TDC system design and initial evaluation** based on the models from step (1), using an *iterative* application of an appropriate discrete-time design methodology, coupled with closed loop sensitivity analysis based on Monte-Carlo stochastic simulation.
(3) **Implementation, testing and evaluation** of the control system on the real process; in the case of self-tuning or self-adaptive control, employing on-line versions of the recursive estimation algorithms used in step (1).

The TDC design philosophy is best exemplified in a new control systems design procedure developed at Lancaster over the past few years. This is based on the definition of a special *Non-Minimum State-Space (NMSS)* representation of systems whose dynamic behaviour can be characterised by transfer functions in either the backward shift (z^{-1}) or the discrete-differential (δ) operator. Although simple to implement in practice, the resultant *Proportional-Integral-Plus (PIP)* controller is attractive because it exploits the power of State Variable Feedback (*SVF*). Moreover, since it has been developed directly within the context of glasshouse horticulture, we believe that the *PIP* approach has particular potential for application to horticultural and agricultural systems.

In two other papers presented at this Workshop (Chotai and Young, 1991; Chotai et al, 1991c), we explore the practical potential of *PIP* by describing how fixed gain and adaptive *PIP* systems have been applied to the control of environmental temperature in a glasshouse and nutrient levels in a *Nutrient Film Technique (NFT)* system used in glasshouse horticulture. The present paper outlines the theoretical background to the various identification, estimation and control procedures on which these *PIP* system designs are based and briefly describes a *Computer Aided Control System Design (CACSD)* package designed to facilitate the practical application of these procedures.

PIP vs Conventional Control System Design

Most conventional control systems used in horticulture and agriculture are either analog or utilise some form of *Direct Digital Control (DDC)*. In the latter case, the control system design is normally in the form of digitised versions of *Proportional-Integral-Derivative (PID)* or *PI* controllers, which are tuned manually on commission. Such tuning is quite often based on trial-and-error procedures and rarely involves any reference to mathematical models of the system to be controlled. Moreover, the controllers may well require manual re-tuning if the system dynamics change over time. Such conventional controllers can also suffer from various problems: for example, since they operate at fairly rapid sampling frequencies, they can experience stability problems if the controlled system is characterised by sizable pure time delays.

Unlike conventional *DDC* controllers, the *PIP* system utilises the alternative *TDC* design philosophy: namely, it *actively* exploits the digital mechanisation directly to achieve better control performance and it is *not* based on a digitised continuous-time design. This involves several innovations: first, the design is always based on a mathematical model of the system, normally obtained by prior modelling using general computer programs such as the "Identification Toolbox" in *Matlab* (Moler et al, 1987), or the *micro-CAPTAIN* package (Young and Benner, 1989); secondly, because it utilises state variable feedback, the *PIP* controller is inherently more powerful and flexible than the conventional *PI* and *PID* alternatives in achieving such objectives as closed loop pole assignment or optimal control; thirdly, the *PIP* controller has a structure which avoids common control system problems, such as integral wind-up; and finally, it is relatively robust to pure time (transportation) delays because, in the z^{-1}operator version, it uses control sampling intervals that would be considered quite coarse by conventional standards.

Of course, these advantages are obtained at some cost; not in terms of implementational complexity, since the *PIP* system can be very little more complicated than the conventional three term controller, but in relation to the design process itself. In particular, this involves the use of more sophisticated theoretical analysis and access to specially designed *CACSD* programs that assist in the application of this analysis to practical design problems. In subsequent Sections of this paper, we outline this theoretical background and describe briefly the *CACSD* program developed at Lancaster.

TF MODELS IN THE z^{-1} AND δ OPERATORS

In all model-based control system design procedures, the form of the models and the associated theoretical background is of paramount importance. Most continuous-time procedures are based either on a Laplace transform transfer function model or state-space differential equations. It is not surprising, therefore, that TDC designs utilise the discrete-time equivalents of these models. It is possible to unify the model analysis in terms of a general operator (see e.g. Goodwin et al, 1988; Middleton and Goodwin, 1990). However, it is felt that the analysis is more transparent if we consider separately the two major forms of the transfer function model in the discrete-time domain: namely, the well known transfer function in the backward shift (z^{-1}, or equivalently the forward shift, z transform) operator; and the alternative, but less well known, delta (δ or discrete differential) operator transfer function.

Backward Shift (z^{-1}) TF Model

The general, discrete-time, z^{-1} operator transfer function (TF) representation of an *nth* order single input, single output (*SISO*), discrete-time system, with a sampling interval of Δt time units, is normally written in the following form,

$$y(k) = \frac{B(z^{-1})}{A(z^{-1})} u(k) \qquad (1)$$

where $A(z^{-1})$ and $B(z^{-1})$ are the following polynomials in z^{-1},

$$A(z^{-1}) = 1 + a_1 z^{-1} + \dots + a_n z^{-n}$$

$$B(z^{-1}) = b_1 z^{-1} + b_2 z^{-2} + \dots + b_m z^{-m}$$

In general, no prior assumptions are made about the nature of the transfer function $B(z^{-1})/A(z^{-1})$, which may be marginally stable, unstable, or possess non-minimum phase characteristics. However, if the input-output behaviour of the

system is characterised by any pure time delay of τ sampling intervals, then this is accommodated by assuming that the first τ coefficients of the $B(z^{-1})$ polynomial, i.e.b_1, b_2, $\dots$, b_τ, are all zero.

The Discrete Differential (δ) Operator TF Model

A interesting alternative to the z^{-1} operator TF model is the following "discrete differential operator" model, which was revived recently under the title delta (δ) operator by Goodwin and his co-workers (see Middleton and Goodwin, 1990),

$$y(k) = \frac{B(\delta)}{A(\delta)} u(k) \qquad (2)$$

where $A(\delta)$ and $B(\delta)$ are polynomials of the following form,

$$A(\delta) = \delta^p + a_1 \delta^{p-1} + \dots + a_p$$
$$B(\delta) = b_1 \delta^{p-1} + \dots + b_p$$

with the index $p = \max(n,m)$ and the δ operator, for the sampling interval Δt, defined as follows in terms of the forward shift operator z,

$$\delta = \frac{z-1}{\Delta t} \; ; \; i.e. \; \delta x(k) = \frac{x(k+1)-x(k)}{\Delta t}$$

Note that, for convenience, we have retained the nomenclature for the parameters a_i, b_j used in the z^{-1} operator case; but the parameter values will, of course, take on different values (see remark (3) below)

Remarks

(1) As $\Delta t \to 0$, the delta operator reduces to the derivative operator ($s = d/dt$) in continuous time (i.e $\delta \to s$).
(2) Given a polynomial of any order n in the z operator, this will be exactly equivalent to some polynomial in δ, also of order n.
(3) As a consequence of (2), we can easily move between the z and δ operator domains. For example, a δ design can be implemented in practice by converting it back to the z^{-1} domain, so avoiding direct differencing (although this may not be the best form of implementation). Also, the δ operator model coefficients are related to forward z operator coefficients by simple vector matrix equations (see Chotai et al, 1990a)
(4) One attraction of the δ operator model to those designers who prefer to think in continuous-time terms is that it can be considered as a direct approximation to a continuous-time system. For example, it is easy to see that the unit circle in the complex z plane maps to a circle with centre $-1/\Delta t$ and radius $1/\Delta t$ in the complex δ plane; so that, as $\Delta t \to 0$, this circular stability region is transformed to the left half of the complex s plane. For very rapidly sampled systems, therefore, the δ operator model can be considered in almost continuous-time terms, with the pole positions in the δ plane close to those of the 'equivalent' continuous-time system in the s plane; and with the *TF* parameters directly yielding information on factors such as the approximate natural frequency and damping ratio.

MODEL IDENTIFICATION AND PARAMETER ESTIMATION

In control systems design, it makes good sense to use recursive methods model parameter estimation. Not only do these allow for the modelling of nonstationary systems, but they can also be implemented on-line for self-tuning and self-adaptive control applications. Of the many recursive methods that are now available for the estimation of z^{-1} operator transfer function model parameters, only one can also be applied directly to δ operator models. This is the *Simplified Refined Instrumental Variable* (SRIV) procedure (Young, 1984, 1985) which exploits special adaptive prefiltering, both to achieve good estimation performance and, in the δ operator case, to avoid numerical differentiation. As such, the *SRIV* algorithm provides the most obvious vehicle for identification and estimation within the *TDC* context.

The *SRIV* Algorithm for z^{-1} and δ operator *TF* models

If we consider first the z^{-1} operator case, the adaptive prefiltering which characterises the *SRIV* algorithm can be justified qualitatively by considering the following stochastic form of equation (1),

$$y(k) = \frac{B(z^{-1})}{A(z^{-1})} u(k) + e(k) \qquad (3)$$

where $e(k)$ is a zero mean, serially uncorrelated sequence of random variables with variance σ^2; and the *TF* is assumed to be stable, i.e. the roots of the characteristic equation $A(z)=0$ all lie within the unit circle of the complex z plane. This equation can be written in the following alternative vector form, which is *linear-in-the-parameters* $\{a_i, b_j\}$ of the *TF* model,

$$y(k) = z(k)^T a + \eta(k)$$

where,

$$z(k)^T = [-y(k-1), \ldots, -y(k-n) \; u(k-1), \ldots, u(k-m)]$$

$$a = [a_1 \; a_2, \ldots, a_n \; b_1, \ldots, b_m]^T$$

and $\eta(k)$ is a noise variable defined as follows in relation to the original white noise $e(k)$,

$$\eta(k) = e(k) + a_1 e(k-1) + \ldots + a_n e(k-n)$$

Most estimation problems are posed in a manner such that the variable to be minimised has white noise properties. Thus, a sensible error function is the *response* or *prediction error*, $\hat{e}(k)$,

$$\hat{e}(k) = y(k) - \frac{\hat{B}(z^{-1})}{\hat{A}(z^{-1})} u(k)$$

where $\hat{B}(z^{-1})$ and $\hat{A}(z^{-1})$ are estimates of the *TF* polynomials $A(z^{-1})$ *and* $B(z^{-1})$. Unfortunately, this is nonlinear in the unknown parameters and so cannot be posed directly in simple, linear estimation terms. However, the problem becomes *linear-in-the-parameters* if we assume prior knowledge of $A(z^{-1})$ in the form of an estimate $\hat{A}(z^{-1})$: then the error equation can be written in the form,

$$\hat{e}(k) = \frac{1}{\hat{A}(z^{-1})} \left[\hat{A}(z^{-1}) y(k) - \hat{B}(z^{-1}) u(k) \right]$$

which can be rewritten as,

$$\hat{e}(k) = \hat{A}(z^{-1}) y^*(k) - \hat{B}(z^{-1}) u^*(k)$$

where,

$$y^*(k) = \frac{1}{\hat{A}(z^{-1})} y(k) \quad ; \quad u^*(k) = \frac{1}{\hat{A}(z^{-1})} u(k) \qquad (4)$$

are "prefiltered" variables, obtained by passing $y(k)$ and $u(k)$ through the prefilter $1/\hat{A}(z^{-1})$.

With this reasoning in mind, the ordinary recursive IV algorithm (e.g. Young, 1984) can be applied iteratively to estimate the model parameter vector a, with the variables $y(k)$, $u(k)$ and the instrumental variable $\hat{x}(k)$ replaced, at each iteration, by their adaptively prefiltered equivalents $y^*(k)$, $u^*(k)$ and $\hat{x}^*(k)$, respectively, and with the prefilter parameters based on the parameter estimates obtained at the previous

iteration (see Young, 1984, 1985). The main recursive part of the SRIV algorithm takes the form,

$$\hat{a}(k) = \hat{a}(k-1) + g(k) \{ y^*(k) - z^*(k)^T \hat{a}(k-1) \} \qquad (i)$$

where,

$$g(k) = P(k-1)\hat{x}^*(k) [1 + z^*(k)^T P(k-1)\hat{x}^*(k)]^{-1} \qquad (ii) \qquad (5)$$

and

$$P(k) = P(k-1) + g(k) z^*(k)^T P(k-1) \qquad (iii)$$

where $P(k)$ is related to the covariance matrix $P^*(k)$ of the estimated parameter vector $a(k)$ by the equation,

$$P^*(k) = \sigma^2 P(k) \qquad (6)$$

and an estimate $\hat{\sigma}^2$ of the variance σ^2 can be obtained from an additional recursive equation based on the squared values of a suitably normalised recursive innovation sequence (see Young, 1984; p.100; Young et al, 1991b).

This complete recursive algorithm (5) can be considered simply as a modification of the well known *Recursive Least Squares* (*RLS*) algorithm, with the data vector $z^*(k)$ replaced alternately by $\hat{x}^*(k)$. At the jth iteration, the prefiltered instrumental variable $\hat{x}^*(k)$ required in the definition of $\hat{x}^*(k)$ is generated by adaptively prefiltering the output of an "auxiliary model" of the following form,

$$\hat{x}^*(k) = \frac{1}{\hat{A}_{j-1}(z^{-1})} \hat{x}(k) \quad ; \quad \hat{x}(k) = \frac{\hat{B}_{j-1}(z^{-1})}{\hat{A}_{j-1}(z^{-1})} u(k) \qquad (7)$$

adaptive prefilter   *adaptive auxiliary model*

where $\hat{A}_{j-1}(z^{-1})$ and $\hat{B}_{j-1}(z^{-1})$ are estimates of the polynomials $A(z^{-1})$ and $B(z^{-1})$ obtained by reference to the parameter estimates obtained by the algorithm at the end of the *previous* (j-1)th iteration. The prefiltered input and output variables are obtained in a similar manner. Such a *recursive-iterative* or *relaxation* approach normally requires only two to three iterations to converge on sensible estimates of the parameters.

The δ operator version of the *SRIV* algorithm (5) is applied to the vector version of the δ operator *TF* model, i.e.,

$$\delta^p y(k) = z(k)^T a + \eta(k) \qquad (8)$$

where now,

$$z(k) = [\delta^{p-1} y(k), \delta^{p-2} y(k), \ldots, y(k), \delta^{p-1} u(k), \ldots, u(k)]^T$$

with a and $\eta(k)$ defined accordingly. The resulting algorithm is algebraically identical to the z^{-1} version: it is simply necessary to define the data vectors appropriately, i.e.,

$$z^*(k) = [\delta^{p-1} y^*(k), \delta^{p-2} y^*(k), \ldots, y^*(k), \delta^{p-1} u^*(k), \ldots, u^*(k)]^T$$

$$\hat{x}^*(k = [\delta^{p-1} \hat{x}^*(k), \delta^{p-2} \hat{x}^*(k), \ldots, \hat{x}^*(k), \delta^{p-1} u^*(k), \ldots, u^*(k)]^T$$

with $y^*(k)$ in the innovation term $\{y^*(k) - z^*(k)^T \hat{a}(k-1)\}$ replaced by $\delta^p y^*(k)$. Here, the star superscript again indicates that the variables have been adaptively prefiltered, this time by the δ operator prefilter. This prefilter now performs two functions: first it helps to improve the statistical efficiency and yield lower variance estimates; second, the elements of $z^*(k)$ and $\hat{x}^*(k)$ can be obtained directly from the adaptive prefilter, thus avoiding direct multiple differencing of the input and output signals, with its attendant problems of noise amplification. Used in this manner, the prefilters are seen as direct descendants of the "state variable filters" proposed by the first author in the early nineteen sixties for the estimation of continuous-time system models (Young, 1964, 1965, 1966, 1969a, 1969b, 1970)).

Model Order Identification

Model order identification is extremely important in *TF* modelling for control system design. A successful identification procedure based on *IV* estimation (see Young, 1989) is to choose the model which minimises the following identification statistic,

$$YIC = log_e \left\{ \frac{\sigma^2}{\sigma_y^2} \right\} + log_e \{ NEVN \} \qquad (9)$$

where,

σ^2 is the sample variance of the model residuals e(k)
σ_y^2 is the sample variance of the measured system output $y(k)$ about its mean value.

while *NEVN* is the "Normalised Error Variance Norm" (Young et al, 1980) defined as,

$$NEVN = \frac{1}{np} \sum_{i=1}^{i=np} \frac{\hat{\sigma}^2 p_{ii}}{\hat{a}_i^2} \qquad$$

Here, in relation to the *TF* models (1) and (2), *np* is the total number of parameters estimated, i.e. *n+m* for model (1) and *2p* for model (2); $\hat{a}_i^2$ is the estimate of the *ith* parameter in the parameter vector *a*; while p_{ii} is the ith diagonal element of the *P(N)* matrix, where *N* is the sample size (so that $\hat{\sigma}^2 p_{ii}$ is an estimate of the error variance associated with the *ith* parameter estimate after *N* samples).

The first term in (9) provides a normalised measure of how well the model explains the data: the smaller the variance of the model residuals in relation to the variance of the measured output, the more negative the first term becomes. Similarly, the second term is a normalised measure of how well the parameter estimates are defined for the *np*th order model: clearly the smaller the relative error variance, the better defined are the parameter estimates in statistical terms, and this is once more reflected in a more negative value for the term. Thus the model which minimises the YIC provides a good compromise between model fit and parametric efficiency: as the model order is increased, so the first term tends always to decrease; while the second term tends to decrease at first and then to increase quite markedly when the model becomes over-parameterised and the standard error on its parameter estimates becomes large in relation to the estimated values (in this connection, note that the square root of $\hat{\sigma}^2 p_{ii} / \hat{a}_i^2$ is simply the relative standard error on the ith parameter estimate).

PIP CONTROL DESIGN FOR THE z^{-1} OPERATOR MODEL

The special non-minimal state-space (*NMSS*) representation which we associate with the z^{-1} transfer function model (1) is defined by the following state vector,

$$x^T = [y(k) \, y(k-1) \, ... \, y(k-n+1) \, u(k-1) \, ... \, u(k-m+1) \, z(k)]$$

where $z(k)$ is an "integral of error" term at sampling instant k,

$$z(k) = z(k-1) + \{y_d(k) - y(k)\}$$

in which $y_d(k)$ is the reference or command input to the servomechanism system. This integral-of-error term is introduced to ensure type 1 servomechanism performance in the resultant *PIP* control system (Young and Willems, 1972). The *NMSS* representation is then obtained directly in the following form,

$$x(k) = F \, x(k-1) + g \, u(k-1) + d \, y_d(k) \qquad (10)$$

where,

$$F = \begin{bmatrix} -a_1 & -a_2 & .... & -a_{n-1} & -a_n & b_2 & b_3 & ..... & b_{m-1} & b_m & 0 \\ 1 & 0 & .... & 0 & 0 & 0 & 0 & ..... & 0 & 0 & 0 \\ 0 & 1 & .... & 0 & 0 & 0 & 0 & ..... & 0 & 0 & 0 \\ . & . & . & . & . & . & . & & . & . & . \\ 0 & 0 & .... & 1 & 0 & 0 & 0 & ..... & 0 & 0 & 0 \\ 0 & 0 & .... & 0 & 0 & 0 & 0 & ..... & 0 & 0 & 0 \\ 0 & 0 & .... & 0 & 0 & 1 & 0 & ..... & 0 & 0 & 0 \\ 0 & 0 & .... & 0 & 0 & 0 & 1 & ..... & 0 & 0 & 0 \\ . & . & . & . & . & . & . & & . & . & . \\ 0 & 0 & .... & 0 & 0 & 0 & 0 & ..... & 1 & 0 & 0 \\ a_1 & a_2 & .... & a_{n-1} & a_n & -b_2 & -b_3 & ..... & -b_{m-1} & -b_m & 1 \end{bmatrix}$$

$$g = [\, b_1 \, 0 \, ... \, 0 \, 1 \, 0 \, 0 \, ... \, 0 \, 0 \, -b_1 \,]^T$$

$$d = [\, 0 \, 0 \, ... \, 0 \, 0 \, 0 \, 0 \, ... \, 0 \, 0 \, 1 \,]^T$$

The definition of this particular *NMSS* state vector is significant: it means that the State Variable Feedback (*SVF*) control law involves only the *directly measurable* input and output signals, together with their past values, all of which can be stored in the digital computer. As a result, any *SVF* control system design (e.g. pole assignment or optimal) does not need to resort to the complication of a state reconstructor (i.e. observer), since the *effective* output matrix in control terms is the I_{m+n} identity matrix. The conditions for controllability of the *NMSS* model are given by the following theorem.

Theorem (Wang and Young, 1988). Given a single input-single output system described by (1), the non-minimal state space representation (10), as described by the pair {F,g}, is completely controllable if, and only if, the following two conditions are satisfied:

(i)the polynomials A (z^{-1}) and B (z^{-1}) are coprime
(ii) $b_1 + b_2 + ... + b_m \neq 0$

The coprimeness condition is equivalent to the normal requirement that the transfer function model (1) should have no pole-zero cancellations. The second condition avoids the presence of a zero at unity which would cancel with the unity pole associated with the integral action.

The Control Algorithm

In the context of the *NMSS* system representation, the automatic control objective is to design an *SVF* control law with gain vector $k = [\, f_0, f_1, ..., f_{n-1}, g_1, ..., g_{m-1}, k_I\,]$, i.e.,

$$u(k) = -k^T x(k) = -f_0 y(k) - f_1 y(k-1) - ... - f_{n-1} y(k-n+1)$$
$$- g_1 u(k-1) - ... - g_{m-1} u(k-m+1) - k_I z(k)$$

such that either the closed loop poles are at preassigned positions in the complex z-plane; or the system is optimised in an Linear-Quadratic (*LQ*) sense (or Linear-Quadratic-Gaussian (*LQG*) in the stochastic situation).

In the pole assignment case, the closed-loop system block diagram takes the Proportional-Integral-Plus (*PIP*) form shown in Fig.1.The closed-loop *TF* associated with this block diagram can be written,

$$y(k) = \frac{D(z^{-1})}{CL(z^{-1})} \, y_d(k)$$

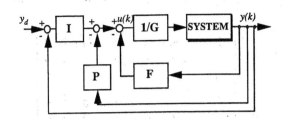

P - Proportional control (gain f_0)
I - Integral control (gain k_I)
F_o - Feedback filter $f_1 z^{-1} + f_2 z^{-2} + ... + f_{n-1} z^{-n+1}$
G - Input filter $1 + g_1 z^{-1} + ... + g_{m-1} z^{-m+1}$

Fig. 1. Proportional-Integral-Plus (PIP) control system.

where,

$$D(z^{-1}) = k_I B(z^{-1})$$

and,

$$CL(z^{-1}) = (1-z^{-1})[G(z^{-1})A(z^{-1}) + F(z^{-1})B(z^{-1}) + f_0 B(z^{-1})] + k_I B(z^{-1})$$

The closed-loop characteristic polynomial $CL(z^{-1})$ can now be expanded and the coefficients for like powers of z^{-1} equated to those the desired closed loop characteristic polynomial,

$$d(z^{-1}) = 1 + d_1 z^{-1} + d_2 z^{-2} + ... + d_{m+n} z^{-(n+m)}$$

which is chosen to ensure the assignment of the closed loop poles to designer-specified positions in the complex z plane. This results in the following set of linear, algebraic equations,

$$\sum(n,m) \cdot k = S_c \cdot f \qquad (11)$$

where k is the SVF control gain vector, i.e.,

$$k = [f_0, f_1, ..., f_{n-1}, g_1, ..., g_{m-1}, k_I]$$

while S_c and f are the following vectors,

$$S_c = [d_1, d_2, ..., d_{n+m-1}, d_{n+m}]^T$$
$$f = [a_1 - 1, a_2 - a_1, ..., a_n - a_{n-1}, -a_n, 0, 0, ..., 0]^T$$

and $\Sigma(n,m)$ is the $(n+m)x(n+m)$ matrix shown in Fig.2 (see Young et al.; 1987b),

Provided the controllability conditions of the above theorem are satisfied, this set of linear, simultaneous equations can be solved to yield the unique set of SVF control gains which define the vector k.. In the LQ or LQG situation, the SVF gains are computed to minimise the following quadratic cost function,

$$J = \sum_{i=1}^{i=\infty} x(k)^T Q x(k) + q_u u(k)^2 \qquad (12)$$

where Q is an $(n+m)x(n+m)$ diagonal matrix, with elements defined as follows,

$$Q = diag.[q_1 q_2 ... q_n q_{n+1} q_{n+2} ... q_{n+m-1} q_{n+m}]$$

with,

$$q_1 = q_2 = ... = q_n = q_y; \quad q_{n+1} = q_{n+2} = ... = q_{n+m-1} = q_u; \quad q_{n+m} = q_z$$

Fig.2 The $\Sigma(n,m)$ matrix

Here q_y, q_u and q_z are the *partial* weightings on the output, input and integral-of-error variables in the NMSS vector x. These partial weightings are defined as,

$$q_y = \frac{W_y}{n} \; ; \; q_u = \frac{W_u}{m} \; \text{ and } \; q_z = W_z \qquad (13)$$

so that the total weightings on the output $y(k)$, input $u(k)$ and integral-of-error $z(k)$ are W_y, W_u, and W_z. These three weighting variables are then chosen in the usual manner to achieve the desired closed loop performance. The optimum SVF gains for the selected weighting values are obtained by computing the steady state solution of the associated, discrete-time, matrix Riccati equation, given the NMSS system description (F, g) and weighting matrices (Q, R). Note that, in this case, $R = q_u$ because of the special, NMSS formulation.

PIP CONTROL DESIGN FOR THE δ OPERATOR MODEL

Following a similar design philosophy to that used for the z^{-1} operator model, the δ operator TF model can be represented by the following NMSS equations,

$$\delta x(k) = F x(k) + g v(k) + d y_d(k) \qquad (14)$$

where,

$$g = [b_1 \; 0 \; ..... \; 0 \; 1 \; 0 \; 0 \; ..... \; 0 \; 0 \; 0]^T$$
$$d = [0 \; 0 \; ..... \; 0 \; 0 \; 0 \; 0 \; ..... \; 0 \; 0 \; 1]^T$$

and,

$$F = \begin{bmatrix} -a_1 & -a_2 & .... & -a_{P-1} & -a_P & b_2 & b_3 & ..... & b_{P-1} & b_P & 0 \\ 1 & 0 & .... & 0 & 0 & 0 & 0 & ..... & 0 & 0 & 0 \\ 0 & 1 & .... & 0 & 0 & 0 & 0 & ..... & 0 & 0 & 0 \\ \cdot & & & & & & & & & & \\ 0 & 0 & .... & 1 & 0 & 0 & 0 & ..... & 0 & 0 & 0 \\ 0 & 0 & .... & 0 & 0 & 0 & 0 & ..... & 0 & 0 & 0 \\ 0 & 0 & .... & 0 & 0 & 1 & 0 & ..... & 0 & 0 & 0 \\ 0 & 0 & .... & 0 & 0 & 0 & 1 & ..... & 0 & 0 & 0 \\ \cdot & & & & & & & & & & \\ 0 & 0 & .... & 0 & 0 & 0 & 0 & ..... & 1 & 0 & 0 \\ 0 & 0 & .... & 0 & -1 & 0 & 0 & ..... & 0 & 0 & 0 \end{bmatrix}$$

In this formulation, however, the control variable is denoted by $v(k)$, which is defined as follows in terms of the control input $u(k)$,

$$v(k) = \delta^{P-1} u(k)$$

with the associated state vector $x(k)$ defined as,

$$x(k) = [\delta^{P-1} y(k), \delta^{P-2} y(k), ..., \delta y(k), y(k), \delta^{P-2} u(k),$$
$$\delta u(k), u(k), z(k)]^T$$

In these equations, $z(k)$ is, once again, the "integral of error" state, which is now defined in terms of the the inverse delta operator, or digital integrator δ^{-1}, i.e.,

$$z(k) = \delta^{-1} \{y_d(k) - y(k)\}$$

and, as before, $y_d(k)$ is the reference or command input at the kth sampling instant.

The Control Algorithm

As in the z^{-1} operator case, the SVF control law is defined in terms of the state variables. In this δ operator situation, however, these are the output and input and their discrete differentials up to the appropriate order, as well as the integral of error state $z(k)$. This control law can be written in the form,

$$v(k) = -k^T x(k)$$

where now,

$$k^T = [f_{p-1} \ f_{p-2} \ \cdots f_0 \ g_{p-2} \ g_{p-3} \ \cdots g_0 \ k_I]$$

is the SVF control gain vector for the δ operator model NMSS form.

The block diagram of this δ operator PIP control system is similar to the z^{-1} operator system in Fig.1, with appropriate changes in the form of the control filters F and G. In order to avoid the numerical differentiation inherent in these δ operator filters and the consequent danger of noise amplification, however, the controller will normally be converted into an alternative, practically realisable, form prior to implementation. This can be achieved by either (a) converting it back to the z^{-1} domain; (b) manipulating it into the form of realisable ratios of δ operator polynomials; or (c) invoking the separation theorem and introducing an NMSS observer (i.e. a state reconstruction filter or Kalman filter in the stochastic case), with the SVF control law implemented directly in δ operator terms and the state estimates replacing the discrete differentials.

THE TDC COMPUTER AIDED CONTROL SYSTEM DESIGN (CACSD) PACKAGE

The efficacy of any modern control systems design procedure depends to a large extent on the availability of suitable computer software which can assist the control systems analyst in the various stages of the design process. Our major objectives in developing such a *Computer Aided Control System Design (CACSD)* package (see Tych et al, 1991a,b) for PIP control system design have been to produce a user-friendly, but powerful, set of integrated design tools, which are as portable as possible and will also prove popular to the widest possible cross-section of the control and systems community.

Of course, it is not possible to guarantee that such objectives will be met completely but, by integrating the whole package, which we have called TDC, within the popular Matlab

program (Moler et al, 1987), we believe we have designed a system which will go some way to satisfying the current tastes of many control systems analysts. Moreover, in shielding the user from the Matlab command line by means of a well designed but simple Graphical User Interface (GUI), while still allowing easy access to Matlab if this is desired, we believe we have provided a good combination of a well designed, systematic procedure for day-to-day design studies, and a flexible tool for more novel research exercises.

The Graphical User Interface (GUI)

The GUI of the package takes full advantage of the Matlab graphical interactive capabilities. It is organized as a two level interface, with the main menu calling lower level menus. The fact that the user interface is almost entirely written in Matlab's macro language, constrains the flexibility and "look" of the GUI to some extent but gives portability to most of the computing platforms where Matlab has been implemented. As shown in Fig.3, the GUI shields the user not acquainted with Matlab from its command line, whilst giving easy access to Matlab's commands, other toolboxes, and the workspace, if this is required.

Fortran / C Routines

In the main SUN version of the TDC package some numerical computation (e.g. the recursive identification and estimation; and Monte Carlo closed loop sensitivity analysis) is performed by external, compiled programs written in either *Fortran* or *C*. These programs then communicate with Matlab by means of disk files, so speeding up the calculations, often to a considerable degree. In some cases, where a user interface more flexible than that provided by Matlab is required, the programs also take advantage of the SUN's' interactive facilities. Clearly, these latter components of the TDC system are not portable to other machines, where alternative arrangements prove necessary. However, such special facilities have been kept to a minimum.

Internal Organisation of the Package

The TDC package is designed as a quasi-object-oriented, high level specialized programming language using the Matlab environment. The commands are issued using either a system of menus and input panels, as shown in Fig.4, or via the Matlab command line. The two main classes of objects defined in the system are *data* and *systems*. *Data* are defined as one or more series of samples, while a *system* consists of *blocks* (with parameters) and structure (connections between the blocks, inputs and outputs).

Control System Design and Evaluation Tools

The control system design is implemented according to the TDC approach based on PIP control system designs. A high level macro **control** leads the user through the design procedure, which includes the following options,

❑ choice between pole assignment and LQ-optimal control;
❑ optional cancellation of numerator effects for minimum phase systems;
❑ graphical specification of the closed loop pole locations or entry of LQ cost function weights;
❑ standard evaluation (time and frequency responses, pole locations) of closed loop behaviour;
❑ Monte Carlo (MC) closed loop sensitivity analysis based on the uncertainty associated with the open loop model parameter estimates.

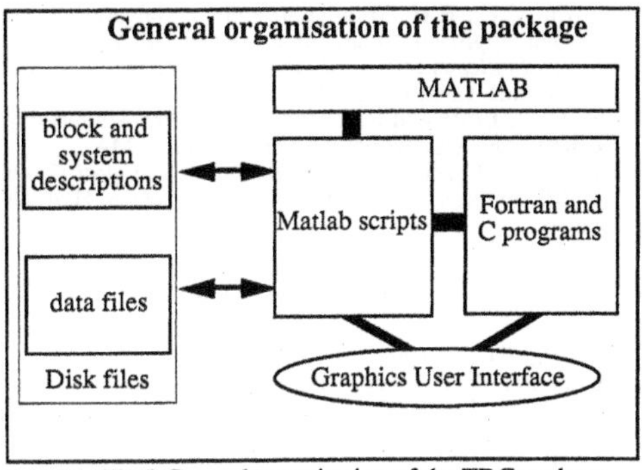

General organisation of the package

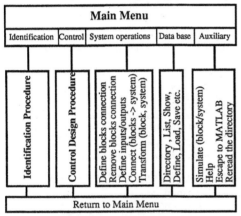

Fig.3 General organisation of the TDC package

Structure of the package

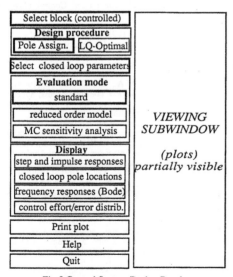

Fig.4 Structure of the TDC package

The associated interactive control system design panel is shown in Fig.5. Buttons are highlighted to indicate the current status (i.e. the selections made by the user); and the graphical output appears in the adjacent graphical window.

Fig.5 Control System Design Panel
in *TDC* program package

A PRACTICAL EXAMPLE: OPTIMAL *PIP* CONTROL OF A PILOT SCALE *NFT* FLOW SYSTEM

Other publications (Young et al, 1987a,b, 1988, 1991a; Young, 1989; Chotai and Young, 1991), have described our research on the modelling and *PIP* control of *NFT* systems based on z^{-1} operator models. Here, we will consider the alternative δ operator approach, based on the same experimental data set used in these other studies and shown in Fig.6. This was obtained from a physical, pilot scale model of the *NFT* flow system which is dynamically similar to the full scale process but operates on a much shorter time scale, so facilitating experimentation.

For a sampling interval Δt defined as unity, for convenience, the δ operator *SRIV* algorithm in the *TDC* package identifies a 3rd order model (i.e. $p=3$ in equation (2)), with either zero or unity pure time delay. In the former case, the estimated model has the following non-minimum phase (*nmp*) form,

$$y(k) = \frac{-0.0228\,\delta^2 + 0.0219\,\delta + 0.1662}{\delta^3 + 1.1156\,\delta^2 + 0.6271\,\delta + 0.00411}\,u(k) + \xi(k)$$

where $\xi(k)$ represents the noise on the system, including the effects of a sinusoidal leak introduced to simulate nutrient uptake by the plants. The covariance matrix $P^*(N)$ associated with the parameter estimates in this model has the form,

$$10^{-4}\begin{bmatrix} 24.10 & 14.91 & 0.1094 & -0.3926 & -0.2933 & 0.4180 \\ 14.91 & 27.41 & 0.1946 & 0.5077 & -0.6210 & 0.7586 \\ 0.1094 & 0.1946 & 0.0017 & 0.0039 & -0.0046 & 0.0058 \\ -0.3926 & 0.5077 & 0.0039 & 0.1610 & 0.0197 & 0.0145 \\ -0.2933 & -0.6210 & -0.0046 & 0.0197 & 0.0819 & -0.0177 \\ 0.4180 & 0.7586 & 0.0058 & 0.0145 & -0.0177 & 0.0217 \end{bmatrix}$$

where the sample size $N=160$. This model has similar dynamic characteristics to the previously estimated z^{-1} operator models, except that these former models had an explicit time delay of one sampling interval: here the *nmp* characteristics are effectively acting as a surrogate for this time delay, and the model was selected here in order to demonstrate that the *PIP* design procedure is applicable to *nmp* systems.

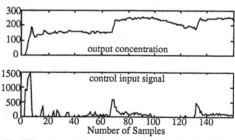

Fig.6 Experimental results from a scale model of an *NFT* flow system (a) output dye concentration; (b) control input signal to peristaltic pump

The initial *PIP* design results for this model, as obtained from the *TDC* package, are given in Fig.7. Here the control system is designed on an optimal LQ basis, with $q_y=1.0$; $q_u=0.5$ and $q_z=0.1$. The closed loop step responses, and pole locations are shown in 7(a) and (b), respectively. Parts (c) and (d) of the Figure illustrate the sensitivity of the design to the uncertainty on the parameter estimates: these were obtained using Monte Carlo stochastic simulation analysis, with the control system gains set at the nominal values obtained in the LQ design (and based on the nominal parameter estimates given above), but with the model parameters for each stochastic realisation selected from a Gaussian distribution with mean defined by the nominal parameter estimates and covariance defined by the associated covariance matrix $P^*(N)$ given above.

The resulting ensemble of step/impulse responses and "stochastic root loci" given in Figs. 7(c) and (d), respectively, all indicate that, while the nominal design has very good response characteristics, as shown in Figs.7(a) and (b), it is a rather sensitive to the inherent uncertainty in the model. In particular, there is clearly some tendency for the oscillatory mode (which arises from the recirculation in the flow system) to become excited during a small proportion of the stochastic realisations. This suggests that this initial design should be rejected and the design procedure repeated, with modified weightings in the performance index, in order to obtain greater stability margins and less sensitivity to uncertainty. Improved designs such as these will be discussed at the Workshop.

CONCLUSIONS

This paper has outlined the *True Digital Control (TDC)* design philosophy and described briefly the major aspects of the *Proportional-Integral-Plus* design procedure for single input-single output systems described by transfer functions in either the backward shift or the delta operator. Since this procedure is based on *Non-Minimum State-Space (NMSS)* concepts, it can be extended straightforwardly to multivariable and stochastic systems (Wang, 1988; Chotai, 1990b). And, when combined with on-line, recursive estimation techniques, such as the *SRIV* algorithms described in this paper, it provides an excellent basis for the design of self-tuning and self-adaptive control systems (e.g. Young et al, 1988, 1991a,b; Young, 1989; Chotai et al, 1991a). Finally, it should be emphasised that the research described here has been carried out in the context of glasshouse horticulture and, therefore, it has particuarly good potential for application to horticultural and agricultural systems.

ACKNOWLEDGEMENTS

The research described in this paper was supported by the U.K. Science and Engineering Research Council (*SERC*) under grant no. GR/E 8156.2. The authors are grateful for this support and also to colleagues at the *AFRC* Engineering Institute, Silsoe, who collaborated in the research programme.

REFERENCES

Chotai, A., P.C. Young and W.Tych (1990a). A non-minimum state space approach to true digital control based on the backward shift and delta operator models, appears in M.H. Hamza (ed.) *Proc. IASTED Conference*, Acta press: Calgary, 1-4.

Chotai, A., P.C.Young, and C.L Wang., (1990b) True digital control of multivariable systems by input/output state variable feedback, *Report No.TR86/(1990), CRES, University of Lancaster*.

Chotai, A., and P.C. Young. (1991) Self-adaptive and self-tuning control of a Nutrient Film Technique (NFT) system, to appear in *IFAC Workshop on Mathematical and Control Applications in Agriculture and Horticulture Matsuyama, Japan*.

Chotai, A., P. C. Young and M. A. Behzadi. (1991a). The self-adaptive design of a nonlinear temperature control system, *Proc. I.E.E.*, Pt.D., Special Issue on *Self Tuning Control*, 41-49.

Chotai, A., P.C. Young and W.Tych (1991b). A Non-Minimum State Space approach to True Digital Control

Based on δ Operator Models, IEE *Conference Publication No. 332, Control 91*, 567-573

Chotai, A., P.C. Young, P. Davis and Z. Chalabi, (1991c) True Digital Control of Glasshouse Systems, to appear in *IFAC Workshop on Mathematical and Control Applications in Agriculture and Horticulture, Matsuyama, Japan*.

Goodwin, G.C., R.H .Middleton, and M. Salgado(1988) A unified approach to adaptive control, in Warwick, K. (ed.): *Implementation of self-tuning controllers*, Peter Peregrinus, London, 126-139.

Moler C., J. Little and S.Bangert (1987). *Pro-Matlab Users Guide*, The Mathworks Inc., South Natick, MA.

Middleton R.H. and G.C.Goodwin (1990). *Digital Control and Estimation - a Unified Approach*, Prentice Hall, New Jersey.

Tych, W., A.Chotai and P.C.Young (1991a). Computer aided design package for true digital control (TDC) systems, *IEE Conference Publication No. 332, Control 91.*, 288-293.

Tych, W., P.C. Young, and A. Chotai (1991b) TDC - A computer aided design package for True Digital Control , to appear, *Proc. IFAC Symposium on Computer Aided Design in Control Systems, Swansea*.

Wang, C. L. (1988) *New methods for the direct digital control of discrete-time systems*, PhD Thesis, University of Lancaster.

Wang, C.L., and P.C. Young (1988), Direct digital control by input-output, state variable feedback: theoretical background, *Int. J. Control*, **47**, 97-109.

Young, P.C. (1964) In-flight dynamic checkout, *I.E.E.E. Trans. on Aerospace*, AS2, 1106-1111.

Young, P.C. (1965) The determination of the parameters of a dynamic process, *Radio and Electronic Engineer*, **29**, 345-362.

Young, P.C. (1966) Process parameter estimation and self adaptive control; appears in P.H. Hammond (ed.). *Theory of Self Adaptive Control Systems*, Plenum Press, New York, 118-140.

Young, P.C. (1969a) *The Differential Equation Error Method of Process Parameter Estimation*, Ph.D Thesis, Department of Engineering, University of Cambridge.

Young, P.C. (1969b, 1970) An instrumental variable method for real time identification of a noisy process, *Proc. IFAC Congress*, Warsaw (also appears in *Automatica*, **6**, 271-287).

Young P.C. (1984). *Recursive Estimation and Time Series Analysis*, Springer Verlag, Berlin.

Young, P.C.(1985). The instrumental variable method: a practical approach to identification and system parameter estimation, in Barker, H.A. and Young, P.C. (Eds.), *Identification and System Parameter Estimation*, Pergamon, Oxford, 1-16.

Young, P.C.(1989). Recursive estimation, forecasting and adaptive control, in C.T. Leondes (ed.), *Control and Dynamic Systems*, **30**, Academic Press, San Diego, 119-166.

Young, P.C.(1991) Simplified Refined Instrumental Variable (SRIV) estimation and True Digital Control (TDC): a tutorial introduction, to appear, European Control Conference, Grenoble.

Young P.C., and S. Benner (1989). *microCAPTAIN 2.0 User Handbook*, CRES, Lancaster University.

Young, P.C., and J. Willems (1972) An approach to the linear multivariable servomechanism problem, *Int. J. of Control*, **15**, 961-979.

Young, P. C., Behzadi, M.A., and Chotai, A. (1988) Self tuning and self adaptive PIP control systems; appears in K. Warwick (ed.). *Implementation of Self-Tuning Controllers*. Peter Perigrinus: London, 220-259

Young P.C., A.Chotai and W.Tych (1991a) True Digital Control: a unified design procedure for linear sampled data systems, to appear in K. Warwick [ed.], *Advanced Methods in Adaptive Control for Industrial Applications*, Springer Verlag, Berlin.

Young, P.C., A. Chotai, and W. Tych, (1991b) Identification, estimation and control of continuous-time systems described by delta operator models, to appear in N.K. Sinha and G.P. Rao (eds.) *Identification of Continuous Time Systems*, Kluwer Academic Publishers, Dordrecht.

Young, P. C., M. A. Behzadi, A. Chotai and P. Davis. (1987a). The modelling and control of nutrient film systems; appears in J. A. Clark, K. Gregson and R. A. Scafell (ed.). *Computer Applications in Agricultural Environments*. Butterworth: London, 21-43.

Young P.C., M.A. Behzadi, C.L.Wang and A.Chotai (1987b). Direct digital and adaptive control by input-output state variable feeback pole assignment, *Int. J. of Control*, **46**, 1867-1881.

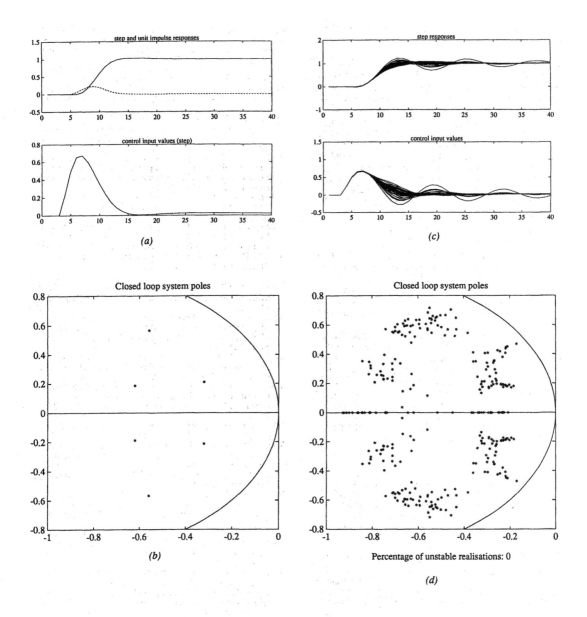

Fig.7 Closed-loop design evaluation: (a) step/impulse
responses (b) pole locations (c) ensemble of step responses;
(d) stochastic root loci.

APPLICATION OF SIGNAL ANALYSIS AND NONLINEAR SYSTEM IDENTIFICATION METHODS TO MODELLING DRY MATTER PRODUCTION IN WINTER WHEAT

Z. S. Chalabi* and P. W. Gandar**

*AFRC Institute of Engineering Research, Wrest Park, Silsoe, Bedford, MK45 4HS, UK
**DSIR Fruit and Trees, Palmerston North, New Zealand

Abstract

A procedure is outlined for the use of stochastic signal analysis and nonlinear system identification methods to model dry matter production in winter wheat crops. In this model, intercepted radiation is treated as the system input and crop dry weight is the output. Intercepted radiation depends on incident solar radiation and leaf area per unit ground area (LAI). It is shown that observed temporal patterns of LAI and dry weight can be represented by empirical orthogonal functions derived from their respective auto-covariance functions. Standard nonlinear system identification procedures can then be used to identify linear and quadratic kernels linking intercepted radiation as input to dry weight as output.

Introduction

A variety of mathematical models have been used to model dynamic agricultural processes. These range from mechanistic models, where strong assumptions are made about processes, through to highly empirical models where assumptions are weak. Relationships between models and data also take a variety of forms. Empirical models are derived directly from data sets and are therefore constrained by the dictates of data. In general, large data sets are required to help narrow the choice between alternative empiricisms. Mechanistic models, on the other hand constitute hypotheses which need to be tested against data. In practice, this approach is often imprecise for two main reasons. First, hypotheses may not be well formulated despite use of equations, because large numbers of assumptions are used. Second, explicit relationships between models and observations are lacking so that testing agreement between the two involves subjective assessments.

The systems approach stands in contrast with empirical and mechanistic approaches. Here, the central hypothesis is mathematical rather than mechanistic. A mathematical relationship or map between 'input' and 'output' variables is assumed to exist. This relationship must then be 'identified' using observations of input and output variables. The identification procedure involves construction of a dynamic differential map which links input and output variables and which is based directly and explicitly on observations.

The objective of this study is to outline some procedures used in developing a systems model for dry matter production in a winter wheat crop. The results of this analysis will be reported elsewhere. Our concern here is to set out the computational procedures used for system identification.

Dry Matter Production

Crops grow by intercepting solar radiation and using this energy to convert carbon dioxide into dry matter. Intercepted solar radiation is therefore often used as a basis for predicting dry matter production and crop yield. The usual form of prediction model employs daily values of solar radiation and a constant factor to convert this into estimates of daily matter production (Monteith, 1977). If $W(t)$, $L(t)$, $I(t)$ and $U(t)$ denote respectively total crop dry weight [g m^{-2}], projected leaf area index [-], incident radiation [MJ m^{-2} d^{-1}] and intercepted radiation [MJ m^{-2} d^{-1}] all defined at t days from sowing, then dry matter production rate can be linked to incident radiation and leaf area index using:

$$\frac{dW(t)}{dt} = a \, U(t) \tag{1}$$

$$U(t) = I(t) \, [1 - e^{-k \, L(t)}] \tag{2}$$

where a is the 'efficiency' of crop production and k is the extinction coefficient of the crop. Equation (1) defines a linear relationship between dry matter production and intercepted radiation and is based on strong physiological assumptions. Tests show that this model is not satisfactory for winter wheat because of nonlinearities arising from grain filling and leaf senescence and from perturbing effects of other environmental variables (Green, 1987). To account for these factors, additional terms or functional dependencies must be introduced in (1) and (2). We sought to avoid this ad hoc approach through use of system identification approach.

Mathematical Strategy

Linear and nonlinear system identification techniques have been used to model engineering and biological systems. These systems can be identified if sufficient data are collected in properly designed experimets. For example, 'white noise' disturbances can be applied experimentally and used to identify nonlinear systems uniquely (Billings, 1980). This approach is not feasible for agricultural field experiments because inputs cannot be controlled to generate 'white noise' disturbances and because outputs can rarely be sampled with sufficient frequency. Indeed it seems unlikely that crop physiologists will be able to arrange field experiments to collect data in a suitable form for direct use of the identification techniques that are standard in other disciplines. This means that an extra step is required before identification methods can be applied to data from field experiments. It is necessary first to determine statistical dynamic models of the input and output variables in order to interpolate these variables between sample observations. Standard procedures of systems identification then follow using inputs and outputs based on the statistical models of the variables.

Data Structure

Data were obtained from six selected nitrogen treatments within a multidisciplinary field experiment designed to examine factors affecting the yield of winter wheat (cv. Avalon) at Rothamsted Experimental Station, Harpenden, United Kingdom. Above-ground dry matter production and projected leaf area index were sampled on 12 and 10 occasions respectively from sowing on 18 September 1986. Replicated nitrogen treatments were considered as separate realizations so that there were, in total, 18 realizations per ensemble for W(t) and L(t). Incident radiation, I(t), was measured daily. In the following, I(t) and L(t) are treated as inputs (combined via equation (2)) and W(t) is the output variable.

Pattern Representation

Complete statistical descriptions are required to quantify the observed ensembles of temporal patterns of projected leaf area index and dry matter production. In practice the mean and auto-covariance of these processes are sufficient statistical descriptors if, to a first order approximation, the processes are assumed to be Gaussian-Markov. The objective is to determine optimal orthogonal series representaion for these patterns. An optimal orthogonal representation of a Gaussian-Markov processes is given by the generalized Karhunen-Loeve theorem (Watanabe,1965). If X(t) denotes one of the observed processes (either L(t) or W(t); I(t) is adequately represented by a Fourier expansion), then the set of optimal orthonormal basis functions $\{\theta_i(t); i=1...n\}$, is the set of eigenfunctions of the auto-covariance kernel R(t,s) of X(t) in the Fredholm integral equation

$$\int_{t_s}^{t_h} R(t,s)\theta_i(s)\,ds = \lambda_i^2\,\theta_i(t) \qquad (3)$$

in which $(\lambda_i^2$, $i=1...n)$ are the eigenvalues of R(t,s). R(t,s) is defined as

$$R(t,s) = \overline{X(t)X(s)} - \overline{X(t)}\ \overline{X(s)} \qquad (4)$$

where the bar symbol denotes ensemble averaging. X(t) is then modelled between sowing (t_s) and harvest times (t_h) by

$$X(t) = \Sigma_i\zeta_i\,\theta_i(t) \qquad (5)$$

where ξ_i is a random variable with mean zero and variance λ_i^2.

If X(t) is a non-stationary Gaussian-Markov process, its auto-covariance kernel is of the form (Kailath,1966)

$$R(t,s) = r[\min(t,s)]\ g[\max(t,s)] \qquad (6)$$

where min(.,.) and max(.,.) denote respectively the minimum and maximum of their arguments, and r and g are continuous functions such that f=r/g is a monotonic increasing continuous function. Using (6), equation (3) can be solved by successive differentiation into a Sturm-Liouville equation (Kailath,1966; Chalabi & Souckova,1984):

$$\frac{d}{dt}\left[\frac{1}{\frac{df}{dt}}\ \frac{d}{dt}\left(\frac{\theta(t)}{g(t)}\right)\right]$$
$$+\lambda^{-1}g(t)\theta(t) = 0 \qquad (7)$$

with the boundary conditions:

$$\frac{\theta(t_s)}{r(t_s)} = \frac{g(t_s)\ \frac{d\theta(t)}{dt}\big|_{t_s} - \frac{dg}{dt}\big|_{t_s}\theta(t_s)}{g^2(t_s)\ \frac{df}{dt}\big|_{t_s}} \qquad (8)$$

and

$$\frac{g(t_h)\ \frac{d\theta(t)}{dt}\big|_{t_h} - \frac{dg(t)}{dt}\big|_{t_h}\theta(t_h)}{\frac{df}{dt}\big|_{t_h}} = 0 \qquad (9)$$

Equations (7) to (9) solve (3) when R(t,s) is given by the general form of (6).

Analyses carried out to determine the auto-covariance of L(t) and W(t) showed that minimum least-squares estimates were obtained in the forms

$$R_L(t,s) = c\ e^{-\alpha|t-s|}\ e^{-\beta s} \qquad (10)$$

and

$$R_W(t,s) = \gamma\ s\ e^{-\eta|t-s|} \qquad (11)$$

for L(t) and W(t) respectively.

Model of Projected Leaf Area Index

It follows from equation (3) that the optimal representation of L(t) is given by an orthogonal expansion in terms of the eigenfunctions of

$$\int_{t_s}^{t_h} c\ e^{-\alpha|t-s|}\ e^{-\beta s}\ \phi_L(s)\,ds$$
$$= \lambda_L^2\ \phi_L(t) \qquad (12)$$

which is a Fredholm integral equation. Analytical methods are available for solving specific forms of the Fredholm equation (Juncosa, 1945; Miller and Zadeh, 1956; Price, 1956). These methods are less general than those given in equations (7) to (9) and hence lead to specific types of Sturm-Liouville equations. The form of equation (12) is similar to that solved by Juncosa (1945). If

$$t_h - t_s > \max(\alpha^{-1}, \beta^{-1}) \qquad (13)$$

then, following Juncosa (1945), we differentiate equation (12) twice with respect to time and re-arrange to give

$$\frac{d^2\phi_L}{dt^2} + \left[\frac{2\alpha c}{\lambda_L^2}e^{-\beta t}-\alpha^2\right]\phi_L(t) = 0 \qquad (14)$$

Substituting

$$\phi_L(t) = \psi\left(\frac{\sqrt{2\alpha c}}{\lambda_L}e^{-\frac{\beta t}{2}}\right) \qquad (15)$$

and

$$\tau = \frac{\sqrt{2\alpha c}}{\lambda_L}e^{-\frac{\beta t}{2}} \qquad (16)$$

into (14) gives Bessel's differential equation of fractional order α:

$$\tau^2\frac{d^2\psi}{d\tau^2} + \tau\frac{d\psi}{d\tau} + (\tau^2-\alpha^2)\psi = 0 \qquad (17)$$

It follows that the eigenfunctions of $R_L(t,s)$ are

$$(\phi_L)_n = e^{-\frac{\beta t}{2}}J_\alpha(\sigma_n e^{-\frac{\beta t}{2}}) \qquad (18)$$

318

and the corresponding eigenvalues are

$$(\lambda_L)^2{}_n = \frac{2\alpha c}{\sigma^2{}_n} \qquad (19)$$

where $J_\alpha(.)$ is Bessel's function of order α and $\{\sigma_n; n=1,2...\}$ are the roots of

$$J_{\alpha-1}(\sigma_n) = 0 \qquad (20)$$

Equations (19) and (20) give the necessary and sufficient conditions for (18) to be the non-trivial solutions of (12).

Model of Dry Matter Production

The optimal orthogonal expansion of $W(t)$ is given in terms of the eigenfunctions of $R_W(t,s)$ as specified by (11). A Sturm-Liouville equation defining the eigenfunctions is obtained by substituting

$$r(t) = t \, e^{\beta t} \qquad (21)$$

$$g(t) = e^{-\beta t} \qquad (22)$$

and

$$f(t) = \frac{r(t)}{g(t)} = t \, e^{2\beta t} \qquad (23)$$

into equation (7) to give

$$\frac{d}{dt}[(1+2\beta t)^{-1} e^{-2\beta t}\frac{d}{dt}\Gamma(t)]$$
$$\lambda_W^{-1} e^{-2\beta t} \Gamma(t) = 0 \qquad (24)$$

where

$$\phi_W(t) = \Gamma(t) \, e^{-\beta t} \qquad (25)$$

For simplicity of notation, we have assumed that γ in equation (11) is unity. The resultant eigenvalues need then to be scaled by $\gamma^{-1/2}$.

Equation (24) cannot be solved analytically. However, there are procedures for solving it numerically (Birkhoff & Rota, 1962; Godart, 1966). Following Birkhoff and Rota (1962), we apply Prufer transformation to equation (24) to give

$$(1+2\beta t)^{-1} e^{-2\beta t} \frac{d\Gamma}{dt} =$$
$$x(t) \cos(y(t)) \qquad (26)$$

and

$$\Gamma(t) = x(t) \sin(y(t)) \qquad (27)$$

Substitution of equations (26) and (27) into equation (24) gives the pair of differential equations:

$$\frac{dy}{dt} = [\lambda_W e^{-2\beta t}] \sin^2 y$$
$$+ [1+2 \beta t] e^{2\beta t} \cos^2 y \qquad (28)$$

and

$$\frac{dx}{dt} = \frac{1}{2}[(1+2\beta t) e^{2\beta t} - \lambda_W e^{-2\beta t}]$$
$$. x \sin(2y) \qquad (29)$$

If equation (28) is solved for y, then Γ is determined explicitly from y

$$\Gamma(t) = x(t_s) \, e^{\Psi(t)} \sin(y(t)) \qquad (30)$$

where

$$\Psi(t) = \frac{1}{2}\int_{t_s}^{t} [(1+2\beta s) \, e^{2\beta s}$$
$$-\lambda_W \, e^{-2\beta s}] \sin(2 \, y(s)) \, ds \qquad (31)$$

Equation (28) could be solved iteratively using the approximation method of Godart (1966). The set of eigenfunctions of $R_W(t,s)$ are then given by substituting (30) into (25).

System Identification

The use of mathematical analysis to identify nonlinear systems is an area of active research. Over the years many computational methods have been developed to identify the structure of nonlinear systems using Volterra and Wiener kernels (Ku & Wolf, 1966; Ku, 1983; Billings, 1980).

A nonlinear system is defined uniquely by an infinite set of kernels. However, the output response can usually be approximated using a small subset (linear, quadratic, cubic, ...) of these kernels. We assume, as a first approximation, that dry matter production can be modelled by a second order nonlinearity, i.e.

$$W(t) = \int_\Omega h_1(\tau_1) \, U(t-\tau_1) \, d\tau_1 + \int_\Omega\int_\Omega h_2(\tau_1,\tau_2)$$
$$. \, U(t-\tau_1) \, U(t-\tau_2) \, d\tau_1 d\tau_2 + \epsilon(t) \qquad (32)$$

where $h_1(.)$ and $h_2(.,.)$ are the first and second order kernels, ϵ is the error in the model and Ω defines the domain of integration, $[t_s, t_h]$. $h_1(.)$ and $h_2(.,.)$ are determined by minimizing the mean-square error

$$\min_{h_1,h_2} \overline{e^2(t)} =$$
$$\min_{h_1,h_2} \overline{[W(t) - \int_\Omega h_1(\tau_1) \, U(t-\tau_1) \, d\tau_1 -}$$
$$\overline{\int_\Omega\int_\Omega h_2(\tau_1,\tau_2) \, u(t-\tau_1, t-\tau_2) \, d\tau_1 d\tau_2]^2} \qquad (33)$$

It can be shown using calculus of variation arguments (Katzenelson and Gould, 1962) that h_1 and h_2 are given by the solution of two simultaneous integral equations:

$$\int_\Omega h_1(p) \, \overline{U(t-p) \, U(t-\tau_1)} \, dp =$$
$$\overline{W(t) \, U(t-\tau_1)}$$
$$-\int_\Omega\int_\Omega h_2(p,q) \, \overline{U(t-p) \, U(t-q) \, U(t-\tau_1)} \, dpdq \qquad (34)$$

and

$$\int_\Omega\int_\Omega h_2(p,q) \, \overline{U(t-p) \, U(t-q) \, U(t-\tau_1)}$$
$$\overline{U(t-\tau_2)]} \, dpdq =$$
$$\overline{W(t) \, U(t-\tau_1) \, U(t-\tau_2)} -$$
$$\int_\Omega h_1(p) \, \overline{U(t-p) \, U(t-\tau_1) \, U(t-\tau_2)} \, dp \qquad (35)$$

Since the factors under the ensemble average operator can be computed numerically using the orthogonal expansions of $W(t)$ and $L(t)$, only h_1 and h_2 are unknown in equations (34) and (35). Interpretation of the resultant linear kernel is straight-forward. However, interpretation of the quadratic kernel in the context of dry matter production will require further development and analysis. Hung et al. (1977) provide mathematical interpretation of quadratic kernels in terms of combination of linear elements. However, it is important that the system kernels are interpreted in physiological as well as in mathematical terms.

Conclusion

This study is a first attempt to use system identification techniques to analyze an agricultural process. The procedures, outlined in this paper, have been applied to modelling dry matter production in winter wheat and the results will be described elsewhere. These procedures appear to hold promise for wider application in agricultural systems and that they warrant further investigation. The power of the methodology would be reinforced if future experiments were designed to collect data in forms more approriate to this style of analysis.

Acknowledgements

The authors are grateful to G.F.J. Milford for permission to use data from the winter wheat experiment.

References

Billings,S.A. (1980) Identification of nonlinear systems - a survey. IEE Proc. 127, D, 6, 272-285.

Birkhoff,G. & Rota,G.C. (1962) Ordinary Differential Equations.
Ginn and Co., Boston and New York.

Chalabi,Z. & Souckova,S. (1985) Pattern analysis of Bekesy audiograms. J. Acoust. Soc. Am. 77, 1185-1191.

Godart,M. (1966) An iterative method for the solution of eigenvalue problems. Mathematics of Computation 20, 339-406.

Green,C.F. (1987) Nitrogen nutrition and wheat growth in relation to absorbed solar radiation. Agric. Forest Met. 41, 207-248.

Hung,G. ; Stark,L. & Eykhoff,P. (1977) On the interpretation of kernels. I. Computer simulation of responses to impulse pairs. Annals of Biomedical Engineering 5, 130-143.

Juncosa,M. (1945) An integral equation related to Bessel functions. Duke. Math. J. 12, 465-471

Kailath,T. (1966) Some integral equations with non-rational kernels. IEEE Trans. Inform. Theory IT-12, 442-447.

Katzenelson,J & Gould,L. (1962) The desiogn of nonlinear filter and control systems. Part I. Information and Control 5, 108-143.

Ku,Y.H. & Wolf,A.A. (1966) Volterra-Wiener functionals for the analysis of nonlinear

Miller,K.S. & Zadeh,L.A. (1956) Solution of an integral equation occuring in the theories of prediction and detection IRE Trans. Inform. Theory IT-2, 72-75.

Monteith,J.L. (1977) Climate and the efficiency of vrop production in Britain. Phil. Trans. R. Soc. Lon. (B), 277 - 294.

Price,R. (1956) Optimum detection of random signals in noise, with application to scatter-multipath communication. Part I. IRE Trans. Inform. Theory IT-2, 125-135.

Watanabe,S. (1965) Karhunen-Loeve expansion and factor analysis- theoretical remarks and applications. In Trans. 4th Prague Conf. Inform. Theory, Stat. Decision Functions and Random Processes, 635-660.

PROCESS IDENTIFICATION, A MARKOV PARAMETER APPROACH

T. Backx

Datex/IPCOS, P.O. Box 3377, 5203 DJ 's-Hertogenbosch, The Netherlands
and
Measurement and Control Group, Faculty of Electrical Engineering, Eindhoven University of
Technology, P.O. Box 513, 5600 MB Eindhoven, The Netherlands

Abstract. The modelling of dynamics of complex MIMO processes by means of
process identication techniques is discussed. A multistep identification
approach is presented, which only requires rough a priori knowledge on the
process dynamics to come to linear, time invariant, multi input multi
output, discrete time models of low complexity that approximately describe
the main dynamical transfer characteristics of a process with high
accuracy. The approach does not require structural identification. The
results of an application to the modelling of spray dryer are presented.

Keywords. Identification, modelling, model reduction, multivariable
systems, process control, spray dryer

INTRODUCTION

Social developments over the past
decade have resulted in an
increased awareness of the need for
attention of our environment.
Superfluous use of raw materials
and deposition of waste materials
in our environment are subject to
discussion and criticism nowadays.
This increased attention for the
careful use of natural resources
and reuse of waste materials
requires improved control of
processes. Processes have to be
operated in such a way that only
those products are produced, which
are asked for. Reuse of materials
usually implies that processes have
to be operated with a broader input
material specifications. As a
consequence more demands are put on
the control of the processes with
respect to accuracies and
flexibility. The required advanced
control techniques need accurate
models of the process dynamics.
Two approaches may be used for
modelling system dynamics:
Mechanistic Modelling and System
Identification. The application of
both techniques is required in
general to come to models that
accurately describe relevant
dynamics of complex processes.
Mechanistic modelling techniques
are required for getting a good
understanding of the mechanisms
that govern process dynamics and to
find appropriate inputs for
controlling the process. System
identification techniques may be
used to come to models that
accurately describe the main
dynamical properties of the true
process.

During the sixties and early seventies a manifold of identification
techniques have been developed,
which can be used for the
identification of Single Input
Single Output (SISO) systems
(Åström, 1971; Eykhoff, 1974). The
early developed techniques are
mainly based on classical
statistical approaches and
concentrate on the estimation of
unknown model parameters in known
model structures.

Later developments of the identification techniques involved the
determination of appropriate model
structures and extended model
complexity from SISO systems to
Multi Input Multi Output (MIMO)
systems. A lot of research is done
on the development of theory that
gives a detailed understanding of
the available techniques both for
SISO and for MIMO systems (Ljung,
1983; Van den Hof, 1986, 1988;
Söderström, 1987; Ljung, 1987;
Janssen, 1988). As a result of
these developments identification
tools have been developed, which
are presently available for common
use (Ljung, 1986). Although a
detailed understanding has been
obtained on the properties of identification techniques by these
developments, the techniques appeared not to be so very well
suited for approximate modelling of
dynamics of complex MIMO processes.

This is mainly caused by the requirement that true systems are part of the model sets used for identification in combination with the use of model sets of low complexity. Model sets applied for identification are mainly Finite dimensional, Linear, Time Invariant, Causal, Discrete time (FDLTICD). True systems usually don't show such a nice behaviour and therefore cannot be described exactly by a model of the FDLTICD set.

To enable identification of complex MIMO systems a different approach has been developed in parallel (Richalet, 1978; Hajdasinski, 1978; Backx, 1985, 1987, 1989; Swaanenburg, 1985; Zhu, 1990; Heuberger, 1991). In this appoach it is no longer assumed that true systems belong to the model sets used for system identification. These techniques appear to be applicable for approximate modelling of main dynamics of complex MIMO systems.

In the sequel an overview will be given of a robust multistep Markov parameter based identification approach. Properties of the models obtained in the various steps are described. The techniques discussed will be demonstrated on an application to the modelling of the dynamics of a spray drying process.

A MARKOV PARAMETER BASED IDENTIFICATION STRATEGY

Main problem with the use of system identification techniques for approximate modelling of main system dynamics is the lack of sufficiently detailed a priori knowledge on the process characteristics. Preliminary process investigations have shown some aspects of the dynamics (e.g. most appropriate inputs and outputs, non-linearities, largest and smallest relevant time constants, static gains, ...), but the information gathered from these analyses is mostly not sufficient for immediately chosing the right compact model sets. To overcome this problem a strategy has been developed consisting of several subsequent identification steps. In this approach the a priori knowledge required for each step is available from previous steps. Preliminary process investigations consisting of rough mechanistic modelling and dedicated experiments

directed to the analysis of dynamical properties have to give sufficient a priori knowledge on the main process dynamics for application of the first process identification step, which only requires limited a priori knowledge (cf. Backx, 1989). The subsequent identification steps of the approach consist of:

.Finite Impulse Response (FIR) estimation
.Fit of an initial Minimal Polynomial Start Sequence of Markov parameter (MPSSM) model
.Least Squares estimation of an MPSSM model
.Approximate realization of the MPSSM model with a minimal State Space (SS) model

The FIR model:

$$y_k - \sum_{i=0}^{N} M_i u_{k-i} \qquad (1)$$

just requires a priori knowledge on the fastest and slowest relevant process dynamics. In this expression M_i are the Markov parameters, u_i are the input signal samples and y_i are the output signal samples. (N+1) is the number of samples of the FIR. The fastest process dynamics determine the required sampling time. The length (N+1) of the impulse responses is determined by the slowest relevant system dynamics. This knowledge is easily obtained during preliminary process investigations. The FIR model parameters can be estimated by minimization of a squared output error criterion:

$$\min_{M_i} arg \ v - \min_{M_i} arg \ \|Y - \hat{Y}(M_i)\|_2 \qquad (2)$$

In this expression Y are the measured output signal samples:

$$Y - [y_k \ y_{k+1} \ \cdots \ y_{k+l}]$$

$\hat{Y}(M_i)$ denotes the outputs generated by the FIR model:

$$\hat{Y} - [M_0 \ M_1 \ \cdots \ M_N] \begin{bmatrix} u_k & u_{k+1} & \cdots & u_{k+l} \\ u_{k-1} & u_k & \cdots & u_{k+l-1} \\ . & . & \cdots & . \\ u_{k-N} & u_{k-N+1} & \cdots & u_{k+l-N} \end{bmatrix}$$

$$- M\Omega$$

The function to be minimized is a quadratic function of M_i as can easily be seen from eq. (2). As a consequence the estimation of the FIR model parameters implies the least squares solution of an overdetermined set of linear equations. This problem has a unique solution and no detailed a priori knowledge on initial model parameters is required for this step.

The FIR model, although it always can be found, is a rather complex model. The number of parameters of the model is quite high: $N_{par} = (N+1)pq$ (a system with p inputs and q outputs). To reduce model complexity a simpler model -an MPSSM model- is fitted to this FIR model. This MPSSM model is given by:

$$y_k = \sum_{i=0}^{\infty} F_i(a_j, \tilde{M}_j | j=1,2,\ldots,r) u_{k-i} \qquad (3)$$

In this expression F_i is given by:

$$F_i = \begin{cases} \tilde{M}_0 & i=0 \\ \tilde{M}_i & 1 \le i \le r \\ -\sum_{j=1}^{r} a_j F_{i-j} & i > r \end{cases}$$

The MPSSM model is an Infinite Impulse Response (IIR) model. The reconstruction of the impulse responses is done by a recurrent relation. The first r elements of the impulse response are considered to be unique. Subsequent impulse response elements are considered to be linear combinations of the first r elements. Minimal polynomial coefficients a_i give the depencies between elements. The number of degrees of freedom of the MPSSM model is $N_{par}=(r+1)pq+r$. The complexity of the MPSSM model will be much lower than the complexity of the FIR model in general, while the degree r of the minimal polynomial will usually be small compared to N.

To find an initial MPSSM model an appropriate estimate for the degree r has to be determined first. As r indicates the number of independant Markov parameters, the elements of the FIR obtained from the first identification step can be analysed for dependencies. This can be done by writing the columns of the Markov parameters into a vector:

$$vct(M_i) = [m_{i,1}^T \ m_{i,2}^T \ \ldots \ m_{i,p}^T]^T \qquad (4)$$

In eq. (4) $m_{i,j}$ denotes the j-th column of the i-th Markov parameter. An appropriate value of r can be found by analysis of the singular values of the following finite block Hankel matrix:

$$H = \begin{vmatrix} vct(M_1) & vct(M_2) & .. & vct(M_{N/2}) \\ vct(M_2) & vct(M_3) & .. & vct(M_{N/2+1}) \\ . & . & .. & . \\ vct(M_{N/2+1}) & vct(M_{N/2+2}) & .. & vct(M_N) \end{vmatrix}$$

$$= U \Sigma V^T \qquad (5)$$

This singular value decomposition of H gives the singular values σ_i, which indicate the dependencies of complete Markov parameters. The entries of the Hankel matrix H resulted from the FIR parameter estimation. This implies that the parameters will be contaminated with noise. As a result the singular values of the Hankel matrix will represent both relevant system dynamics and noise. Due to the ordering of singular values the first singular value will be largest. The FIR used for this computations corresponds with a very high order model. The true system dynamics usually can be very well represented by a relatively low order model. It may therefore be expected that the singular values will tend to some noise level (cf. Backx, 1987) given by the inaccuracies of the FIR estimation. An appropriate value of r is found by looking at the course of R:

$$R = \sigma_{i+1}/\sigma_i \qquad (6)$$

This ratio R will tend to one due to stabilization to the noise level, if sufficient properly estimated initial Markov parameters have been used for determination of the degree of r. A small value before an increase to about 1 of the ratio R indicates that r=i may be a good estimate for the degree of the minimal polynomial.

Once the degree r has been chosen, an initial MPSSM model can be fitted to the FIR by making use of Gerth's algorithm (Gerth, 1972). This is a two step approach where on the basis of an available FIR model first minimal polynomial coefficients are estimated and in a second step with the coefficients a_i start sequence Markov parameters are fitted to this FIR. Two sets of overdetermined linear equations have to be solved in least squares sense for this purpose:

$$Ga - v \qquad (7)$$

$$H\tilde{M} - M_v \qquad (8)$$

$$G = \begin{vmatrix} vct(M_1) & vct(M_2) & .. & vct(M_r) \\ vct(M_2) & vct(M_3) & .. & vct(M_{r+1}) \\ . & . & .. & . \\ vct(M_{N-r}) & vct(M_{N-r+1}) & .. & vct(M_{N-1}) \end{vmatrix}$$

$$a^T - -[a_r \ a_{r-1} \ .. \ a_1]^T$$

$$M_v^T - [vct(M_1) \ vct(M_2) \ .. \ vct(M_N)]$$

$$\tilde{M}^T - [vct(M_1) \ vct(M_2) \ .. \ vct(M_r)]$$

$$E - [0 \ 0 \ .. \ 0 \ 1]^T$$

$$A = \begin{bmatrix} 0 & 0 & .. & 0 & -a_r \\ 1 & 0 & .. & 0 & -a_{r-1} \\ . & . & .. & . & . \\ 0 & 0 & .. & 1 & -a_1 \end{bmatrix}$$

$$H^T - [I \ AE \ A^2E \ .. \ A^{N-r}E]$$

The initial MPSSM model can be used as a starting value for the estimation of a final MPSSM model by minimization of the following squared output error criterion:

$$\min_{a_j, \tilde{M}_j} \arg V_1 - \min_{a_j, \tilde{M}_j} \arg \|Y - \hat{Y}(F_1(a_j, \tilde{M}_j))\|_2 \qquad (9)$$

The criterion function V_1 is a quadratic function of the start

sequence of Markov parameters $\tilde{M}_j$

and a high order polynomial function of the minimal polynomial coefficients a_j. With some mathematical manipulations the

implicit solutions for $\tilde{M}(a_j)$ can

be found and substituted in V_1. As a result numerical minimization is only required towards the r minimal polynomial coefficients.
No structural identification is required for the estimation of the MIMO MPSSM model. The MPSSM model has one drawback however. The eigenvalues of the model, given by the zeros of the minimal polynomial:

$$f(\lambda) - \lambda^r + a_1\lambda^{r-1} + .. + a_r - 0 \qquad (10)$$

have multiplicity min(p,q). These eigenvalues are distinct, but do not contribute equally to the systems input/output behaviour. To further reduce complexity of the model and to come to a minimal representation of the model the approximate realization technique of

Pernebo and Silverman (Pernebo, 1982) is applied. For this purpose the MPSSM model is one to one translated into a balanced state space realization (A,B,C,D). The approximate realization is obtained from the following Lyapunov equations:

$$\begin{bmatrix} A_{11} & A_{12} \\ A_{21} & A_{22} \end{bmatrix} \begin{bmatrix} P_1 & 0 \\ 0 & p_2 \end{bmatrix} \begin{bmatrix} A_{11}^T & A_{21}^T \\ A_{12}^T & A_{22}^T \end{bmatrix}^-$$

$$\begin{vmatrix} p_1 & 0 \\ 0 & p_2 \end{vmatrix}^+ \begin{vmatrix} B_1 \\ B_2 \end{vmatrix} [B_1^T \ B_2^T] - 0 \qquad (11)$$

$$\begin{bmatrix} A_{11}^T & A_{21}^T \\ A_{12}^T & A_{22}^T \end{bmatrix} \begin{bmatrix} Q_1 & 0 \\ 0 & Q_2 \end{bmatrix} \begin{bmatrix} A_{11} & A_{12} \\ A_{21} & A_{22} \end{bmatrix}^-$$

$$\begin{bmatrix} Q_1 & 0 \\ 0 & Q_2 \end{bmatrix}^+ \begin{bmatrix} C_1^T \\ C_2^T \end{bmatrix} [C_1 \ C_2] - 0 \qquad (12)$$

The matrices P and Q satisfy the following relation:

$$P - Q - \Sigma_1 \qquad (13)$$

with Σ_1 a diagonal matrix, which

contains the Hankel singular values of the system. The blocks P_1 and Q_1 are assumed to contain only those Hankel singular values, which are related to relevant input/output behaviour of the system. The reduced order minimal SS model (A_{11}, B_1, C_1, D) obtained from eq. (11) and (12) is the model of low complexity that approximates the relevant true process dynamics very well.

SOME PROPERTIES OF THE VARIOUS MODELS

The applied input test signals u and the process output noise n have to satisfy the following conditions:

$$E[u_i u_j^T] - \begin{array}{cc} \sigma^2 I_p & i-j \\ 0 & i \neq j \end{array} \qquad (14)$$

$$E[u] - 0 \qquad (15)$$

$$E[u_i n_j^T] - 0 \qquad \forall i, j \qquad (16)$$

Furthermore the datasets used for identification have to contain sufficient samples to guarantee that the cross products of input signal samples and noise samples become negligibly small. If these conditions are satisfied the following holds:

$$\min_{M} arg V = \min_{M} arg \|Y - M\Omega\|_2 =$$

$$= \min_{M} arg \|M_{true} - M\|_2 \qquad (17)$$

and

$$\min_{a_i, \tilde{M}_i} arg V_1 = \min_{a_i, \tilde{M}_i} arg \|Y - F(a_i, \tilde{M}_i) \Omega\|_2 =$$

$$= \min_{a_i, \tilde{M}_i} arg \|M_{true} - F(a_i, \tilde{M}_i)\|_2 \qquad (18)$$

The models give a good representation of the main process dynamics.

AN APPLICATION

The above described process identification techniques have been successfully applied for the modelling of the dynamics of many different MIMO processes that cannot be described exactly by an FDLTICD model. As an example the results are shown of an application to the modelling of a spray dryer, which is used for drying milk to milkpowder.

The spray dryer is a two input two output system (cf. Fig. 1). The inputs are the slurry flow and the temperature of the air used for drying the slurry. The two outputs are the humidity and the temperature of the material after drying. This process is a distributed parameter system. The slurry is sprayed into the spray drying tower by a nozzle. The globules fall down and are in the mean time dried by the hot air that is blown through the drying tower. The impulse responses of the system are shown in Fig. 2. As can be seen from this figure the process shows quite some interaction. Validation results, obtained by comparison of the outputs of both the process and the model obtained after excitation of the inputs with some test signal, are shown in Fig. 3. As can be seen the model fits very well to the true process behaviour. This model has been used for the design of an internal model based multivariable control system.

CONCLUSIONS

In this paper a multistep identification approach is described that is well suited for modelling of the dynamics of complex MIMO processes as e.g. a distributed parameter type systems. The techniques discussed enable straightforward accurate modelling of the main dynamics of such a process although no detailed a priori knowledge of the process dynamics are available. The MIMO identification techniques do not require an explicit structural identification. The method has been successfully applied to model the transfers of many industrial processes. As an example of such applications the results of an application to the modelling of the dynamics of a spray drying tower are discussed.

REFERENCES

Åström, K.J. and P. Eykhoff (1971). *Automatica*, Vol. 7, pp. 123-162

Backx, T. (1985). *Proc. 17-th JAACE Symp. on Stoch. Syst. Theory and its Appl.*, Kyoto, Japan

Backx, T. (1987). *PhD thesis*, Eindhoven University of Technology, Eindhoven, The Netherlands

Backx, T. and A. Damen (1989). *Journal A*, Vol. 30, pp. 3-12, 33-43

Eykhoff, P. (1974). *System Identification Parameter and State Estimation* John Wiley & Sons, New York

Gerth, W. (1972). *PhD thesis*, Technischen Universität Hannover, Hannover, Germany

Hajdasinski, A.K. (1978). *Report Technical University Eindhoven*, TUE 78-E-88, Eindhoven, The Netherlands

Heuberger, P. (1991). *PhD thesis*, Delft University of Technology, Delft, The Netherlands

Janssen, P. (1988). *PhD thesis*, Eindhoven University of Technology, Eindhoven, The Netherlands

Ljung, L. and T. Söderström (1983). *Theory and Practice of Recursive Identification*, The MIT Press, Cambridge, Massachusetts

Ljung, L. (1986). *System Identification Toolbox: Manual*, The Mathworks, Inc., Sherborn, Massachusetts

Ljung, L. (1987). *System Identification - Theory for the User* Prentice-Hall, Englewood Cliffs, New Jersey

Pernebo, L. and L.M. Silverman (1982). *IEEE Trans. A.C.*, Vol. AC-27, pp. 382-387

Richalet, J. (1978). _Automatica_,
   Vol. 14, pp. 413-428
Söderström, T. and P. Stoica
   (1987). _System Identification_,
   Prentice Hall, Englewood Cliffs,
   New Jersey
Swaanenburg, H.A.C. (1985). _Proc._
   7-th IFAC Symp. on Ident. and
   Syst. Par. Est., York, England
Van den Hof, P. and P. Janssen
   (1986). _IEEE Trans. A.C._, Vol.
   AC-32, pp. 89-92
Van den Hof, P. (1988), _Proc. 8-th_
   IFAC Symp. on Ident. and Syst.
   Par. Est., Beijing, China
Zhu, Y.C. (1990), _PhD thesis_,

   Eindhoven University of
   Technology, Eindhoven, The
   Netherlands

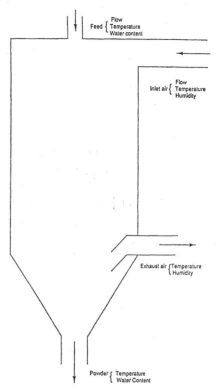

Fig. 1

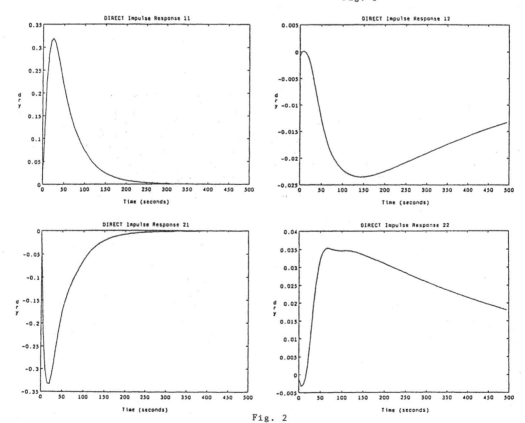

Fig. 2

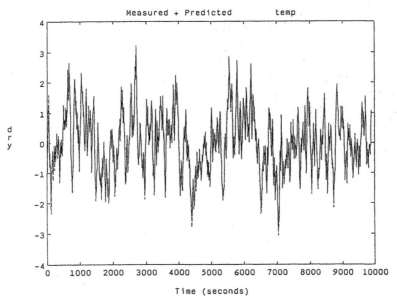

Fig. 3

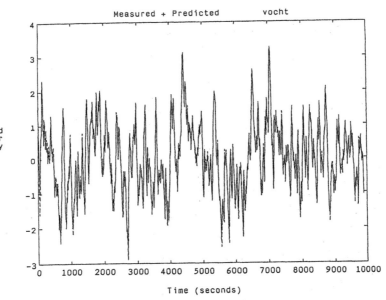

Fig. 3

INTERACTIVE DECISION SUPPORT SYSTEM FOR MULTIOBJECTIVE PROGRAMMING IN AGRICULTURE

M. Tanaka and T. Tanino

*Dept. of Information Technology, Faculty of Engineering, Okayama University,
Okayama 700, Japan*

Abstract. This paper describes a framework of software for an interactive decision support system for designing input values in complicated multiobjective system such as environmental control system of plant factories or greenhouses. The software includes an interactive multiobjective programming and interactive estimation of unknown parameters in linear and nonlinear functions, so that the decision maker can decide his input values without the knowledge of multiobjective decision theory and estimation theory in detail. By introducing state variables, the decision maker can construct a hierarchical model. For multiobjective programming, we apply the interactive ε-constraint method that seems to be most appropriate for the decision maker of environmental control among various methods.

Keywords. Multiobjective optimization; decision theory; parameter estimation; computer software; agriculture; linear systems; nonlinear systems.

INTRODUCTION

The main objectives of plant factories or greenhouses are to increase the production, to control the plant growth, and so on. To achive these objectives, the environmental factors have to be controlled. For the environmental control for plants, we have to supply carbon dioxide, water, fertilizer, adjust the thermal level and shading rate, for example. To consider this complicated system, the total control system should be described by hierarchical structure (Tantau, 1990). On the other hand, the cost for environmental control have to be as small as possible. There may exist some other objectives in the enviromental control systems. These objectives are the functions of the variables that we can manipulate. There exist a lot of studies of mathematical modelling of the input-output relations in environmental control (e.g. Enoch, 1978; Liebig, 1989). Based on theoretical studies, various functions will be developed.

Multiobjective programming (e.g. Sawaragi, Nakayama and Tanino, 1985) deals with problems where conflicting objectives exist. The solution depends on the preference of the decision maker (DM). Brief explanations of principal methods will be given. But all these methods require the explicit expression of the functions between the input and output variables. This paper includes the estimation of the unknown parameters of functions. The software is termed as Interactive Decision Support System (IDSS), where the methodology for multiobjective programming is based on the ε-constraint method (Haimes, Hall and Freedman, 1975; Nakayama, Tanino and Sawaragi, 1980; Shimizu, 1982).

The feature of IDSS is that it includes

- interactive multiobjective programming scheme

- interactive parameter estimation scheme for linear and nonlinear models

- state variable formulation to enable hierarchical structure.

We define state variables as follows. Some of the state variables can be the input or objective variables themselves, but they are usually intermediate ones between the input and objective ones, and are inevitable in considering the system. State variables do not usually appear in the literature on multiobjective programming, but they will help us to deal with this kind of complicated system. A prototype system is coded by Matlab system.

MATHEMATICAL MODEL

Here, we define input variables to be the ones we can manipulate; for example, the amount of carbon dioxide per hour, the time of lightening, etc. The input variables are denoted as $\mathbf{u} = (u_1, u_2, \cdots, u_m)$, and

$$\mathbf{U}_o = \{\mathbf{u} : u_i^- \leq u_i \leq u_i^+, \quad i = 1, 2, \cdots, m\} \qquad (1)$$

is the initial feasible domain to be taken. We also define state variables $\mathbf{x} = (x_1, x_2, \cdots, x_n)$ as

$$x_i = x_i(\mathbf{u}), \; i = 1, 2, \cdots, n.$$

This also can have a constraint

$$X = \{\mathbf{x} : x_i^- \le x_i \le x_i^+, \quad i = 1, 2, \cdots, n\} \qquad (2)$$

from its physical sense. The intermediate values such as photosynthesis rate, temperature on leaves, carbon dioxide concentration in the air are defined as state variables. In the above expression, u_i^-, u_i^+, x_i^- and x_i^+ are fixed.

Let $\mathbf{U}_x$ be the set of $\mathbf{u}$ that satisfy (2); i.e.,

$$\mathbf{U}_x = \{\mathbf{u} : \mathbf{x}(\mathbf{u}) \in \mathbf{X}\}. \qquad (3)$$

<u>Definition</u> (feasible solution)
The set $\mathbf{U} = \mathbf{U}_o \cap \mathbf{U}_x$ is termed as "feasible solutions"(Fig. 1).

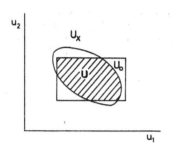

Fig. 1. Feasible solutions.

Suppose we have r objectives:

$$\mathbf{f}^T = (f_1, f_2, \cdots, f_r)$$

where the superscript ^{T} denotes the transpose of a vector. Define $\mathbf{f}$ to be the function of $\mathbf{u}$ and $\mathbf{x}$, i.e.,

$$f_i = f_i(\mathbf{u}, \mathbf{x}(\mathbf{u})), \quad i = 1, 2, \cdots, r \qquad (4)$$

<u>Definition</u> (feasible region of objective function)
The set $\mathbf{F} = \{\mathbf{f}(\mathbf{u}) : \mathbf{u} \in \mathbf{U}\}$ is called "feasible region of objective functions". Without loss of generality, we can suppose that all $\{f_i, \ i = 1, 2, \cdots, r\}$ are functions to be minimized.

PARAMETER ESTIMATION

The functions $x_i(\mathbf{u})$ and $f_i(\mathbf{u}, \mathbf{x}(\mathbf{u}))$ are not always completely known. IDSS includes parameter estimation schemes both for linear and nonlinear models. Here, we will investigate interactive parameter estimation schemes where the errors are treated as unknown deterministic values. The DM will choose a parameter vector in the feasible solutions (this value can be a statistically estimated value), and can look at the input-output behaviour by using the selected parameters. The reason for taking this approach is that the amount of data available for parameter estimation is usually small, so the statistically estimated parameters are not always reliable. Another reason is that we can find outliers by interactive method even from a small amount of data.

<u>Linear Model</u>

We will use a common formulation for the functions $x_i(\mathbf{u})$ and $f_i(\mathbf{u}, \mathbf{x})$ by a linear model

$$y(k) = \theta^T \mathbf{q}(k) + \varepsilon(k), \quad k = 1, 2, \cdots, N_1 \qquad (5)$$

$$\varepsilon^- \le \varepsilon(k) \le \varepsilon^+ \qquad (6)$$

where $\mathbf{q}(k)$ is the ℓ_1-dimensional input vector, θ is the ℓ_1-dimensional unknown parameter, $y(k)$ is the scalar output value, and $\varepsilon(k)$ is the modelling error. The integer k indicated the number assigned to the tuples of input-output data. It is not difficult to suppose that $\theta_i \ge \theta_i^-, i = 1, 2, \cdots, \ell_1$, where θ_i^- is a known value. For the function $x_i = x_i(\mathbf{u})$, $\mathbf{q}(k)$ and $y(k)$ correspond to $\mathbf{u}$ and x_i respectively, and for the model $f_i = f_i(\mathbf{u}, \mathbf{x})$, they correspond to f_i and $(\mathbf{u}, \mathbf{x})$ respectively. Usually, $\mathbf{q}$ consists of some of the elements of $\mathbf{u}$ and/or $\mathbf{x}$, not all. If no statistics of $\varepsilon(k)$ are known, one of the most popular approach for estimating θ is to take the least square estimates:

$$\hat{\theta} = \left(\sum_{k=1}^{N_1} \mathbf{q}(k)\mathbf{q}(k)^T\right)^{-1} \left(\sum_{k=1}^{N_1} \mathbf{q}(k)y(k)\right). \qquad (7)$$

Note that it is necessary that

$$rank\{\mathbf{q}(1), \mathbf{q}(2), \cdots, \mathbf{q}(N_1)\} = \ell_1$$

for the unique solution of $\hat{\theta}$ to exist. If the amount of data N_1 is small, the estimate does not always take a reasonable value. But, the estimate (7) does not suggest any alternatives if the DM does not like the estimated value of the parameter.

Suppose $\varepsilon(k) \in [\varepsilon^-, \varepsilon^+]$ where ε^- and ε^+ are pre-assigned values. Then the possible set of θ that satisfies (5) with the error-constraint (6) can be written as

$$\mathbf{s}^T \mathbf{q}(k) \ge y(k) - \varepsilon^+ - \bar{\theta}^T \mathbf{q}(k), \quad k = 1, 2, \cdots, N_1 \quad (8)$$

$$\mathbf{s}^T \mathbf{q}(k) \le y(k) - \varepsilon^- - \bar{\theta}^T \mathbf{q}(k), \quad k = 1, 2, \cdots, N_1 \quad (9)$$

$$s_1, s_2, \cdots, s_\ell \ge 0 \qquad (10)$$

$$\theta = \bar{\theta} + \mathbf{s} \qquad (11)$$

The smallest set of θ that satisfies (6),(8)-(11) is the convex polyhedron, but IDSS shows the minimum and maximum values of each parameter instead to let the DM recognize the feasible solutions of each parameter quickly (Fig. 2).

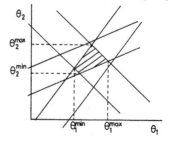

Fig. 2. Parameter region in linear model.

The minimum and maximum values of θ can be computed by solving $2\ell_1$ times simplex method, where the objective function is the solution of the next problem.

Problem P_1:

$$\min s_i \quad \text{or} \quad \max s_i$$

subject to (8)-(10), for $i = 1, 2, \cdots, \ell_1$.

When the DM chooses a value for the parameter, IDSS verifies if all the inequalities hold. If not, IDSS notifies the DM. It is not a difficult task for the DM to find a parameter value that satisfies all the conditions, because the feasible parameter set is a convex polyhedron.

Nonlinear Model

The nonlinear model treated here is

$$y(k) = \varphi(\theta, \mathbf{q}(k)) + \varepsilon(k), \quad k = 1, 2, \cdots, N_2 \qquad (12)$$

where $\varepsilon(k) \in [\varepsilon^-, \varepsilon^+]$, and the meaning of the variables is the same as in the linear model. Define Θ to be the set of ℓ_2-dimensional parameter vector θ that satisfies all the inequalities

$$y(k) - \varepsilon^+ \le \varphi(\theta, \mathbf{q}(k)) \le y(k) - \varepsilon^-, \quad k = 1, 2, \cdots, N_2. \quad (13)$$

Since the function φ is nonlinear and convexity is not assumed, it is difficult to describe the shape of feasible parameter region parametrically. Therefore, we adopt the following algorithm.

Step 1 Find $\theta^{(0)}$ that satisfy all the inequalities (13). Set $j = 0$.

Step 2 Search extremal points of Θ by fixing $\ell_2 - 1$ parameter values of $\theta^{(j)}$.

Step 3 The DM decides the next searching point $\theta^{(j+1)}$ by changing one of the parameters.

Step 4 If the DM is satisfied with $\theta^{(j+1)}$ then stop; otherwise $j := j + 1$ and go to step 2.

For the problem in Step 2, we have to solve

Problem P_2:

$$\max \theta_i \quad (\text{or} \quad \min \theta_i)$$

subject to

$$y(k) - \varepsilon^+ - \varphi(\theta, \mathbf{q}(k)) \le 0, \quad k = 1, 2, \cdots, N_2 \qquad (14)$$

$$\varphi(\theta, \mathbf{q}(k)) - y(k) + \varepsilon^- \le 0, \quad k = 1, 2, \cdots, N_2 \qquad (15)$$

for all $i \in [1, \ell_2]$, where

$$\theta_m = \theta_m^{(j)}, \quad m \in [1, \ell_2] \setminus \{i\}.$$

By the above algorithm, it is clear that the yielding $\theta^{(j)}$ is guaranteed to be in the feasible domain Θ (Fig. 3).

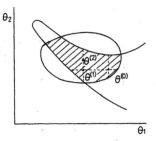

Fig. 3. Parameter estimation in nonlinear model.

MULTIOBJECTIVE PROGRAMMING

When the functions $\{x_i(\mathbf{u}), i = 1, 2, \cdots, n\}$ and $\{f_i(\mathbf{u}, \mathbf{x}), i = 1, 2, \cdots, r\}$ are given or decided by the parameter estimation schemes explained in the previous section, we have the following multiobjective programming problem:

minimize

$$\mathbf{f}(\mathbf{u}) = (f_1(\mathbf{u}), f_2(\mathbf{u}), \cdots, f_r(\mathbf{u}))$$

subject to $\mathbf{u} \in \mathbf{U}_o \cap \mathbf{U}_x$.

<u>Definition</u> (Pareto-optimality)

$\mathbf{f}^*$ is "Pareto-optimal" if and only if

$$\forall \mathbf{f} \in \mathbf{F}, \quad \mathbf{f}^* - \mathbf{f} \in \Re_+^r \quad \Rightarrow \quad \mathbf{f}^* = \mathbf{f}$$

where $\mathbf{F}$ is the feasible region of objective functions and $\Re_+^r$ is the nonnegative orthant in r-dimensional space

Thus, our problem is to choose $\mathbf{f}$ on the Pareto-optimal surface, and also $\mathbf{u}$ correspondingly.

Next, we consider the preference of the decision maker. The decision maker has a preference order for the set of Pareto-optimal solutions in $\mathbf{F}$. For denoting the preference explicitly, we define the utility function as follows.

Let $\mu(\mathbf{f})$ be a utility function of the DM. We suppose that the DM's preference satisfies the weak order (i.e. asymmetric and negatively transitive) (Sawaragi, Nakayama, and Tanino, 1985) to guarantee that there exists a unique indifference curve for any alternative $\mathbf{f}$. When the utility function $\mu(\mathbf{f}(\mathbf{u}))$ is known, the multiobjective programming problem reduces to a nonlinear optimization problem given by

Problem P_3:

$$\max_{\mathbf{u}} \mu(\mathbf{f}(\mathbf{u}))$$

subject to $\quad g_i(\mathbf{u}) \le 0 \quad i = 1, 2, \cdots, 2(m + n)$

where

$$g_i(\mathbf{u}) = u_i^- - u_i, \quad i = 1, 2, \cdots, m$$

$$g_{i+m}(\mathbf{u}) = u_i - u_i^+, \quad i = 1, 2, \cdots, m$$

$$g_i(\mathbf{u}) = x_i^- - x_i(\mathbf{u}), \quad i = 1, 2, \cdots, n$$

$$g_{i+n}(\mathbf{u}) = x_i(\mathbf{u}) - x_i^+, \quad i = 1, 2, \cdots, n$$

But we will not deal with this kind of problem any more, because the multiobjective programming problem arises due to the difficulty of the unification of the various objectives. This case is not essentially a multiobjective programming problem.

331

Methodologies

Here we list some principal multiobjective programming methods considered before, with brief explanations.

(a) Weighted Sum
The objective function is the weighted sum, i.e.

$$\min_{\mathbf{u}\in U} \mathbf{w}^T\mathbf{f}(\mathbf{u}), \quad w_i \geq 0, \ i = 1, 2, \cdots, r; \sum_{i=1}^{r} w_i = 1$$

Using this function, the DM cannot attain Pareto-optimal solutions at duality-gap part by using any $\mathbf{w}$.

(b) Goal Programming (Charnes and Cooper, 1961) By using ℓ_p norm,

$$\min_{\mathbf{u}\in U} ||\mathbf{f}(\mathbf{u})-\bar{\mathbf{f}}||_p^{\mathbf{w}} = \begin{cases} \left(\sum_{i=1}^{r}\left|w_i(f_i(\mathbf{u})-\bar{f}_i)\right|^p\right)^{1/p} & 1 \leq p < \infty \\ \max_{i\in[1,r]}\left|w_i(f_i(\mathbf{u})-\bar{f}_i)\right| & p = \infty \end{cases}$$

is computed, where $\bar{\mathbf{f}}$ is the goal fixed by DM. This method may yield a solution that is not Pareto-optimal.

(c) Compromise Programming (Yu and Leitmann, 1973)
This is the case when $\mathbf{f}$ is given by

$$\bar{f}_i = \min_{\mathbf{u}\in U} f_i(\mathbf{u}) \quad i = 1, 2, \cdots, r$$

in the goal programming.

(d) Interactive Frank-Wolfe Method (Geoffrion, Dyer and Feinberg, 1972)
This method is based on the maximization of utility function that is not given explicitly. The critical steps are
(Step 1)

$$\max_{\mathbf{u}\in U} \nabla_{\mathbf{u}}\mu(\mathbf{f}(\bar{\mathbf{u}})) \cdot \mathbf{u}$$

where $\bar{\mathbf{u}}$ is a given feasible solution $\bar{\mathbf{u}} \in U$. To evaluate $\nabla_{\mathbf{u}}\mu(\mathbf{f}(\bar{\mathbf{u}}))$, the marginal rates of substitution is used, which is given interactively.
(Step 2)

$$\max_{0\leq\alpha\leq 1} \mu(\mathbf{f}(\bar{\mathbf{u}} + \alpha(\mathbf{u}^* - \bar{\mathbf{u}})))$$

where $\mathbf{u}^*$ is the solution of step 1. Since μ is not given explicitly, the DM solves step 2 by looking at the value of $\mathbf{f}$ numerically or graphically.

(e) ε-constraint Method (Haimes, Hall and Freedman, 1975)
Problem $P_\varepsilon(\bar{\mathbf{e}}_k)$:

$$\min_{\mathbf{u}\in U} \ f_k(\mathbf{u}) \tag{16}$$

$$\text{subject to} \quad f_j(\mathbf{u}) \leq \varepsilon_j \quad j \in N_k(r) \tag{17}$$

$$N_k(r) = \{1, 2, \cdots, r\}\backslash\{k\}$$

This is the minimization problem of $f_k(\mathbf{u})$, where all the other objective functions have upper constraints (Fig. 4). Define

$$\bar{\mathbf{e}}_k = (\varepsilon_1, \cdots, \varepsilon_{k-1}, \varepsilon_{k+1}, \cdots, \varepsilon_r)$$

$$U_k(\bar{\mathbf{e}}_k) = \{\mathbf{u} : f_j(\mathbf{u}) \leq \varepsilon_j, \ j \in N_k(r)\} \tag{18}$$

$$S_k = \{\bar{\mathbf{e}}_k : U_k(\bar{\mathbf{e}}_k) \neq \emptyset\} \tag{19}$$

$$AS_k = \{\bar{\mathbf{e}}_k : \bar{\mathbf{e}}_k \in S_k, \ f_j(\mathbf{u}^*(\bar{\mathbf{e}}_k)) = \varepsilon_j, \ j \in N_k(r)\} \tag{20}$$

where $\mathbf{u}^*(\bar{\mathbf{e}}_k)$ is the solution of the problem $P_\varepsilon(\bar{\mathbf{e}}_k)$. (19) is the necessary condition for $P_\varepsilon(\bar{\mathbf{e}}_k)$ to have feasible solutions.
<u>Theorem 1</u> (Sakawa, 1986). Let

$$\bar{\mathbf{e}}_k^* = (\varepsilon_1^*, \cdots, \varepsilon_{k-1}^*, \varepsilon_{k+1}^*, \cdots, \varepsilon_r^*)$$

and suppose that $\mathbf{u}^*$ is the unique solution of $P_\varepsilon(\bar{\mathbf{e}}_k^*)$. Then $\mathbf{u}^*$ is Pareto-optimal.

<u>Theorem 2</u> Any Pareto-optimal solution $\mathbf{u}^* \in U$ is the solution of $P_\varepsilon(\bar{\mathbf{e}}_k)$, where $\bar{\mathbf{e}}_k \in AS_k$.

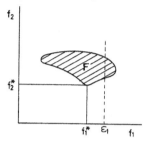

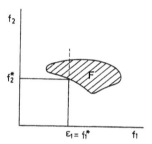

Fig. 4. ε-constraint method for two cases.

(f) Surrogate Worth Trade-Off Method (Haimes, Hall and Freedman, 1975)
This method is based on ε-constraint method. When the problem $P_\varepsilon(\bar{\mathbf{e}}_k)$ gives a (Pareto-optimal) solution and all the ε-constraints are active, the corresponding Lagrange multiplier gives the trade-off rate :

$$\lambda_{1i} = -\frac{\partial f_1(\mathbf{u})}{\partial f_i(\mathbf{u})}, \quad i = 2, \cdots, r$$

Prepare a small set $\{\bar{\mathbf{e}}_1\}$. For each $\bar{\mathbf{e}}_1$, computer shows $\{\lambda_{1i}, i = 2, \cdots, r\}$ to the DM and asks to compare to his/her surrogate worth. If all $\{\lambda_{1i}, i = 2, \cdots, r\}$ coincide with his/her surrogate worth, then the solution of the ε-constraint problem is the final solution.

(g) Satisficing Trade-Off Method (Nakayama, 1984)
Firstly, the DM sets up the ideal point $\mathbf{f}^*$. Here, f_i should be a very small value, e.g.,

$$f_i^* \leq \min_{\mathbf{u}\in U} f_i(\mathbf{u}) \quad i = 1, 2, \cdots, r$$

and the DM also sets up aspiration levels $\hat{\mathbf{f}}(> \mathbf{f}^*)$. These two points are used for the problem defined as

$$\min_{\mathbf{u}\in U} ||\mathbf{f}(\mathbf{u}) - \mathbf{f}^*||_\infty^{\mathbf{w}}$$

or, equivalently

$$\min_{\mathbf{u}\in U} \max_{i\in[1,r]}\{w_i(f_i(\mathbf{u}) - f_i^*)\}$$

where

$$w_i = \frac{1}{\hat{f}_i - f_i^*}$$

Satisficing Trade-Off method is to repeat revising $\hat{\mathbf{f}}$ and solving the min-max problem until the DM is satisfied with the solution.

IDSS SOFTWARE

IDSS is composed of three modules. Here we will give brief explanations to the components.

A. Variable Definition

(A1) Input Variables

IDSS asks the names of input variables u_i and the lower and upper limits u_i^-, u_i^+ which the input variables can take.

(A2) State Variables

IDSS asks the names of state variables x_i, the lower and upper limits x_i^-, x_i^+ if limited, and the functions $x_i(\mathbf{u})$. Here, if the function includes unknown parameters, the unknown parameter should be written as #a1#, for example. If the function includes undefined input variables, IDSS calls the function (A1).

(A3) Objective Functions

IDSS asks the names of objective functions f_i, and the functions $f_i(\mathbf{u}, \mathbf{x}(\mathbf{u}))$. If the function includes unknown parameters, the format in (A2) is also applied. When undefined input or state variables exist, IDSS calls the functions (A1) or (A2).

B. Parameter Estimation

(B1) Experiment Data

Tuples of data are prepared in this function.

(B2) Linear Model

Define the lower and upper limits ε^-, ε^+ of error ε which the DM considers to be appropriate. IDSS computes the least square estimate also. Looking at the lower and upper limit of the parameters that are computed by solving $2\ell_1$ LP problems P_1 and the least square estimate, the DM chooses a point. Then IDSS shows a table of input variables, output variable, and the estimate of output variable using those parameters for the DM to check validity. If the DM is not satisfied, IDSS repeats this interactive procedure.

(B3) Nonlinear Model

Define the lower and upper limits ε^-, ε^+ of error ε which the DM considers to be appropriate. Following the algorithm (steps 1-4), the DM selects a point.

C. Multiobjective Programming

(C1) Forming $\mathbf{f}(\mathbf{u})$

Substituting $x_i(\mathbf{u})$ into $f_i(\mathbf{u}, (\mathbf{x}))$, the IDSS constitutes $f_i(\mathbf{u})$. This step is automatically done when the DM enters into module "C".

(C2) Component-wise Pareto-optimal Solution Region

This function first computes

$$\tilde{f}_i(i) = f_i^- = \min_{\mathbf{u}\in U} f_i(\mathbf{u}) \quad i = 1, 2, \cdots, r$$

and define $\mathbf{u}^*(i)$ to be the corresponding input vector. Then using $\mathbf{u}^*(i)$, IDSS computes

$$\tilde{f}_j(i) = f_j(\mathbf{u}^*(i)), \quad j \in N_k(r), \quad i = 1, 2, \cdots, r.$$

Now we can find that all the Pareto-optimal solutions are within a hypercuboid

$$f_i^- \le f_i \le f_i^+, \quad i = 1, 2, \cdots, r$$

where

$$f_i^- = \tilde{f}_i(i),$$
$$f_i^+ = \max_{j=1,\cdots,r} \tilde{f}_i(j).$$

See Fig. 5 for two dimensional case. This is also automatically computed after (C1).

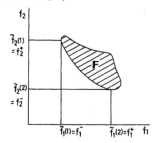

Fig. 5. Component-wise Pareto-optimal solution region.

(C3) Interactive ε-Constraint Method

This is basically the same to the algorithms presented by Nakayama, Tanino, and Sawaragi (1980), Shimizu (1982). The algorithm we use is described as follows.

Step 1 IDSS shows tuples $(f_1^-, f_1^+), (f_2^-, f_2^+), \cdots, (f_r^-, f_r^+)$ to the DM.

Step 2 The DM selects the index $k \in \{1, 2, \cdots, r\}$ for the corresponding function to be minimized. According to the requirement of IDSS to enter the value $\varepsilon_j \in [f_j^-, f_j^+]$, $j \in N_k(r)$, the DM inputs them. If the DM is from Step 3, he is required to input only one bound he can mostly accept relaxation.

Step 3 IDSS computes f_k^*, $\mathbf{u}^*(\bar{\mathbf{e}}_k)$ by the ε-constrained method, and also $\{f_i(\mathbf{u}^*(\bar{\mathbf{e}}_k)), i \in N_k(r)\}$. Note that $f_i(\mathbf{u}^*(\bar{\mathbf{e}}_k))$ may be smaller than ε_i. If the DM is satisfied with the solution, then stop; otherwise go to Step 2.

CONCLUSIONS

We developed an interactive decision support system for multiobjective programming. The estimation of unknown parameters in linear and nonlinear models is also formed to be interactive, so that the user can make a model with his/her favourite value. This paper dealt with only continuous variables. But, it is not difficult to apply IDSS for the system with discrete input variables by fixing the variable and try several times. Although this work was motivated by agricultural problem, it can be used in other field without any modification.

REFERENCES

Charnes, A., and W. W. Cooper (1961). *Management Models and Industrial Applications of Linear Programming*. John Wiley & Sons, New York.

Enoch, H. Z. (1978). A theory for optimization of primary production in protected cultivation. *Acta Horticulturae, 76*, 32-57.

Geoffrion, A. M., J. S. Dyer, and A.Feinberg (1972). An interactive approach for multi-criterion optimization with an application to the operation of an academic department. *Management Science, 19*, 357-368.

Haimes, Y. Y., W. A. Hall, and H. T. Freedman (1975). *Multiobjective Optimization in Water Resources Systems*. Elsevier, Amsterdam.

Liebig, H. -P. (1989). Growth and yield models as an aid for decision making in protected crop production control. *Acta Horticulturae, 260*, 99-113.

Nakayama, H. (1984). Proposal of satisficing trade-off method for multiobjective programming. *Trans. Soc. of Instrument and Control Engineers, 20*, 29-35 (in Japanese).

Nakayama, H., T. Tanino, and Y. Sawaragi (1980). An interactive optimization method in multicriteria decisionmaking. *IEEE Trans. SMC-10*, 163-169.

Sakawa, M. (1986). *Optimization of Nonlinear Systems*. Morikita Publisher, Tokyo (in Japanese).

Sawaragi, Y., H. Nakayama, and T. Tanino (1985). *Theory of Multiobjective Optimization*. Academic Press, London.

Shimizu, K. (1982). *Theory of multiple objectives and conflicts*. Kyoritsu Publisher, Tokyo (in Japanese).

Tantau, H. -J. (1990). Automatic control application in greenhouse. *11th IFAC World Congress, 11*, 302-305.

Yu, P. L., and G. Leitmann (1974). Compromise solutions, domination structures and Sulukvadze's solution. *J. Opt. Th. & Appl., 13*, 362-378.

MATHEMATICAL SIMULATION OF THE BROILER PRODUCTION - A TOOL FOR MODEL AIDED CONTROL OF THE PROCESS

K. Großkopf*, D. Heide**, M. Flechsig*** and E. Matthäus[†]***

**Friedrich-Schiller-University Jena, Faculty of Biology, Dept. of Environment and Nutrition,
Gustav-Herrmann-Str. 24, Stadtroda, D-6540 Germany*
***University of Leipzig, Faculty of Agricultural Sciences, Institute of Agricultural Farm Economics,
Schillerstr. 6, Leipzig, D-7010 Germany*
***Institute of Ecosystem Research, Magdalenenstr. 17-19, Berlin, D-1130 Germany

Abstract. Based on the causal factor-product-relationships and inter-relationships acting in forming the yield and expenditure in broiler production a model for analysing the behaviour and controlling the broiler production process has been developed using the simulation system SONCHES[TM]. The model maps different aspects - the metabolism and process within a single animal submodel, a description of intraspecific inter-actions contained in a stock submodel, the interaction of the biological system with its physical environment within the poultry house climate submodel and an economic evaluation of yield, expenditures and control activities within an economic submodel. Various possibilities for model validation and management preparing simulation experiments are presented.

Keywords: Agriculture; system analysis; modeling; computer simulation; production control.

PROBLEM FORMULATION

Animal production can be considered as a very complex dynamic system characterized by tight interactions of biological, technological and economic processes. Its organisation includes from a cybernetical point of view the purposeful, economically effective technological use and control of biological basic processes (e.g. growth). For better understanding and for solving this task the use of adequate models of the process is required, ingenious and effective. With their help consequences of changes in quality, quantity, time of application and evaluation of production factors may be predicted in a simulative way by computer experiments off-line or in connection with adequate hardware an on-line process control may be carried out within the framework of production control systems (Kuhlmann, 1987; Korver, Arendonk, 1988).
In every case such models have to reflect the functional structure of the processes controlled, i.e. the causal factor - pro-duct - relationships and interrelation-ships acting in forming the yield and expenditure in order to compare the be-haviour of the real system (production process) with the simulated one of the model. An adequate process model has been developed successively for broiler pro-duction (Grosskopf, 1988; Grosskopf and others 1988 and 1990) using the modelling and simulation system SONCHES[TM] (Knijnen-burg and others, 1984; Wenzel and others, 1985; Matthäus and others, 1985).
In order to concentrate on the most impor-tant and essential relationships of the reality and taking into account the avail-able knowledge on the quality and quantity of causal relationships in broiler produc-tion the enormous number of potential factors of influence on this aim is re-duced to the following ones:

- genotype (genetically determined reten-tion ability of material (protein lipid) and growth ability),
- sex of the chicken,
- time and ontogenetic age of the chicken,

respectively
- food (limited to the energy and protein content of the ration),
- stocking density,
- thermal environment (limited to the temperature and relative air humidity).

Obviously, within the system of causal relationships built up from these factors the biological processes connected with metabolism and growth represent the basic processes of the whole real system and, hence, of the whole model. In practice, however, these processes are inseparably connected with the technological ones, which have to be reflected by the model. Among them controlling of climatic environment of the chicken has a special, because relatively autonomous, function.

MODEL DESCRIPTION

The structure of the simulation model is adequate to the functional structure of the real broiler production process defining (see Fig.1)

- a broiler compartment for mapping the growth of a spatially limited stock,
- a poultry house compartment for mapping the climate inside the the poultry house
- an economy compartment for evaluating outputs and inputs of the broiler and poultry house compartment to calculate costs, proceeds, gross margin, profit and economic efficiency.

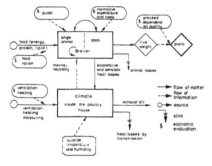

Fig.1. Structural diagram of the whole broiler model

Broiler Compartment

The core of the broiler compartment and, hence, the base for reflecting all the other processes is represented by mapping the growth of a single animal in dependence on the first four factors of influence mentioned above under the condition of an optimal environment (i.e. influences of the stock and thermal environmental factors on yield and expenditure formation are excluded). The argument for reducing the stock to a single animal mainly results from the necessity to trace back growth first of all to its physiological basic processes. This includes the demand to eliminate intraspecific interrelationships within the broiler compartment. Growth is realized due to intake of energy

and essential components (e.g. amino acids) with the food, their conversion in the metabolism and - resulting from this - their retention as body own matter (lipid, protein) (see Fig. 2).

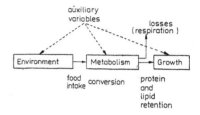

Fig.2. Basic processes of the single animal growth model

For modelling of these complex basic processes a great number of additional variables is required. The starting point consists in the hypothesis that the retention ability of material and the growth ability of an animal are genetically determined (Emmans, 1981 and 1987, Emmans and other 1983). This first of all requires a mathematical description of the genetically determined growth velocity for the live weight (DW in g/chicken/d) as a function of the actual live weight of the chicken (W in g/chicken). Three different growth functions fitted to experimental values are available and may be used alternatively:

- Gompertz-function

$$DW = W*(A-K*ln(W)) \qquad (1.1)$$

with A = 0.29269
K = 0.03426

- Bertalanffy-function

$$DW = 3*W*(-B+AC*W^{-1/3}) \qquad (1.2)$$

with B = 0.023819
AC = 0.42982

- Evolon-function (Mende, 1985)

$$DW = C*(W-A)^k**(B-W)^l \qquad (1.3)$$

with C = .0001606
A = 0
k = 0.56
B = 6832
l = 0.79

The connection of the genetically determined live weight gain with a genetically determined protein retention per unit of live weight gain (DPDW) and a normative lipid retention per unit of protein gain (DLDP) results in a protein retention ability (DP1).

$$DP1 = DW * DPDW \quad [g/chicken/d] \quad (2.1)$$
whereas

$$DPDW = .24 * (W/W_{max})^{p*xp} \quad [g/g]$$

with pexp = 0.06

$$W_{max} = \begin{cases} e^{A/K} & \text{(Gompertz)} \\ (AC/B)^3 & \text{(Bertalanffy)} \\ B & \text{(Evolon)} \end{cases} \quad (2.2)$$

and in a lipid retention characteristic (DL).

$$DL = DW * DPDW * DLDP \quad (3.1)$$
$$[g/chicken/d]$$

whereas

$$DLDP = Q * P^D \ [g/g] \quad (3.2)$$

with Q = 0.23
    D = 0.15
    P = actual cumulative protein retention [g/chicken].

Using the three normative rates for live weight (DW), protein (DP1) and lipid (DL) described, the daily food energy (DE) and food protein requirement (FPR) can be derived because the chicken fed ad libitum aims to feed itself in such a way that it can realize its genetical growth pattern.

$$DE = DW * DPDW * CEPN + DL * CEL + CEM * W^{0.75} \quad (4)$$
$$[kJ \ m.E.^{[1]}/chicken/d]$$

$$FPR = \frac{DP1}{PCR} + \frac{(CPM * W^{2/3})}{PCRM} \quad (5)$$
$$[g/chicken/d]$$

with
CEPN - energy requirement per unit of protein retention under the condition of a genetically determined growth,
CEL - energy requirement per unit of lipid retention,
CEM - energy requirement for maintenance metabolism per unit of metabolic live weight,
CPM - net protein requirement for maintenance metabolism per unit of metabolic live weight,
PCR - protein conversion rate in protein gain dependent on the energy and protein content of the ration and on time,
PCRM - protein conversion rate in the protein maintenance metabolism.

These requirements are met by the daily food intake (FOOD in [g/chicken/d]). Different control functions were developed and can be used alternatively, so making allowance for the different opinions on the control mechanism for food intake under ad libitum feeding.

$$FOOD = \frac{DE}{FEC} \quad (6)$$

(Food intake is controlled only by the energy requirement.)

or

$$FOOD = Max(\frac{DE}{FEC}, \frac{FPR}{FPC}) \quad (7)$$

(Food intake is controlled by energy as well as by protein requirement according to the limiting one.)

with
FEC - energy content of the ration
FPC - crude protein content of the ration

In any case an upper bound for food intake (MFI) caused by a limited intake capacity of the digestive tract and by its passage ability is involved

$$FOOD = Min \ (FOOD, \ MFI) \quad (8)$$

The real daily food intake determines the development of the live weight of the chicken, so becoming the central state variable (abundance) of the whole broiler compartment. The live weight of the chicken (W) is influenced by its real daily live weight gain which is calculated as the growth rate of the compartment abundance. At first cumulative protein (P) and lipid retention (L) declared as state variables has to be calculated because the growth of an animal cannot be interpreted and mapped simply as live weight gain[1]. Hence, the broiler compartment is described vectorially by the state variables W, L and P. Lipid and protein gain for their part depend on the actual available food energy and food protein for growth - which are the differences between the adequate global intake by the chicken and its respective maintenance requirement -, on the ratio between both values and moreover they are limited by genetically determined values (environmental variables and fixed parameters) to an upper (DP1, maximal lipid content of the whole body) and/or to a lower bound (protein and lipid content of the whole body which are mortally for the chicken).
The extension of the single animal growth model to a stock model is compulsory for mapping the following processes significant in practice:

- animal losses,
- effect of stocking density on food intake, and
- effect of temperature inside the poultry house on food intake, on growth of the single animal and on the amount of heat production and the structure of heat losses of a stock.

----
[1] From the point of view of physiology of human nutrition it is of utmost importance to produce live weight for slaughtering with a low content of fat but with a content of protein as high as possible. Besides, the efficiency of food nutrient conversion depends on the structure of live weight gain.

Modelling of this is realized with strong references to variables of a single animal growth model

- modifying these variables under the influence of thermal environment and/or
- using these variables for calculating new ones representing inputs of other compartments and/or outputs of the whole system.

Poultry House Compartment

The climate inside the poultry house is modelled by a special poultry house compartment. Its task consists in ensuring the maintaining of a physiologically and economically admissible range of temperature and relative air humidity inside the poultry house independent on outside climatic conditions. This makes the basis for determining suitable control strategies of the climate. The control variables of the climate are temperature (TEMPI) and relative air humidity inside the poultry house (PHII). These are state vaiables of the air which should be controlled properly by suitable variables independent from each other like enthalpy (EI) and absolute water content of the air inside the poultry house (WI). The calculation of the enthalpy and the absolute water content from air temperature and relative air humidity is realized outside of SONCHES by special FORTRAN-subroutines (ENTHAL and WATER) stored in the SONCHES user library. The vapour saturation values of the air in dependence on its temperature needed for calculation of WI are stored as a table function in the table library. In the case of simulation these values are read and interpolated, if necessary.

Factors of influence on the control variables of the climate inside the poultry house are:

- vapour (TOVAP) and heat losses by the stock (HLLAT, HLSEN) which are available as outputs of the broiler compartment;
- heat losses by transmission (HLTRAN) which are calculated as an additional state variable of the poultry house compartment;
- temperature and relative air humidity of the ventilated air in connection with the ventilation with unchanged or treated (heated, moistured) outside air.

The balances resulting from this are:

$$EI/\Delta t = ((HLSEN + HLLAT - HLTRAN)/MI) + (MV/MI) * (EV - EI) \qquad (9)$$
$$[kJ/kg * d]$$

$$WI/\Delta t = TOVAP/MI + (MV/MI) * (WV - WI) \qquad [g/kg * d] \qquad (10)$$

with
MI - mass of air inside the poultry house,
MV - mass of the ventilated air,
WV - absolute water content of the ventilated air,
EV - enthalpy of the ventilated air.

The outputs of the poultry house compartment which are essential from the point of view of the whole model are:

- Mass and volume of the ventilated air per time-step of the poultry house and the broiler compartment;
- Change of enthalpy of the outside air and the daily expenditure of heating energy required for keeping the normative ranges of temperature and relative air humidity inside the poultry house.
- Increase of water content of the outside air and the daily expenditure of water for moisturing required for keeping the normative ranges of temperature and relative air humidity inside the poultry house;

Economy Compartment

The completation for the scientific part of the model with an economy compartment which evaluating the output and input of the broiler and poultry house compartment to calculate costs, proceeds, profit and economie efficiency is necessary to assess financially the consequences of interactions and connections between several production factors. It is the presupposition for the economic contemplation of the whole broiler production process.

The functions of the economy compartment are:

- to assess financially the output of the other compartments of the model
  - food costs (CFOOD),
  - animal costs (CAN),
  - costs for electrical energy (CELTOT)
  - costs for heating (CHEAT)
  - the financial proceeds (PROCB)
- to determine the amount of costs by standards for which are the exact calculation is very difficult but the influence on the economic result is insignificant
  - costs for drugs and other materials (CMED)
  - former costs (CUNI)
- to asses some index numbers additionally depends on the number of animal places and the number of animals
  - costs for labour (CLAB)
  - costs for illumination (CILUM)
  - costs for repair (CREP) and capital (CCAP)
- to calculate cost - proceed - relationships for the determination of the efficiency of the whole process
  - profit (PR)
  - operating profit (DB1, DB2) [1]
  - production index (PI)
  - break-even-point (BEP)

---

[1] The operating profit 2 (DB2) includes the costs for labour the operating profit 1 (DB1) not. These are the conditions to consider differ conditions.

These four main points are shown by the structure of the reference numbers in fig. 1.

In following there will be described many selected reference numbers.

The calculation of the food costs was realised by assessing the food consumption of the day with the unit price for food of the ration. The algorithm allows to use two-stage-ration. This kind of ration is common in broiler feeding.

The food costs represent approximately 50% of the whole costs.

The broiler production process is characterized as a process with a high demand for energy . Especially the climatic requirements of the animals demand to use air-conditioning.

The use of ventilation increases the electricity costs (CVENT). They are calculated by a comparison between the ventilation capacity of the air-conditioning (VENTMAX, $[m^3/h]$) and the ventilation rate which is needed per day (VLDAY, $[m^3/h]$).

The calculation is based on the assumption, that a fixed number of homogeneous ventilators (NMFG) are linked. So these ventilators are uniformly regulated. The algorithm for the calculation of a ventilator group is:

$$VENTMAX = NUMFG * VENTMAXFG \qquad (11.1)$$
$$[m^3/h]$$

The next ventilator group is switched on when the capacity of the air-conditioning (VENTMAX) of the active ventilator group is less than the ventilation rate which is needed per day.

The algorithm (11.2.) is defined as :

$$IF \ (VLDAY/24 ..LE. \ IZ * VENTMAX)$$
$$IZ = IZ +1 \qquad (11.2.)$$

The equation for calculations the electricity consumption (ELCON, kWh) of one ventilator group is the multiplication of the maximum consumption of electricty per ventilator and the number of ventilators per group.

$$ELCON = NMFG * ELMAXFG \qquad (11.3.)$$

under consideration of the proportions:

$$\frac{\text{maximum electroenergy consumption} \quad [kW]}{\text{seeked electroenergy consumption} \quad [kW]} =$$

$$\left[\frac{\text{maximum capacity of the ventilator group } [qm/h]}{\text{capacity under the maximum } [m^3/h]} \right]^3 \qquad (11.4)$$

That means, notated in an algorithmn:

$$ELCON = ELCON / (VENTMAX / (VLDAY / 24 - ((IZ-1) * VENTMAX)))^3$$

$$ELCON = ELCON + (IZ-1) * (NMFG * ELMAXFG)$$
$$ELCON = ELCON * 24$$

$$CVENT = ELCON * CKWH * UNIT / AF \qquad (11.5)$$

Because only variables are used for the calculation of the costs for electricity it is possible to consider differential air-conditionings.

Besides the calculation of several costs it is important to calculate the financial proceed for an exactly computation of the profit.

The calculation of the financial proceeds is based on the leaving weight production (LWP, [kg/qm]) and the weight per animal (W, [g/animal]). The leaving weight production and the weight per animal are calculated in the scientific part of the model. According to the "Handelsklassenverordnung für Geflügelfleisch vom 20.4.1983" the classification of broiler carcasses is done in three categories (classes) A, B, C. Animals which have a minimum weight of 1200 g can be classified in the class "EXTRA".

Some of the produced animals are not classified because they are less than the minimum weight (MW, [g]). These are the "Zählverluste" (ASDZ, [animals/m²]). An other part of the produced animals has damages of the skin or the skeleton. These animals are rejected (ASDV, [animals/m²]). In the economy compartment there was defined a minimum weight (WM, [g]) of 900 g/animal.

Literature studies and analysis in slaughterhouses have shown, that the count losses and the number of rejected animals depend on the slaughter weight. In determined weight ranges there are relatively constant percentages. For instance in the range of 1270 to 1600 g slaughterweight per animal there are the count losses nearly 1,5 % and the number of rejected animals nearly 3 % of the produced animals. (ASD, [animals/m²])

The calculation of the count losses is:

$$ASDZ = ASD * 0.015 \qquad (12.1)$$
$$[animals/m^2]$$

And the number of rejected animals is:

$$ASDV = (ASD - ASDZ) * 0.03 \qquad (12.2)$$
$$[animals/m^2]$$

The price for carcasses depends on the weight of the animal (W, [g/animal]) in the classes. In class A there are five price classes at the moment.

The calculation of the profit (PR) (difference between sum total costs (CTOT) and financial proceed (PROCB)) is carried out in three stages

1. operating profit 1
   ("profit" incl. fixed costs)

$$DB1 = PROCB - CTOT + CGFIX \qquad (13.1.)$$

2. operating profit 2
   ("profit" incl. fixed costs and labour)

   DB2 = DB1 + CLAB   (13.2.)

3. profit as the difference between financial proceeds and total costs

   PR = PROCB - CTOT   (13.3.)

The fixed costs are the sum of the costs
of capital (CCAP) and maintenance (CREP).
A special reference number characterises
the economic efficiency of the broiler
production process. The production index
(PI) is calculated as the quotient of the
intersection of the weight production
(W(O) - WO) of the usuable animals (PUA,
kg/animal) and the product from productions days (DAY, [d]) and food utilization
rate (FA, [kg food/kg live weight produc-
tion])

$$PI = (W - WO * PUA / DAY * FA) * 1000$$
$$(14.1.)$$

The calculation of the number of useable
animals from the produced animals is the
difference between the number of animals
on the beginning (ASD1, [animals/m^2]) and
the number of produced animals:

PUA = 1 - ((ASD1 - ASD + ASDZ +
       ASDV/ASD1)   (14.2.)

The economic model allows also to calculate the break-even-point (BEP):

BEP = PROCB / CTOT   (15)

Controlled by a parameter (PUNIT) all the
simulated financial variables (costs,
proceed, profit) may be calculated alternatively in relation to the following
bases (UNIT)

- per slaughtert animal
- per m^2 poultry house in a production
  cycle
- per kilogram live weight
- per kilogram live weight gain.

Simulation Experiments

The described model structure allows simulation experiments whose results can be
used for the preparation and accomplish-
ment for economic decisions.
The determination of widely variable inputs, which varied allows manifold possibilities for experiments on different
questions.
It is possible to differ between four main
points

- the agreement of the model with the
  reality (model validation)
- the effects of varied scientific inputs
  for scientific and economic output
- the consequences of varied economic
  frame conditions (for instance prices)
  for the efficiency of the whole broiler
  production process and based on this

- the conclusion for technical and organi-
  zational formation of the broiler
  production process under consideration
  of determined economic frame conditions
  but also for the determination of urgent
  economic frame conditions which guarantee a desired productivity.

The wide spectrum of using the process
simulation model shows the following
questions:

1. What are the consequences of changing
   the energy and protein content of the
   food ration for the growth rate of the
   chicken, for the amount of the structure of gain of matter (lipid, protein)
   of the body and especially for the
   efficency of the process (proceeds
   PROCB, foodcosts CFOOD, profit PR)? An
   example is shown in fig.3 to 5.
2. What are the consequences of changing
   the number of housed chickens per unit
   area as to states of affairs relevant
   for the economy of the process (live
   weight production W, proceeds, of the
   relation between special costs and fix
   costs)?
3. What are the consequences of changing
   the official food prices for the costs
   of feeding and on following the whole
   efficiency of the process?
4. Which air - conditioning process proves
   to be effective under the special climatic conditions?
5. Which prices should consumers pay for
   the broilers when the weight of the
   animals is over 1600 g and for the
   production process should be effective?

In the following there is shown the effect
of changing the energy and protein content
(FEC2, FPC2) of the food stuff 2 (from the
11th day up to the end of the production
period) for the efficiency of the whole
process. A simulation study was performed
for

- case I: FPC2 = 23%   FEC2 = 12.3 MJ/kg
- case II: FPC2 = 28%   FEC2 = 12.3 MJ/kg
- case III: FPC2 = 23%   FEC2 = 20.0 MJ/kg

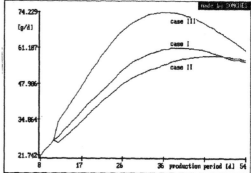

Fig. 3. Daily growth rate per animal

The live weight production is higher on a
high energy content in the ration as a
high protein content and a low level of

energy in the food.

The reason is, that the growth of protein has been a priority over the substance growth of fat (the substance growth of fat over the minimum).

So on the first the animal satisfies its protein requirement with the food consumption. A low protein level in the ration brings a higher food consumption and so a higher energy consumption of the animal.

It has already been shown that the payment based on the live weight of the animals, does not depend on the body composition. Heavy animals bring a higher price independent of the protein-fat-relation in the carcasses (fig. 4) and so a higher profit (fig. 5).

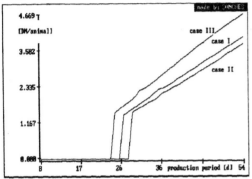

Fig. 4. Proceed per animal

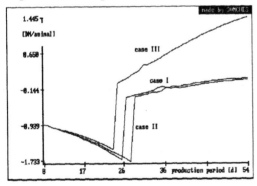

Fig. 5. Profit per animal

Based on the simulation results the following conclusions for the organisation of production are possible:

- under consideration of the actual economic frame conditions feed rations with a sufficient protein content and a high energy content allow the largest positive difference between proceeds and costs.
- using foodstuffs with a higher energy content brings shorter productions periods because the necessary slaughter weight - the highest prices are realised today for animals with 1400 g live weight - can be achieved earlier. The number of production passages per year can be rised. The productivity and so the profit/m² unit area can be raised.

References

Emmans, G.C. (1981). Computer simulation in poultry nutrition. Proceed., 3rd. European Symp. on Poultry Production, pp. 91 - 103

Emmans, G.C. (1987). Growth, body composition and feed intake. World's Science Journal, London 43, pp. 208 - 227

Emmans, G.C.; Whittemore, C.T. (1983). Simulation models in nutritional management of livestock. In: Feed information and animal production, Proceed., Second Symp. of the International Network of Feed Information Centres, 1983, pp. 323 - 227

Großkopf, K. (1988). Grundlagen zur mathematischen Simulation der Broilerproduktion - ein Beitrag zu einer rechnergestützten Produktionskontrolle und -steuerung in der Tierproduktion. Dissertation B, Universität Leipzig.

Großkopf, K.; E. Matthäus and M. Flechsig (1988). Modelling of growth processes in broiler production using the simulation system SONCHES. In: System analysis and simulation, Proceed., of the 3rd International Symp. held in Berlin, September 12 -16, 1988. Schriftenreihe Mathematical Research, Vol. 47, Akademie Verlag, Berlin, pp. 241 - 244

Großkopf, K. and E. Matthäus (1990). Mathematical Simulation of the Broiler Production Process. Systems Analysis-Modelling-Simulation 7, pp. 473 - 491

Knijnenburg, A. and others (1984). Concept and usage of the interactive simulation system for ecosystems SONCHES. Ecol. Mod. 26, pp. 51 - 76

Korver, S. and J.V.S. van Arendonk (1988). Modelling of live stock production systems. Kluwer Academic Press Publishers, Dordrecht/Boston/ London. Kuhlmann, F. (1987). Steuerung und Regelung biologischer Produktionssysteme. Spiegel der Forschung/Wissenschaftsmagazin der Univ. Giessen, 6, pp. 14 - 18

Matthäus, E. and others (1985). Use of task facilities of the simulator "SONCHES" for agroecosystem model investigations and control. In: Sydow and others (Ed.). Systems analysis and simulation, 28. Akademie Verlag, Berlin. pp. 133 - 138

Mende, W. (1985). The Evolon Model and its Application to Natural Processes. In: Ebeling, W.; M. Peschel (Eds.). Approach to cooperation and Competition in Dynamic Systems, Akademie Verlag, Berlin. pp. 261 - 271

Wenzel, V. and others (1985). SONCHES - an interactive simulation language for design, validation and usage of ecological models. In: Sydow and others (Ed.). Systems analysis and simulation, 28. Akademie Verlag, Berlin. pp. 129 - 132

KALMAN FILTER NEURO-COMPUTING FOR BIOLOGICAL SYSTEM MODELS USING NEURAL NETWORKS

H. Murase and N Honami

Dept. of Agricultural Engineering, University of Osaka Prefecture, 4-804, Mozuume, Sakai, Osaka, Japan 591

Abstract. The hierarchical neural network can be used to model biological systems such as plant growth, photosynthesis, evapotranspiration, etc. The development of back-propagation algorithm for neuron training has made it possible to use the layered network for simulating such non-linear systems. Modeling of such biological systems using the neural network often requires large number of layers and units in the network architecture because of the complexity of the system. The back propagation algorithm, however, often fails to achieve satisfactory identification of the system in the sense that output error minimization characteristics of the steepest descent scheme of the back propagation algorithm does not fit the problems involving large number of estimation parameters (synapse weights).An attempt of implementing Kalman filter algorithm (Kalman,1960) in the procedure for training the neural network was made and evaluated. The performance of Kalman filter neuro-computing algorithm was compared to commonly used back propagation algorithm. Simulation of growth of radish sprouts under influence of changes in temperature and concentration of nutrient solution was attempted by two different neural network models, i.e., Kalman filter model and back propagation model. Results revealed a superior effectiveness of Kalman filter algorithm over the standard back propagation algorithm in the parameter estimation.

Keywords. Artificial Intelligence; Neural nets; Kalman filter; Nonlinear systems; System identification.

INTRODUCTION

Artificial intelligence (AI) is concerned with developing software systems that are capable of performing works that one would describe as intelligent if human did them. Neural nets are currently one of the hottest AI research areas, after having a period of inconspicuous research development for over a decade (Tello 1989). In many of the other AI research area, neural nets related studies are being conducted because neural modeling is a technique that can in principle be directed at a variety of different applications. One of most attractive aspects is that there are essential mechanisms that allow for incrementally improved performance. The neural modeling technique does not confine itself in AI research area. It has opened up a new way of describing nonlinear systems in the form that computers can handle.

Most of constitutive elements of biological functions and/or ecosystem are nonlinear in nature. Many attempts of nonlinear analysis in biological systems have been made for quite a number of years now, but because of the difficulty of the problem, mostly they ended up with the simplification of the problem by linear assumptions. The hierarchical neural net can be applicable to simulating nonlinear phenomena often found in the biological systems (Hoshi, et al. 1990) The learning process of the hierarchical neural net can be used as an algorithm for the nonlinear maltivariate analysis. Inversely the Kalman filter, which can be effectively utilized to such types of problems as nonlinear maltivariate analysis, inverse analysis and so forth, can possibly be an algorithm to train the neural net (Murase and Koyama 1990).

In this study, an attempt of implementing the Kalman filter algorithm in the procedure for training the neural network was made and evaluated.

NEURON TRAINING

Figure 1 illustrates an example of a simple hierarchical neural net architecture. The neural net consists of an input edge layer, a middle layer and an output edge layer.

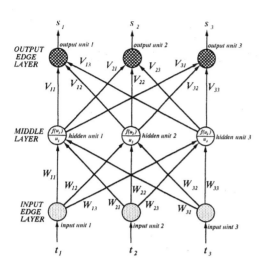

$$s_j = \sum_{i=1}^{m} V_{ij} f(u_i) \qquad (2)$$

The output signal is given in a vector form as $\mathbf{S}=\{s_1, s_2, \cdots\cdots, s_n\}$. After all what this neural network does is to perform a nonlinear transform on $\mathbf{T}$ as expressed in Eq.3.

$$\mathbf{S} = F(\mathbf{T}) \qquad (3)$$

Once those nonlinear functions of hidden units are specified, the behavior of the network can be identified by determining all synapse weights contained in the network. The neuron training is a procedure to determine optimal values of synapse weights by adjusting them step by step using known input data and their associated output data called teacher's signal. The most common algorithm for this training procedure is the gradient descent method of back propagation, which is a version of the hill climbing strategy. This technique can be considered as one of techniques used for parameter identification problems. Since the Kalman filter has many achievements on the engineering problems of parameter identification , it is worth trying to apply Kalman filter for the neuron training.

Fig. 1. Hierarchical Neural Net Architecture.

In general the hierarchical neural net may have middle layers as many as it needs. In this example the input edge layer contains three input units. Each input unit is stimulated by input signal. The middle unit has also three hidden units. Three output units form the output edge layer. Each layer may contain desired number of units in general. The input units project to the hidden units, and the hidden units project to the output units. There are no direct connections between the input units and the output units. The units are connected each other with an adjustable weight called synapse weight. The training procedure adjusts the weight so that each output unit gives a desired answer signal. Mathematical explanation of the process of signal (stimulus) transfer in the network might help understand the hierarchical neural net. The stimulus $\mathbf{T}$ as an input signal can be expected in a vector form as $\mathbf{T}=\{t_1, t_2, \cdots\cdots, t_n\}$. In Fig.1, n is equal to 3. i-th component of the stimulus $\mathbf{T}$,i.e., t_i comes out from the input unit i is transferred to a hidden unit j through the synapse weight W_{ij}. Since each hidden unit has a summation function operating on inputs, the total input u_j received by the hidden unit becomes

$$u_j = \sum_{i=1}^{n} W_{ij} t_i \qquad (1)$$

The hidden unit i has also a transfer function that performs nonlinear transform on the total input u_i and then gives an output $f(u_i)$. $f(u_i)$ becomes the next input fed into the output unit j, which has a summation function, through another synapse weight V_{ij}. The total input received by the output unit j becomes directly its output s_j expressed as

KALMAN FILTER

The Kalman filter technique will be adopted for the neuron training. A linear discrete system can be described by the following two equations.

1) State equation

$$\{x\}_{k+1} = [A]\{x\}_k \qquad (4)$$

$\{x\}$:state variable vector.
$[A]$:system matrix.
k:discrete time.

2) Observation equation

$$\{y\}_k = \{h(x)\}_k \qquad (5)$$

$\{y\}$:observation vector

In this study synapse weights should be considered as state variables. The discrete time can be taken as the iterative step, i.e., training iteration. Since synapse weight in this problem are independent upon time, $[A]$ in Eq.4 must be the unit matrix. The state equation for this problem can be written as

$$\{W\}_{k+1} = [I]\{W\}_k \qquad (6)$$

where, $\{W\}$ is a state variable vector of which components are synapse weights to be determined. Equation 5 can be considered as a nonlinear observation equation.

Equation 5 may be expressed in simpler form using the sensitivity matrix [H] as given by Eqs.7,8,and 9.

$$\{p\}_k=[H]\{W\}_k \qquad (7)$$

where,

$$\{p\}_k=\{q\}-\{F(\widehat{W},t)\}_k+[H(\widehat{W},t)]_k\{\widehat{W}\}_k \quad (8)$$

$$[H_{ij}]=\frac{\partial F_i}{\partial W_j} \qquad (9)$$

$\{q\}$:teacher's signal.
$\widehat{W}$:state variable estimated at one step prior to the present iteration.

The observation vector $\{p\}$ and matrix[H] can be evaluated using a priori information.

NUMERICAL EXPERIMENT

Numerical experiments have been conducted by considering the hierarchical neural net illustrated in Fig.2. A three layered network is used for this test. Each layer consists of two units. Hidden unit $H1$ and $H2$ have the same type of transfer function called logistic function described by Eq.10.

$$f(x)=\frac{1}{1+e^{-x}} \qquad (10)$$

There are 8 synapse weights to be trained in this network. Eqs. 11 and 12 are the transfer functions of hidden unit $H1$ and $H2$, respectively.

$$F_1(t_1,t_2)=\frac{W_5}{1+e^{-(W_1\,t_1+W_3\,t_2)}}+\frac{W_6}{1+e^{-(W_2\,t_1+W_4\,t_2)}} \quad (11)$$

$$F_2(t_1,t_2)=\frac{W_7}{1+e^{-(W_1\,t_1+W_3\,t_2)}}+\frac{W_8}{1+e^{-(W_2\,t_1+W_4\,t_2)}} \quad (12)$$

The sensitivity matrix [H] ((2xn)x8), where n indicates a number of data sets used as teacher's signal, can be determined by differentiation of Eqs. 11 and 12 with respect to synapse weights W_1~W_8. The elements H_{11} and H_{12} of the matrix [H], for example, are as follows

$$H_{11}=\frac{\partial F_1(t)}{\partial W_1}=\frac{W_5\,t_1e^{-(W_1\,t_1+W_3\,t_2)}}{(1+e^{-(W_1\,t_1+W_3\,t_2)})^2}$$

$$H_{12}=\frac{\partial F_1(t)}{\partial W_2}=\frac{W_6\,t_1e^{-(W_2\,t_1+W_4\,t_2)}}{(1+e^{-(W_2\,t_1+W_4\,t_2)})^2}$$

$$(13)$$

Teacher's signal for the neuron training were generated by using Eqs.11 and 12 with prescribed synapse weights W_1 through W_8 such as 0.1, 0.2, 0.3, 0.4, 0.5, 0.6, 0.7, 0.8, respectively. Input values were generated by using RND function of the computer language BASIC. Twenty sets of training data (teacher's signal) were provided. Influence of Kalman filter parameters (variance of observation vector

and diagonal elements of the estimation error) were examined. The total output error (e) was calculated using the following formula (Eq.14), where o_i is a calculated output signal using i-th data set and q_i is i-th teacher's signal

$$e=\sum_{i=1}^{n}(o_i\text{-}q_i)^2 \qquad (14)$$

Twenty-four combinations of four different values (1×10^0, 1×10^{-2}, 1×10^{-5}, 1×10^{-9}) for the variance of observation vector and six magnitudes (0.01, 0.1, 1, 10, 100, 1000) for the diagonal elements of the estimation error matrix were assigned an tested. Convergent characteristics of state variables (synapse weights) were also examined.

A comparison of the performance of this Kalman filter neuron training algorithm (KFA) and a standard back propagation algorithm (BPA) was made. Forty data sets as teacher's signal were prepared. Those data were also generated using Eqs. 11 and 12 with random number inputs between 0 and 1. Those data, however, contain a noise in the output signal. The format of the noise was $\pm10\%$ random fluctuation of exact output value. In this numerical test, a random number between 0 and 1 was assigned to each initial value of state variables.

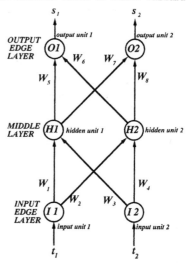

Fig. 2. Neural Network used for Numerical Experiment.

RESULTS OF NUMERICAL TEST

Table 1 capsulizes the test results regarding influence of the Kalman filter parameters on the output error. After 50 times of iteration, the sum of square error of output signal reached a value between 1×10^{-3} and 1×10^{-13} depending upon the combination of the Kalman filter parameters. It seems that when the variance of observation vector is smaller value of diagonal element of estimation error matrix results in smaller error.

Figure 3 shows the convergence of eight synapse weights. All estimates became very close to the target values after 50 times of iteration.

The quick reduction of output error can be observed in Fig.4. The first five iterations practically gave correct output signal with the error (e) less than 5.0×10^{-10}.

Table 2 highlights the outstanding performance of the Kalman filter neuron training algorithm (KFA). BPA has a option to choose one minimization scheme from either the steepest descent method or conjugate gradient method. The final output error was defined as such a manner that iterative calculations no longer give appreciable reduction in the sum of square error of output signals. The steepest descent method of BPA attained no acceptable final output error within 30 minutes of calculation. BPA with the conjugate gradient method also gave about three times larger value of final output error than that calculated by KFA. The Kalman filter parameters chosen for this comparison were the followings; the variance of observation vector was 1×10^{-5}, and the diagonal element of estimation error matrix was 10.

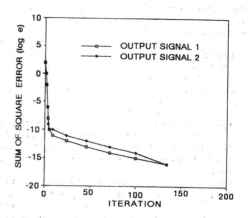

Fig. 4. Convergence of Output Error.

TABLE 2 Comparison of Neuron Training Algorithms

NEURON TRAINING ALGORITHM	SUM OF SQUARE ERROR OF OUTPUT SIGNAL	ITERATION	CALCULATION TIME
Kalman Filter	3.386×10^{-2}	50	5 min
Stpst. Desc. Mtd.	9.070×10^{0}	7576	30 min
Conj. Grad. Mtd.	1.080×10^{-1}	2886	30 min

GROWTH MODEL OF RADISH SPROUTS

Radish sprouts are popular vegetables in Japan. They are used for salad. They can be used as a garnish. The commercial production of radish sprouts is mostly hydroponic in the simple plastic-film-greenhouse.

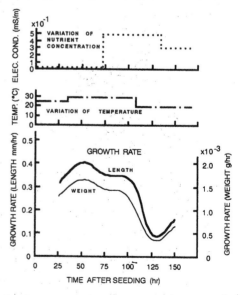

Fig. 5. Variation of Growth Rate of Radish Sprout due to Changes in Environmental Factors.

TABLE 1 Influence of Kalman Filter Parameters on The Output Error after 50 Iterations

VARIANCE OF OBSERVATION VECTOR	DIAGONAL ELEMENT OF ESTIMATION ERROR MATRIX					
	0.01	0.1	1	10	100	1000
	SUM OF SQUARE ERROR OF OUTPUT SIGNAL					
1×10^{0}	1×10^{-3}	1×10^{-3}	1×10^{-5}	1×10^{-6}	1×10^{-7}	1×10^{-8}
1×10^{-2}	1×10^{-5}	1×10^{-7}	1×10^{-9}	1×10^{-9}	1×10^{-9}	1×10^{-9}
1×10^{-5}	1×10^{-7}	1×10^{-9}	1×10^{-9}	1×10^{-13}	1×10^{-12}	1×10^{-12}
1×10^{-9}	1×10^{-13}	1×10^{-12}	1×10^{-12}	1×10^{-8}	1×10^{-7}	1×10^{-6}

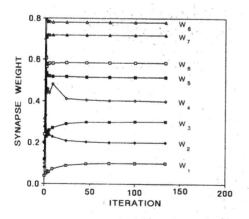

Fig. 3. Convergence of Synapse Weights.

The growth rate of radish sprouts is highly depending upon temperature of growing environment and the concentration of nutrient solution. Figure 5 shows an experimental result indicating that the growth rate of radish sprouts responds sensitively to changes of ambient temperature and concentration of hydroponic solution. The growth rate of a radish sprout is also affected by its growth stage. The growth of radish sprouts can be considered as a nonlinear system. An identification of this nonlinear system was attempted by making use of a neural network model.

Figure 6 illustrates the three layered hierarchical neural network used for this problem. Input signals are ambient temperature, concentration of nutrient solution measured by electric conductivity, and time after seeding. Output signals are growth rates of hypocotyl length and hole weight of an averaged single radish sprout. Teacher's signals were obtained from experimental data shown in Fig.5. Forty sets of teacher's signal were used for the neuron training. The Kalman filter algorithm and standard back propagation algorithm (steepest descent method) were both tested to make a comparison of their performance of neuron training in nonlinear system identification. For reference a multiple regression model expressed by Eq.15 was also shown.

$$\begin{Bmatrix} y_1 \\ y_2 \end{Bmatrix} = \begin{bmatrix} A_{11} & A_{12} & A_{13} \\ A_{21} & A_{22} & A_{23} \end{bmatrix} \begin{Bmatrix} x_1 \\ x_2 \\ x_3 \end{Bmatrix} + \begin{Bmatrix} C_1 \\ C_2 \end{Bmatrix} \quad (15)$$

where, y_1=length of hypocotyl, y_2=hole weight of sprout, x_1=ambient temperature, x_2=electric conductivity(concentration of hydroponic solution), x_3=time, A_{ij}=regression coefficients, and C_i=Constants.

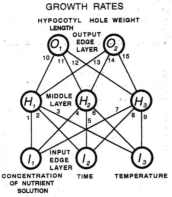

GROWTH RATES

AFFECTING PARAMETERS

Fig. 6. Neural Network Model for Radish Sprout Growth.

Radish sprouts were grown in a growth chamber. The temperature in the growth chamber was kept at 20 °C. The concentration of nutrient solution was also kept at 0.2 in the electric conductivity scale mS/m. The elongation of hypocotyl and the increase of hole weight of a single sprout were measured every 12.5 hours continuously for about 6 days after seeing. Figure 7 shows the plots of measured growth pattern, data calculated based upon neural network models with synapse weights determined by two different training algorithms, and multiple regression curves. The neural network model of which synapse weights were determined by the Kalman filter can simulate very closely the experimentally obtained growth pattern of radish sprouts. The back propagation scheme, however, fails to fit the actual growth pattern.

The total output error on the kalman filter calculation converged to 3.3×10^{-3} after 120 times of iteration. The final output error of the back propagation became 3.9×10^{-2} after 10000 times of iteration. The Kalman filter parameters, i.e., the variance of observation vector and diagonal elements of the estimation error were 0.001 and 1, respectively.

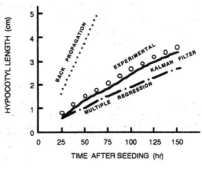

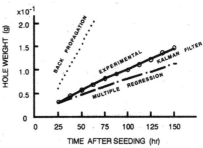

Fig. 7. Simulation Results in Comparison to Experimental Data.

CONCLUSIONS

From this study, it was concluded that the Kalman filter algorithm is applicable and effective to determination of synapse weights of hierarchical neural nets. It was also concluded that thoroughly trained hierarchical neural network models may be utilized to simulate the behavior of nonlinear biological system. The Kalman filter is one of effective algorithms for the thorough training of neural network as revealed in this study.

REFERENCES

Hoshi, T., M.Hirafuji, and T.Honjyo (1990). Expertsystems for biotechnology and agriculture.Corona Pub.Co., Ltd. Tokyo.

Kalman, R. E. (1960). A New Approach to Linear Filtering and Prediction Theory. Trans. ASME, J. Basic Eng., vol. 82D, no.1, 95-108.

Murase, H., S. Koyama, and R. Ishida (1990). Pasokonniyoru Keisanrikigaku. Jyungyakukaisekinyumon.(Introduction to inverse analysis). Morikitashuppan. (Morikita Pub. Co., Ltd.). Tokyo.

Tello, E. R. (1989). Object - oriented programing for artificial intelligence. Addison-Wesley Pub. Co. Inc. New York.

SENSORS AND INTELLIGENT SENSING SYSTEMS

H. Yamasaki

*Dept. of Mathematical Engineering and Information Physics,
Faculty of Engineering, University of Tokyo, 7-3-1, Hongo, Bunkyo-ku, Tokyo, 113, Japan*

Abstract.   Recent progress of technology of sensor and sensing system
is discussed placing stress on intelligent sensors and intelligent
sensing systems. Measurement and control are considered as information
flow between different world.   Roles of machine intelligence are
described. Recent topics; effective expansion of application and human
friendly system, both can be realized only by support of powerful machine intelligence.

Keywords.   Sensors; sensing systems; intelligent machine; machine
intelligence; measurement system; man-machine system; system structure.

INTRODUCTION

Rapid progress of measurement and control
technology is widely recognized. The progress is supported by development of new
sensors and related information processing.
As a result, fields of application of
measurement and control are rapidly expanded. Typical examples of newly developed
fields are bio related areas. Biomedical
applications are the most eminent. Plants
production in agriculture is also eminent
and promising area in which technical
innovation can be triggered by modern
measurement and automatic control technologies.

These newly developed applications require
new functions and advanced performances to
sensors and related signal processing.
These new functions and advanced features
are realized by machine intelligence and
systematization for enhancement of sensing
performances.

In this paper, recent progress of sensing
technology is discussed placing stress on
intelligent sensor and intelligent sensing
systems.

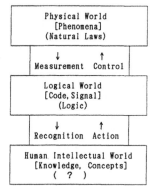

Fig.1 Information Flow between
Three Different Worlds

MEASUREMENTS, CONTROL AND MAN-MACHINE COMMUNICATIONS

Information Flow in Three Different Worlds

Let us discuss information flow. This
information flow includes not only communications between objects and measurement
and control system but also man-machine
communications.
Importance of systematization in information handling in the area of various industry is now widely recognized.
A role of information is very important
for realization of advanced function of
various systems. In addition, information
flow in communication between three different worlds which are shown in Fig.1 is
the most important to realize human friendly systems in the industry.

The three different worlds of information
shown in Fig.1 are as follows;

[1] Physical world. It is ruled by natu-
    ral laws, in which information is
    transmitted as physical signal.
[2] Logical world. It is ruled by logic,
    in which information is described by
    logical codes.
[3] Human intellectual world. It is the
    internal mental world of human. Infor-
    mations are transferred as knowledges
    and concepts in this world. What kind
    of laws dominate this world?
    Author can not answer the question at
    this moment. This is the one of the
    our old but new problems left unsolved.

Information obtained by measurement is
extracted from the physical world and
transferred into the logical world. In the
logical world informations are recognized
by human through man-machine interfaces
which display the measured and calculated
results. Thus the information is transferred into human intellectual world then
constitutes a part of knowledge and concepts in the human world.

Mass of obtained information segments are structured then become a part of knowledge. Assembled knowledges constitute concepts. Group of concepts constitute a field of science and engineering.

In contrast, man's behaviors are based on obtained informations and knowledge. Man's intention are expressed and transmitted to the system by actions through man-machine interfaces. Logical control systems regulate its objects in the physical world based on the command. We can define "Control" as the information flow in which human desire is realized in the physical world through the logical world.

Between physical world and logical one, informations are transmitted through the measurement systems and control systems. Between logical world and human internal world the communication are carried out through human recognition and action.

Sensors in Sensing Systems

Sensing systems perform information transfer from physical world into man-machine interfaces. An intelligent sensing system is the system of which the border between human and logical world is shifted to human side. A part of human intelligence for recognition is replaced by machine intelligence.

Sensors are the front end elements of sensing systems and they extract information of the objects in the physical world and transmit the physical signal. Sensors act as interfaces between physical world and logical world.
An intelligent sensor is the sensor of which border to logical world is shifted to logical world. A part of the signal processing done in the logical world is replaced by information processing within the sensors.

NEEDS FOR PRESENT SENSOR TECHNOLOGY

Present sensor technology is built to meet past various needs for sensors. Development of sensor technology is essentially needs-oriented. So we can forecast the future of sensing technology by discussing present needs for sensors.

[1] Sensing of normal states of objects can be achieved with high accuracy and sensitivity. However detection of abnormalities and malfunctions are poorly developed.
Needs for fault detection and prediction are widely strongly recognized.

[2] Physical or chemical quantities at single point can be sensed accurately by present sensing techniques. However, sensing of multi-dimensional states are difficult. For example; widely spread states or objects of which characteristic parameters are spatially dependent and also time dependent. Sensing of these objects are very difficult, if spatial and temporal resolution of measurement are required.

[3] Well defined physical quantities can be sensed easily with accuracy and sensitivity. However, quantities which are sensed by human sensory systems are not clearly defined, then they can not be sensed by simple present sensors.

Difficulty in above described items share common feature. That is the difficulty in simple model building of the objects.
The problem can be solved by combining outputs of multiple or different kind of sensors. This is called sensor signal fusion or integration. Another solution is adding intelligence to sensors for improvement of their performances.

STRUCTURE OF INTELLIGENT SENSING SYSTEMS

The human sensory system has a hierarchical structure, so a hierarchical system was created to realize advanced sensing functions.
Sensor signal integration and sensor signal fusion are adopted as the basic concepts to design the adaptive intelligent sensing system.

Sensors incorporated with dedicated signal processing functions are called intelligent sensors or smart sensors. The roles of the intelligent sensors are to enhance design flexibility and realize new sensing functions, and additional roles are to reduce loads on central processing units and signal transmission lines by distributed information processing in the lower layer of the system.

Intelligent sensing systems have a hierarchical structure having multi-layer as shown in Fig.2.
On the upper layer the most highly intelligent information processing is done. The function of processing is centralized like human brain. Processed information is abstract and dependent on knowledge, and it is independent on operating principle and physical structure of sensors. In contrast, information processing in the lower layer is strongly dependent on sensor's underlying principles and structures. A group of intelligent sensors on the lower layer collect informations from the external objects like our distributed sensory organs. Signal processing of these intelligent sensors is done in distributed and parallel manner.

There are intermediate signal processing functions in the middle layer. One function of the intermediate signal processing is the integration of signals from multiple intelligent sensors. When the signals come from different type of intelligent sensors, the function is referred to as sensor signal fusion which is described later. Another function is tuning of parameters of the sensors to optimize the total system performance.

In general, the property of information processing done in each layer is more directly oriented to hardware in the lower layer and less hardware oriented in the upper layer. For the same reason, algorithm of information processing is more flexible and need more knowledge in the upper layer and less flexible and less knowledge in the lower layer.

Upper layer	[KNOWLEDGE PROCESSING]
TOTAL CONTROL Concentrated central processing (Digital serial processing)	

Middle layer [INFORMATION PROCESSING] INTERMEDIATE CONTROL, TUNING & OPTIMIZATION OF LOWER LEVEL Distributed processing & control (Digital & Analog)	INTERMEDIATE CONTROL OF LOWER INTERFACES Distributed processing (Digital & Analog)
Lower layer [SIGNAL PROCESSING] SENSING & ACTUATING [INTELLIGENT SENSORS] Distributed parallel processing (Analog)	MAN-MACHINE INTERFACE DISPLAY & CONTROL [INTELLIGENT TRANS- DUCERS] Distributed processing

Fig.2 Hierarchical Structure of Intelligent Sensing System

Processing in each layer can be characterized as follows: signal processing in the lower layer, information processing in the middle layer and knowledge processing in the upper layer.

Machine Intelligence in man-machine interfaces

The intelligent sensing systems have two different types of interface. The first one is an object interface including sensors and actuators. The second interface is a man-machine interface. Man-machine interface has also hierarchical structure as shown in Fig.2. Intelligent transducers are in the lower layer. An advanced function of information processing for integration and fusion is in the middle layer. Machine intelligence in the middle layer harmonize the performance difference between man and machine to reduce load of man. It can check human errors and suggest reasonable action in emergency. It is useful for realization of human friendly system.

In general, machine intelligence in these interfaces translates information throughput and adjusts the difference between different worlds.

ROLES OF INTELLIGENCE IN SENSORS

The sensor intelligence performs distributed signal processing at the lower layer in the hierarchy of the sensing system.

The role of signal processing function in intelligent sensors can be summarized as follows;

 1) Signal enhancement for extraction of useful features of the objects.
 2) Reinforcement of inherent characteristics of sensor devices.

The important role of the sensor intelligence is to improve the signal selectivity of individual sensors in the physical world. This includes simple operations of output from multiple sensor devices for feature extraction. However, this does not include optimization of device parameters or signal integration from multiple sensor devices, because this requires knowledge of the sensor devices and the object.

ROLES OF INTELLIGENCE IN MIDDLE LAYER

One of the roles of the intelligence in the middle layer is to extract essential feature of the object. The other role is to organize multiple output from the lower layer and to generate intermediate output. In the processing of the middle layer, the output signals from multiple sensors are combined or integrated. The extracted features are utilized for recognition of the situation by upper layer intelligence. These processing are done in the logical world.

Sensor Signal Fusion and Integration

Signals from sensors of different measurands are combined and the results give us new useful information. Ambiguity or imperfection in the signal of a measurand can be compensated by another measurand. This is sensor signal fusion. The processing creates a new phase of information.

Optimization

Another function of the middle layer is parameter tuning of the sensors to optimize the total system performance. Optimization is done based upon the extracted feature and knowledge about target signal. The knowledge is given by the upper layer as a form of optimization algorithm.

Example: Intelligent Microphone

An intelligent adaptive sound sensing system was developed by the authors. [Takahashi & Yamasaki(1990)], [Yamasaki & Takahashi(1990)].

The system receives the necessary sound from a signal source out of various noise environments with improved S/N ratio. The location of the target sound source is not given but some feature of signal is given to discriminate the signal out of noise. The feature is given as a cue signal.

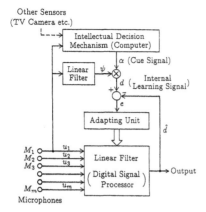

Other Sensors
(TV Camera etc.)

(a) Block Diagram of Intelligent
Adaptive Microphone

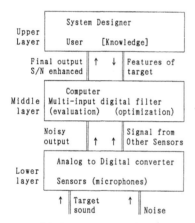

(b) Diagram of Information Flow

Fig.3 Intelligent Adaptive Sound Sensing System

The system consists of one set of multiple
microphone and multiple input linear filters and a self-learning system for adaptive filtering. The adaptive sensing
system has three layers structure as shown
in Fig.3.
The lower layer includes multiple microphone and A/D (analog to digital) converters. The middle layer has a computer and
a multiple input linear filter which combines multiple sensor signal and generates
a integrated output. The performance of
the filter is tuned and controlled by the
computer in the middle layer. The filter
plays the role of an advanced signal processing of the middle layer described in
this paper. The computer has self-learning function and optimizes the filter
performance on the basis of knowledge
given by an external source which is in
the higher layer (in this case, the source
is man).

The output from the optimized filter is
used as the final output of the system.
Max. S/N improvement of 18dB is obtained
experimentally. A video signal of object's
movement can be utilized as a cue signal.
In the case, the system is an example of
sensor fusion of visual and auditory signals. Possible application of the system
is detection of abnormal sound from failed
machines or human voice in a noisy environment.

APPROACHES TO REALIZATION
OF SENSOR INTELLIGENCE

There are three different approaches to
realize sensor intelligence.
[Yamasaki(1984),(1988),(1990)]

  1) Use of specific functional materials
     (Intelligent materials).
  2) Use of functional mechanical structure
     (Intelligent structure).
  3) Integration with computers.

In the first and second approach, signal
processing is analog signal discrimination. Only useful signal is selected and
noise or undesirable effects are suppressed. Thus signal selectivity is realized
by the specific materials or specific
mechanical structure.

In the first approach, unique combination
between object material and sensor material contributes to realize the almost
ideal signal selectivity. The typical
example of the sensor materials is enzyme
fixed on the tip of bio-sensors.

In the second approach, propagation of
optical and acoustic wave can be controlled by the specific shape of the wave
media. Also reflection of these waves are
controlled by the surface shape of the
reflectors. A lens and a concave mirror
are examples; only light which is emitted
from a certain point in the object space
can be concentrated at a certain point in
the image space, and the effect of stray
light can be rejected on the image plane.

The hardware of these analog processing is
relatively simple and reliable and its
processing time is very short due to the
intrinsically parallel processing. But the
algorithm of analog processing is usually
not programmable and it is difficult to
modify once it is designed.

The third approach is most popular and it
is usually represented as the integration
of sensor devices and micro-processors.
The algorithm is programmable and it can
be changed even after it is designed.

Typical examples of the technical approaches are described in the separate litera-
tures.[Yamasaki(1988),(1990)]
These range from single chip sensing devices integrated with microprocessors to
big sensor arrays utilizing synthetic
aperture techniques, and from two dimensional functional materials to human sensory systems.

CONCLUSIONS

Future Concepts of
Intelligent Sensing Systems

Rapid progress in fabrication techniques for ultra large scale integration circuit devices will realize a highly integrated advanced sensing system. Fig.4 shows a concept of an intelligent area image sensing system integrated on single chip device. A concept for future design of micro intelligent sensors can be seen in the figure.[Kataoka(1982)]

The device consists of multi-layer structure, each layer performing a different function based on physical properties of layer materials. A number of light sensing devices are arrayed two-dimensionally on the top layer, signal transmission devices are built in the second layer, memories are in the third, computing devices are in the fourth, and power supplies are in the bottom layer.
Image processing such as feature extraction and edge enhancement can be performed in the three-dimensional multi-functional structure. This is just like the retina of our eyes. More advanced function, for example, an object identification can be possible utilizing given knowledge in the intermediate layer. Then it can execute a part of functions of brain.

Similar configuration is available for tactile imaging system.

The important feature of sensory systems of man and animals is such integration of multi-sensor signals and distributed signal processing.
It can be said that a future concepts or targets of our intelligent sensing systems are the sensory systems of man and advanced animals.

It is important to note that hierarchical structure is essential for advanced information processing and different roles should be reasonably allocated to each layer. The roles of machine intelligence can not be exaggerated in each layer and interface for efficient information flow.
As shown in Fig.5, machine intelligence translates information throughputs and adjust the difference in paradigm between different worlds.

REFERENCES

Kataoka S.: Sensors & human society, J. IEEJ Vol.102(5) 345-348, (1982)

Mackintosh International, Sensors, Vol.24 (1981)

Takahashi K. and Yamasaki H., Self-Adapting Multiple Microphone System, Sensors & Actuators Vol.22 A3, 610-614 Elsevier Sequoia (1990)

Yamasaki H. and Takahashi K.: An Intelligent Adaptive Sensing System, F. Harashima(ed.) Integrated Micro-Motion Systems . 257-277 Elsevier Science (1990)

Yamasaki H.: Approaches to Intelligent Sensors, Proc. 4th Sensor Symposium, 69-76, IEEJ (1984)

Yamasaki H.: From Sensor Devices To Intelligent Sensing Systems, Y. Hashimoto et al ((ed.)) Measurement Techniques in Plant Science, 25-42 Academic Press (1990)

Yamasaki H.: Integrated Intelligent Sensors, H.Yamasaki (ed.) Computrol NO.21, 2-11, Corona Pub.(1988)

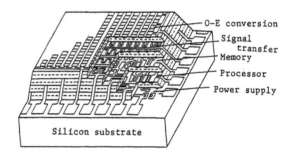

Fig.4 Design Concept of Integrated Image Sensing System having Multi-layer Structure on Silicon Substrate

Exter.	Intelligent Sens. System	Intell. Man-Machine Interface	Human
Physical W	M.Intelligence	Logical W.	M.Intelligence	Senses	Intell. W.

Fig.5 Machine Intelligence Adjust Difference in Paradigm of Different World.

DEVELOPMENT OF AN ION CONTROLLED
FEEDING METHOD IN HYDROPONICS

A. Okuya and T. Okuya

Yokogawa Electric Corporation, Nishi-shinjuku 1-25-1, Shinjuku-ku, Tokyo 163, Japan

Abstract. An ion controlled feeding method in hydroponics for accurate nutrient control was developed by calculating the ion concentrations absorbed by crops. Experiments have been carried out in the commercial greenhouse. Cherry tomatoes were grown on a hydroponic culture. EC (Electrical Conductivity), pH, mixing tank level, and water flow were monitored. The feeding and irrigation were controlled by a computer. Other environments such as temperature were controlled by the same computer. The concentrations of ions in the mixing tank were analyzed at a laboratory. When EC set value was below the measurement value, four kinds of stock solutions were transported to the mixing tank. The adding ratio of each stock solution was calculated by the absorbed concentration of each ion and the difference between the set value and the measurement value. Applying this method enables to establish the automatic control of nutrient solution to meet the changing nutritional pattern of the crop.

Keywords. Agriculture; Control systems; Weather control; Hydroponics; Nutrient

INTRODUCTION

The computers which control greenhouse climate are now available in the market and have been recognized to make highly intensive greenhouse cultivation. Many studies have been carried out to investigate environmental conditions in greenhouse (i.e. Bot, 1983). And more sophisticated control systems such as a an expert system with a crop model has been developed (Kano and Shimaji, 1988). However these studies were mainly concerned with the above-ground environment, not the root environment.

In Japan, several kinds of hydroponic cultures in greenhouse are widespread and the respective area is increased to more than 300 ha. Especially for salad vegetable such as lettuces, tomatoes and cucumbers are grown on that system.

In most of the hydroponic cultures, only EC, pH, the temperature of the solution and irrigation are controlled as the root environment. The popular method of making fertilizer in these systems is transporting the two kinds stock solutions (named A and B stock solutions) to a large mixing tank simultaneously at a ratio dependent on the desired EC. In this kind of system, the ratio among ions of the stock solution transporting to the mixing tank is usually the same throughout the cropping period. However, the ion concentrations absorbed by the crop are different depending on the growth stages and are influenced by the environment such as solar radiation and air temperature. Thus, the balance of the ion concentration in the mixing tank may become not suitable for the growth of

crops particularly in the system recirculating the nutrient solution through the system.

One of disadvantages of recirculating the nutrient solution is the risk of spreading root-infecting pathogens all through the system. Therefore, excess nutrient solution which is not absorbed by crops (drain water) is not recirculating but flows directly into the soil in some kinds of hydroponic cultures, such as rockwool culture widespread in the Netherlands.

For economical reasons (i.e. no waste of water and nutrients) and to avoid pollution of soil and surface water, it is preferable to adopt recirculating system which collects and reuses the drain water. Therefore, a equipment for disinfection of drain water by heat treatment has been studied (Runia, van Os and Bollen, 1988). Thus, the recirculating system of hydroponic system has been much important.

Recently systems of multi-head injectors to control the concentration of individual nutrient were developed (Bauerle and co-workers, 1988; Papadopoulos and Liburdi, 1989). These systems were used for hydroponic cultures in which nutrient solution was not recirculated.

Accordingly, a control system was developed which comprises the ion control feeding method adopted for recirculating hydroponic system. Not only fertigation but also above-ground environment was controlled by the same computer. The objectives of the investigation was to develop the system and to test in a commercial greenhouse.

MATERIALS AND METHODS

Theory

In this system, five kinds of fertilizers were used. One of these fertilizers was used for adding the micronutrients. Four kinds of fertilizers contained main nutrients (called main fertilizers), as shown in TABLE 1.

TABLE 1 The Kinds of Ions Contained in Main Fertilizers

Main fertilizer	Ion	
1	Ca^{2+}	NO_3^-
2	K^+	NO_3^-
3	Mg^{2+}	SO_4^{2-}
4	NH_4^+	PO_4^{3-}

The method of adjusting the adding ratio of the main fertilizers was given by the following algorithm: Firstly, the absorbed concentrations of each ion were calculated by the following formula (NO_3^- is used as an example).

$$IT_{NO3} = (IL_{NO3}*V + A_{NO3} - AB_{NO3}) / (V + AW - T) \qquad (1)$$

where

IT_{NO3} : Ion concentration of NO_3^- in the time (me/l)

IL_{NO3} : Ion concentration of NO_3^- in the mixing tank at the last sampling time (me/l)

V : Volume of the mixing tank (l)

A_{NO3} : Amount of NO_3^-, added to the mixing tank from last sampling time to this sampling time (me)

AB_{NO3} : Absorbed amount of NO_3^- by whole crops in the greenhouse (me)

AW : Amount of water, added to the mixing tank from last sampling time to this sampling time (l)

T : Transpiration amount of whole crops from last sampling time to this sampling time (l)

The level of transpiration amount of crops be able to estimate by means of a model based on climatic factors (Okuya A. and T. Okuya, 1988). This method was particularly useful for irrigation against the hydroponic system in which nutrient was not recirculated. In this investigation, the nutrient was recirculated, the reduction amount of the nutrient in the mixing tank was equivalent to the transpiration amount by crops. As the level of the mixing tank was controlled at the same level, the amount of water added to the mixing tank was equal to the reduction amount of the nutrient solution. Then above expression was simplified as follow by assuming AW is equal to T.

$$IT_{NO3} = (IL_{NO3}*V + A_{NO3} - AB_{NO3}) / V \qquad (2)$$

Then, the absorbed amount of NO_3^- by whole plants in the greenhouse (AB_{NO3}) was expressed as follow;

$$AB_{NO3} = IL_{NO3}*V + A_{NO3} - IT_{NO3}*V \qquad (3)$$

The absorbed concentration of NO_3^- was expressed as follow;

$$C_{NO3} = AB_{NO3} / AW \qquad (4)$$

Where

C_{NO3} : Absorbed concentration of NO_3^- (me/l)

The absorbed concentration of other ions were calculated by the same method.

Secondly, the adding ratio among the four main fertilizers were calculated by the following methods. If the nutrient requirements of crops was equal to the absorbed concentration of each ion, the adding ratio among the main ions could be equal to the ratio of absorbed concentration of each ion. Each of the main fertilizers contained two kinds of main ions as shown in TABLE 1. For example, the main fertilizer No.1 contained Ca^{2+} and NO_3^-, the adding value was calculated using the related values of Ca^{2+}.

$$V\_1 = C_{Ca} / S1_{Ca} \qquad (5)$$

Where

V_1 : Adding value of the main fertilizer No.1

C_{Ca} : Absorbed concentration of Ca^{2+} (me/l)

$S1_{Ca}$: Concentration of Ca^{2+} in the main fertilizer No.1 (me/l)

The adding values of the main fertilizer No.2, No.3 and No.4 were calculated as equations (6), (7) and (8) respectively.

$$V\_2 = C_K / S2_K \qquad (6)$$

$$V\_3 = C_{Mg} / S3_{Mg} \qquad (7)$$

$$V\_4 = C_{PO} / S4_{PO} \qquad (8)$$

Thus, the adding ratio among the main fertilizers was expressed as follow.

$$V\_1 : V\_2 : V\_3 : V\_4 \qquad (9)$$

To protect the crop from unsuitable concentration, the adding concentration ratio of each ion calculated by the above adding ratio can be made to remain with allowable limits, as shown in TABLE 2. The ion of NO_3^- is used for standard.

TABLE 2 The Limits of the Adding Concentration Ratio

Ion	Upper limit	Lower limit
NH_4^+	0.3	0.0
PO_4^{3-}	0.5	0.1
K^+	1.0	0.1
Ca^{2+}	1.0	0.1
Mg^{2+}	0.5	0.1

The adding ratio of micronutrients fertilizer was not given by the above algorithm. It was added proportionally to the mixing tank with the amount of water added to the mixing tank.

Hardware Configuration

The design of this feeding system is illustrated in Fig.1. The hydroponic system contained five stock tanks for concentrated solution of each fertilizer. Each concentrated solution was injected to the mixing tank by dosimetric pumps. Water was added to the mixing tank in case the surface of the water was below the target depth.

EC and pH of the mixing tank was measured in every fifteen minutes while the nutrient solution was recirculating. When EC set value was below the measurement value, four kinds of stock main fertilizer solutions in each stock tanks were injected to the mixing tank according to the values of adding ratio. When pH set value was below or above the measurement value, alkaline or acid was injected to the mixing tank as long as difference between the EC set value and the measurement value was smaller than 0.05 mS/cm.

The nutrient solution in the mixing tank was circulating through the mixing tank and the beds uncontinuously by circulating pump. The suspend and running minutes of the circulating pump set up by a operator were revised by a computer automatically, according to solar radiation.

All the measuring and the control, including above-ground environment were conducted by the greenhouse environmental control system, YEWFARM (a product of Yokogawa Electric Corporation). There were two computers in YEWFARM, line controller and line computer.

Line controller independently measured the environmental conditions and controlled the equipment in the greenhouse. Line controller consists of main unit which communicates with line computer via ML-bus and sub unit for input/output expansion.

Line computer, which has a 40 MB hard disk, a floppy disk and a 14" color CRT, employs a MC68020 CPU (16.7 MHz). It is capable of running plural BASIC programs simultaneously on its real-time operating system: one for collecting the environmental data and storing them in the hard disk, the other for displaying measured data and editing target values.

In the greenhouse, there was an interface panel containing input/output signal conditioning modules, which connected to the sub unit for input/output expansion of line controller via twisted-pair cable. All control devices and sensors were connected to these modules. When a control device was to be operated, an output signal was transmitted from line controller to the signal conditioning module.

The concentrations of ion in the mixing tank were analyzed by Ion Chromatographic Analyzer (Model IC-200, Yokogawa Electric Corporation) every one week or two weeks at the laboratory.

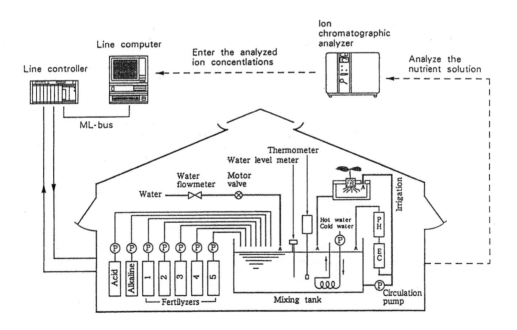

Fig. 1. Schematic diagram of the system.

357

Software Configuration

After the operator entered the analysis concentration of each ion to the line computer from key-board, the absorbed concentrations of each ion and the value of adding ratio of the four main fertilizers were calculated. The inputs data and the calculating results were displayed on the screen of the computer similar to Fig. 2. The value of each adding ratio was transported to the line controller via ML-bus to be used for controlling EC of the nutrient solution in the mixing tank. Until the next entry of the analysis data was conducted, the main fertilizer was added to the mixing tank using the same ratio.

When the measurement values such as EC, pH were exceed or less than the limit values set by the operator, an alarm was given to warn against mechanical or electrical failure in the system.

Experiment

The ion controlled feeding method was introduced in a commercial greenhouse in Itsukaichi, Tokyo in the winter of 1990. The floor area is 806 m^2, crops are cherry tomatoes with planting density of 2.2 plant/m^2, transplanted on 8th of August, 1990. The hydroponic system contained five stock solution tanks, an acid tank and an alkaline tank of 200 litres each, and a mixing tank of 2000 litres.

RESULTS AND DISCUSSION

The analyzed value and the absorbed concentration of seven kinds of ions in the mixing tank during three months are shown in Fig's 3 and 4.

In the hydroponic cultivation, to achieve maximum

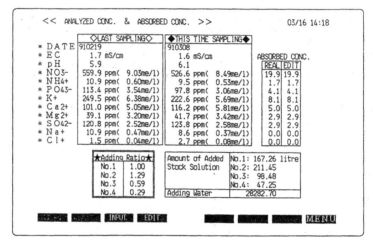

Fig. 2. The example of the user interface.

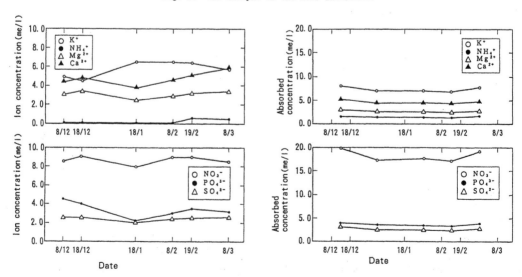

Fig. 3. Changes of the ion concentrations.

Fig. 4. Changes of the absorbed concentrations.

yields and crop quality, it is important to maintain the plants at optimum nutrient status. To meet this target, it it necessary to monitor the nutritional status of the nutrient solution regularly. So, the service of analyzing the nutrient solution of hydroponic cultivation is being conducted by some organizations and private companies for their customers. To reflect these analyzed concentration of the nutrient solution to maintain the nutrient at optimum status, it is necessary to know the absorbed concentration of the crops. However, as indicated in Fig's 3 and 4, changing patterns of the absorbed concentration were not the same as the pattern of the analyzed values. This means the difficulties to maintain the nutrient solution at optimum status with the analyzed concentrations only, particularly in recirculating system, if growers manage the nutritional maintenance manually. So, the control system is needed which measuring and logging the concerned data and decided the strategy of nutrient maintenance with the analyzed concentrations effectively.

The adding ratios during the same period are shown in TABLE 3. The values on 8th of Dec. were the initial values recommended for tomatoes by a fertilizer manufacturer in Japan. The main fertilizers were transported to the mixing tank according with the initial values from the transplanting date to 18th of Dec.

The adding ratios of the main fertilizers No.1, No.3 and No.4 were not changed largely from the initial value after the system was installed. The value for fertilizer No.2 was less than the initial value. This is because that the absorbed concentration of potassium was less than expectation of the initial value.

The adding ratio calculated by the system hardly changed excepting the values of fertilizer No.2, because the limitation shown in TABLE 2 was not so restrict and the climate during the period was stable.

TABLE 3 Adding Ratio of the Main Fertilizers

Date	Adding ratio			
	1	2	3	4
Dec. 8	1.00	1.96	0.59	0.26
Dec. 18	1.00	1.28	0.58	0.29
Jan. 18	1.00	1.25	0.59	0.29
Feb. 8	1.00	1.28	0.59	0.28
Feb. 19	1.00	1.28	0.58	0.28
Mar. 8	1.00	1.29	0.59	0.29

CONCLUSION

A precise control method was developed with the recirculating hydroponic system to maintain the status of the nutrient solution reflecting the changing nutritional pattern of crop. Its application to the commercial cultivation enable to reduce the amount of nutrient and water, furthermore the pollution of soil and surface water could be avoided.

REFERENCE

Bauerle, W., T. Short, E. Mora, S. Hoffman and T. Nantais(1988). Computerized Individual-nutrient Fertilizer Injector:The System. HortScience, 23(5), 910.

Bot, G. P. A.(1983). Greenhouse Climate: from Physical Processes to a Dynamic Model. Ph.D. dissertation, Wageningen Agricultural University.

Kano, A. and H. Shimaji(1988). Greenhouse Environmental Control System with a Crop Model and an Expert System. Acta Horticulturae, 230, 229-236.

Okuya, A. and T. Okuya(1988). The Transpiration of Greenhouse Tomato Plants in Rockwool Culture and its Relationship to Climatic Factors. Acta Horticulturae, 230, 307-311.

Papadopoulos, A. P. and N. Liburdi(1989). The "Harrow Fertigation Manager" - A Computerized Multifertilizer Injector. Proceedings of International Symposium on Growth and Yield Control in Vegetable Production, Berlin, GDR.

Runia,W. Th., E. A. van Os and G. j. Bollen(1988). Disinfection of drainwater from soilless cultures by heat treatment. Netherlands Journal of Agricultural Sciences, 36, 231-238.

Copyright © IFAC Mathematical and Control Applications
in Agriculture and Horticulture, Matsuyama, Japan 1991

AN ISFET-BASED NUTRIENT SENSOR FOR
ROCKWOOL CULTURE

I. Amemiya, H. Yagi, T. Sakai

*Research and Development Center, Toshiba Corp., 1 Komukai Toshiba-cho, Saiwai-ku,
Kawasaki 210, Japan*

Abstract. An application of ISFETs (ion-sensitive field-effect
transistors) to nutrient solution control systems in rockwool culture has
been investigated. Measuring the ion concentration of a solution in
rockwool is difficult and has not been carried out in practice. The
authors anticipated that an ISFET, with its advantages of small size and
low output resistance, could be used to measure the ion concentration of
such a solution in rockwool. In this study, the authors examined the basic
characteristics of ISFET-based nutrient sensor probes in rockwool. The
characteristics examined were as follows: 1) Sensitivity characteristics
for certain ion concentration changes (over a 10^{-4}-10^{-1}M NO_3^- range).
2) Reproducibility of the output when they were inserted into rockwool
repeatedly. 3) Output dependence on the rockwool water content (over a 20-
80% range). 4) Output dependence on the distance between the ISFET and a
reference electrode (over a 10-80 cm range). 5) Long-term stability of the
sensitivity and output baseline in rockwool (over 2 weeks). These experiments have shown that the sensors have sufficient stability for use in
this kind of measurement. The authors have found that ISFETs are able to
measure the ion concentrations for a nutrient solution in rockwool
culture.

Keywords. Sensors; Electric sensing devices; Ion-sensitive field-effect
transistor; Monitoring; Microcomputer-based control; On-line operation;
Remote control; Agriculture; Rockwool culture.

INTRODUCTION

Nutrient solution control in nutriculture
is important for promoting plant growth and
efficient production. In particular, ion
concentration control of a nutrient solution is one of the dominant factor
(Hashimoto and co-workers, 1988).

Recently, rockwool culture has been actively studied as a new promising nutriculture
method. Rockwool culture is a system using
rockwool, which is an inorganic fiber block
having an over 90% pore rate, as the plant
growth medium. Rockwool is usually kept
under a half dry condition, and almost no
solution flows out from the rockwool block.
Therefore, it is difficult to measure the
nutrient solution both inside and outside
of the rockwool. In order to control the
solution, it is necessary to set an ion
sensor directly inside the rockwool and to
keep it in contact with the solution in the
rockwool efficiently. For this purpose, ion
sensors are desirable to be small in size.
However, conventional ion-selective electrodes (ISEs) are relatively large. Also,
when they are made smaller, they have a

problem in regard to a high electrode
resistance causing noise. Therefore, an ion
sensor for directly measuring the solution
in rockwool has not yet been developed.

On the other hand, the authors have been
investigating ISFET (ion-sensitive field-
effect transistor) devices, using an SIS
(silicon-insulator-silicon) substrate
(Sakai and co-workers, 1987, 1990). ISFETs
have an ion-sensitive membrane on the
MOSFET gate insulator instead of its metal
electrode. The potential difference between
an ion-sensitive membrane and a solution
leads to a channel resistance change.
Therefore, the ion concentration in a
solution can be measured as the drain
current change in the ISFET. In general,
the interface potential change is measured
directly, using a source-follower circuit.

ISFETs have the following advantages compared with conventional ISEs: 1) The small
size. 2) Low output resistance. 3) All
solid state structure. 4) Multiplexing
ability. Based on these features, the
authors anticipated that ISFETs could be
used to measure the ion concentration of

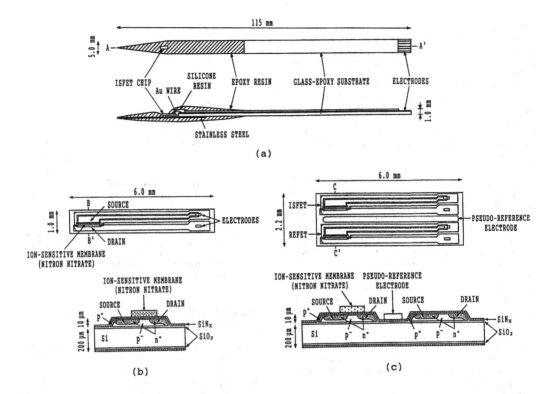

Fig.1 Structure of sensor probe and ISFET chips: (a) sensor probe,
(b) single-type ISFET chip, (c) differential-type ISFET chip.

a solution in rockwool on a real time basis, and examined the basic characteristics in rockwool during the present study.

EXPERIMENTS

The structure of an ISFET-based nutrient sensor probe is shown in Fig. 1(a). The size was approximately 115 x 5 x 1 mm. An ISFET chip was mounted on the edge of a glass-epoxy substrate and electrically connected to lead tracks by wire bonding. A sharp-pointed stainless steel plate was attached to the glass-epoxy substrate to protect the chip from damage during insertion. The parts, excepting a sensitive area of the chip, were encapsulated with silicone rubber and epoxy resin. The resin, which surrounded the chip, was piled up a little higher than the chip surface to protect the ion-sensitive membrane from friction with the rockwool.

In general, ISFETs are operated with a reference electrode, the same as conventional ISEs. The surface potential of the ion-sensitive membrane is detected by the voltage between the ISFET and the reference electrode.

On the other hand, an operating method using two ISFETs and a differential cir-

cuit, making the best use of the ISFET's advantages, has been suggested. The differential-type ISFET contains an ISFET with an ion-sensitive membrane and a REFET (reference field-effect transistor) with an ion-insensitive membrane on the same substrate. The surface potential change for the ion-sensitive membrane is obtained by measuring the differential output between the ISFET and the REFET. Although REFET is still under development (Bergveld and coworkers, 1989), the differential-type ISFET has advantages of cancelling output disturbance due to the common phase external noise and the thermal drift in FETs. Therefore, in addition to a single-type ISFET, a differential-type one was also tried for use in the present experiments.

The chip structures for single-type and differential-type ISFETs are shown in Figs. 1(b) and 1(c), respectively. The size of a single-type chip was 6.0 x 1.0 x 0.2 mm. The size of a differential-type chip was 6.0 x 2.2 x 0.2 mm. The sensitive area on each FET was 1.0 x 0.05 mm. These chips were fabricated using an SIS substrate, and the FETs were formed in silicon islands on one side of the substrate. The silicon nitride covering the gate surface acts as a pH-sensitive membrane. A nitron nitrate (50 wt%) bound by epoxy resin was cast on the ISFET gate region as a nitrate ion-

sensitive membrane. The membrane was cured
at 100°C for 90 minutes. A pH-ISFET (a bare
ISFET) was used in the REFET, owing to
using a pH-constant buffer solution. An
Ag/AgCl electrode was used as the reference
electrode paired with a single-type sensor
probe. The tip of the reference electrode
was minimized to about 2 mm in diameter, to
enable inserting it into the rockwool. The
distance between the sensor probe and the
reference electrode was usually kept at
approximately 2 cm.

In the present experiments, the nutrient
solution conditions for rockwool in culturing a plant were assumed to be as follows:
1) Ion concentration changes between a one
hundredth standard composition and a ten
times one (over about a 10^{-4}-10^{-1}M range).
2) pH changes between 4 and 8. 3) Water
content changes between 80% and 20%. The
following characteristics of the sensor
were examined inside the rockwool to verify
whether or not the sensor probe could
correctly and stably measure such a solution in rockwool:

1) Sensitivity characteristics for
   nitrate ion concentration changes.
2) Output reproducibility for repeated
   insertions.
3) Output dependence on the water
   content.
4) Output dependence on the distance
   between the ISFET and a reference
   electrode.
5) Long-term stability of sensitivity and
   output baseline.

A rockwool (Nichias Corp.) which was dried
at 60°C for 8 hours after being boiled for
1 hour was used. The volumetric water
content of the rockwool was calculated from
its weight assuming that gravity is one.
The detailed conditions and procedure for
these examinations are described with the
result in a later section.

RESULTS AND DISCUSSION

1. Sensitivity Characteristics in Rockwool

The sensitivity characteristic of the
sensor probes in rockwool for nitrate ion
concentration is shown in Fig. 2. The
solutions used were standard phosphate
buffer solutions (pH 6.86) containing from
10^{-4} to 10^{-1}M nitrate ions. They were poured
into rockwool blocks (30 x 30 x 30 mm) with
a 60% volumetric water content.

In Fig. 2, the horizontal-axis shows the
nitrate ion concentration for the solution
extracted from rockwool. The plotted values
were average outputs for five tests, which
were within a 6 mV range. As a result, the

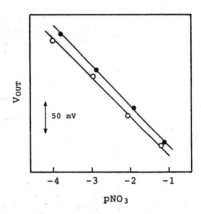

Fig.2 Sensitivity characteristic of sensor probes
in rockwool. Open circles are for a single-
type probe, and solid circles are for a
differential- type probe; the solution was a
standard phosphate solution (pH 6.86).

output for each type of ISFET probe showed
a linear relationship with the concentration. The sensitivities for both sensor
probes were approximately 60 mV/pNO₃. This
sensitivity value was almost the same as
that for an ideal Nernst equation. As a
result, it became obvious that ISFET-based
sensor probes were able to measure the ion
concentration in rockwool.

A difficulty encountered during this experiment was that the correct concentration of the solution in rockwool was not
obvious. In a preparatory experiment, the
pH value of the solution without a buffer
action before pouring it into rockwool was
compared with that after extracting it from
rockwool, and the pH value of the solution
was found to change owing to the rockwool.
In particular, the pH value for a weak acid
solution changed to a neutral region.
Changes in other ions were relatively
smaller. This pH change of the solution
influenced the REFET output, which was
desirable to be kept constant, and led to
a differential output variation. Therefore,
the relationship between the sensor output
in rockwool and the extracted solution
concentration was examined. The reason was
that the contents of the extracted solution
were almost the same as the solution contents in rockwool. The sensitivity characteristics in rockwool were examined in this
way, and good results were obtained as
described above. Moreover, the slight
deviation in plots from the line were
considered to be caused by the concentration difference between the solution in
rockwool and the extracted solution.

2. Output Reproducibility for Repeated Insertions

The sensor output reproducibility was checked under the severest conditions in the assumed range. That is, the volumetric water contents were 20%, 50%, and 80% in rockwool, and the one thousandth concentration solution was for a standard phosphate buffer solution containing 10^{-4}M nitrate ions.

The output variation range for both probes during ten insertions into rockwool blocks containing such a solution is shown in Table 1. The output of a single-type probe stayed within a 4.0 mV range. This output reproducibility is thought to be sufficient for nutrient sensors. The output of a differential-type probe varied largely compared with the single-type one. This result was due to the REFET (pH-ISFET) output as a result of monitoring each ISFET output. Two reasons for the REFET's output variation were considered. One is the pH change of the solution in the rockwool. The severance of the rockwool fiber by probe insertion causes the rockwool and the binder to dissolve into the solution. The dissolving of these materials might lead to a pH change of the solution, and such a pH change might be larger in a lower water content. The other reason is the difference in the height of the sensitive membranes between the ISFET and the REFET. In a differential-type probe, the ISFET had an ion-sensitive epoxy membrane (0.05-0.10 mm thick) on the gate insulator and the REFET had no thick membrane. This difference was disadvantageous for the REFET to contact the solution in the rockwool. These reasons could explain the fact that the REFET output reproducibility was low under low water content conditions.

As a result, it became obvious that the sensor probe could measure the solution with reproducibility, even under severe conditions, excepting for the differential-type probe at the lowest water content condition.

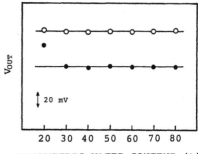

Fig.3 Sensor probe output dependence on water content in rockwool. Open circles are for a single-type probe, and solid circles is a differential-type probe; the solution contained 10^{-4}M NO_3^- in 3×10^{-4}M ionic strength, approximately.

varied from 80% to 20% was examined. The solution which was used in the previous experiment was used again. The water content was adjusted as follows: First, the rockwool block (30 x 30 x 30 mm) was sunk in the solution. Second, the rockwool containing a sufficient amount of solution was put in a basket which was attached to the head of a spinner and had many holes in it to facilitate draining. Then, the water content was adjusted by dehydration through high speed spinning.

The results are shown in Fig. 3. The output stayed within a 2.5 mV range for the single-type probe. This variation was thought to include the drift depending on the surface condition of the ion-sensitive membrane. The output stayed within a 3 mV range in a water content range from 80% to 30% for the differential-type. However, a shift at 20% water content was produced by the output change of the REFET (pH-ISFET). The mechanism for this was thought to be the same as in the case of the previous experiments. Therefore, it was considered that the sensor probe could measure the solution independently from the water content in the assumed range.

4. Output Dependence on Distance to Reference Electrode

It is necessary in a single-type sensor to check the output dependence on the distance between the probe and the reference electrode. The result of this experiment is shown in Fig. 4. The maximum distance of 80 cm was determined from the length of the rockwool slab (75 x 150 x 910 mm). The solution which was used in the previous experiment to realize the severest condition was poured into the rockwool slab uniformly in volumetric water content of 20%.

TABLE 1   Output Variation Range for Sensor Probes While Inserted Ten Times

	Volumetric water content		
	80%	50%	20%
single-type	3.5	4.0	3.0
differential-type	12.0	6.0	28.0

(mV)

3. Output Dependence on Water Content

The output change for sensor probes when the volumetric water content of rockwool

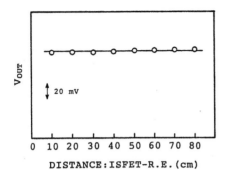

The output of the probe stayed within a 3 mV range when the distance changed from 10 cm to 80 cm. This output variation was considered to be small enough, and it is considered obvious that the distance between the probe and the reference electrode, that is the length of the electrical path in rockwool, does not influence measurements in the slab.

5. Long-Term Stability of Sensitivity and Output Baseline

The long-term stability of the sensitivities and output baseline for a sensor probe settled in rockwool were examined under conditions resembling an actual culturing circumstance. Two different nitrate ion concentration solutions which had the same pH value were poured reciprocally on the sensor probe in rockwool.

The response characteristic for a differential-type probe when the solution was dropped at a speed of 1 ml/min is shown in Fig. 5. The output signal included no noise, and it took a few minutes for the output to become stable after changing the

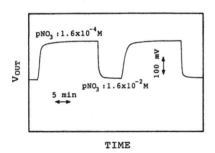

Fig.5 Typical differential-type probe output to nitrate ion concentration change in rockwool.

solution. This response was considered to depend on the exchange speed of the solution, and the net response speed of the sensors must be within a few seconds.

The result for a differential-type sensor probe is shown in Fig. 6. The sensitivity varied initially from 54 mV/pNO_3 to 58 mV/pNO_3 after 4 hours. After that, it became stable and the difference between the value after 4 hours and that after 380 hours was only 2.5 mV/pNO_3. The absolute output voltage for each solution varied 20 mV initially after 30 hours. After that, it drifted at a 0.2 mV/hour rate. As a result of this experiment, it is thought that rockwool does not markedly influence the sensor. If the ion concentration range is from $10^{-4}M$ to $10^{-1}M$, calibration once a day enables measuring the solution with a 2.7% full scale accuracy.

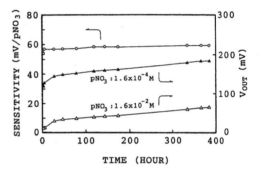

Fig.6 Long-term stability in sensitivity and output baseline for differential-type probe in rockwool. Open circles show sensitivity, and open triangles show output voltage for $1.6 \times 10^{-2}M$ NO_3^-, solid triangles show voltage for $1.6 \times 10^{-4}M$ NO_3^-.

CONCLUSION

The application of an ISFET for use in a nutrient solution control system for rockwool culture, an ISFET-based nutrient sensor, has been investigated. The new sensor probes using single-type and differential-type ISFETs were fabricated in a form enabling insertion into rockwool.

As a result of experiments in the sensitivity characteristic to nitrate ions, the probes have been found to be able to measure the solution contained in rockwool. The output of the probes had sufficient reproducibility and stability for distance changes from the reference electrode under the severest conditions expected in culturing plants. Moreover, the output hardly depended on the water content change, and the sensor maintained a high long-term stability in rockwool. These results suggest the possibility of using an ISFET-based nutrient sensor, and suggest several methods for using the sensor probe. For example, from the result of the distance dependence experiment, a number of sensor probes in rockwool cubes on the same slab can be operated with a common reference electrode in the slab. The sensor probe can be moved to any part of the rockwool, with the reference electrode being fixed.

A block diagram for a nutrient solution control system using an ISFET as a nutrient sensor is shown in Fig. 7. If it is possible to monitor the solution concentration in rockwool by such an ISFET, feedback from the solution information in rockwool to a microcomputer would enable a close-by control of plant growth.

On the other hand, the advantage of a differential-type ISFET was not shown in the present experiments. However, its advantage is thought to become important under actual culturing conditions, which include biophysical noise from plants. Thus, a differential-type probe should still be researched. Moreover, the probe structure for protecting the soft ion-sensitive membrane, such as PVC, from friction during insertion should be improved.

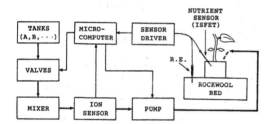

Fig.7 Block diagram of nutrient solution control system for rockwool culture, using an ISFET-based nutrient sensor.

ACKNOWLEDGEMENTS

The authors would like to thank Mr. H. Watake and Mr. S. Yamaguchi for valuable discussions of an ion sensor in nutriculture.

REFERENCES

Hashimoto,H., T.Morimoto, T.Fukuyama, H.Watake, S.Yamaguchi, and H.Kikuchi (1988). Identification and control of hydroponic system using ion sensors. Acta Horiculture, 245, 490-487.

Sakai,T., H.Hiraki, S.Uno, O.Takikawa, M.Katsura, and M.Shimbo (1987). Ion sensitive FET with silicon-insulator-silicon structure. Proc. 4th Int. Conf. Solid-State Sensors and Actuators, Tokyo, Japan, 1987, (Transducers '87), 711-714.

Sakai,T., I.Amemiya, S.Uno, and M.Katsura (1990). A backside contact ISFET with a silicon-insulator-silicon structure. Sensors and Actuators, B1, 341-344.

Bergveld,P., A.van den Berg, P.D.van der Wal, M.Skwronska-Ptasinska, E.J.R. Sudholter, and D.N.Reinhoudt (1989). How electrical and chemical requirements REFETs may coincide. Sensors and Actuators, 18, 309-327.

Copyright © IFAC Mathematical and Control Applications
in Agriculture and Horticulture, Matsuyama, Japan 1991

DYNAMICS OF PLANT WATER RELATIONS AS AFFECTED BY EVAPORATIVE DEMAND

M. Kitano and H. Eguchi

Biotron Institute, Kyushu University 12, Fukuoka 812, Japan

Abstract. Evaporative demand was evaluated physically by using environmental factors
of radiant flux density, air temperature, humidity and wind velocity. The evaluated
evaporative demand explained quantitatively effects of the respective environmental
factors on evaporation. Furthermore, a heat flux control method developed for on-line
measurement of stem water flux was applicable to dynamic analysis of transpiration
stream in a cucumber plant (*Cucumis sativus* L.) responding to the evaporative demand.
The evaporative demand and the stem base water flux closely related to stomatal
response and leaf growth through dynamics of plant water balance. These facts suggest
that plant water relations affected by environment can be better understood by on-
line measurements of the evaporative demand and the transpiration stream.

Keywords. Plant water relations; evaporative demand; environment control; dynamic
response; on-line operation; flowmeters; agriculture; biology.

INTRODUCTION

In plant hydraulic system, transpiration stream
from roots to leaves is induced by evaporative
demand of environment, and plant water relations
are affected by the evaporative demand through
plant-environment water balance(e.g. Kramer,1983).
The evaporative demand depends on environmental
factors complicatedly in processes of radiative
and convective transfers. However, there remain
difficulties of quantitative evaluation of the
evaporative demand. The present paper deals with
physical evaluation of the evaporative demand by
using environmental factors and also deals with
analysis of relationship between the evaporative
demand and the transpiration stream with reference
to plant water relations.

EVAPOLATIVE DEMAND

In general, water vapor difference (W_L-W_A) between
leaf intercellular spaces and ambient air has been
used as a measure of evaporative demand (e.g.
Assmann and Grantz,1990; Farquhar and Cowan,1974;
Grantz and Meinzer,1990; Kitano and Eguchi,1989b).
In fact, W_L-W_A is the driving force for leaf-to-
air vapor transfer and can be a relative measure
under a constant air current. W_L-W_A, however, is
not necessarily adequate as a quantitative
measure, because W_L-W_A is difficult to explain the
respective effects of environmental factors on
evaporative demand explicitly. In particular, W_L-
W_A can not reflect the effects of air current.
Furthermore, for evaluation of W_L-W_A, it is
essential to measure leaf temperature which varies
unsteadily with environmental factors and stomatal
responses. Therefore, evaporative demand was newly
evaluated by using only physical environmental
factors on the basis of leaf heat balance.

Evaluation of Evaporative Demand

Definition. Leaf heat balance can be expressed as

$$aR_S+2\varepsilon_L\varepsilon_A\sigma T_A{}^4=2\varepsilon_L\sigma T_L{}^4+2C_P\rho G_{AH}(T_L-T_A)+\lambda E_A \quad (1)$$

where R_S is the sort wave radiant flux density, a
the short wave absorption coefficient of a leaf,
ε_L and ε_A the respective emissivities of leaf and
environment, σ the Stephan-Boltzmann constant, T_A
and T_L the respective temperatures of air and
leaf, C_P the specific heat of air at constant
pressure, ρ the density of air, G_{AH} the boundary
layer conductance for heat transfer, λ the latent
heat of vaporization of water, and E_A the
transpiration rate per unit leaf area.
Subtracting $2\varepsilon_L\sigma T_A{}^4$ from both sides of Eq.(1) and
linearizing $T_L{}^4-T_A{}^4$ as $4T_E{}^3(T_L-T_A)$, Eq.(1) can be
rewritten as

$$aR_S-2\sigma\varepsilon_L(1-\varepsilon_A)T_A{}^4=2C_P\rho(G_R+G_{AH})(T_L-T_A)+\lambda E_A \quad (2)$$

where G_R is the radiative transfer conductance $(G_R=4\varepsilon_L\sigma T_E{}^3/C_P\rho)$, and T_E should be between T_A and T_L
(Campbell,1977). Transpiration rate (E_A) can be
written as

$$E_A = n\, G_L(W_L-W_A) = n\, G_L\{SVD+\Delta(T_L-T_A)\} \quad (3)$$

where G_L is the leaf conductance, SVD the
saturation deficit of air, Δ the slope of the
saturation vapor density curve, and n the constant
($n=2$ in an amphistomatous leaf and $n=1$ in a
hypostomatous leaf). Leaf conductance (G_L) can be
written by stomatal conductance (G_S) and boundary
layer conductance (G_{AV}) for vapor transfer as

$$G_L = \frac{G_S \cdot G_{AV}}{G_S+G_{AV}} \quad (4a)$$

From Eqs.(2) and (3),

$$E_A = \frac{2C_P\rho G_E\,SVD+\Delta\{aR_S-2\sigma\varepsilon_L(1-\varepsilon_A)T_A{}^4\}}{\{(2/n)\gamma(G_E/G_L)+\Delta\}\,\lambda} \quad (5)$$

where G_E is the parallel conductance of G_R and G_{AH}

$(G_E = G_R + G_{AH})$, and γ the thermodynamic psychrometer constant ($\gamma = C_p \rho / \lambda$). Thus, evaporation rate per unit area of an evaporating surface was defined as evaporative demand (ED), which can be written by substituting G_{AV} for G_L into Eq.(5) as

$$ED = \frac{2C_p \rho G_E SVD + \Delta\{aR_s - 2\sigma\varepsilon_L(1-\varepsilon_A)T_A^4\}}{\{(2/n)\gamma(G_E/G_{AV})+\Delta\}\lambda} \quad (6)$$

ED means the evaporative demand when stomatal resistance does not exist (i.e. $G_s = \infty$). That is, ED does not depend on plant physiological function but on physical conditions of environment.

Boundary layer conductance. For evaluating ED from Eq.(6), it is essential to determine boundary layer conductances for heat transfer (G_{AH}) and for vapor transfer (G_{AV}). In general, G_{AH} has been evaluated for either forced convection or free convection by using each standard formula (Monteith,1973). Standard conductances for forced convection (G_{FO}) and free convection (G_{FR}) can be written as (Dixon and Grace,1983; Grace and co-workers,1980)

$$G_{FO} = \frac{0.66 \, \kappa^{0.67}}{d^{0.5} \, \nu^{0.17}} U^{0.5} \quad (7)$$

$$G_{FR} = \frac{b \, \beta^{0.25} \, g^{0.25} \, \kappa}{d^{0.5} \, \nu^{0.5}} (T_L - T_A)^{0.25} \quad (8)$$

where U is the wind velocity, d the characteristic dimension of a leaf, κ the thermal diffusivity of air, ν the coefficient of kinematic viscosity of air, β the coefficient of thermal expansion of air, g the acceleration of gravity, and b the constant. The value of b is 0.50 and 0.23 on the respective upper and lower leaf surfaces (Monteith, 1973).

In the leaf boundary layer, however, mixed convection (forced convection + free convection) became a dominant form of leaf-to-air convection at lower U and higher $T_L - T_A$ (Kitano and Eguchi,1991). Thus, G_{AH} for mixed convection was evaluated by a parallel model as (Kitano and Eguchi,1989b)

$$G_{AH} = G_{FO} + G_{FR} \quad (9)$$

G_{AV} for vapor transfer was evaluated by using the conductances for heat transfer and Lewis number (Le) as

$$G_{AV} = G_{FO}/Le^{0.67} + G_{FR}/Le^{0.75} \quad (10)$$

The buoyancy effect in the mixed convection was evaluated for a leaf temperature (T_{LP}) calculated from Eq.(2) and (3) by substituting G_{AV} for G_L as

$$T_{LP} = T_A + \frac{(2/n)\gamma(G_E/G_{AV})}{(2/n)\gamma(G_E/G_{AV})+\Delta}$$
$$\times \{\frac{aR_s - 2\sigma\varepsilon_L(1-\varepsilon_A)T_A^4}{2C_p \rho G_E} - \frac{SVD}{(2/n)\gamma(G_E/G_{AV})}\}$$

$$(11)$$

Physical properties. The radiative conductance (G_R), which depends on T_E as $G_R = 4\varepsilon_L \sigma T_E^3/C_p\rho$, was also evaluated by using T_{LP} estimated by Eq.(11). That is, T_E was given by $T_E = (T_A + T_{LP})/2$. Furthermore, respective values of physical properties of air such as Δ, γ, $C_p\rho$, ν and κ were given at temperature of T_E. Short wave absorption coefficient (a) of a leaf was given as 0.5, which can be useful as a good estimate under solar radiation (Gates,1980). Long wave emissivity (ε_L) of a leaf was adopted as 0.97, which was the mean value measured in 34 plant species by Idso and co-workers (1969). Emissivity (ε_A) of environment was given as $\varepsilon_A = \varepsilon_L$ and $\varepsilon_A = 0.674 + 0.007 T_A$ (Gates,1980) under respective covered and uncovered conditions.

Evaporative Demand and Evaporation

Relationship between ED and evaporation rate (E_{AM}) in a model leaf with a wetted filter paper was examined under different conditions of T_A, SVD, U and aR_s in an air controlled wind tunnel. A cucumber leaf with d of 16.8 cm was modeled by using a pair of 0.1 mm thick brass sheets between which a micro heater of 0.1 mm constantan wire was sandwiched (Kitano and Eguchi,1991), and a wetted filter paper was mounted on the brass sheets. That is, the model leaf resulted in a hypostomatous leaf. The model leaf was suspended horizontally in the working section of the wind tunnel by using a frame. Through a brass rod, the frame was connected to an electronic balance (PM1200, Mettler Instrumente AG) installed at the outside of the wind tunnel, and E_{AM} of the model leaf was evaluated from the weight change. The model leaf was heated at desired aR_s by applying direct electric current to the micro heater through a regulated power supply.

In the respective effects of environmental factors of T_A, SVD, U and aR_s, positive correlations were clearly found among ED, E_{AM} and each environmental factor. On the other hand, $W_L - W_A$ was not observed in positive correlation with E_{AM} under the effects of T_A and U (Kitano and co-workers,1990). Figure 1 shows the relationship between ED and E_{AM} in the model leaf under different 220 conditions of T_A of 15-35°C, SVD of 5-35 g m^{-3}, U of 7.5-120 cm s^{-1} and aR_s of 0-800 W m^{-2}. A stable linear relationship was clearly found between ED and E_{AM} even under different environmental conditions. This result indicates that ED of Eq.(6) can evaluate evaporative demand quantitatively. Figure 1, furthermore, shows that the evaporative demand on an amphistomatous leaf did not reach to twice of the demand on a hypostomatous leaf.

Thus, evaporative demand can be evaluated by ED in terms of physical environmental factors. The ED evaluation is free from measurement of leaf temperature which varies unsteadily with environmental factors and stomatal responses.

TRANSPIRATION STREAM

Transpiration stream in a stem is located between water uptake in roots and transpiration in leaves. For on-line measurement of water flux in an intact stem, we have developed a heat flux control (HFC) method (Kitano and Eguchi,1989a).

Heat Flux Control Method

Figure 2 shows a schematic diagram of the HFC system. A main heater (HM) of 1 cm width and two subheaters (HU and HD) of 5 mm width are thin flexible heaters made of a manganin wire. These heaters are mounted on the outer circumference of the stem, where the respective positions of HU and HD are about 5 mm upward and downward from HM. The outer surface of HM is covered with a plastic tape (PT$_1$) of about 2mm thickness, and the temperature difference $T_s - T'_s$ between the outer surfaces of HM and PT$_1$ is detected by a copper-constantan thermocouple of 0.1 mm diameter. PT$_1$ is, furthermore, covered with the plastic tape (PT$_2$) for preventing rapid change in T'_s influenced by the incident radiation and the ambient air temperature. The stem temperature differences of $T_u - T_d$, $T_u - T'_u$ and $T_d - T'_d$ are detected by the thermocouples inserted into the

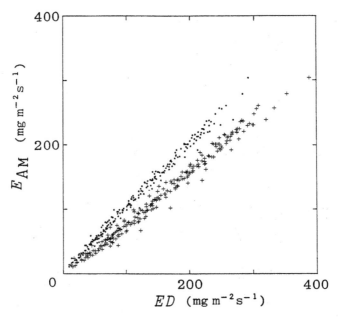

Fig. 1. Relationship between evaporative demand (ED) and evaporation rate (E_{Am}) in a model leaf under different 220 conditions of ambient air temperatures of 15-35℃, saturation deficits of 5-35 g m⁻³, wind velocities of 7.5-120 cm s⁻¹ and applied heats of 0-800 W m⁻². ED was evaluated for a hypostomatous leaf (●) and an amphistomatous leaf (+).

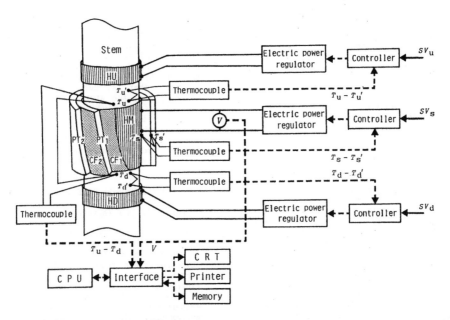

Fig. 2. Schematic diagram of the HFC method for on-line measurement of water flux in stem: HM, main heater for controlling $T_s - T'_s$; HU and HD, respective subheaters for controlling $T_u - T'_u$ and $T_d - T'_d$; PT_1, plastic tape mounted on HM; PT_2, plastic tape mounted on PT_1; CF_1 and CF_2, copper foils covering the respective surfaces of HM and PT_1; T_s, surface temperature of HM; T'_s, surface temperature of PT_1; T_u and T_d, stem temperature at the respective upper and lower ends of the part heated by HM; T'_u and T'_d, stem temperatures at points about 2 mm apart from the respective T_u and T_d measuring points; SV_s, SV_u and SV_d, set values for the respective controls of $T_s - T'_s$, $T_u - T'_u$ and $T_d - T'_d$; V, direct current voltage applied to HM for controlling $T_s - T'_s$. (after Kitano and Eguchi, 1989a)

stem. When the heat Q is continuously applied to the stem by HM, the water flux (F) in the stem can be expressed on the basis of stem heat balance as

$$F = \frac{q_f}{C_w(T_u-T_d)} = \frac{Q-(q_s+q_u+q_d)}{C_w(T_u-T_d)} \qquad (12)$$

wher Q is the heat of HM, q_f the heat flux transferred with water flux in the stem, q_s the heat flux diffused from the outer surface of HM to the surroundings, q_u and q_d the respective upward and downward heat fluxes through the stem by heat conduction, and C_w the specific heat of water. The heat Q can be given by the direct current voltage (V) and the electric resistance (R) of HM as $Q=V^2/R$. The respective heat fluxes of q_s, q_u and q_d are proportional to temperature differences of $T_s-T'_s$, $T_u-T'_u$ and $T_d-T'_d$ (Sakuratani,1981). However, these heat fluxes of q_s, q_u and q_d are difficult to evaluate accurately in real time because of difficulties in determinations of the respective proportional constants and difficulties in accurate measurements of the related temperature differences which are very small and vary with F in complex manners.

In the HFC method, the respective heat fluxes of q_s, q_u and q_d are kept constant without evaluations. For keeping the respective fluxes constant, the related temperature differences of $T_s-T'_s$, $T_u-T'_u$ and $T_d-T'_d$ are controlled at the respective set values of SV_s, SV_u and SV_d by manipulating the respective heaters of HM, HU and HD, where a constant SV_s of about $0.2℃$ and constant SV_u and SV_d of about $0℃$ are preset in the respective feedback control systems with PI (proportional plus integral) control action. In the feedback control systems, the signals of the respective temperature differences are amplified by low drift amplifiers. Thus, the respective fluxes of q_s, q_u and q_d are always kept constant irrespective of variation in F, and then $q_s+q_u+q_d$ in Eq.(12) is kept constant at all times. From the stem heat balance when $F = 0$ (i.e. $q_f=0$), the constant $q_s+q_u+q_d$ can be given as

$$q_s + q_u + q_d = Q_0 = \frac{V_0^2}{R} = \text{const.} \qquad (13)$$

where Q_0 and V_0 are the heat and the voltage of HM when $F=0$. Therefore, the water flux F can be obtained from

$$F = \frac{Q - Q_0}{C_w(T_u-T_d)} = \frac{V^2 - V_0^2}{R\,C_w(T_u-T_d)} \qquad (14)$$

V_0 can be evaluated before the measurement with an enough accuracy by assuming that V_0 is about 95% of V when the transpiration rate becomes remarkably low in the dark (the real V_0 can be obtained by cutting the stem after the measurement). The main heater voltage V and the amplified signal of T_u-T_d are transmitted to CPU, and the water flux F is evaluated on-line by Eq.(14) without any calibrations. In the HFC system, the applied heat of HM is manipulated according to F, and the temperature of the heated stem results in nearly constant at about $3.5℃$ higher than the ambient air temperature. Consequently, the indeterminable heat leak other than q_s, q_u and q_d is almost constant, which is involved in Q_0. Characteristics of the HFC system are $\pm 5\%$ in accuracy, 1 mg s^{-1} in resolution and 1 min in time constant.

Stem Water Flux and Transpiration

Water flux (F) in the stem base and transpiration rate per plant (E_M) in a 10 leaf stage cucumber plant(*Cucumis sativus* L.) were measured by the HFC

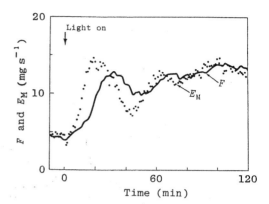

Fig. 3. Time course patterns of stem water flux and transpiration rate under a step input of light: F, water flux measured by the HFC method at the stem base; E_M, transpiration rate per plant measured by weighing the plant and the pot. (after Kitano and Eguchi, 1989a)

method and the weighing method, respectively. Figure 3 shows responses of F and E_M to a step input of light. F and E_M increased after lighting, where rise in F lagged behind that in E_M. The difference between the respective peak times was 13 min. The amount of water transpired excessively before the lag F caught up with E_M 30 min after lighting, was about 4.5g which corresponded to about 5% of the total water content in the stem. Thereafter, F and E_M oscillated with a period of about 40 min and settled about 80 min after lighting where F and E_M reached to the same level. The oscillation of F lagged and damped as compared with that of E_M. This lag and dampening response of the stem water flux was brought by the lag water uptake in roots and the hydraulic buffer action of stem water storage. That is, during the lag time in root water uptake, the stored water in stem internal tissues was supplied for rapid increase in leaf transpiration.

Thus, dynamics of water balance in a whole plant can be analyzed quantitatively by applying the on-line system of HFC method.

PLANT RESPONSES TO EVAPORATIVE DEMAND

Measurements in a Phytotron

Cucumber plants were potted in vermiculite moistened with nutrient solution and grown at T_A of 23 ℃, RH of 70% and U of 30 cm s^{-1} in a phytotron glass room, and the plant of healthy growth at the 10 leaf stage was used for the experiment on transpiration stream, stomatal response and leaf expansion affected by the evaporative demand. In the on-line data processing system used, signals from the respective sensors were transmitted to the computer through interfaces at a 1 min interval.

ED on a whole plant. Evaporative demand per unit area was evaluated by ED of Eq.(6). Therefore, the evaporative demand on a whole plant (SED) can be obtained by multiplying ED by the total leaf area (ΣLA) of the plant as

$$SED = ED \times \Sigma LA \qquad (15)$$

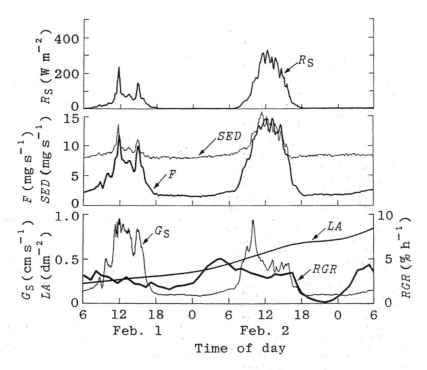

Fig. 4. Diurnal variations of short wave radiant flux density (R_s), evaporative demand on a whole plant (SED), water flux in the stem base (F), stomatal conductance (G_s), area (LA) and relative growth rate (RGR) of a younger leaf.

Area (LA) of each leaf was evaluated from length (LL) of the midrib on the basis of a LA-LL relation obtained as $LA = 1.265 LL^2 - 4.784 LL + 12.79$, where LL of each leaf was measured by a rule at every 9:00 a.m. For evaluating G_{av} and G_{aH} in ED, leaf characteristic dimension (d) on the chord basis was given by using a d-LL relation obtained as $d = 0.852 LL - 1.27$, and the mean of d values in respective leaves was adopted.

Stem water flux. By using the on-line system of HFC method, water flux (F) in the stem base was measured. F was treated as transpiration rate per plant, and the transpiration rate (E_{aH}) per unit leaf area was obtained by dividing F by ΣLA.

Stomatal conductance. Stomatal conductance (G_s) was evaluated by

$$G_s = \cfrac{1}{\cfrac{W_L - W_A}{E_{aH}} - \cfrac{1}{G_{av}}} \qquad (16)$$

where W_L was calculated from T_L which was measured by 10 thermocouples inserted into leaves (Kitano and Eguchi, 1989b).

Leaf expansion. In a developing younger leaf, the midrib length (LL) was measured by using an on-line differential transformer: Both ends of the midrib of the leaf were held by small clips which were connected to the differential transformer system through stretched threads, and change in LL (i.e. distance between the clips) was transformed into change in output voltage of the system. LA of the leaf was evaluated by using the LA-LL relationship mentioned above, and a relative growth rate (RGR) was calculated on the area basis as $\triangle LA/LA \times 100$.

Plant Responses

Figure 4. shows diurnal variations of R_s, SED, F, G_s, LA and RGR in the phytotron glass room on a cloudy day (Feb.1) and a fair day (Feb.2). SED and F varied with R_s of solar radiation. In the dark during the nighttime and under lower R_s in the daytime, the level of F was appreciably lower as compared with that of SED, and G_s in the cloudy daytime was kept at higher level. On the other hand, under higher R_s around the midday of the fair daytime, F increased to the higher level of SED, and G_s was extremely depressed. This G_s depression under the equivalence relation between higher SED and F in the fair daytime can be considered to be caused by turgor loss of leaves. The relative growth rate (RGR) in the younger leaf was approximately constant in the cloudy day. After the midnight (Feb.2), RGR began to increase and reached to the maximum before the sunrise. This increase in RGR can be considered to be brought by turgor increase in the leaf under lower SED and F in the dark. After the sunrise of the fair day, RGR decreased gradually and remarkably dropped to nearly 0 at night. This drop of RGR can be considered to relate to the water stress and the G_s depression induced under the equivalence relation between higher SED and F in the fair midday, because such RGR drop was not found in the cloudy day safe from the water stress. Thus, SED and F closely related to stomatal response and leaf expansion through plant-environment water balance.

The relationship between SED and F depends largely on leaf conductance (G_L) of Eq.(4a), which can be rewritten as

$$G_L = \frac{G_S}{1 + \dfrac{G_S}{G_{RU}}} = \frac{G_{RU}}{1 + \dfrac{G_{RU}}{G_S}} \qquad (4b)$$

G_L and F are regulated by G_S when G_S is smaller than G_{RU}, and consequently F is kept lower than SED. These relations were found under dark at night and under lower R_S in the cloudy daytime. On the other hand, G_L and F are regulated by G_{RU} when G_S increases over G_{RU}, and F closely approaches SED with the G_S increase. These relations were found under higher R_S and SED around the fair midday when water stress was induced. These facts suggest that plants are susceptible to water stress under the condition that F becomes nearly equivalent to higher level of SED.

CONCLUSION

ED defined in Eq.(6) was useful to quantify the evaporative demand of environment. The stem base water flux measured by the on-line HFC system was reliable for dynamic analysis of water balance in a whole plant. In the phytotron glass room, the evaporative demand varied with solar irradiance, and water stress was induced when the stem base water flux increased to higher level of the evaporative demand around the fair midday. The evaporative demand and the stem water flux closely related to dynamics of plant water balance. Thus, on-line measurements of the evaporative demand and the stem water flux can give explanatory information on plant water relations.

REFERENCES

Assmann, S.M., and D.A. Grantz (1990) Stomatal response to humidity in sugarcane and soybean: Effect of vapour pressure difference on the kinetics of the blue light response. *Plant,Cell Environ.*,13,163-169.

Campbell, G.S. (1977) *An Introduction to Environmental Biophysics*. Springer-Verlag, New York.

Dixon, M., and J. Grace (1983) Natural convection from leaves at realistic Grashof numbers. *Plant,Cell Environ.*,6,665-670.

Farquhar G.D.,and I.R. Cowan (1974) Oscillation in stomatal conductance: The influence of environmental gain. *Plant Physiol.*,54,769-772.

Gates, D.M. (1980) *Biophysical Ecology*. Springer-Verlag, New York.

Grace, J., F.E.Fasehun,and M.Dixon (1980) Boundary layer conductance of the leaves of some tropical timber trees. *Plant,Cell Environ.*,3, 443-450.

Grantz, D.A., and F.C. Meinzer (1990) Stomatal response to light and leaf-air water vapor pressure difference show similar kinetics in sugarcane and soybean. *Plant Physiol.*,81,865-868.

Idso, S.B., R.D. Jackson, W.L. Ehrler, and S.T. Michell (1969) A method for determination of infrared emittance of leaves. *Ecology*, 50, 899-902.

Kitano, M., and H. Eguchi (1989a) Quantitative analysis of transpiration stream dynamics in an intact cucumber stem by a heat flux control method. *Plant Physiol.*, 89, 643-647.

Kitano, M., and H. Eguchi (1989b) Dynamic analysis of stomatal responses by an improved method of leaf heat balance. *Environ. exp. Bot.*, 29, 175-185.

Kitano, M., M. Hamakoga, and H. Eguchi (1990) Physical evaluation of effective evaporative demand with reference to plant water relations. *Biotronics*, 19, 109-119.

Kitano, M., and H. Eguchi (1991) Buoyancy effect on forced convection in the leaf boundary layer. *Plant, Cell Environ.*, 14 (*in press*)

Kramer, P.J. (1983) *Water Relations of Plants*. Academic Press, New York.

Monteith, J.L. (1973) *Principles of Environmental Physics*. Edward Arnold, London.

Sakuratani, T. (1981) A heat balance method for measuring water flux in the stem of intact plants. *J. Agric. Met.Tokyo*, 37, 9-17.

MEASUREMENT OF BIOELECTRIC POTENTIAL
ON THE SURFACE OF <u>SPINACH</u> LAMINA

T. Uchida, Y. Nakanishi and T. Sakano

Shikoku Research Institute Inc., Takamatsu 761-01, Japan

Abstract. In order to develop the environmental control system based on plant
responses, the bioelectric potential transition patterns on <u>Spinach</u> lamina in
response to light-on condition were mesured easily as an useful plant physiological information with our felt electrode. Our study on bioelectric potential
showed changes in pattern specifically with respect to environmental and growth
conditions such as temperature, humidity, CO_2 concentration, light intensity,
growth stage and dying process. It was found that these transient pattern
changes of bioelectric potential can be simulated by the exponential function.
Our study also suggests the possibility of expressing relations between culture
environmental conditions and growth conditions by means of function parameters.

Keywords. Environment control; Biocommunications; Sensors; Biocybernetics;
Electrophysiology.

INTRODUCTION

The bioelectric potential of plants seems to
have close relations with their growth, photo-
synthsis, metabolism and other life activities.
In the past, studies on bioelectric potential
of intact plants have focused on looking at
cell-level electric phenomena by using micro-
electrodes. Recently, efforts have come to be
done to understand macroscopic electric phenomenon of plants in a culture state in a non-
destructive manner. Systems to measure such
phenomenon stably for an extended period
without contact to the plants concerned are
also being developed (Hirao, 1988; Kano, 1988;
Sekiyama, 1984, 1989). One of such efforts
is SPAEC (Speaking Plant Approach to Environmental Control), which is intended to incorporate such information into a culture system
and thereby build a more sophisticated culture
environmental control system (Hashimoto, 1981,
1989).
At plant factories and other facilities,
research is being done for intensive cultivation. But almost no effort has so far been
made to measure the bioelectric potential
of soft vegetables. In view of this, in an
effort to determine patterns of bioelectric
potential information capable of application
to culture environmental control, we studied
bioelectric potential transition patterns on
the surface of <u>Spinach</u> laminae by turning the
light on and off. We simulated the transition

patterns by using the exponential function and
thereby determined how their growth is related
to environmental conditions by using the parameters of the function.

MATERIAL AND METHOD

Sample

We used <Spinacia oleacea × hybrida "Lead">
for our lamina surface bioelectric potential
measurement. Samples were nursed and grown in
an almost fixed environmental condition in our
research plant factory. Before measuring their
bioelectric potential, we placed them in the
measurement system for more than three days
for acclimatization.

Measurement System

As shown in Fig. 1, we used a system comprising a bio-signal amplification circuit and a
recorder to measure the potential difference
between the surface of the sample lamina and
the culture solution around the roots. For
the signal amplification circuit, we used
Analog Devices' AD524 amplifier for high
precision measurement.
To set an environment for our experiments, we
used a hydroponics test system (made by
Shimazu Seisakusho Ltd.) consisting of a
chamber capable of controlling the air
temperature, humidity and CO_2 concentration

around the shoots of samples, a culture
solution control unit capable of controlling
the water temperature and the solution level,
and an artificial lighting device. The
lighting device used a metal halide lamp
(Mitsubishi Electric's BOC lamp). We also
provided the system with a function to
measure the speeds of photosynthesis and
transpiration based on the differences of CO_2
concentration, temperature and humidity
between the inlet and outlet of the chamber.

Erectrode to Measure Lamina Surface Electric Potential

We used a surface electrode to measure the
electric potential on the lamina surfaces of
intact plants in a non-destructive manner
stably and macroscopically for an extended
period. As shown in Fig. 2, the electrode,
similar in structure to the felt electrode
developed by Hirao and Arai (1988), was
designed to measure plant surface electric
potential from a system comprising a sample,
electrolyte, a silver wire (Ag-AgCl) and a
signal amplification circuit. We used KCl for
electrolyte, with a gel included solution
(for Nihon Koden Kogyo's P-150 electrode)
added for improved water retention and
prolonged use of the electrode. The reference
electrode was installed in the solution
around the roots of the samples.
Our method has a number of advantages over the
use of a microelectrode for cell membrane
potential measurement: it can be installed
easily (simply by clipping a leaf with electrodes), dose not require the electromagnetic
shielding of the sample (only shielding from
the electrode to the signal amplification
circuit is required), simplifies the sample
measurement system, and can measure data
continuously for more than a month.

Measuring Method

We measured transition changes in the potential on the surface of sample laminae by
turning the artificial light source (a metal
halide lamp) on and off in constant environmental conditions (temperature, humidity, CO_2
concentration, light intensity, water temperature, etc.)
Measurement was done in time periods of four
hours during the daytime and another four
hours in the nighttime in which transition
changes in the lamina surface potential,
photosynthesis and transpiration are made
sufficiently stable.
As the culture solution, we used a mix of
Otsuka House Fertilizers' No. 1 and No. 2 in
the ratio of 3 to 2 with an electric condition
speed of 1.8mS/cm.

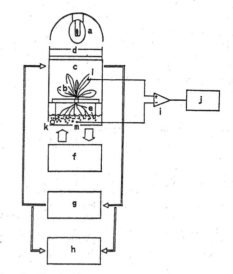

Fig. 1. Block diagram of measurement system.
a: Lamp, b: Plant, c: Shoot chamber,
d: Shading panel, e: Root chamber,
f: Nutrient solution control system,
g: Air control system, h: Measurement
system of CO_2 uptake and transpiration,
i: Amplifier, j: Recorder, k: Bubbler,
l: Electrode for measurement,
m: Reference electrode.

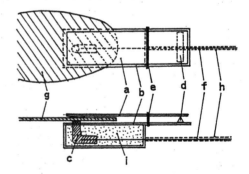

Fig. 2. Construction of the measurement electrode on the surface of plant.
a: Supporting plate of sample,
b: Plastic box, c: Felt, d: Fulcrum,
e: Spring, f: Electric wire (Ag-AgCl),
g: Sample, h: Electric shield, i: Gel
included solution (KCl).

RESULTS AND EVALUATION

Measurements of Lamina Surface Electric Potential

Simultaneous measurement on three places of
the same sample. Fig. 3 gives sample lamina
surface electric potentials measured on the
first to three leaf blades (the leaf blade

developed first was made the first leaf) of
the same sample. At different places of
measurement, transition changes in potential
with the turning on/off of the light source
showed resemblance, though the levels of
change varied to some extent. After the work
done by Hirao and Arai (1988), we assigned
the numbers n-1, n-2, n-3 and n-4 in a time
series manner to each peak measured when the
light source was on and identified individual
potential transition patterns with the
corresponding peak heights and positions.

Measurements of samples at different growth
stages. Fig. 5 gives sample lamina surface
electric potential measurements for samples
at different stages of growth. The difference
with the stage of growth was particularly
notable at the height of peak n-4. The earlier
the stage of growth, the higher peak n-4.

Measurements with changes in temperature.
Fig. 6 shows sample measurements made by
changing the temperature at the inlet of the
chamber from 25°C to 20°C to 25°C. Peak
positions varied significantly with changes in
temperature. The higher the temperature, the
faster and sharper changes in the pattern.

Measurement with changes in relative humidity.
Fig. 7 shows sample measurements made by
changing the relative humidity at the inlet of
the chamber from 40% to 50% to 60% and to 70%.
Changes in the relative humidity caused
changes in the heights at n-2 and n-4. There
was the tendency that the higher the relative
humidity, the higher the peaks at n-2 and n-4.

Measurements in the process of leaf dying.
Fig. 8 gives sample measurements in the
process of leaf dying.
In the process of dying, while peak n-1 showed
no change, peaks n-2, n-3 and n-4 lowered in
height gradually as the samples went dying.

Measurements with changes in CO_2 concentration.
Fig. 9 shows sample measurements where the CO_2
concentration at the inlet of the chamber was
changed from 50ppm to 100ppm to 400ppm and to
600ppm.
The height of peak n-3 varied significantly
with changes in CO_2 concentration. The higher
the CO_2 concentration, the deeper the peak at
n-3.
Also, the tendency was that the lower the CO_2
concentration, the larger the amplitude and
the longer the cycle of electric potential
vibration on the surface of the sample leaf
blades measured in the daytime.

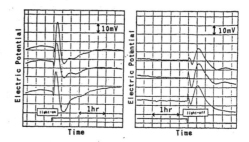

Fig. 3. Comparison of bioelectric responses on
three Spinach laminae.
Air temperature   : 20°C,
Relative humidity   : 60%,
CO_2 concentration   : 600 ppm,
Water temperature   : 20°C
Light intensity   : 190 μE/m²sec.

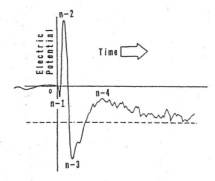

Fig. 4. Bioelectric potential transition pattern on the surface of Spinach lamina.
n-1 ~ n-4: Four characteristic peak
when the light source was on.

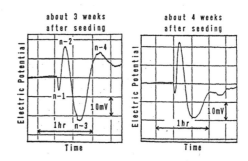

Fig. 5. Effect of growth on the bioelectric
response of Spinach lamina.
Air temperature   : 20°C,
Relative humidity   : 60%,
CO_2 concentration   : 600 ppm,
Water temperature   : 20°C
Light intensity   : 190 μE/m²sec.

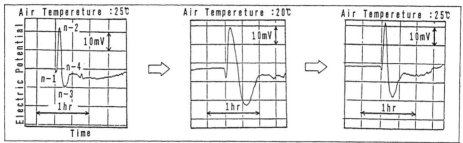

Fig. 6. Effect of air temperature on the bioelectric response of <u>Spinach</u> lamina.
Relative humidity: 60%, CO_2 cocentration: 600ppm, Water temperature: 20℃, Light intensity : 190 μ E/m²sec.

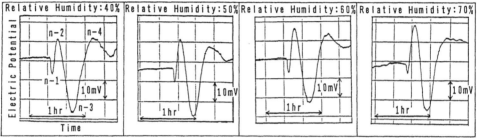

Fig. 7. Effect of relative humidity on the bioelectric response of <u>Spinach</u> lamina.
Air temperature: 20℃, CO_2 cocentration: 600ppm, Water temperature: 20℃, Light intensity: 190 μ E/m²sec.

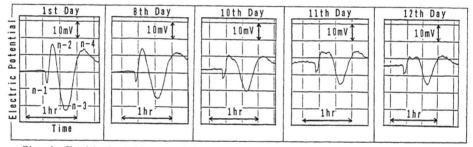

Fig. 8. The bioelectric potential patterns on Spinach lamina during process of dying.
Air temperature: 20℃, Relative humidity: 60%, CO_2 cocentration: 600ppm, Water temperature: 20℃, Light intensity: 190 μ E/m²sec.

Measurements with changes in light intensity.
Fig. 7 gives sample measurements with the light intensity level changed from 110 to 190 to 340 μ E/m²sec.
As with changes in the CO_2 concentration, changes in light intensity were reflected significantly in changes in the height of peak n-3. The higher the light intensity level, the more apparent the vibration of electric potential on the surfaces of the sample laminae measured in the daytime.

Relationship between Photosynthesis and Electric Potential on Lamina Surface

From the measurements made in the process of leaf dying, it seems that while changes in peak n-1 were passive and independent of life activity, changes in peaks n-2, n-3 and n-4 were active and related to life activity. It can also be judged that from the measurements made by changing the CO_2 concentration and light intensity, peak n-3 was related to photosynthesis and that from the data obtained by changing the relative humidity, peaks n-2 and n-4 had to do with plant physiology in relation to water.

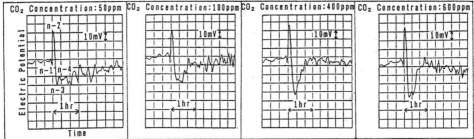

Fig. 9. Effect of CO_2 cocentration on the bioelectric response of <u>Spinach</u> lamina. Air temperature: 20℃, Relative humidity: 60%, Water temperature: 20℃, Light intensity: 190 μE/㎡sec.

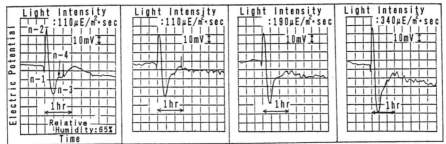

Fig. 10. Effect of light intensity on the bioelectric response of <u>Spinach</u> lamina. Air temperature: 20℃, Relative humidity: 60%, CO_2 cocentration: 600ppm, Water temperature: 20℃.

Fig .11 shows the relationship between the height of peak n-3 and the rate of photosynthesis with light intensity changed from 110 to 190 to 340 μE/㎡sec and the relative humidity within the range from 45% to 70%. there was strong correlations between the two factors.

Modeling of Transition Changes in Bioelectric Potential with Respect to Light

To use transition changes of bioelectric potential with respect to light to monitor the state of plant growth, it is necessary to have some appropriate model to represent changes in the transition pattern. Hodgkin and Huxley analyzed active membrane potential levels by using a membrane-equivalent capacitor-resistance circuit (Tasaki, 1968). In like manner, we assumed the model of Eq. (1) for changes in the bioelectric potential transition pattern V(t) with respect to the stepped input of light.

$$V(t) \equiv \Sigma_{n=1}^{6} A_n \cdot \{\exp(-\alpha_n \cdot t) - \exp(-\alpha_{n+1} \cdot t)\} , \quad (1)$$

$$\alpha_n \equiv \alpha / n \quad (n=1\sim6) , \quad \alpha_7 \equiv 0$$

Fig. 12(A) gives the result of a simulation of changes in the stages of growth shown in Fig. 5. In the simulation only the weighting

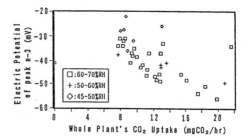

Fig. 11. Relation between whole plant's CO_2 uptake and the height of peak n-3 on the bioelectric response.
Air temperature   : 20℃
CO_2 concentration : 600ppm
Water temperature : 20℃.

factor $A_4 \sim A_6$ was changes with the time constant left unchanged.
Fig. 12(B) gives the result of a simulation of changes with respect to temperature shown in Fig. 6. In this simulation, only the time constant α was changes with the weighting factor left unchanged.
Fig. 12(C) gives the result of a simulation of changes with respect to relative humidity shown in Fig. 7. In the simulation, coefficients A_5 were changed.
It suggests the relationship with the water condition of plants.
Fig. 12(D) gives the result of a simulation of changes with respect to light intensity

and CO_2 concentration shown in Fig. 8 and 9,
respectively. In the simulation, coefficients
A_4 and A_5 were changed.
It suggests the relationship with the rate of
photosynthesis.
We plan to further examine these relations
and develop models for plants' internal water
conditions and rate of photosynthesis.

CONCLUSION

Our study on bioelectric potential on laminae
of Spinach in culture found that the electric
potential showed changes in pattern specifically with respect to environmental and
growth conditions such as temperature,
humidity, CO_2 concentration, light intensity,
growth stage and dying process. The height of
peak n-3 was related to the rate of
photosynthesis. It was found that these
transient pattern changes of bioelectric
potential can be simulated by the exponential
function given by Eq. (1). Our study also
suggests the possibility of expressing
relations between culture environmental
conditions and growth conditions by means of
function parameters.
The bioelectric potential measurement system
used in our study is a simple sensor
comprising a measurement electrode and signal
amplification circuit and does not require
providing electromagnetic shielding for the
plants to be measured. It can therefore be
used in practical culture facilities. Our
method of measuring transition changes in
electric potential on lamina by turning the
light source on and off has the advantage of
obtaining patterns of data easily. It can be
integrated in culture support systems
(Nakanishi, 1989) using supplement lamps in
culture facilities or the like and used to
evaluate and control culture environmental
and growth conditions.

Our thanks go to Professor Y.Hashimoto,
College of Agriculture, Ehime University, who
gave us valuable advice in preparing this
article, and to Dr. N. Arai and Dr. T. Hirao,
National Institute of Agrobiological resources,
Professor Y. Kano, Tokyo University of
Agriculture and Technology, and Dr. T.
Sekiyama, Central Research Institute of
Electric Power Industry (CRIEPI), who
extended us advice in pushing this study.

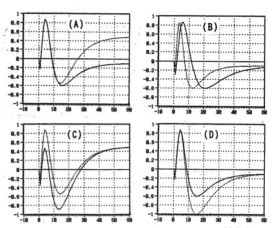

Fig. 12. The result of simulation of the
bioelectric potential patterns.
(A):Changes in the stage of growth,
(B):Changes with respect to temperature,
(c):Changes with respect to relative
humidity,
(d):Changes with respect to light
intensity and CO_2 concentration.

REFERENCES

Hashimoto, Y., T. Morimoto, and S. Funada
(1981). Computer processing of speaking
plant for climate control and computer
aided plantation. Acta Horticult., 115,
27-33.

Hashimoto, Y., H. Nishina, H. Nonami, and
Y. Yi (1989). Dynamical Approach to
Physiological Plant Ecology. Laboratory
of Agricultural Environmental Engineering,
College of Agriculture, Ehime University
Press.
Hirao, T., and N. Arai (1988). Development of
measurement system of bioelectric
potentials in plant. Bull. Natl. Inst.
Agrobiol. Resour., 4, 65-92.
Kano, Y., S. Hasebe, and K. Omasa (1988).
Measurement of electric potential of
living plant by double layer electrodes.
Environ. Control in Biol., 26(2), 79-82.
Nakanishi, Y., and T.Uchida (1989). Develop-
ment of a stand-alone computer integrated
cultivation support system. Environ.
Control in Biol., 27(4), 145-147.
Sekiyama, T., and H. Hanyu (1984). Continuous
measurement of electrical characteristics
in plant (1). Abiko research laboratory,
CRIEPI Rep., 484002.
Sekiyama, T., H. Hanyu, and H. Saiki (1989).
Continuous measurement of electrical
characteristics in plant (3). Abiko
research laboratory, CRIEPI Rep., U89023.
Tasaki, I.(1968). Nerve Excitation, A Macromolecular Approach. Charles C Thomas
Publisher.
Uchida, T., Y.Nakanishi, and M.Urata (1990).
Studies on measurement of bioelectric
potential of intact plant for the environmental control (part 1). Environ. Control
in Biol., 28(3), 113-118.

MEASUREMENT OF PLANT PHYSIOLOGICAL
INFORMATION FOR THE PLANT FACTORY

K. Iwao

*Electrotechnology Applications Research and Development Center, Chubu Electric Co., Inc.,
Ohdakacho, Midori-Ku, Nagoya 459, Japan*

Abstract. These studies were carried out for the development of
non-invasive and continuous monitoring technique of plant growth
in the plant factories. In this kind of facilities, the
environmental factors such as ambient temperature, relative
humidity, light intensity, pH of nutrient solution are measured
and reported with high accuracy. But, data on plant growth are
not measured almost or at all even nowadays.

We have, therefore, tried to apply the method of image
processing and obtained sufficient results.

Those data were applied to the following jobs.

(1) Evaluation of the seedling growth before transplanting into
the growth unit

(2) Estimation of the date of harvest for each kind of
vegetables

(3) Evaluation of the spacing methods

As for leaf area (A) of the seedling of salad, its growth
curve was expressed as $A \ (mm^2) = 13.8 \cdot \exp(0.51 \cdot T)$, $N=90$, $R=0.89$.
Where, T is number of the days after the gemination. The
relationship between its fresh weight W (mg) and leaf area A
(mm^2), $W=1.7+0.21 \cdot A$, $N=90$, $R=0.99$. The evaluation (1),(3) and
estimation (2) were done by these mathematical expressions.

Keywords. Image processing; plant growth; plant factory;
non-invasive and continuous monitoring; quality control.

INTRODUCTION

In the plant factories, the machinery and controls are strengthened
heavier than that of ordinary green
houses. The environmental factors
such as ambient temperature relative
humidity, light intensity, pH of
nutrient solution are measured and
output with high accuracy.

Monitoring of plant growth as the
production control, however, have not
been considered and tried, almost or
at all. The reason may be that it is
very difficult to measure the growth
mon-invasively and continuously.

Those facillities are often
featured to be able to produce some
kinds of vegetables uniformly
according to the schedulle. But, in
fact, it is impossible to realize the
features mentioned above.

We have, therefore, tried to apply
the technique of image processing
based on a personal computer
(PC-9801RX) for the continuous
monitoring of plant growth in a plant
factory. This paper report the
methods and results of the study.

METHHODS AND APPARATUS

1.Materials

The following vegetables are
tested. They have grown in the
artifical light type plant factory
"Ele-farm" which we are developing.

(1) Salad
·Kuroha-wear-head

(2) Lettuce
·Manoa lettuce
·Green wave

2.Environmental conditions
(1)Light source
   ·Fluorecent lamps
(2)Light intensity and period
   ·22klx & 16hr

(3)Temperature and humidity
   ·Lighting period 25C, 75%
   ·Dark period   20C, 75%

(4)Nutrient solution
   ·Ohtsuka house No.1 & No.2
   ·EC   1.3 mS
   ·pH   5.5
   ·System DFT

3.Measurement by image processing

Images of vegetables were recorded by a video tape recorder for 10 seconds at 11:00 every day, for 10 days. The recorded images were processed by the image processer TOSPIX-I and its transverse area, perimeter, maximum length, width, ratio of circle, equivalent circle diameter and area ratio were measured (see Appendix). The schema of the apparatus was shown in Fig.1.

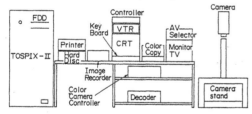

1 Image input   : Camera , Camera Controller , Decoder , VTR
2 Image processor : TOSPIX-II
3 Image output   : Monitor , Hard Copy , CTR Image Recorder , Printer , VTR

Fig.1 Apparatus for image input and image processing

Next, those data were analized statistically in regression line, and which some of them can be the index of growth of fresh weight or not.

4.Cultivation apparatus

Fig.2 shows the configuration of our plant factory "Ele-form" where the material plant grew up. The growing period (29 days) was divided into 4 sections, that is, 2 days in germination in water, 9 days in seedling unit-1. The growth data

shown in this paper are mainly obtained in this unit.

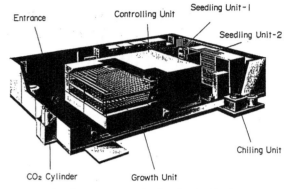

Fig.2 Structure of plant factory "Ele-farm"

RESULTS AND DISCUSSION

Fig.3 shows the growth curve of salad seedling by fresh weight during 9 days after germination. The number of materials were 10 plants every days.

In this process (seedling unit-1), the salad seedling grew by about 50 times (about 5~240mg/plant).

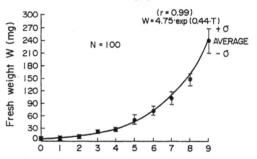

Fig.3 Growth curve of salad seedling by fresh weight
(Kuroha-wear-head, in seedling unit-1)

The growth characteristics of weight W fitted exponential curves well, that is, $W = 4.75 \cdot \exp(0.44 \cdot T)$. The coefficient of correlation R was 0.99.

On the other hand, the transverse area A as one of the measurements by image processing was expressed as $A = 13.8 \cdot \exp(0.51 \cdot T)$, $R = 0.98$ as shown in Fig.4. It grew by about 80 times (about $14 \sim 1100 mm^2$). The material plants were the same in Fig.2.

Other image processing items such

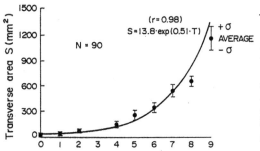

Fig.4 Growth crve of salad seedling
by transverse area
(Kuroha-wear-head, in seedling
unit-1)

as Perimeter, Length, Width,
Equivalent circle diameter also
fitted exponential curves. For
example, Fig.5 shows the growth curve
by perimeter. It increased by about
14 times (about 16 ~ 230mm). The
exponent K was 0.31 and was
considerably smaller than that of
arer, because of its dimension
(area:2 dimension, perimeter:1
dimension). Besides, area ratio
decreased gradually from about 3.0 to
1.0.

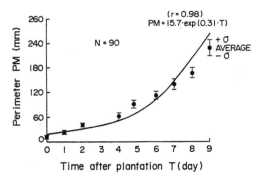

Fig.5 Growth curve of salad seedling
by perimeter (Kuroha-wear-head,
in seedling unit-1)

Fig.6 shows the correlation
cofficients of several image items
described above. Area's correlation
was 0.99, the biggest among them, so
we decided to use the area as the
growth index.

Fig.7 shows the relationship
between fresh weight W and area A. It
was expressed as a linear regression
line, W=1.69+0.21 A, R=0.99. This
result suggested that area is the

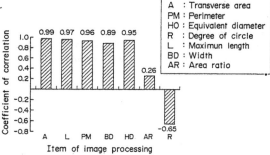

Fig.6 Comparison among coefficient of
correlations of various growth
curves by image processing

most reliable index instead of fresh
weight for the non-invasive growth
monitoring. The correlation
coefficient R decreased according to
the growth because of the
super-position of leaves. For
example, R was 0.97 in seedling
unit-2 and 0.89 in growth unit.

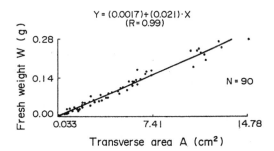

Fig.7 Relationship between fresh
weight(W) and transverse area
(A) of salad seedling

Fig.8 shows the two conceptions of
the spacing according to the
vegetobles growth. One is "Length" L,
another is "Equivalent circle
diameter" HD. L is effective in order

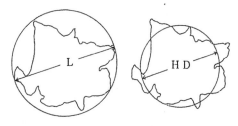

(a)Maximum length   (b)Equiv. circle dia.
Fig.8 Spacing depended on maximum
length L or equivalent circle
diameter HD

to avoid superposition of the leaves, that is, the decrease of photosynthesis. But, it needs more space for cultivation.

From the economical point of view, the spacing depended on HD would be realistic and efficient.

Fig.9 shows the growth curve by HD and our spacing curve in all growing period of salad. Our spacing could follow the increase of HD for 23 days after germination.

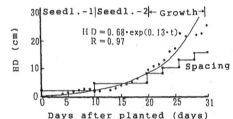

Fig.9 Growth curve by equivalent circle diameter HD and our spacing curve (Each plot is the average of 40 plants/day)

As for the growth curve by L, our spacing could cover only for 13 days after germination, which meant in the seedling unit-2.

Spacing for the vegetables seedling should be determined by the exponential functions for each species.

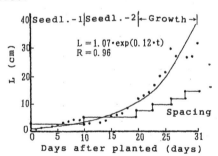

Fig.10 Growth curve by maixmum length L and our spacing curve (Each plot is the average of 40 plants/day)

CONCLUSIONS

(1) We have developed the plant growth monitoring system by image processing with a personal computer for the plant factories.

(2) The leaf area was regarded as the most useful index of vegetable growth as a non-invasive method.

(3) It was linear to the fresh weight with high coefficient of correlation.

(4) The scheduled harvest was achieved by these techniques.

(5) Spacing should be determined by the exponential expressions for each crops.

REFERENCE

MATSUI Tsuyoshi and Hiromi EGUCHI. (1976). Computer Control of Plant Growth by Image processing I, Mathematical Representation of Relation between Growth and Photographs of Plants. Environ.Control in Biol., 14, 1-7.

IMAGE DIAGNOSIS OF PHOTOSYNTHESIS IN WATER-DEFICIT PLANTS

K. Omasa*, S. Maruyama**, M. A Matthews*** and J. S Boyer****

*National Institute for Environmental Studies, Tsukuba, Ibaraki 305, Japan
**National Institute of Agrobiological Resources, Tsukuba, Ibaraki 305, Japan
***University of California, Davis, California 95616, USA
****University of Delaware, Lewes, Delaware 19958, USA

Abstract. An image instrumentation system was introduced for quantitative analysis of the rapid changes in intensity of chlorophyll fluorescence during dark-light transitions, which are sensitive indicators of various reactions of photosynthesis. This system provides information not only on localized differences in photosynthetic activity of the whole leaf *in situ*, but also on the inhibition site under various conditions. Using this system we analyzed the effects of slight water deficit on photosynthesis of attached leaves. Spatial differences in the inhibition of photochemical reactions on the leaf was shown. This system is useful for the analytical diagnosis of various stress-actions on plants *in situ*.

Keywords. Biology; Image processing; Instrumentation; Photosynthesis; Water deficit.

INTRODUCTION

The recent advances in imaging of physio-ecological functions of intact plants are remarkable (Omasa and Aiga, 1987; Hashimoto *et al.*, 1990; Omasa, 1990). For example, thermal imaging methods reveal spatial differences in stomatal responses to environmental stimuli and differences in gas exchange between leaves and the surrounding atmosphere. Remote-controlled light microscopes are used to observe heterogeneous movements of neighboring stomata of attached leaves. Advances in computed tomography (CT) such as portable X-ray CT, magnetic resonance imaging, and positron CT provide the information about mass transfer and metabolism inside the plants as well as morphological characteristics.

Visible light is absorbed by leaf tissues and used to photosynthesis, photomorphogenesis and other reactions. Therefore, light that is reflected or emitted from leaves gives information about these reactions. For example, rapid changes in intensity of fluorescence emitted from chlorophyll (Chl) *a* during a dark-light transition reflect the various reactions of photosynthesis, especially photosynthetic electron transport (Papageorgiou, 1975; Lichtenthaler, 1988).

Recently, Omasa *et al.* (1987) developed a new instrumentation system for analyzing chlorophyll *a* fluorescence quantitatively. This system gave information not only about the localized differences in photosynthetic activity on whole leaves *in situ* but also on the site of inhibition. In the present paper, we analyzed the stress imposed on the photosynthetic system by water deficits in

attached leaves employing an improved system for removing noise in playback images.

CHLOROPHYLL *a* FLUORESCENCE

When a dark-adapted leaf is illuminated with actinic blue-green light, the intensity of chlorophyll *a* fluorescence shows complicated changes with time during the illumination (Fig. 1). Although the actinic light may differ in intensity, fluorescence generally shows a fast rise (O), an intermediary level (I) with a small decline (D), then a gradual increase to a maximum (P), and again a slow decline with a secondary maximum (M) via a minimum (S), before reaching the steady-state level (T). These changes are called chlorophyll fluorescence induction (CFI) or the Kautsky effect (Kautsky *et al.*, 1960).

Since chlorophyll fluorescence is the re-emission of the light energy which is trapped by antenna chlorophyll and not utilized for the photochemical reaction (Duysens and Sweers, 1963; Murata *et al.*, 1966), the fluorescence intensity virtually reflects the magnitude of the photochemical reaction in a complementary fashion. Fluorescence intensity is low when the primary electron acceptor of photosystem II (Q_A) exists in the oxidized state, and high in its reduced state in the fast phase of fluorescence transients, because fluorescence intensity is regulated by the availability of Q_A.

The fluorescence transient from O to P takes a maximum of several seconds when the leaf is illuminated with moderate light, and the transient is called the fast phase. The fast phase

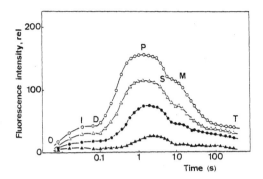

Fig. 1. The CFI curves of a healthy cucumber leaf *in situ* under different intensities of actinic blue-green light. The CFI was measured at the same position on the leaf. Before measurement, the leaf was dark-adapted for 30 min. Intensity of actinic light (μmol photons $m^{-2}s^{-1}$) : (o) 200, (Δ) 150, ($\bullet$) 100, and ($\blacktriangle$) 50. Light intensity was varied using ND filters. Environmental conditions: air temperature, 25.0°C; RH, 70% (Omasa *et al.*, 1987).

of the fluorescence transient (OIDP) is closely correlated with redox reactions of photosystem II (PS II); OI represents the photoreduction of Q_A by PS II reaction center, ID dip represents rapid oxidation of Q_A by photosystem I (PS I) (Munday and Govindjee, 1969), and DP indicates the photoreduction of Q_A by the PS II reaction linked to the water-spitting enzyme system (Munday and Govindjee, 1969). The fluorescence transient from P to T requires more than minutes, and is called the slow phase. The fluorescence intensity in the slow phase is not as simple as in the fast phase; the change includes two components, photochemical quenching (closely correlated with the oxidation state of Q_A) and non-photochemical quenching (mainly due to trans-thylakoid pH gradient). It may also reflect interaction between electron transport and carbon fixation (Walker, 1981; Sivak and Walker, 1985).

IMAGE INSTRUMENTATION SYSTEM

Ordinarily TV cameras and recording systems are not suitable for a quantitative analysis of CFI in small areas of attached leaves, because of their low sensitivity, large after-image, poor image quality, the presence of AGC (automatic gain control) function, and the indistinctness in timing of the playback image. Also, common tungsten and fluorescent lamps cannot be used as light sources for CFI imaging, because of the unevenness and fluctuation in light intensity.

These problems were overcome by the following system (Fig. 2). A highly sensitive CCD imager with uniform sensitivity and an after-image suppression was selected for a TV camera (SONY XC-47, improved type). The image quality was improved by use of a digital time base corrector (FOR.A FA-300), optical video disc recorder (OVDR) (SONY LVR-5000 + LVS-5000), and the preprocessing was improved by using a high-speed TV image processor (KCR nexus 6800). The use of OVDR instead of a video tape recorder (VTR) (Omasa *et al.*, 1987) especially decreased noise caused by abrasion of the VTR's head. The AGC function, which changes the relationship between the fluorescence intensity and the gray level of the recorded image, was removed from the TV camera. The timing of the playback image was exactly defined by the use of a shutter synchronized to the TV signal and by the time code recorded in each image. The unevenness in light source intensity was improved by the use of two xenon lamp (CERMAX LX-300F) projectors, and by attaching a special neutral density (ND) filter with concentric circles of different density.

After the plant had adapted to the dark for 20 min, the CFI was begun by irradiating the whole leaf with two beams of blue-green light (380-620 nm) from the projectors, with filters (Corning 4-96 + two heat-absorbing filters + a special ND filter) via the shutter opening. The fluorescence image was continuously measured at a TV field interval of 1/60 s by the TV camera equipped with an interference filter (Vacuum Optics Co. Japan IF-W, 683 nm; half-band width, 10 nm) and a red cut-off filter (Corning 2-64, >650 nm)

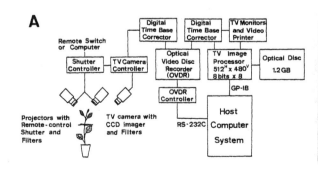

Fig. 2. Image instrumentation system for quantitative analysis of CFI in attached leaves. A, block diagram; B, photograph.

under intensities of actinic light from 50 to 200 µmol photons m^{-2}s^{-1}. Although the ordinary 50 mm lens (Nikon NIKKOR 1:1.2) gives a resolution of 0.3 mm, the use of a close-up lens (Nikon Micro-NIKKOR 105 mm 1:2.8 + PK-13) made it possible to measure within a resolution of about 15 µm. A series of images were recorded, with time code, by the OVDR. The OVDR image was digitized by a video A/D converter after it was played back to a still image. A series of the digitized images (512H x 480V 8 bits) was stored on an optical disc (National DU-15). A host computer was used to control the TV image processor and the OVDR.

The CFI curves and image intensity of characteristic transient levels (I,D,P,S,M,T) were calculated by the TV image processor on the basis of a series of images after preprocessing for shading correction and noise removal. These were represented by scales which correspond to the A/D conversion level. The relationship between the fluorescence intensity and the A/D conversion level of OVDR image showed a positive linear correlation. The image shading was corrected by calculating the ratio of an original image to a specific image (shading master), obtained by measuring a uniform light of definite intensity, because the shading was mainly caused by the lens and optical filters of the TV camera. The noise was removed by the use of a spatial smoothing filter, and the averaging of images digitized from a still OVDR image. For example, the image quality could be improved to within 1 % standard deviation by the shading correction and the smoothing of 3 x 3 pixels after an averaging of 10 images, when the image resolution was 280 lines. The after-image for the TV camera was about 4 % at 30 ms after shutter opening, and decreased to 0.3 % at 50 ms.

RESPONSES OF CFI TO WATER DEFICIT

Water deficit inhibits photosynthesis in attached leaves. The decline in photosynthesis involves stomatal closure restricting the entry of CO_2 (Brix, 1962; Barrs, 1968) and direct inhibition of photosynthetic apparatus (Boyer, 1971; Keck and Boyer, 1974; Matthews and Boyer, 1984).

Recent evidence indicates that under some conditions photosynthetic activity does not always show an uniform distribution over the leaf area but varies spatially, and sometimes forms a "patchy" over the leaf (Omasa et al., 1981a,b, 1987; Terashima et al., 1988; Omasa and Shimazaki, 1990). Hashimoto et al. (1984) and Downton et al. (1988) have demonstrated non-uniform stomatal closure induced by water deficit. Sharkey and Seemann (1989) reported non-uniform CO_2 uptake in intact leaves, however, spatial differences in the inhibition of photochemical reactions of water-deficient leaves have not been presented.

Govindjee et al. (1981) and Ögren and Öquist (1985) have shown that severe water deficit causes changes in CFI of leaves. However, effects of slight water deficit have not been clear. Therefore, changes in CFI of attached sunflower leaves during a temporary water deficit were investigated

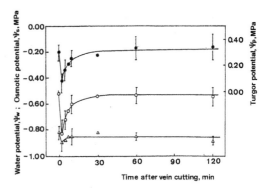

Fig. 3. Changes in water status of attached sunflower leaves after vein cutting. o, Water potential; Δ, osmotic potential; ●, turgor potential. These potentials were measured in downstream regions near the cutting line shown in Fig. 4. Environmental conditions: air temperature, 25.0°C; RH, 60%; light intensity, 230 µmol photons m^{-2}s^{-1}.

using our image instrumentation system.

Figure 3 shows changes in water status of attached sunflower leaves after vein cutting. The water potential decreased from -0.52 MPa to -0.83 MPa within 2.5 min after the cutting, then it gradually recovered to -0.60 MPa within 10 min. As the effect of the treatment on osmotic potential was relatively small, the larger part of the changes in water potential was attributable to turgor potential. The abrupt changes in the turgor potential from 0.30 MPa to 0.07 MPa after vein cutting was presumably caused by the removal of a main water source. The quick recovery of turgor was due to the entry of water from other tissues in response to the water potential difference created after the cutting.

Figure 4 shows changes in CFI curves after cutting veins in an attached sunflower leaf and Fig. 5 the intensity images of characteristic transient levels (D,P,S,M,T). In leaf regions (cf. 5 in Fig. 4B) near the midvein, some distance from the cut veins, the CFI clearly presented the typical IDPSMT transients similar to those appearing at any location of the leaf before the vein cutting. Since the CFI observed upon dark-light transition of the leaf reflects the partial reactions of photosynthesis, we can detect the alteration in photosynthetic apparatus by the vein cutting from the changes in CFI curves.

The effects of the temporary water deficit caused by vein cutting on the CFI curves differed in time after the cutting. When the time lapse was 30 min (4 in Fig. 4) after the cutting (involving dark during 20 min), the amplitude of the ID-dip and the DP-rise was decreased, and the appearance of peak P was advanced slightly. Also, the amplitude of the PS-decline was decreased, but the MT-decline was increased. These phenomena appeared in the surrounding regions of the cut vein. At an elapsed time of 60 min, the amplitude of the

MT-decline was recovered. When the elapsed time was 90 and 150 min, a severe decline of the DP-rise was extended as "patchiness" to regions near the cut vein although the appearance of patchiness was variable in the experiments. Thereafter, the appearance of peak P was late and the amplitude of the PS- and MT-decline was reduced.

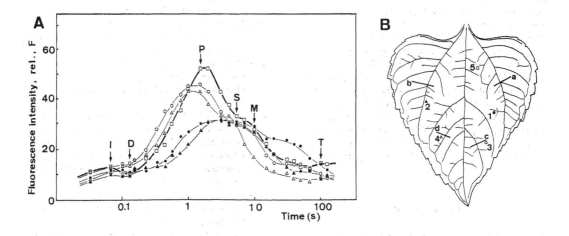

Fig. 4. Changes in CFI curves after cutting veins in an attached sunflower leaf. The veins were cut in order of cut line a to d shown in B (line a was cut at 150 min before the CFI measurement; line b at 90 min; line c at 60 min; line d at 30 min), and then the leaf was exposed to 230 μmol photons m⁻²s⁻¹ light intensity. After dark adaptation for 20 min, CFI of the leaf was measured under 100 μmol photons m⁻²s⁻¹. The CFI curves in A correspond to site 1 to 5 (●, site 1; ▲, site 2; o, site 3; Δ, site 4;□, site 5). Environmental conditions: air temperature, 25.0°C; RH, 60%.

Fig. 5. Intensity images of characteristic transient levels (I,D,P,S,M,T). The transient levels correspond with I, D, P, S, M, and T of CFI curves in Fig. 4.

As mentioned above, a loss of the ID dip in CFI suggests the decline of oxidation of Q_A by PS I (Munday and Govindjee, 1969). Since the DP rise in CFI reflects photoreduction of Q_A through reductant from H_2O (Munday and Govindjee, 1969), a diminished rise of DP was consistent with inactivation of the water-splitting enzyme system. Since PS decline involves energy-dependent quenching (Krause et al., 1981), the suppression of PS decline suggested the depression of formation of the trans-thylakoid proton gradient, probably due to the inactivation of the water-splitting enzyme. However, the possibility that the PS decline was affected by the inhibition of electron flow from Q_A to PS I cannot be excluded because the PS decline partly reflects the oxidation of Q_A by PS I (Bradbury and Baker, 1981). The increase of MT decline just after the vein cutting was probably due to a delay of formation of the trans-thylakoid proton gradient in addition to other unidentified reactions in chloroplasts (Papageorgiou, 1975; Bradbury and Baker, 1981).

CONCLUSION

A spatial variety of changes in CFI to slight water deficit was shown by our image instrumentation system. Since the CFI provides the information not only on the photosynthetic activity but also on the inhibition site in the photosynthetic system, we could use it to analyze mechanisms of inhibition of photosynthetic electron transport caused by slight water deficiency. This system is useful for the analytical diagnosis of various stress-actions on plants in situ.

REFERENCES

Barrs, H.D. (1968). Effect of cyclic variations in gas exchange under constant environmental conditions on the ratio of transpiration to net photosynthesis. Physiol. Plant., 21, 918-929.

Boyer, J.S. (1971). Nonstomatal inhibition of photosynthesis in sunflower at low leaf water potentials and high light intensities. Plant Physiol., 48, 532-536.

Bradbury, M., and N.R. Baker (1981). Analysis of the slow phases of the in vivo chlorophyll fluorescence induction curve. Changes in the redox state of photosystem II electron acceptors and fluorescence emission from photosystems I and II. Biochim. Biophys. Acta, 635, 542-551.

Brix, H. (1962). The effect of water stress on the rates of photosynthesis and respiration in tomato plants and loblolly pine seedlings. Physiol. Plant., 15, 10-20.

Downton, W.J.S., B.R. Loveys, and W.J.R. Grant (1988). Non-uniform stomatal closure induced by water stress cause putative non-stomatal inhibition of photosynthesis. New Phytol., 110, 503-509.

Duysens, L.M.N., and H.E. Sweers (1963). Mechanism of two photochemical reactions in algae as studies by means of fluorescence. In Jpn. Soc. Plant Physiol. (Ed.) Studies on Microalgae and Photosynthetic Bacteria. Univ. Tokyo Press. pp.353-372.

Govindjee, W.J.S. Downton, D.C. Fork, and P.A. Armond (1981). Chlorophyll a fluorescence transient as an indicator of water potential of leaves. Plant Sci. Let., 20, 191-194.

Hashimoto, Y., P.J. Kramer, H. Nonami, and B.R. Strain (Eds.) (1990). Measurement Techniques in Plant Science. Academic Press. pp.343-431.

Kautsky, H., W. Appel, and H. Amann (1960). Die Fluoreszenzkurve und die Photochemie der Pflanze. Biochem. Z., 332, 277-292.

Keck, R.W., and J.S. Boyer (1974). Chloroplast response to low water potentials. III. Differing inhibition of electron transport and photophosphorylation. Plant Physiol., 53, 474-479.

Krause, G.H., J-M. Brisntais, and C. Vernotte (1981). Two mechanisms of reversible fluorescence quenching in chloroplasts. In G. Akoyunoglou (Ed.) Photosynthesis. I. Photophysical Processes - Membrane Energization. Balaban International Science Services. pp.575-593.

Lichtenthaler, H.K. (Ed.) (1988). Applications of Chlorophyll Fluorescence. Kluwer Academic Publishers.

Matthews, M.A., and J.S. Boyer (1984) Acclimation of photosynthesis to low leaf water potential in sunflower. Plant Physiol., 74, 161-166.

Munday, C.J., and Govindjee (1969). Light-induced changes in the fluorescence yield of chlorophyll a in vivo III. The dip and peak in Chlorella pyrenoidosa. Biophys. J., 9, 1.

Murata, N., M. Nishimura, and A. Takamiya (1966). Fluorescence of chlorophyll in photosynthetic systems. II. Induction of fluorescence in isolated spinach chloroplasts. Biochim. Biophys. Acta, 120, 23-33

Ögren E., and G. Öquist (1985). Effects of drought on photosynthesis, chlorophyll fluorescence and photoinhibition susceptibility in intact willow leaves. Planta, 166, 380-388.

Omasa, K. (1990). Image instrumentation methods of plant analysis. In H.F. Linskens, and J.F. Jackson (Eds.) Modern Methods of Plant Analysis, New Series, Vol.11. Springer-Verlag. pp.203-243.

Omasa, K., and I. Aiga (1987). Environmental measurement: Image instrumentation for evaluating pollution effects on plants. In M.G. Singh (Ed.) Systems & Control Encyclopedia. Pergamon Press. pp.1516-1522.

Omasa, K., Y. Hashimoto, and I. Aiga (1981a). A quantitative analysis of the relationships between SO_2 or NO_2 sorption and their acute effects on plant leaves using image instrumentation. Environ. Control Biol., 19, 59-67.

Omasa, K., Y. Hashimoto, and I. Aiga (1981b). A quantitative analysis of the relationships between O_3 sorption and its acute effects on plant leaves using image instrumentation. Environ. Control Biol., 19, 85-92.

Omasa, K., and K. Shimazaki (1990). Image analysis of chlorophyll fluorescence in leaves. In Y. Hashimoto et al. (Eds.) Measurement Techniques in Plant Science. Academic Press. pp.387-401.

Omasa, K., K. Shimazaki, I. Aiga, W. Larcher, and M.

Onoe (1987). Image analysis of chlorophyll fluorescence transients for diagnosing the photosynthetic system of attached leaves. *Plant Physiol.*, *84*, 748-752.

Papageorgiou, G. (1975). Chlorophyll fluorescence: An intrinsic probe of photosynthesis. In Govindjee (Ed.) *Bioenergetics of photosynthesis*. Academic Press. pp.319-371.

Sharkey, T.D., and J.R. Seemann (1989). Mild water stress effects on carbon-reduction-cycle intermediates, ribulose biphosphate carboxylase activity, and spatial homogeneity of photosynthesis in intact leaves. *Plant Physiol.*, *89*, 1060-1065.

Sivak, M., and D.A. Walker (1985). Chlorophyll a fluorescence: can it shed light on fundamental questions in photosynthetic carbon dioxide fixation? *Plant Cell Environ.*, *8*, 439-448.

Terashima, I., S-C. Wong, C.B. Osmond, and G.D. Farquhar (1988). Characterization of non-uniform photosynthesis induced by abscisic acid in leaves having different mesophyll anatomies. *Plant Cell Physiol.*, *29*, 385-394.

Walker, D.A. (1981). Secondary fluorescence kinetics of spinach leaves in relation to the onset photosynthetic carbon assimilation. *Planta*, *153*, 273-278.

Copyright © IFAC Mathematical and Control Applications
in Agriculture and Horticulture, Matsuyama, Japan 1991

IDENTIFICATION OF THE WATER STATUS OF
THE PLANT - IDENTIFICATION OF AR MODEL

T. Torii, T. Ochiai, T. Okamoto and O. Kitani

*Dept. of Agricultural Engineering, Faculty of Agriculture, University of Tokyo, 1-1-1 Yayoi,
Bunkyo-ku, Tokyo, Japan 113*

ABSTRACT

To control the water stress to the plant requires one of the advanced
technologies for the greenhouse and the field cultivation. The highest
quality of the crops is obtained when the optimal water stress is provided. In this study, the plant's responses to the randomly varying two
levels of light intensity were measured, and the frequency characteristics of the plant were calculated. For the mathematical model of the
plant, auto-regressive model (AR model) was assumed, and its parameters were identified. From this model, the simulation of the experiment
and the frequency characteristics were calculated. The results of the
simulation corresponded to those of the experiment. The step response
of the lighting to the plant was also measured, and the frequency response functions were calculated. The frequency characteristics under
random lighting condition, from step response, and by AR model showed
good agreement. These results showed that the AR model would be a possible model to explain the plant responses.

Keywords. Agriculture, Identification, Modeling, Greenhouse Automation,
Simulation, AR model

INTRODUCTION

Automation to control of supplying water and
giving water stress to the crops require an important technologies to get crops of good quality.
To enable this, the instrumentation and the modeling of the water status of the plant is important
issue. The responses of the plant to the water
stress were studied, and a mathematical model was
identified. The water stress was given by providing varying levels of light intensity. The changing stem diameters were measured to monitor the
water status of the plant. There were some previous reports about identification of plant responses (Hashimoto, 1985)(Morimoto, 1988)(
Youlin, 1988)(Morimoto, 1989). In this study,
auto-regressive model was assumed for the mathematical model of the plant, and its parameters
were identified.

Modeling of AR model

The input-output relation of AR model is following equation (Kido,1985) .
$$Y(z)=H(z)X(z)+N(z) \qquad (1)$$
where
$H(z)$ transfer function
$Y(z)$ output signal
$X(z)$ input signal
$N(z)$ noise
and
$$H(z)=\frac{1}{1+\displaystyle\sum_{i=1}^{N}a_i z^{-1}} \qquad (2)$$

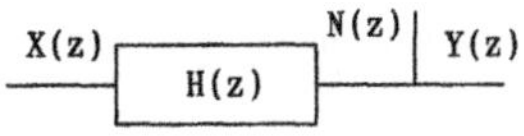

Fig. 1 Input-output relation of AR model

To estimate the parameters of AR model, the relation between input and output of the model is
reduced to Yule-Walker's equation.

$$\begin{bmatrix} R_0 & R_1 & & R_{N-1} \\ R_1 & R_0 & & R_{N-2} \\ \cdot & \cdot & & \cdot \\ R_{N-1} & R_{N-2} & & R_0 \end{bmatrix}\begin{bmatrix} a_1 \\ a_2 \\ \cdot \\ a_N \end{bmatrix}=-\begin{bmatrix} R_1 \\ R_2 \\ \cdot \\ R_N \end{bmatrix} \qquad (3)$$

where
R_0 R_1 R_{N-1} autocorrelation function
The parameters of auto-regressive model a_1 and
the dimension of the models N were calculated. The
Levinson-Durbin's algorithm was used to determine
parameters of a_1, and A.I.C (Akaike Information
Criteria) optimum dimension of N. Frequency characteristics of the model was given as follow.

$$H(z)=\left| \frac{1}{1+\displaystyle\sum_{i=1}^{N}a_i z^{-1}} \right|^2 \qquad (4)$$

To simulate the model, $H(z)$ was reduced to the
following equation.
$$H(z)=1+b_1 z^{-1}+b_2 z^{-2}+ \ldots \qquad (5)$$
where
$b_0=1$ $b_1=a_1 b_0$ $b_2=a_2 b_0+a_1 b_1$ ..
The input-output relation at time domain was
shown next .

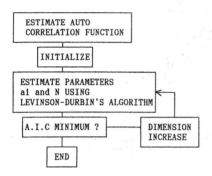

Fig. 2. Flowchart of the estimation of AR parameters.

$$y(t) = h(t) * u(t)$$

$$= 1 + \sum_{i=1}^{N} b_i u(t - i\tau) \qquad (6)$$

* convolution
τ sampling time

Experiment

Control system of the Chamber

The test apparatus of the environment-controlled chamber are illustrated in Fig. 3. The temperature and humidity were controlled by P.I.D. temperature controller. In this study, the temperature was kept at 30 °C and the humidity, 60 %. The light shield plate was used to control the intensity of the illumination, and solid state relays as switches.

Instrumentation system

Transpiration and Absorption of the root were measured by the electric balance. The water level of the pot was kept at a same level by Mariot's

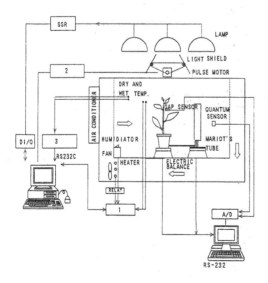

Fig. 3 Sketch of test apparatus.
   1. P.I.D. temperature controller
   2. Pulse motor controller
   3. Data logger

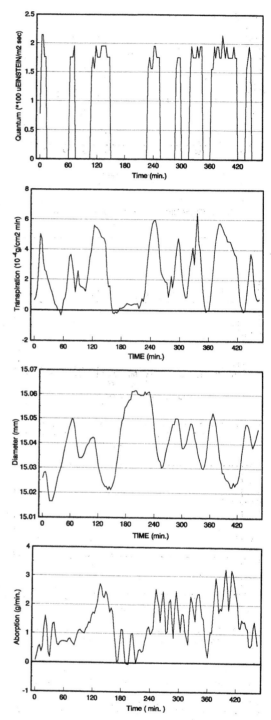

Fig. 4 Responses of the plant under random lighting conditions. The variation of the light intensity, transpiration, diameter changing, and root absorption, from above.

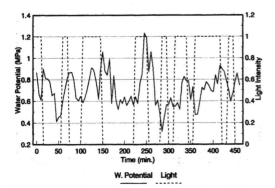

Fig. 5 Changing of the water potential

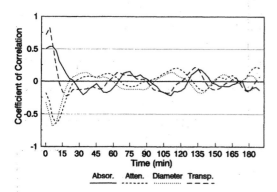

Fig. 6 Cross correlation function with light intensity and transpiration, changing stem diameter, and root absorption.

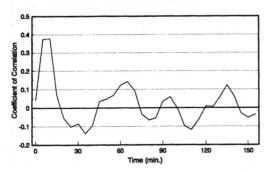

Fig. 7 Cross correlation function with light intensity and leaf water potential

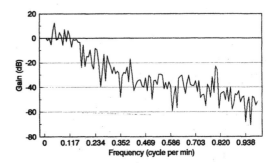

Fig. 8 Cross spectrum with light intensity and transpiration

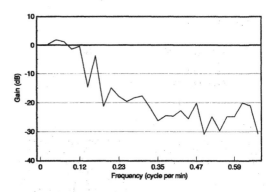

Fig. 9 Frequency characteristic of transpiration calculated from step response

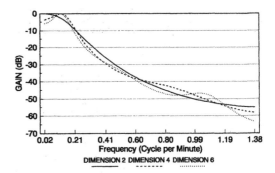

Fig. 10 Frequency characteristic of transpiration calculated by AR model. Dimension of the AR model which was 2, 3, and 6 was calculated.

tube. The variation of the diameter was measured by eddy-current displacement sensor. The leaf temperatures were measured by data logger using thermocouples. These data were saved to the personal computers.

Plant materials

Tobacco plant (Nicotiana tabacum cv. BY 4) with water culture was used in this study.

Input signals

As input signals, the light intensity of varying two levels were provided. Pseudo random signal determined by M sequence was applied (Sato,1967).

Input signal 1 and -1 by M sequence correspond to the intensity of illumination 15 kLx, 0 Lx respectively. The dimension of M sequence was 5, number of output 31, and the basic clock pulse interval was 15 minutes.

Water potential

Water potential was measured in a chamber at 20 °C and at about 100% humidity. Decagon SC-10A Psychrometer was used.

RESULT AND DISCUSSION

Cross correlation function

The result of the experiment by pseudo random signals are shown at Fig. 4. These data were measured at 3 minutes interval. The result of water potential is shown in Fig. 5, measuring at 5 minutes intervals. Cross spectrum and cross correlation function of the light intensity were calculated with transpiration , diameter variation, and root absorption using Fast Fourier Transfer (Fig. 6). Time lag of transpiration was about 3 minutes, root absorption was 3 to 6 minutes, and diameter variation was 6 to 9 minutes. Cross correlation function with light intensity and water potential are at Fig. 7 . Time lag was about 5-10 minutes which was almost the same as observed in the case of the diameter changes. This result suggests that the difference of time lag between transpiration and root absorption might have caused the fall of the leaf water potential, the changing of stem diameter.

Frequency characteristics

Frequency characteristics were calculated by three different method: cross spectrum with light intensity and transpiration, shown in Fig. 8; frequency characteristics gained from step response, shown in Fig. 9; frequency characteristics gained from AR model, shown in Fig. 10. The results obtained from these three method were of good agreement. The frequency characteristics was flat until 0.15 cycle per minute, which declined

afterward. This result coincided with the previous report (Hashimoto,1982).

Identification of AR model and simulation

The identified parameters are listed in table 1. To make sure of the suitability of AR parameters, the simulation was calculated. The result of the simulation of transpiration is shown in Fig. 11, and the stem diameter changes in fig. 12. The changing cycles of both the experiment and the simulation followed close patterns, with some differences in the actual values. The variation of light intensity might be too strong to estimate plant as linear system.

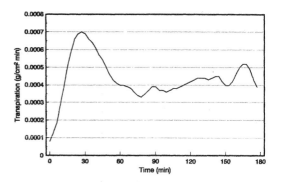

Fig. 13 The step response of the transpiration

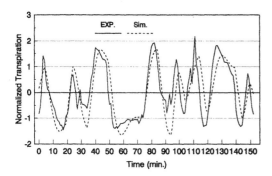

Fig. 11 Simulation of the transpiration under random lighting conditions. The dimension of the AR model was 6.

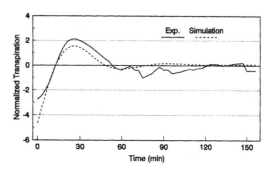

Fig. 14 The simulation of the step response of transpiration.

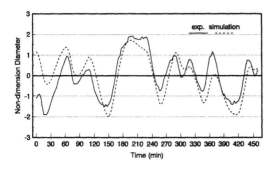

Fig. 12 Simulation of the changing diameter under random lighting conditions. The dimension of the AR model was 3.

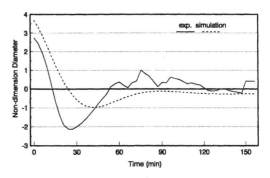

Fig.15 The simulation of the step response of changing diameter.

392

Table 1. Identified AR parameters

dimension	transpiration	diameter
1	1.43607	1.45294
2	-0.699047	-0.26639
3	0.385029	-0.24693
4	-0.453785	
5	0.349369	
6	-0.170439	

Further, the step responses were simulated with these identified parameters. The result of the simulation and experiment which also showed close pattern of changing cycles is shown in Fig. 13 , with some differences in the damping characteristics. These phenomena might be also caused by the reason mentioned before.

Conclusion

The plant responses were observed under random lighting conditions, and the system parameters of the plant were calculated. Step responses of the plant were also measured. To make sure of the suitability of the parameters, these simulations were applied to those experiment.
1. The frequency characteristics of the plant responses were about 1.5 cycle per minute.
2. By any of the three methods, the frequency characteristics of the plant were found to be almost the same.
3. The simulation by the AR model corresponded with the experiment data. Therefore, AR model would be a possible model to explain plant responses.

REFERENCES

Hashimoto, Y., Morimoto T., and Fukuyama T., (1982). Identification of water deficiency and photosynthesis in short-term plant growth under random variation of the environment. Proc. 6th IFAC Symposium on Identification and System Parameter Estimation, Pergamon Press, Oxford. 1559-1564.

Hashimoto, Y., Morimoto T., (1985). Identification of water relations and CO_2 uptake in physiological ecological processes in a controlled environment (I). Proc. 7th IFAC Symposium on Identification and System Parameter Estimation, Pergamon Press, Oxford. 1677-1681,

Morimoto, T., Fukuyama, T., Youlin, Y., and Hashimoto, Y. (1988). Identification of physiological dynamics in hydroponics.
Proc. 8th IFAC/IFORS Symposium on Identification and System Parameter Estimation, Pergamon Press, Oxford.

Morimoto T., and Fukuyama T., Hashimoto, Y. (1989). Growth diagnosis and optimal environmental control of Tomato plant cultivation in hydroponics(I). Environ. Control in Biol., 27, 137-143

Youlin, Y., Hashimoto, Y. (1988). Parametric Model of Photosynthesis of Lettuce as Affected by the Pulsed Light. Environ. Control in Biol., 26, 119-127

Kido, k (1985). Digital Signal Processing, Maruzen, Tokyo. 204-212

Sato, I., (1967). Pseudo random sequence. Correlation function and spectrum, University of Tokyo Press, 170-181

MEASUREMENT OF ELONGATION RATE OF PLANT

H. Shimizu and S. Oshita

Faculty of Bioresources, Mie University, Kamihama-cho, Tsu, Mie, Japan 514

Abstract. A noncontact measurement system for a plant elongation was developed using a image processing technique. Theoretical basis was presented to determine space coordinates of a point on a plant by a set of two dimensional images.

A plant was set up on a turn table, and a computer acquired a set of images before and after rotating the turn table. The digitized images were transformed to thinned images. Three dimensional coordinates of points on the thinned line were computed and by the connection of these points, the representative length of a plant was determined. The elongation rates of rice plant leaves were measured with the system developed and it was confirmed that a elongation rate was able to be measured with noncontact.

Keywords. Agriculture ; image processing ; elongation rate of plant ; noncontact measurement.

INTRODUCTION

Analysis of plant reaction closely related to growth environment is one of the fundamental studies concerning growth phenomena in plant. In these phenomena, an elongation of plant or plant parts is one of important expressions of plant reaction. Formerly mechanical instruments, such as the auxanometer or kymographion, were used for the measurement of the plant elongation, recently electric devices have been widely used.

Klueter(1967) developed the system for continuous detection, measurement and recording the movements of plants. It was consisted of an angle transducer connected to the plant through a variable-length rod. Aimi(1969) reported the electric growth recorder based on conversion of displacement into electric current through the ratio-transformer. This apparatus was equipped with a compensating mechanism for the temperature change. The leaf movement of *Phaseolus* was shown together with records of some environmental factors. Warner(1971) investigated the latent times for growth responses of individual pea plants to auxin, cytokinin, ethylen and abscisic acid using the electric

position-sensing device. Miyazato(1976) measured the micro growth rate in the cereal plant with the electric strain meter. Linear Variable Differential Transducer (L.V.D.T.) have been used to record the elongation rate (Barlow,1972; Gallagher,1976; Christ,1978a, 1978b). Barlow et al. monitored the rapid change in leaf elongation rate that occurred after a plant was released from stress. Gallagher et al. designed an auxanometer to measure the extension rates of grass leaves in the fields, and constructed it with L.V.D.T.. Chrsit reported that the elongation rate of monocotyledonous leaves was determined during the whole growing period by means of L.V.D.T. connected to a computer.

Their ways are, however, contact method. In measuring the elongation of plant, attention has to be paid to make influence upon the plant itself minimum as well as the growth environment. Hence, it is ideal to measure without contact to a plant. This paper describes a noncontact method to measure the elongation rate of monocotyledonous leaves with the use of image processing technique, a measurement system developed and an example of results.

FORMULATION

Theoretical Equation

A method to determine a three dimensional position of a point from data in images is as follows:

It is assumed that a point P is located above a turn table, and that a system of coordinates is a three dimensional rectangular coordinate whose origin is consistent with the center of the turn table. Figure 1 shows a geometrical arrangement between a point P and a turn table.

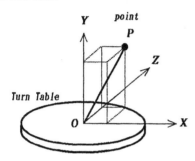

Fig. 1. Geometrical arrangement between point P and turn table

Figure 2 shows a top view and a front view of the turn table. The front view is only acquired with the system developed in this study. X and Y values are directly obtained from images. Therefore, if Z value is calculated, the three dimensional coordinates of the point P can be obtained. Let P be the point before rotation, P' be the point after rotation, α be the angle between the Z axis and line segment PO, θ be the rotational angle of the turn table, and X_1, X_2 be X values of point P and P'. X_1, X_2 are, respectively,

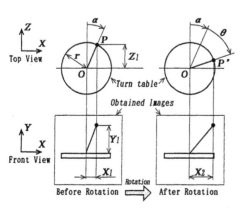

Fig. 2. Data set extracted from two images

expressed by the followings.

$$X_1 = r\cos\alpha$$
$$X_2 = r\cos(\alpha + \theta)$$

Transforming with the addition theorem

$$X_2 = X_1 \cdot \cos\theta + Z_1 \cdot \sin\theta$$

where

$$Z_1 = r\sin\alpha$$

Z_1 is given by

$$Z_1 = \frac{X_1 \cdot \cos\theta - X_2}{\sin\theta} \qquad (1)$$

where

$$\theta \neq n \cdot \pi \quad (n = 0, 1, 2, \cdots)$$

Thus, the three dimensional coordinates of point P (X_1, Y_1, Z_1) is obtained.

Optimum Angel of Rotation

Eq.(1) is the function of X_1, X_2 and θ.

$$Z_1 = f(X_1, X_2, \theta)$$

Since only θ is able to be set in the desired value, it is necessary to find the optimum angle to make an error in measuring minimum. If X_1 and X_2 have the error ε_1 and ε_2 respectively, and θ has no error, Z value is given by

$$Z' = f(X_1 + \varepsilon_1, X_2 + \varepsilon_2, \theta) \qquad (2)$$

Expanding Eq.(2) in Taylor series, and disregarding terms of above second orders,

$$Z' = f(X_1, X_2, \theta) + \sum_{i=1}^{2} \frac{\partial f}{\partial X_i} \varepsilon_i$$

the propagation of errors to Z_1 is expressed by

$$\delta = \sum_{i=1}^{2} \frac{\partial f}{\partial X_i} \varepsilon_i \qquad (3)$$

It is possible to minimize the propagation error δ when the absolute values of partial differential coefficient in Eq.(3) become to be minimum.

The above partial differential coefficients are the function of θ, and that are

$$\frac{\partial f}{\partial X_1} = \cot\theta \qquad (4)$$

$$\frac{\partial f}{\partial X_2} = -\sec\theta \qquad (5)$$

The absolute values of Eq.(4) and (5) become to be minimum when θ is equal to a multiple of $\pi/2$. Therefore, the optimum angle of rotation is $\pi/2$.

SYSTEM DESCRIPTION

Non contacting measurement of plant elongation rate was carried out with two systems. Namely, one is "Plant Image Recording System" that recorded plant image on video tape every set time. Another was "Image Data Processing System" that obtained plant digital image from video tape, and processed to digital thinned images by image processing technique, and calculated the plant elongation rate.

Plant Image Recording System

The block diagram of plant image recording system is shown in Fig.3.

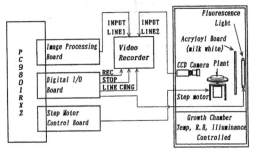

Fig. 3. Plant image recording system

The system was based on a 16-bit computer (NEC, PC9801RX2) equipped with image processing board(512 × 512[pixel],256[gray level]) (PHOTORON ,FRM1-512), digital I/O board (CONTEC ,W- BUS [98]) and stepping motor control board (VEXTA ,K7100-PC). A turn table consisted of stepping motor(ORIENTAL MOTOR,EM569-NA), a CCD camera (378000 pixel) (SONY, XC-77) and a fluorescence light(HITACHI,FML27EX-N), controlled by the computer, were set up in a growth chamber (temperature, relative humidity and illuminance

controlled) (KOITO,M20). The turn table was located between the CCD camera and the fluorescence light.
A plant to be measured its elongation was set up on the turn table, it was illuminated with the 27W fluorescence light from behind.
The timing table of measurement is shown in Fig.4. The video recorder started and the marking pattern and other information for the identification of subsequent images was recorded on the video tape over first 4 seconds. This information given by the computer noted the recording date and time, the beginning date of measurement and a marking pattern. The marking pattern was used to process the recorded images automatically. The image of recording information are shown in Fig.5.

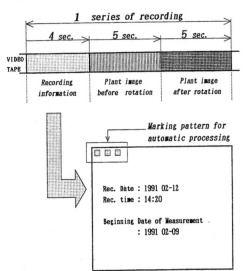

Fig. 5. Information image

Next, the fluorescence light was turned on, input-line of the video recorder was changed. Then a plant image on the turn table was taped. This plant image was a silhouette since the plant was illuminated from behind. After 5 seconds, the turn table was rotated in 90 degrees, and the plant image was taped again for 5 seconds. This series of operations mentioned above was controlled by the computer, and were executed every 10 minutes during day and night. When the recording speed was selected in 2 cm/second with the use of 750 feet length videotape, the plant images were able to be continuously taped for 5 days because the recording time for one measurement was about 15 seconds.

Image Data Processing System

The block diagram of image data processing system

VIDEO RECORDER			TURN TABLE	FLUORESCENCE LIGHT
STATE	LINE1	LINE2		
REC 4 sec.	INF. IMAGE			
5 sec.		PLANT IMAGE		O N
1 sec.			ROTATE IN 90 DEG.	
5 sec.		PLANT IMAGE		
STOP				O F F

Fig. 4. Timing table of measurement

397

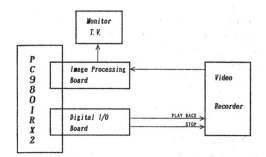

Fig. 6. Image data processing system

is shown in Fig.6.

This system, the modification of the plant image recording system, was consisted of the computer, image processing board, digital I/O board, the video recorder and a monitor TV. The timing table of data processing is shown in Fig.7.

The computer watched the recorded images to detect the marking pattern of the recording information. Detecting it, the computer acquired the plant image before rotation into image processing board after 6 seconds, and stopped the video recorder. The outline(described later) of plant before rotation was extracted from the acquired image. The video recorder was played back again during 5 seconds, and the plant image after rotation was acquired. This outline was also extracted in the same way, and the video recorder was stopped. The outline data were used to determine a thinned line with which a plant length was calculated. A thinned line is described later. It took about 45 seconds to perform above operations. Data processing was fully automated with the use of the marking pattern in the recording information.

VIDEO RECORDER	IMAGE PROCESSING BOARD	COMPUTER
Play Back	Watching	
	Detecting Making Pattern	
	Wait for 6 sec.	
	Acquired Plant Image Before rotaion	
Stop	Extracting outline 1	
Play Back		
	Wait for 5 sec.	
	Acquired Plant Image After rotation	
Stop	Extracting outline 2	
		Thinning 1
		Thinning 2
		Calc. of Plant Length
		Save Outline Data & Length Data

Fig. 7. Timing table of data processing

Outline

Normally outline was an external shape subjected to thresholding. But outline obtained in this way did not completely show the shape of object. Then the method shown in Fig.8 was developed for extracting an exact outline from the plant digital image. It was assumed that gray level of a pixel represented the light intensity at the center of the pixel, and that a threshold was put between gray levels of two neighbor pixels. Then a line segment connecting two pixel centers was drawn, a point on the segment indicated a certain gray level between the two. Finally, the threshold was found on the segment. Position of the threshold found in this way formed the outline of the object. The way in case that threshold equaled 100 was shown in Fig.8.

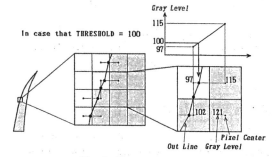

Fig. 8. Extraction of outline

Thinning

Thinning technique for plant leaf was developed, because various proposed thinning methods had usually a defect that a pointed end diminish in length after thinning. One cycle of the developed procedure is explained as follows(Fig.9).

1. Elongate the center line by 5 dots from P.
   (The upper end of the thinned line is named present point P.)
2. Draw line segment AB at a right angles
3. Make midpoint M in line segment AB
4. Draw line segment PM.
   (Line segment PM was newly created as the thinned line in this cycle.)
5. In the next cycle, M becomes the present point P.

This cycle was repeated with the exception that:
1. In the first cycle of thinning, present point P is assigned to the midpoint of horizontal segment line and the

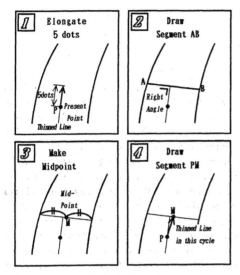

Fig. 9. Thinning

direction of elongation is vertical.

2. When the newly created thinned line crosses the outline, the point of intersection is the terminal of the thinned line.

Corresponding Points in Two Images

In two thinned images before and after rotation, two points that had the same Y value indicated the same point on the plant(Fig.10).

Calculation of Plant Length

The X and Y value of a point on the center line were obtained from the thinned image. The Z value (depth value) was computed with Eq.(1). The X values substituted for Eq.(1) were those of corresponding points in thinned images before and after rotation. Thus, three dimensional data of a point on the

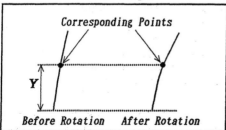

Fig. 10. Corresponding points in two images

thinned line of plant were acquired , and these three dimensional data were computed at intervals of 5 dots. A plant length(L) was approximately calculated by adding up a distance between two adjacent three dimensional points(Eq.6).

$$L = \Sigma dL_i \qquad (6)$$

where

$$dL_i = ((X_i - X_{i-1})^2 + (Y_i - Y_{i-1})^2 + (Z_i - Z_{i-1})^2)^{0.5}$$

The rate of elongation was computed for a time interval of 60 minutes. The plant length, that was calculated from the images obtained every 10 minutes, was compared with the value obtained 60 minutes earlier, and the difference was the value of elongation rate in mm/hr.

RESULTS AND DISCUSSION

A measurement accuracy of this system was examined, and is shown in TABLE 1. A manual precision stage(reading precision:0.0005mm) on which a needle was fixed was set up on the turn table. The stage was moved by 1mm, and a movable distance of the needle was measured five times. The image resolution was 0.097mm/dot in this case. Therefore, a maximum measured distance was 50mm. TABLE 1 shows that measuring accuracy is almost within an error of 0.050mm. Since the subject of measurement was a elongation rate of seedling of rice plant in this study, it was sufficient to measure within range of 50mm.

Continuous measurements of elongation rate were made with the use of above mentioned system. Measured plant was a seedling of rice plant, and a elongation rate of primary leaf was measured. It was treated with brine assortment, disinfection, soak in water and seeding on a mat for seedling culture in a laboratory dish. It was kept under the conditions of 30℃, 50%R.H. and 12h day with a illuminance of about 9000lx. A typical example is given in Fig.11.

TABLE 1 Measurement Accuracy

Moved distance of stage (mm)	Measured length of needle (mm)				
1	1.021	1.038	1.012	0.950	0.979
2	2.000	2.008	2.070	2.010	1.999
3	3.006	3.014	3.007	3.017	3.020
4	4.046	4.024	3.945	4.027	3.959
5	5.011	5.041	5.014	5.009	4.997
6	6.051	6.018	6.006	6.051	6.006
7	7.039	7.015	7.006	7.003	7.030
8	8.006	7.004	7.988	8.000	8.006
9	8.996	8.976	8.991	9.014	9.021

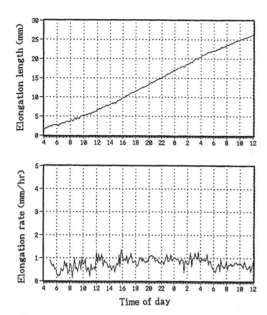

Fig. 11. Elongation of a primary leaf of rice plant

It was observed that the elongation rate of the primary leaf of rice plant was fairly constant during day and night. This suggests that the elongation rate did not obviously react upon lights because a chloroplast in a leaf was in the making.

Since it took about 45 seconds on an average to acquire plant length from one series of recording, it was necessary about 100 minutes to process data obtained for one day. Though it was feel too long for computation time, it did not keep us busy for the processing was automatically performed by the computer.

But, some problems still remain as follows. Threshold level needed to be changed between day and night, because of a influence of the lights equipped with the growth chamber. It was difficult to make the elongation length of plant a line of continuity at the time of illumination condition changed. For that reason, it is necessary to develop some reasonable method in determining the threshold level. Some influence was exerted upon a plant because the fluorescence light was turned on for 11 seconds at every measurement. One way of solving this problem is to use of a light source with no effect upon photosynthesis. Moreover, the shaking of plant due to the rotation caused the measurement errors. On this subject, it is conceivable to apply a mirror system instead of the turn table. Therefore, further improvements in the system are necessary to measure the elongation rate exactly.

CONCLUSION

1. An automatic measurement system for elongation rate of plant with noncontact was designed and constructed with the use of image processing technique.

2. An image processing algorithm was developed to extract the outline of plant from a plant image, create the thinned line from the outline, determine the three dimensional coordinates of a point on the thinned line, and calculate the length of plant.

3. The algorithm performed satisfactorily in the system, and it was demonstrated that the elongation rate of rice plant was automatically measured with noncontact.

4. Further improvements were necessary to measure a real rate of elongation with no influence upon a plant, and a fine change in elongation rate.

REFERENCE

Aimi, R. (1969). Electric growth recorder. Plant & Cell Physiol. 10, 707–710.

Barlow, E.W.R., L. Boersma, (1972). Growth Response of Corn to Changes in Root Temperature and Soil Water Suction Measured with a LVDT. Crop Science, 12, 251–252.

Christ, R.A. (1978a). Elongation Rate of Wheat Leaves —Elongation Rates during Day and Night—. J. of Exp. Bot., 29(110), 603–610.

Christ, R.A. (1978b). Elongation Rate of Wheat Leaves —Effect of Sudden Light Change on the Elongation Rate—. J. of Exp. Bot., 29(110), 611–618.

Gallagher, J.N., P.V. Biscoe, R.A. Saffell, (1976). A Sensitive Auxanometer for Field Use. J. of Exp. Bot., 27(99), 704–716.

Klueter, H.H., W.A. Bailey, (1967). To record plant growth. Agricultural Engineering, Dec. 1967, 720–721.

Warner, H.L., A.C. Leopold, (1971). Timing of growth regulator responses in peas. Biochemical and Biophysical Research Communications, 44(4), 989–994.

DISTRIBUTION OF ELECTRIC POTENTIAL OF PLANT SURFACE MEASURED BY NEW POTENTIAL METER

Y. Kano, Y. Hamauzu, I. Kawaguchi and Y Kida

Tokyo University of Agriculture and Technology, Nakamachi 2-24-16, Koganei, Tokyo 184, Japan

ABSTRACT Measurement of electric potential on a plant surface required high impedance of measurement device, because the potential on a plant surface has high impedance as voltage supply. It also required stability and not to disturb the state of living samples. The new modified amplifier using the MOSFET which has high input impedance and is modified by a new method and solve these issues in measurement of electric potential on a plant surface.

Characteristics of MOSFET is changed when the voltage is given to its substrate. It is tried to use MOSFET for DC-AC converter by supplying a square wave voltage to substrate as the second electrode in order to decrease the drift.

As a conclusion, the output voltage of drain is in proportion to the gate voltage and is modified by substrate voltage by giving AC or square wave voltage to the substrate of MOSFET. Output voltage is in proportion to input voltage linearly from -0.7 [V] to 0.7 [V] and the drift is very small.

The distribution of relative electric potential on the surface of cucumber is measured. It shows high potential at the top and the end of the cucumber comparing the center of it. But the distribution of the potential along the axis of cucumber becomes flat with time. At the last day of measurement, there is not difference of potential between the top, the end and the center of cucumber.

KEYWORDS. Agriculture; Amplifier; Biocommunications; Ecology; Electric sensing device; Evaluation of freshness.

INTRODUCTION

The nature around our life has been changing, that is, the air pollution has been serious problem by combustion of fuel for rapid progress of industry. The earth temperature has been rising according to the density of CO_2, and acid rain has began to appear by NO_x or SO_x. These problems, have been global problem, and will henceforth become more and more serious.

In this study, a new method of measuring and valuating plants is suggested at a viewpoint from electrical engineering. It is tried to measure the electric potential on the surface of plants and organism. These potential and its current is very small, because these are depend on ionic activity and organism has very high resistance. Therefore, the instrument measuring electric potential on surface of plant, is required high input impedance and stability to decrease the drift. Usual method of measuring the electric potential of organism has been contacted on

it or pricking a very small and thin
electrode into it. Our purpose is to
measure electric potential without
injuring the object. This new method
basically has not effected on the organism.

As it has high input impedance,
MOSFET is used as a modulator by the
new method that changes a static input
signal into square pulse signal. The
static electric potential is measured by
this new modulating method. And for new
application, the electric potential on
the surface of cucumber is measured by
this method.

NEW DC AMPLIFIER USING MOSFET

When static electric potential is
measured, a DC amplifier is used for
detection and amplification of potential. A characteristics of these DC
amplifier are required as follows;
1. low drift and low offset
2. low noise
3. high input impedance
4. sensitivity and high stability
5. wide frequency band
6. simple circuit component
But it is very difficult for the amplifier to have all above characteristic.
In this study, the amplifier is required
these characteristics above for detecting electric potential on the surface of
the plant.

The amplifier measuring electric
potential on the surface needs DC
amplifier which has high input impedance
and low drift. It is favorable to measure electric potential by the amplifier
using MOSFET on a serious view of characteristic of high impedance a direct

coupling, but in consideration of drift,
it is not favorable. A modulation type
amplifier is superior in drift. The
amplifier using vibrating capacitance
modulation is non-contact modulation
and has good characteristics, though it
has mechanical modulator.

A new modulation method using
MOSFET in input circuit part is proposed
for keeping high input impedance and
low drift. The structure of MOSFET is
shown in Fig.1. The substrate of MOSFET
is usually connected with a source or
the earth, and most MOSFET has already
substrate connected with source when
they ware made. But the substrate is
seen to be another junction gate controlling channel, and it can be used as
the control-gate for channel as like as
junction type FET. Therefore, the substrate is possible to use a second gate
of FET. [1], [2]

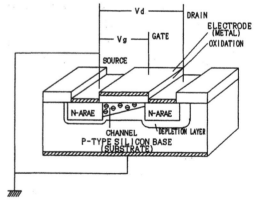

Fig.1 Structure of MOSFET

When negative voltage is supplied
on substrate of n-channel MOSFET, an
invert voltages of p-n-junction between
drain and substrate, and source and
substrate, are increased. And the gap
between these p-n-junction become more
wide growing into the channel, and a
conductance between drain and source

decrease. As the saturation current is controlled by each gate voltage, the output voltage of drain is modified by substrate voltage. At the same time, the gap grows near a drain, channel is pinched off at low voltage of drain is modified by substrate voltage.

When a positive voltage is supplied on substrate of n-channel MOSFET, the gaps between drain and substrate, and substrate and source become thick, but electrons in n-channel between drain and source are absorbed into the substrate. Then the conductance between drain and source increases. Though current begins to flow from substrate to source when a forward voltage (above 0.6V) to p-n-junction on substrate is supplied, the input voltage on gate is also modified by substrate voltage and drain voltage becomes output signal. [3]

As a conclusion, the output voltage of drain is in proportion to the gate voltage and is modified by substrate voltage by giving AC or square wave voltage to the substrate of MOSFET.

THE FREQUENCY RESPONSE
BY CHANGING VOLTAGE
OF SUBSTRATE OF MOSFET

When MOSFET is used as a chopping device, the frequency band is wanted to be wide for amplifier, and then the chopping frequency is also wanted to be more higher. But there is another problem which spike-noises appears by capacitance C_{gd} between gate and drain, and C_{gs} between gate and source when MOSFET is used as a chopping device.

In this circuit, 3SK14 is used for input circuit part, because it is MOS type FET with high input impedance and the terminal electrode of substrate comes out independently from package. In usual chopping type modified amplifiers, transistor or FET are used for switching device, and differential spike-voltage is decreased compensating by a serious and parallel connection of chopper.

The positive voltage is supplied the substrate to decrease the spike-noise. As it is much better than negative voltage, the exciting frequency is decided 3kHz. At this frequency, the wave form is not distorted and the amplifier has wider frequency band. And the exciting voltage is kept constant using zener diode for stability. The results of reformed output voltage modified by positive square voltage is shown in Fig. 2.

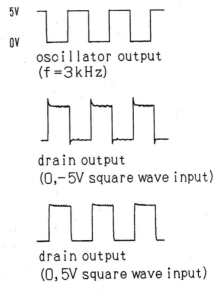

oscillator output
(f = 3kHz)

drain output
(0, −5V square wave input)

drain output
(0, 5V square wave input)

Fig.2 Reformed output voltage

MEASUREMENT OF ELECTRIC POTENTIAL
ON SURFACE OF PLANT

Method of Measuring

The electrode is constructed with detecting electrode and shielding electrode insulating with thin insulator, shown in Fig. 3. When low electric potential on a plant is measured, the electrode is insulated with thin film contacting on the surface of the plant.

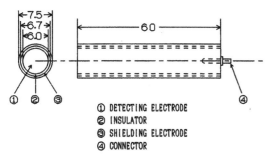

① DETECTING ELECTRODE
② INSULATOR
③ SHIELDING ELECTRODE
④ CONNECTOR

Fig.3 Structure of electrode (Unit: [mm])

In this study, the electric potential on the surface of cucumber is measured because it is easy to get it and to measure electric potential on it. It is measured for 5 days every 24 hours which is decided from decomposition of object. The object has been stored in cool and dark place to neglect the influence of environment. It is measured at every 1 cm along the axial of cucumber on the surface of it. And the distribution of electric potential is shown in Figs. 4, 5, 6. At the same time, distribution of potential is measured at each 120 deg around the axis of cucumber. The instrument is kept in shield box during measurement as this measuring instrument is apt to be considerably influenced by electric field noises, because input impedance is very high.

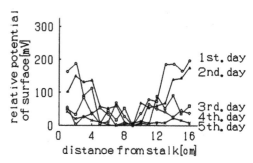

Fig.4 Relative potential of surface on cucumber (around deg.0)

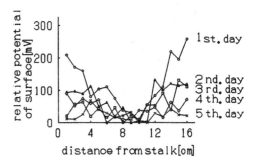

Fig.5 Relative potential of surface on cucumber (around deg.120)

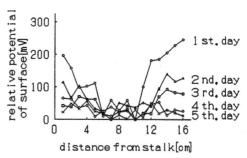

Fig.6 Relative potential of surface on cucumber (around deg.240)

Result

The characteristics of the circuit is shown in Fig. 7, 8. Output voltage is in proportion to input voltage linearly from -0.7 [V] to 0.7 [V] and the drift is very small.

The distribution of relative electric potential on the surface of cucumber are shown in Fig. 4 to 6. And the mean value of around potential of cucumber is shown in Fig. 9, 10. In the measurement, the cucumber is kept at temperature about 7 to 9 Celsius. It shows high potential at the top and the end of the cucumber comparing the center of it. But the distribution of the potential along the axis of cucumber becomes flat with time. At the last day of measurement, there is not difference of potential between the top, the end and the center of cucumber.

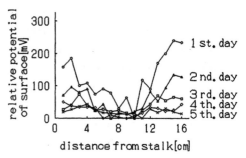

Fig.9 Relative potential of surface on cucumber (average value)

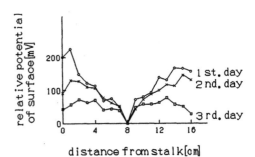

Fig.10 Relative potential of surface on cucumber (average value)

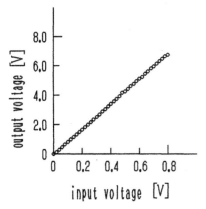

Fig.7 Characteristic of amplifier (positive input)

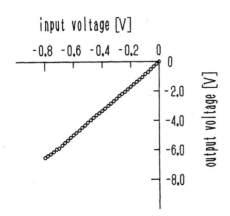

Fig.8 Characteristic of amplifier (negative input)

CONCLUSION

1. It is succeeded to make the modulation type DC amplifier which is able to keep high input impedance without mechanical construction by using MOSFET as the modulating element.

2. The good linearity of the amplifier is obtained for in the input range of -0. 7 [V] to +0. 7 [V].

3. The electric potential on the surface is measured by this amplifier without contact on the surface.

4. The electric potential on the surface of cucumber is measured with electrodes insulated between electrodes and the surface with thin film.

5. The distribution of the electric potential along the axis of cucumber shows high at the top and the end of the cucumber than the center of it.

6. The potential along the axis of cucumber becomes flat with time.

7. There is the correlatively between the deterioration of freshness of fruits or vegetables and the distribution of electric potential on the surface of cucumber.

REFERENCES

(1) Yukio HIRAMOTO, Toshihumi TAKENAKA (1982), Jikken Seibutugakukouza 5 Denkitekikeisokuhou, Maruzenkabushikikaisya

(2) Osamu NISHINO (1969), Jikken Buturigakukouza 5 Denkikeisoku, Kyoritsusyuppan

(3) Yoshiyuki NAGAHASHI (1976), DC anpu no sekkei, CQsyuppansha

(4) P. D. Chapman (1985), A capacitance based ultraprecision spindle error analyser, Butterworth & Co (Publishers) Ltd, pp. 129-137

(5) Toshiharu SHIRAISHI (1985), Measurement of Straightness Error of a Coordinate Measuring Machine by Capacitance Type Sensor

APPLICATION OF IMAGE PROCESSING TO MEASUREMENT OF TOTAL LENGTH IN ROOT SYSTEM

J. Chikushi, S. Yoshida and H. Eguchi

Bioton Institute, Kyushu University 12, Fukuoka 812, Japan

Abstract. Newly developed method of image processing was applied to measurement of the total length in a root system. The image of cucumber root system was taken by a still video camera, and transformed into the thinned root image which was composed of the pixels. The counting operation of respective diagonal and horizontal or vertical connections of pixels was carried out, and the root length was calculated from the counted total numbers of the connections; a reliable correction factor was designed mathematically for diagonal connection. Estimation error was within 2% of actual length, and the correlation coefficient (r) between actual and estimated length was 0.9999 (*P*<0.01). Thus, the present method can be useful for simple and accurate measurement of root length and may help to understand the characteristics of plant root system.

Keywords. Agriculture, Analog-digital conversion, Biology, Computer evaluation, Computer software, Error estimation, Estimation, Image Processing, Reliability, Thinning operation, Root length.

INTRODUCTION

Characteristics of the plant root function closely relate to volume of the root system which can be estimated from total root length (Gardner, 1964). There have been several techniques to measure or estimate root length such as the direct method, the inch counter method, and the line intersect method (Reicosky, Millington and Peters, 1970). In particular the line intersect method developed by Newman (1966) has been widely used in the fields of environmental biology, plant physiology, soil science, and agronomy, *etc.* Furthermore, the method has been improved by using various instruments and devices (Baldwin, Tinker and Mariott, 1971; Barnett and colleagues, 1981; Goubran and Richards, 1979; Marsh, 1971; Nagano and Hara, 1988; Richards and colleagues, 1979; Rowse and Phillips, 1974; Tennant, 1975; Voorhees, Calson and Hallauer, 1980; Wilhelm, Mormann and Newell, 1983).

The present paper deals with root length measurement by image processing in binary image operation, thinning operation, and counting of the connections of picture cells (pixels), and also deals with the application of the method to

measurement of growth of the real root system in cucumber plant.

MATERIALS AND METHODS

Material plant

Cucumber plant (*Cucumis sativus* L. var. Hort. Chojitsu-Ochiai) were grown in a growth chamber at air temperature of 23 °C, relative humidity of 70% and a light intensity of 200 μmol m^{-2} s^{-1} (12h photo period) in hydroponic system. Detached root system was sampled at an interval of 3 days after one leaf stage.

Image processing

A root sample was spreaded in water in order to eliminate the overlapped roots, where the root image was contrasted with black background. Figure 1 shows the image processing system. A still video camera (RC-474, Canon) including the charge-coupled device (CCD) was used for taking the images of the sample with a scale in a floppy disk. Image data were transmitted to the still video recorder (RR-551, Canon) through the floppy disk media. The root image was displayed on a CRT monitor (PC-TV454, NEC) with 3 modes (R, G and B) of 0 to 255 grades in brightness, and was operated to obtain the binary image through a graded filter (split filter) by using the the image processor (Model ED-1382, Edec). The displayed

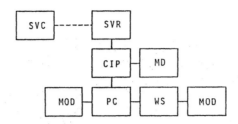

Fig.1. Schematic description of the image processing system:
SVC, Still video camera; SVR, Still video recorder; CIP, Color image processor; PC, Personal computer; MD, Monitor display; WS, Workstation; MOD, Magneto-optical disk.

TABLE 1 *Effects of the Correction of α on Measurement of Thread Length*

Actual length (R) (cm)	$\bar{\alpha}$	Estimate (E1) by $\alpha=\sqrt{2}$ (cm)	$|R-E1|/R$ (%)	Estimate (E2) by $\alpha=\bar{\alpha}$ (cm)	$|R-E2|/R$ (%)
40.2	1.273	42.15	4.85	40.02	0.45
49.4	1.314	51.48	4.21	49.13	0.75
63.9	1.293	66.92	4.72	63.85	0.07
92.2	1.271	95.85	3.95	91.12	1.17
100.7	1.308	104.99	4.26	100.52	0.18
151.6	1.257	156.81	3.43	149.16	1.61
199.7	1.280	209.94	5.12	199.33	0.19

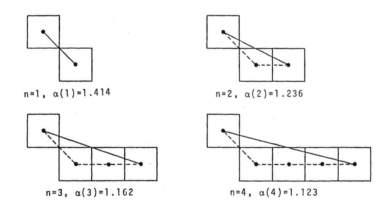

n=1, α(1)=1.414

n=2, α(2)=1.236

n=3, α(3)=1.162

n=4, α(4)=1.123

Fig.2. Schematic diagram of pixel arrangements: n and α(n) are numbers of the pixel and correction factor, respectively.

image consisted of 512 lines x 480 pixels. Binary data of the pixels in one binary image was converted into ASCII data using personal computer (PC9801VX, NEC) for the thinning operation on the basis of the algorithm by Hilditch(Joint System Development Corp.): The root image consisting of some pixels was changed to the lines passing through the centre of the root with one-pixel breadth. The thinning operation was conducted in a workstation (AS3160C, Toshiba) for the high-speed data processing. One-pixel length in the thinned image was evaluated from the interval of the scale marks on the display.

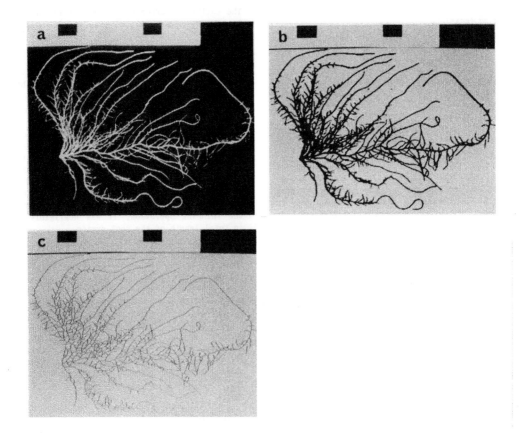

Fig.3. CRT photographs of original image (a), binary image (b), and thinned image (c) of cucumber root system.

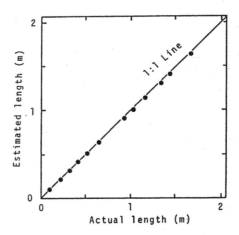

Fig.4. Relationship between actual and estimated thread length.

Counting technique

The total root length in one image can be estimated by using a thinned root image. When the thinned line consists of only vertical or horizontal connections of the pixels, the length of line can be evaluated by the number of pixels multiplied by one-pixel length. On the other hand, when the line declines with 45° to the horizontal, the length can be evaluated by 'the number of pixels - 1' multiplied by $\sqrt{2}$ and one-pixel length, where $\sqrt{2}$ implies the correction factor for estimation length between two pixels connected diagonally. That is, a vertical or horizontal connection becomes shorter than a diagonal one even if the number of pixels is same. So, we counted the total numbers of vertical or horizontal connections H and diagonal connections D respectively, because a root system was composed of various curved lines, i.e. diagonal and vertical or horizontal connections. Thus, total root length was evaluated by using these numbers in this experiment as follows,

$$R = (H + \alpha D)P \qquad (1)$$

where R is the total root length, P is the one-pixel length and α is the correction factor. When $\alpha = \sqrt{2}$ was used for the evaluation of R, the total root length was overestimated by 3-5% of actual length (Table 1). To obtain more accurate estimation, α was corrected further. Figure 2 shows schematic diagrams for correcting of α in various cases of pixel arrangements. The gradient of the actual line of the root is variable, and various arrangements of the pixels were constituted according to degree of the gradient. This pixel arrangement was characterized by numbers n of continuous pixels connected horizontally or vertically. In the case of $n \geq 2$, the length shown in dashed line in Fig.2 is measured to be longer than actual length of the root, and the assumed straight line (solid line) becomes necessarily closer to the actual length. Difference d between these two lines is $(\sqrt{2}+1)-\sqrt{5}$. General d in n pixel connections can be calculated by

$$d(n) = (\sqrt{2} + (n-1)) - \sqrt{(n^2+1)} \qquad (2)$$

The integration of these differences caused distinct deviation from actual value when estimating the total root length. $\sqrt{2}$ in Eq.2 corresponds to α in Eq.(1). When choosing $\alpha(n)$ value to become $d=0$, then general expression of α depending on n is

$$\alpha(n) = \sqrt{(n^2+1)} - (n-1) \qquad (3)$$

From the distribution of frequencies of connected pixels M_n, the weighted mean value $\bar{\alpha}$ of $\alpha(n)$ can be obtained by

$$\bar{\alpha} = \Sigma(\alpha(n)M_n)/\Sigma(M_n) \qquad (4)$$

In this experiment, better estimations were obtained by using $\bar{\alpha}$ in stead of $\alpha = \sqrt{2}$ (Table 1).

The estimation error was within 2% of the actual length.

RESULTS AND DISCUSSION

Figure 3 shows original photograph (a), binary image (b) and thinned image (c) of the root system at 1 leaf stage of cucumber plant. The brightness of the root in the original image varied with the position and the diameter of the root. In the binary image, roots were contracted clearly at even where the brightness was lower. In the thinned image, the tangled roots at the base of the root system were transformed into clearer lines than those in original or binary image. Although partial root image was lost by the thinning operation, it was negligibly small in total length of the root system.

The lengths of white threads were measured for the examination of accuracy in the present method. Figure 4 shows the relationship between actual and estimated values of thread lengths. Plotted marks almost distributed on the 1:1 line and correlation coefficient of the relationship was 0.9999 ($P<0.01$). Table 2 shows the comparison between actual and estimated thread lengths obtained from five samples for each length. The differences between these lengths were remarkably small. Furthermore, both the standard deviation and coefficients of variation were remarkably small. These facts suggest that a total root length can be estimated accurately and reproducibly.

The present method was applied to measurements of real root. Figure 5 shows growths of the root system and leaf expansion in cucumber plants. The root length in a root system was developed exponentially in course of time according to increases in root dry weight and leaf area per plant.

CONCLUSION

The newly developed method by using image processing provided fine results for measurement of thread or root length. There were two advantages in the method for measuring root length; 1) Randomness and direction of root images did not affect the measured root length, and 2) the image processing system available in commerce was able to be used for the measurement without any special instrument. Thus, the present method can be useful for the simple and accurate measurement of root length in various cases and for understanding the characteristics of plant root system.

ACKNOWLEDGEMENT

We thank Joint System Development Corporation for generously allowing us the use of the soft ware package of "SPIDER".

410

TABLE 2 *Accuracy of Thread-length Measurements by the Present Method*

Actual length (cm)	Estimated length (cm)	Standard deviation (cm)	Coefficient of variation (%)
49.4	49.31	0.65	1.31
100.7	100.81	0.37	0.36
151.6	150.20	1.31	0.87
199.7	198.18	0.83	0.42

REFERENCES

Baldwin J.P., P.B. Tinker, and F.H.C. Marriott (1971) The measurement of length and distribution of onion roots in the field and laboratory. *J. appl. Ecol. 8*, 543-554.

Barnett C. E.,R.A. White , A.M. Petrovic, and G.C. Good (1981) An automated apparatus for measuring root length. *Hort Sci. 2*, 140-144.

Gardner W.R. (1964) Relation of root distribution to water uptake and availability. *Agron. J. 56*, 41-45.

Goubran F.H., and D. Richards (1979) The estimation of root length in samples and subsamples. *Plant Soil 52*, 77-83.

Joint System Development Corp. (1987) *SPIDER (Subroutine Package for Image Data Enhancement and Recognition) User's Manual.* Joint System Development Corp., Tokyo. pp. 500-501.

Marsh B.a'B. (1971) Measurement of length in random arrangements of lines (Short communication) *J. appl. Ecol. 8*, 265-267.

Nagano T., and M. Hara (1988) An Application of computer graphics for root length measurement. *J. Agr. Met. 44*, 1-6 (in Japaneese with English summary).

Newman E.I. (1966) A method for estimating root length. *Agron. J. 62*, 451-453.

Reicosky D.C., R.J. Millington, and D.B. Peters (1970) A comparison of three methods of estimating root length. *Agron. J. 62*, 451-453.

Richards D.F.H., W.N. Goubran , W.N. Garwoli, and M.W. Daly (1979) A machine for determining root length. *Plant Soil 52*, 69-76.

Rowse H.R., and D.A. Phillips (1974) An instrument for estimating the total length of root in a sample. *J. appl. Ecol. 11*, 309-314.

Tennant D. (1975) A test of modified line intersect method of estimating root length. *J. Ecol. 63*, 995-1101.

Voorhees W.B., V.A. Calson, and E.A. Hallauer (1980) Root length measurement with a computer-controlled digital scanning microdensiometer. *Agron. J. 72*, 847-851.

Wilhelm W.W., J.M. Mormann, and R.L. Newell (1983) Semi-automated X-Y-plotter-based method for measuring root lengths. *Agron. J. 75*, 149-152.

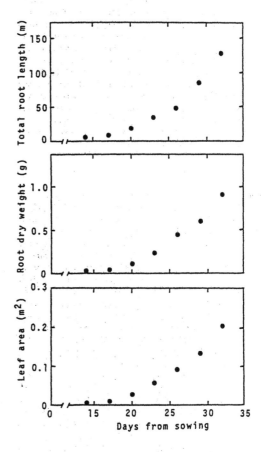

Fig.5. Growths of root length, dry weight of root and leaf area in cucumber plants.

Copyright © IFAC Mathematical and Control Applications
in Agriculture and Horticulture, Matsuyama, Japan 1991

IMPROVEMENT OF TOMATO FRUIT PRODUCTION BY NUTRIENT SUPPLY MANAGEMENT IN GREENHOUSES

H. Nonami, T. Fukuyama, Cho Ill Hwan and Y. Hashimoto

Dept. of Biomechanical Systems, Ehime University, Tarumi, Matsuyama 790, Japan

Abstract. In order to improve the quality of tomato fruit production by avoiding
blossom-end rot, the relationship between the occurrence of blossom-end rot of tomato
fruits and the management of nutrient solution supply in the nutrient-film technique
(NFT) was studied, using a tomato cultivar (Lycopersicon esculentum Mill. cv. Momotaro)
which is sensitive to blossom-end rot. When electric conductivity (EC) of the nutrient
solution was increased from 0.10 $S \cdot m^{-1}$ to 0.40 $S \cdot m^{-1}$, blossom-end rot occurred in all
fruit clusters. Because Ca^{2+} ion concentration in the xylem sap of plants grown at
0.40 $S \cdot m^{-1}$ of EC was much higher than that of the control plants whose fruits did not
suffer from blossom-end rot, Ca deficiency did not seem to be a direct cause of
blossom-end rot in fruits. When plants were subjected to water stress by withholding
nutrient solution supply, photosynthesis rates of plants which had blossom-end rot
became much lower than those of the control plants, although both plants had similar
photosynthetic ability prior to the water stress. Because increases in sugar content
in tomato fruits occurred due to high concentration of the nutrient solution in
combination with water withdrawal, it is possible to improve the quality of tomato
fruits by increasing concentration of the nutrient solution and supplying the nutrient
solution intermittently in NFT.

Keywords. Agriculture; greenhouse; tomato cultivation; nutrient-film technique;
biology.

INTRODUCTION

The management of nutrient solution supply in
hydroponic culture influences growth of crops and
the quality of products. Because hydroponic
culture is known to be a labor-saving culture
method in greenhouses, the practice of hydroponic
culture has been increasing in recent years.
Among the hydroponic culture systems,
nutrient-film technique (NFT) is one of the most-
used method. Because environmental conditions
around roots of crops such as aeration, nutrient
concentrations, nutrient compositions and
temperature can be controlled easily by using
NFT, NFT is used to control growth and production
of horticultural crops in greenhouses, and thus,
NFT is suited for greenhouse automation.
Although NFT allows for flexibility in
controlling the nutrient solution supply,
nutrient supply methods have not been studied
well in terms of the physiological functions of
crops grown in NFT. Because the nutrient supply
is related to water stress and salinity stress in
plants, the way in which the nutrient solution is
supplied to plants determines growth of crops and
quality of products. In order to relate the
nutrient supply methods to growth and fruit
production, we have cultivated tomato plants in
NFT, and have studied root growth and mechanisms
of occurrence of blossom-end rot of tomato
fruits. Blossom-end rot is a widespread
physiological disorder that affects tomato
fruits. This disorder is reported in virtually
every tomato-producing area of the world and
causes serious losses of marketable fruit. This
disorder is commonly attributed to an imbalance
of Ca metabolism in plants. However, detailed
mechanisms of the development of blossom-end rot
in NFT-grown tomato plants have not been

understood. Thus, in the present study, we
attempted to characterize the physiological
properties of tomato plants which yielded fruits
having blossom-end rot. To do so, we grew a
variety of tomato plant which is susceptible to
blossom-end rot under various nutrient supply
methods.

MATERIALS AND METHODS

Plant Material

Tomato (Lycopersicon esculentum Mill. cv.
Momotaro) seeds were sown in well watered
vermiculite on August 11, 1990. Seedlings were
transplanted to NFT beds on September 22, 1990.
Tomato plants were cultivated at either the
standard concentration or elevated concentration
of nutrient solution after anthesis. When the
5th fruit cluster was developed, the upper part
of the stem at the 3rd node from the 5th fruit
cluster was excised in order to enhance fruit
development. Photosynthesis and water status of
a leaf adjacent to the 4th fruit cluster were
measured from December 18 to December 22, 1990.
In order to characterize physiological properties
under water stress, the nutrient solution supply
was stopped at 7 o'clock a.m. on December 19,
1990.

Greenhouse Cultivation System

Figure 1 shows a schematic diagram of the
computer control system of the greenhouse used in
the present study. The temperature of the
greenhouse was maintained by means of a heat pump

413

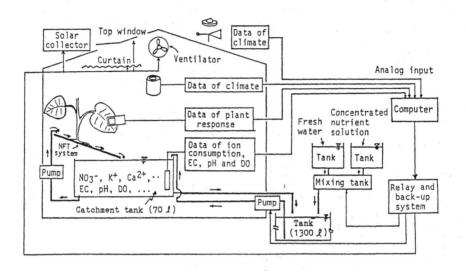

Fig. 1.   Schematic diagram of a tomato cultivation system using NFT.

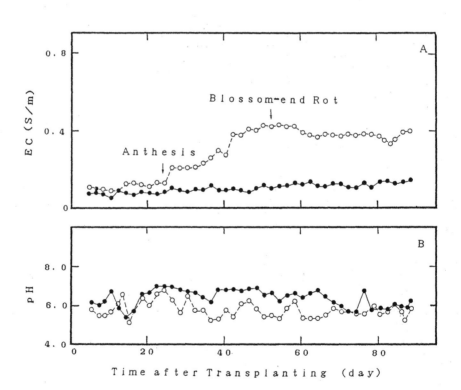

Fig. 2.   Changes in EC (A) and pH (B) in catchment tanks when EC of the nutrient
solution was kept at 0.10 S·m⁻¹ (closed circles) as the control or elevated
to 0.40 S·m⁻¹ after anthesis (open circles) as the treatment.   Time of
anthesis and occurrence of blossom-end rot were shown by arrows (A).

414

(Daikin Heat Pump (Air Cooled Chiller) UWY10J, Daikin Industrial Co.). However, when air temperature in the greenhouse exceeded 30 ℃, cooling was accomplished by a computer-regulated ventilation system. The air temperature in the greenhouse was maintained higher than 20 ℃ throughout the experiment.

Nutrient Supply Systems

The nutrient solution was automatically prepared by mixing fresh water and concentrated nutrient solution, which was a mixture of Otsuka Hydroponic Fertilizer No. 1 and No. 2 at the ratio of 3:2. Nutrient solution with electric conductivity (EC) of 0.10 $S \cdot m^{-1}$ was the control (i.e. the standard nutrient solution concentration for tomato cultivation). When EC = 0.10 $S \cdot m^{-1}$, the concentration of each ion expressed in $mol \cdot m^{-3}$ was as follows; 0.33 NH_4^+, 4.0 NO_3^-, 0.27 PO_4^{3-}, 4.0 K^+, 0.14 Mg^{2+}, 3.5 Ca^{2+}, 0.62 SO_4^{2-}, 0.014 Fe^{3+}, 0.009 Mn^{2+}. The EC of the nutrient solution was adjusted as shown in Fig. 2. After anthesis, the EC was elevated gradually to 0.40 $S \cdot m^{-1}$ in the treated plants. Because blossom-end rot was observed when the EC was elevated to 0.40 $S \cdot m^{-1}$, the EC was maintained at 0.40 $S \cdot m^{-1}$ until the end of the experiment. Through the entire cultivation period, pH was adjusted to 6.0 automatically by adding either H_2SO_4 or $Ca(OH)_2$ solution.

Physiological Measurements

Net photosynthesis rates were measured in a leaf adjacent to the 4th fruit cluster with a small assimilation cuvette and CO_2 analyzer (LCA, ADC Co. Ltd., Hoddesdon, Herts, U.K.).

From the same leaf, a leaf disc was excised for measurements of water status with isopiestic psychrometry (Boyer and Knipling, 1965). A thermocouple chamber was coated with melted and resolidified petrolatum (Boyer, 1967) and loaded with a leaf disc immediately after excision. The water potential was measured by the isopiestic technique (Boyer and Knipling, 1965). This measurement was immediately followed by a measurement of osmotic potential in the same tissue using the same technique after freezing at -70 ℃ and thawing (Ehlig, 1962). Turgor was calculated by subtracting osmotic potential from water potential (Nonami et al., 1987; Nonami and Schulze, 1989).

Growth rates of roots were measured with computer-aided image analysis. The image of a root was fed into a color image unit (Sharp Co.) attached to a computer (X68000pro, Sharp Co.) through a video camera system (TK-910, JVC) with a close-distance focusing lens (Micro-Nikkor, Nikon). Roots and the lens were enclosed in a green plastic container so that growth of roots was not disturbed by illumination necessary to acquire the image of the root with the camera. The green plastic had maximum transmission at 525 nm and negligible transmission below 475 nm and above 575 nm. In order to maximize the image contrast, black was chosen as the color of the plastic sheet used in NFT cultivation so that the image of the root could be acquired with the camera under low illumination. An outline of the root image was identifiable by differences in brightness and hue between the root and NFT sheet, and was defined on the computer monitor with pixels (picture-elements). The outline of

the root image was determined every 3 seconds and the movement of the root cap was memorized as a function of time. Growth rates of the root were calculated by dividing the distance of the root cap displacement by the time interval during the measurement.

When the nutrient supply was stopped on December 19, 1990, a lower stem of a tomato plant was excised and the exudate from the stump was collected in a vial. Concentration of Ca^{2+} ion in the xylem sap and the nutrient solution in the catchment tank was measured with an ion meter (Toshiba Model LQ201).

Sugar content in fruits were evaluated by using a Brix meter (N-20, Brix 0-20%, Atago). Immediately after excising a fruit in half, juice from the center of the fruit was squeezed onto the reflector of the Brix meter. Sugar content was expressed in Brix %.

RESULTS

Effect of Salinity Stress

When concentration of the nutrient solution was increased from EC = 0.10 $S \cdot m^{-1}$ to 0.30 $S \cdot m^{-1}$, the growth of the root was not affected (Fig. 3). However, sudden increases in the concentration from EC = 0.10 $S \cdot m^{-1}$ to 0.60 $S \cdot m^{-1}$ caused inhibition of root growth (Fig. 3). In the present study, the concentration of the nutrient solution was increased gradually as shown in Fig. 2 so that the effects of salinity stress on root growth was avoided.

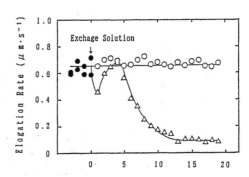

Fig. 3. Elongation rates of roots of plants when the concentration of the nutrient solution was increased from EC=0.10 $S \cdot m^{-1}$ (closed circles) to EC=0.30 $S \cdot m^{-1}$ (open circles) or to EC=0.60 $S \cdot m^{-1}$ (open triangles).

Occurrence of Blossom-end Rot

Although the effect of salinity stress was minimized during tomato cultivation, blossom-end rot occurred in fruits of plants grown in the elevated nutrient solution concentration (see EC= 0.40 $S \cdot m^{-1}$ in TABLE 1). The upper fruit cluster formed higher rates of fruit having blossom-end rot (TABLE 1).

In order to know whether Ca^{2+} ion was absorbed sufficiently by plants having blossom-end rot,

TABLE 1 Percentage of Occurrence of Blossom-end
         Rot in Each Fruit Cluster

Fruit Cluster	EC=0.1 (S/m)	EC=0.4 (S/m)
1st	0 (%)	9.8 (%)
2nd	0 (%)	3.8 (%)
3rd	0 (%)	14.5 (%)
4th	0 (%)	14.1 (%)
5th	0 (%)	32.0 (%)
Total	0 (%)	11.9 (%)

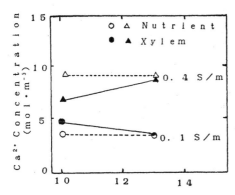

Fig. 4. Ca^{2+} ion concentrations in the nutrient
solution (open symbols) and the xylem
sap (closed symbols) of plants grown in
EC = 0.10 S·m^{-1} (triangles) and EC =
0.40 S·m^{-1} (circles) of the nutrient
solution.

Ca^{2+} ion concentration in the xylem sap was
measured. The Ca^{2+} ion concentration in the
xylem sap was almost similar to that of the
nutrient solution for both the control and
treated plants (Fig. 4). Thus, plants having
blossom-end rot contained much higher Ca^{2+} ion
concentration than the control plants. It is
evident that limitation of absorption ability for
Ca^{2+} ion through roots did not cause the
occurrence of blossom-end rot in fruits in the
present study.

Water Status and Photosynthesis

In order to characterize physiological properties
of plants suffering from blossom-end rot disease,
water stress was applied to both the control and
treated plants simultaneously, and water status
and photosynthesis were measured. Before
applying water stress, both control and treated
plants had similar photosynthesis rates (data is
not shown). After withdrawal of water supply,
both control and treated plants had similar
photosynthesis rates until 12:00 (Fig. 5B), but
plants suffering blossom-end rot began to
decrease photosynthesis rates (Fig. 5B).
Although plants suffering blossom-end rot
exhibited decreased photosynthetic ability, their
leaves had higher turgor than leaves of the
control plants (Fig. 5D), suggesting that plants
suffering blossom-end rot were better able to

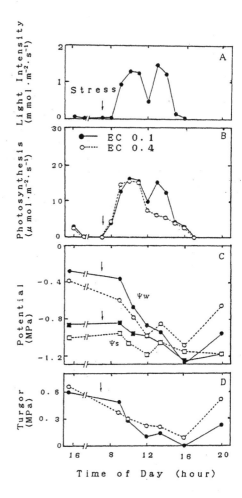

Fig. 5. Diurnal changes in light intensity (A),
photosynthesis rates (B), water (Ψw)
and osmotic (Ψs) potentials (C), turgor
(D) of leaves of plants grown in EC =
0.10 S·m^{-1} (closed symbols) and EC =
0.40 S·m^{-1} (open symbols) of the
nutrient solution after the water stress
was applied.

osmoregulate under water stress. Osmotic
potential of leaves of plants suffering blossom-
end rot was significantly lower than that of the
control plants (Fig. 5C), indicating that more
osmotically active substances were accumulated in
cells of plants suffering blossom-end rot. When
metabolic substances accumulated in too high
concentrations, enzymatic activity may decline
and such inhibitory effects of solute
accumulation may be related to decrease in
photosynthesis rates in plants suffering
blossom-end rot.

Fruit Quality

Fruit quality was evaluated by a Brix meter,
which is a good estimate of soluble sugar in a
fruit. Fruits having blossom-end rot were

416

excluded from this evaluation. Fruits of plants grown in the elevated-nutrient solution concentration had higher sugar content than fruits of the control plants (Fig. 6). After applying water stress by withholding the nutrient solution supply, sugar content of fruits increased in both control and treated plants (Fig. 6). Although excessively increasing the concentration of the nutrient solution causes blossom-end rot in tomato fruits, it is possible to say that supply of appropriately elevated concentration of the nutrient solution in NFT can make tomato fruits sweeter and increase the commercial values of fruits. Furthermore, because water withdrawal caused an increase in sugar content of fruits, it is possible to say that an intermittent nutrient supply may increase water stress on tomato plants and increase sugar contents of fruits, resulting in improved fruit quality.

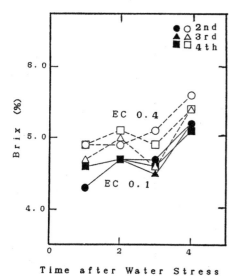

Fig. 6. Sugar content of fruits measured with the Brix meter in the 2nd fruit cluster (circles), the 3rd fruit cluster (triangles) and the 4th fruit cluster (squares) of plants grown in EC = 0.10 S·m⁻¹ (closed symbols) and EC = 0.40 S·m⁻¹ (open symbols) of the nutrient solution after the water stress was applied.

DISCUSSION

It is evident that how nutrient solution is supplied determines the quality of crop produced in the NFT cultivation. A high concentration of the nutrient solution increased the sugar content of tomato fruits, but an excessively high concentration of the nutrient solution caused the blossom-end rot in fruits. The application of water stress by withholding the nutrient solution supply also increased the sugar content of tomato fruits. These observations suggest that intermittent supplies of the nutrient solution in combination with appropriate concentration of the nutrient solution may improve the quality of

tomato fruits under NFT cultivation.

An increase in the nutrient solution concentration causes salinity stress in plants, and intermittent supplies of nutrient solution cause drying stress to plants. Both salinity stress and drying stress cause plants to have water stress. To endure such water stresses, plants osmoregulate so that the cell volume of plants can be maintained (Morgan, 1984). Meyer and Boyer (1981) studied mechanisms of osmoregulation under low water potentials, and found that osmotically active solutes accumulate in order to maintain cell turgor. Osmotically active solutes accumulated during water stress were largely accounted for by the accumulation of sugars and amino acids (Meyer and Boyer, 1981). If water stress is applied appropriately, plants will accumulate sugars due to osmoregulation.

Fukuyama (1990) compared cultivation methods of muskmelon plants grown in greenhouses. When NFT, deep flow technique, rockwool culture, kuntan culture and soil culture were used to grow muskmelon plants, muskmelon fruits produced in NFT were the most commercially-valued (Fukuyama, 1990). When muskmelon plants were cultivated in NFT, intermittent supplies of the nutrient solution were practiced in order to increase the sugar contents of the fruits (Fukuyama, 1990). Because, in the present study, increases in sugar contents of tomato fruits were observed after the nutrient solution supply was stopped, it is also probable that intermittent nutrient solution supplies improves quality of tomato fruits cultivated in NFT.

When the concentration of the nutrient solution was increased too much, blossom-end rot was formed in tomato plants in the present study. It is believed that blossom-end rot is caused by calcium deficiency in tomato plants (Evans and Troxler, 1953; Maynard et al., 1957; Van Goor, 1968; Ward, 1973; Wiersum, 1966). However, in the present study, plants which formed blossom-end rot in fruits had a higher content of Ca²⁺ ion in the xylem sap than the control plants, and thus, it may be possible that blossom-end rot in tomato fruits can be caused by factors other than Ca deficiency.

Evans and Troxler (1953) found Ca concentrations in tomato fruits with and without blossom-end rot to be 0.10-0.13% and 0.17%, respectively. Maynard et al. (1957) found 0.04 and 0.07% Ca for blossom-end rot and normal fruit, respectively. Van Goor (1968) reported 0.03-0.04% Ca for blossom-end rot fruits and 0.09% Ca for normal fruits. Ward (1973) found a normal fruit Ca value to be about 0.08%, whereas that of blossom-end rot to be from 0.02 to 0.07%. Wiersum (1966) gave concentrations of 0.08 and 0.18% Ca for blossom-end rot and normal fruit, respectively. Spurr (1959) did not find a difference between Ca content of blossom-end rot and normal mature fruit. Cerda et al. (1979) reported that the Ca concentration of mature normal fruit varied from 0.039 to 0.079% compared with 0.028 to 0.043% for mature blossom-end rot fruits. The data presented in the published research papers show that there are discrepancies in reported values for Ca in fruit with and without blossom-end rot. Thus, it is very difficult to conclude that Ca deficiency is the only cause of the occurrence of blossom-end rot in tomato fruits.

When plants with blossom-end rot fruits were

subjected to water stress, the photosynthetic ability became much lower than in the control plants, although osmoregulation was active in plants with blossom-end rot fruits (Fig. 5). Because turgor in leaves of plants with blossom-end rot fruits was high, and because photosynthesis was partially active, stomatal inhibition did not seem to cause the partial inhibition of photosynthesis. The cause may be metabolic inhibition in chloroplasts. Rao et al. (1987) grew sunflower plants hydroponically in different concentrations of the nutrient solution, and studied inhibition of photosynthesis under water stress. They found that when plants grown in high concentrations of the nutrient solution were subjected to water stress, concentrations of ions increased too high a level for normal enzymatic activity of ATPase in chloroplasts, resulting in earlier inhibition of photosynthesis compared with plants grown in low nutrient concentrations (Rao et al., 1987). Thus, it is very likely that tomato plants with blossom-end rot had low ATPase activity in cells when the plants were subjected to water stress. Because ATPase is involved in transport of materials through membranes (Serrano, 1989) and cell wall synthesis (Cassab and Varner, 1988; Hayashi, 1989), it is possible that blossom-end rot is caused by enzymatic inhibition related with ATPase due to high concentrations of soluble matters in cells.

CONCLUSION

Water stress induces plants to osmoregulate, resulting in accumulation of sugars in cells. In order to improve the quality of tomato fruits in NFT cultivation by taking advantage of the physiological function of osmoregulation, water stress was applied to tomato plants by increasing concentrations of the nutrient solution and withholding the nutrient supply. When concentration of the nutrient solution was increased too much, blossom-end rot occurred in tomato fruits, resulting in the loss of the commercial value of fruits. When the nutrient solution supply was withheld, sugar accumulation was observed in tomato fruits. To prevent the occurrence of blossom-end rot in tomato fruits, EC of the nutrient solution should not exceed a certain critical level, i.e., EC = 0.40 S·m⁻¹ for the tomato cultivar used in the present study. Intermittent supply of the nutrient solution in NFT may also improve the sugar content of tomato fruits.

REFERENCES

Boyer, J.S. (1967) Leaf water potentials measured with a pressure chamber. Plant Physiol., 42, 133-137.

Boyer, J.S., and E.B. Knipling (1965) Isopiestic technique for measuring leaf water potentials with a thermocouple psychrometer. Proc. Natl. Acad. Sci., 54, 1044-1051.

Cassab, G.I., and J.E. Varner (1988) Cell wall proteins. Annu. Rev. Plant Physiol. Plant Mol. Biol., 39, 321-353.

Cerda, A., F.T. Bingham, and C.K. Labanauskas (1979) Blossom-end rot of tomato fruit as influenced by osmotic potential and phosphorus concentrations of nutrient solution media. J. Amer. Soc. Hort. Sci., 104, 236-239.

Ehlig, C.F. (1962) Measurement of energy status of water in plants with a thermocouple psychrometer. Plant Physiol., 37, 288-290

Evans, H.J., and R.V. Troxler (1953) Relation of calcium nutrition to the incidence of blossom-end rot in tomatoes. Proc. Amer. Soc. Hort. Sci., 61, 346-352.

Fukuyama, T. (1990) Sugar contents, amino acid contents, and taste tests in muskmelon fruits cultivated under NFT, DFT, rockwool culture, kuntan culture and soil culture. Environ. Control Biol., 28, 61-69.

Hayashi, T. (1989) Xyloglucans in the primary cell wall. Annu. Rev. Plant Physiol. Plant Mol. Biol., 40, 139-168.

Maynard, D.N., W.S. Barham, and C.L. McCombs (1957) The effect of calcium nutrition of tomatoes as related to the incidence and severity of blossom-end rot. Proc. Amer. Soc. Hort. Sci., 69, 318-322.

Meyer, R.F., and J.S. Boyer (1981) Osmoregulation, solute distribution, and growth in soybean seedlings having low water potentials. Planta, 151, 482-489.

Morgan, J.M. (1984) Osmoregulation and water stress in higher plants. Annu. Rev. Plant Physiol., 35, 299-319.

Nonami, H., J.S. Boyer, and E. Steudle (1987) Pressure probe and isopiestic psychrometer measure similar turgor. Plant Physiol., 83, 592-595.

Nonami, H., and E.-D. Schulze (1989) Cell water potential, osmotic potential, and turgor in the epidermis and mesophyll of transpiring leaves: Combined measurements with the cell pressure probe and nanoliter osmometer. Planta, 177, 35-46.

Rao, I.M., R.E. Sharp, and J.S. Boyer (1987) Leaf magnesium alters photosynthetic response to low water potentials in sunflower. Plant Physiol., 84, 1214-1219.

Serrano, R. (1989) Structure and function of plasma membrane ATPase. Annu. Rev. Plant Physiol. Plant Mol. Biol., 40, 61-94.

Spurr, A.R. (1959) Anatomical aspects of blossom-end rot in the tomato with special reference to calcium nutrition. Hilgardia, 28, 269-295.

Van Goor, B.J. (1968) The role of Ca and cell permeability in the disease blossom-end rot of tomatoes. Physiol. Plant., 21, 1110-1121.

Ward, G.M. (1973) Causes of blossom-end rot of tomatoes based on tissue analysis. Can. J. Plant Sci., 53, 169-174.

Wiersum, L.K. (1966) Calcium content of fruits and storage tissues in relation to the mode of water supply. Acta Bot. Neerlandica, 15, 406-418.

DEVELOPMENT OF ION SENSORS FOR HYDROPONICS

S. Yamashita, K. Baba, S. Ito and Y Asano

DKK Corporation, 4-13-14 Kichijojikitamachi, Musasinoshi, Tokyo 180, Japan

Abstract. The ion sensors for hydroponics have been developed. The sensor
is the combination type. It consists of the sensing tip, sensor protection
and reference electrode. The basic performance of the sensor for NO_3^-, K^+,
Ca^{2+} is described in detail. As a example, nitrate ion sensor provides the
linear response in the range from 6 to 6,000 mg/l of nitrate ion with the
slope of 55 mV/decade. The results obtained with this sensor gave
reasonably agreement with those by conventional method.

Keywords. process control; monitoring; ion; sensors; hydroponics.

INTRODUCTION

The instrumentation technology in vegetable
production was introduced fairly early days
after the world war II at agricultural
experiment stations and other laboratories.
Studies in these facilities were
accumulated to develop instrumentation
technology attracting public attention
these days. However, the practical
application of this technology is just
started; i.e., there is a great gap between
laboratory level and field level. There are
many hurdles to be overcome.

We are going to explain the direction of
practical application and some topics of
the component sensor for vegetable plant.
Moisture of hydroponics is separately
discussed in an independent section. Since
instrumentation technology is directly
related with control and information
acquisition, it is mainly discussed from
the view point of the on line sensors.

APPLICATION CONDITIONS OF COMPONENT SENSOR IN VEGETABLE PLANT

Compared with ordinary sensors of
temperature, pressure and flow rate, the
component sensor has generally a
complicated and hard-to-handle
constructions. Therefore, to raise its
applicability to a practicable level in a
vegetable plant, the following requirements
should be considered.

(1) Sturdy design durable for the use by
non-skilled person in vegetable plant.
(2) Long term accuracy with maintenance to
some extent.
(3) Sensitivity enough for sensing
biological information.
(4) Low expense.
(5) Capability for small scale production of
many types till the use of the sensor
spreads.

Since we are in the high technology age,
measurement of the object component is
realized with a rather high accuracy. But
the cost reduction to a level on which
farmers can make measurement easily.
Actually, the following steps are thought,
and each step is modified in its own way.

Step1: Experimental plant such as
agricultural station and university.
Step2: Experimental plant for near actual
application.
Step3: Actual plant managed by farmers or
businesses.

When step 3 is reached, the sensor must be
refined so that requirements (1) through
(4) are satisfied. Two types of vegetable
plant available; soil-based plant (advanced
house culture) and hydroponics-based plant.
Component sensors will be gradually
separated into two categories. That is,
however, a future problem.

MAJOR COMPONENT SENSORS USED IN VEGETABLE PLANT

Of course, necessary sensors in this field
are roughly classified into those for
controlling meteorological environment and
those for controlling culture medium. Major
items and their measurement principles are
as tabulated in Table 1.

Many of these are, except some part,
experienced in process industry. These
sensors are now under modification for
agricultural purpose. But, as described
before, agriculture has its own conditions,
and more contrivances must be accumulated
to assort a complete series.

Viewing from general tendency, each
component sensor for vegetable plant must
be furnished with the following functions,
and the present task is to realize the

Table 1 Major Component Sensors for Vegetable Plant and their Working Principles

Use	Item	Principle	Measurement range
Culture liquid/ soil	EC (Conductivity)		0.5～100 mS/cm
	pH	Glass electrode method	5～9 pH
	DO (Dissolved Oxygen)	Galvanic/ Polarographic method	5～20 ppm
	ORP (Oxidation/ Reduction Potential)	Platinum electrode method	±500 mV
	Various ions	Ion Sensor/ Colorimetry	
Meteorology	Ethylene	Infrared absorption method	0～5000 ppm
	CO2	Gas chromato- graphy	―

requirement.

(1) The sensing tip of the component sensor is desirably disposable as a rule for the convenience of user.
(2) If possible, it is preferable that no reagent supplying and no troublesome calibration are required.
(3) When the sensor is inserted into the culture liquid or the soil, some contrivance is required to prevent contamination of sensor surface.
(4) The signal transmitter should be of compact construction, and resistant to high temperature and high humidity environment.
(5) It is recommended that a simple indication related with measured value be provided as the rough standard of the field work.

ION SENSORS AND ION MONITOR IN HYDROPONICS

In hydroponics culture, particularly when the hydroponic liquid is circulated, concentration control of the liquid by the component sensor is essential. Among general items of hydroponic liquid control, basic ones are pH, EC, NO_3^-, K^+, and Ca^{2+}. And dissolved oxygen (DO), PO_4^{3-}, Mg^{2+}, etc. may be added. In any case, the ion sensor is the main stream among component sensors used in hydroponics. The following is the brief description of the sensors.

Ion Sensor in Hydroponics

The ion sensor using an electrode have been put into practice by sensor makers over the past 25 years. Sensors reached to 31 species for selective measurement of various ions have been developed. They are used, like pH measurement with glass electrode, in combination with a reference electrode. Sensors for agriculture use are as shown in Table 2.

These are mostly called liquid membrane type sensors. An organic compound selectively reacts with ions is dissolved in a water-insoluble solvent and mixed with PVC, then shaped in a plate to produce the sensor. When this sensor is immersed in the sample water, the organic compound is dissolved by a little amount to react with specific ions causing a potential proportional to the concentration. The weak point of the ion sensor is the interfering components as shown in the Table 2. The presence of such a interfering component causes an error which seems to exert little effect on actual hydroponic liquid control. Since lead ion sensor is sensitive to PO_4^{3-}, this ion sensor has a possibility to be used as the PO_4^{3-} sensor. Also since the cation sensor is sensitive to Mg^{2+} and Ca^{2+}, Mg^{2+} ion concentration is obtained by subtracting the separately measured Ca^{2+} from Mg^{2+} and Ca^{2+} concentration. Fig. 1 shows the typical combination type NO_3^- sensor for hydroponics. And Fig. 2 shows the comparison of the data measured by sensor method and by conventional method.

HYDROPONIC LIQUID MONITOR IN HYDROPONICS

The noteworthy point of the component

Table 2 Typical Ion Sensors used in Hydroponics

Object	Measurement range	Measured pH	Interfering component
NO_3^-	$60,000 \sim 6$ mg/l	7	Hg^{2+} Br^-, NO_2^-, CN^-
K^+	$4,000 \sim 0.4$ mg/l	6	Cs^+
Ca^{2+}	$4,000 \sim 0.4$ mg/l	6	Zn^{2+}, Pb^{2+}, Sr^{2+}

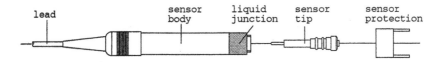

lead   sensor body   liquid junction   sensor tip   sensor protection

Fig. 1 The typical combination type NO_3^- sensor for hydroponics.

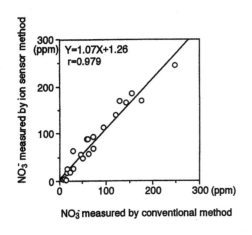

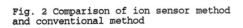

NO_3^- measured by conventional method

Y=1.07X+1.26
r=0.979

Fig. 2 Comparison of ion sensor method and conventional method

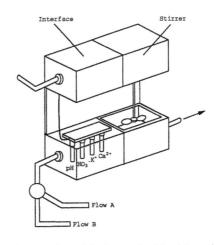

Fig. 3 A example of hydroponic liquid monitor

sensor for hydroponics is the trouble due
to microbes adhered on the sensor surface.
The trouble causes delayed response and
errors. Though the cleaner was used on
trial bases, it was expensive. Thus
occasional cleaning will be the most
practicable. Fig. 3 shows the hydroponic
monitor which collects component sensors in
a compartment for the convenience of
maintenance. Each electrode can be removed
for cleaning in a single action, and
several flow paths can be switched for
controlling several points with one set of
the monitor.

OTHER COMPONENT SENSORS

In recent vegetable plant, kerosine is
burnt for temperature keeping, and for
promoting the crop growth under CO_2 rich
atmosphere, and inexpensive infrared
absorption type CO_2 analyzer came into the
scene. It has been found that ethylene gas
is related with the life of vegetable. The
ethylene sensor utilizing gas chromatograph
and chemiluminescence will be developed
soon.
Further, if physiological measurement
and growth measurement of vegetable become
possible, growth conditions of vegetable
can be adequately performed. These are
important future tasks. We hope the sensor
development goes to a sensor that detects
the disease factor, the most hard problem
in vegetable plant.

SUMMARY

In the near future, more ion sensors for
agriculture will be developed and it is
expected for them to prompt efficiency in
hydroponic culture.

NON DISPERSIVE INFRARED GAS ANALYZER
FOR CO₂ CONTROL

M. Nakano

*Process Control Instruments Dept., Fuji Electric Co., Ltd., Tokyo Factory,
No.1 Fuji-machi Hino-city, Tokyo, Japan*

Abstract. A non-dispersive infrared gas analyzer with single
beam method has been developed. It acculately measures the
concentration of CO₂ gas in the air, and simultaneously outputs
an analog signal and upper/lower alarm for control. This instrument is suitable for use as a CO₂ meter in green houses, building
ventilation systems,CA(Controled Atmosphere) storage facilities,
and so on.

Keywords. Agriculture; air conditioning; Gas analyzer; infrared;
carbon dioxied; energy saving.

INTRODUCTION

Non dispersive infrared (NDIR) gas analyzers have been used to measure various
gas components mainly in process control
system and air polution monitoring system.
And the performance,such as sensitivity,
selectivity or stability, has been
improved remarkably.

On the other hand, a compact and easy
operation analyzer has been required for
new applications.

This paper introduces a CO₂ analyzer for
such applications. The analyzer has the
following features;
1) High stability and easy maintenance
2) Control with upper/lower limit alarm
   functions
3) Compact and lightweight

PRINCIPLE OF OPERATION

The measuring principle is based on the
specific molecular absorption of bands of
infrared beam. The concentration of the
gas is determined by the transmittance of
infrared beam through a measuring cell of
fixed length. The correlation between the
transmittance and concentration is given
by the following Eq.(1).(Lambert-Beer's
law);.

$$I = I_0 \cdot \exp(-kcl) \qquad (1)$$

  k: Absorption coefficient
  c: Gas concintration
  l: lengeh of the measuring cell
  I: Transmitted infrared intencity
 I₀: Original infrared intensity

The analyzer construction is shown in
Fig. 1. The analyzer is composed of an
infrared source, a chopper, a measuring
cell and a detector. This simple construction enables easy maintenance as well
as a compact and lightweght design.

An infrared beam emitted by the infrared
source travels through the measuring cell
filled with the sample gas while being
partially absorbed. Then, the attenuated
beam reaches to the detector which has two
chambers connected by the small channel.
The absorption in the detector causes
pressure rise in both chambers, resulting
a balancing flow in the connecting channel,
since it is designed to cause a different
pressure rise in each chamber. The magnitude of the balancing flow varies corresponding to the concentration of the
measured component in the sample cell.
This slight flow is converted to the electrical signal by the mass-flow sensor.

The mass-flow sensor is similar to a
heated wire wind velocity meter of a very
small size and high sensitivity. High
stability of the analyzer is obtained by
this optical system and mass-flow sensor.

The electrical signal is amplified and
supplied to the indicator and controller.
The controller has lower/upper limit alarm
function and outputs contacct output
signals.

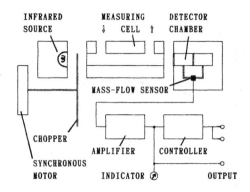

Fig. 1. Single Beam NDIR Analyzer

SPECIFICATIONS

Specifications of the analyzer are as follows;

Measuring method: NDIR single beam method
Measuring gas : CO_2 in air
Measuring range : 0 to 3000ppm or 0 to 5%
Output signal : 4 to 20mA DC or 0 to 100mV DC
Alarm functions : lower/upper limit alarm output (contact output)
Indicator   : CO_2 concentration actual scale
Repeatability : ±1%FS
Zero point drift: within ±10%FS/6 month
Power supply   : 100,115,220V AC,50/60Hz approx.18 VA
Outer dimensions: 257x220x85mm (HxWxD)
Weight   : Approx. 3Kg
Gas sampling   : suction pump,membrane filter built-in

TYPICAL DATA

Tipical data of the analyzer are shown in table 1 and Fig. 2.

APPLICATION

Typical applications of the CO_2 analyzer are as follows;

Agriculture

CO_2 concentration in the greenhouse is increased to improve product quality and productivity by injecting CO_2 from the cylinder or CO_2 generator. Thisis because that plant development is increased by activating photosynthesis with higher CO2 concentration than that of normal air. However,if CO_2 concentration is too high,

TABLE 1   INTERFERENCE (ZERO POINT)

INTERFERENT COMP	INFLUENCE
20℃SAT. H_2O	0.3%FS
0.1% CO	≃ 0
1% CH_4	0.1%FS

RANGE:0-3000ppm

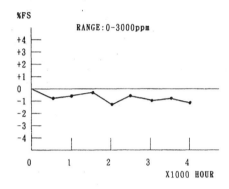

Fig. 2. ZERO DRIFT

the plants are spoiled. On the other hands, lack of CO_2 will cause insufficient growth of the plants. Therefor,CO_2 conecntration is kept to an appropriate value between 600 to 1500ppm,depending on the growth condition and the atmospheric condition such as sunlight, temperature and so on. The CO_2 analyzer is used to monitor and control CO_2 concentration in the green house.

CA (Controled Atmosphere) storage

CA storage is used to keep the freshness of fruits or vegetables during the long term storage. CO_2 and O_2 concentration in the atmosphere are kept at constant value. In case of apple storage, CO_2 and O_2 concentration are controled at applox. 3%. The CO_2 analyzer is suitable for this application.

Air conditioning

An ordinary air conditioning equipment sucks external air at a constant rate and mixes it with recovered air for preventing air pollution in a room,whereafter the air re-heated or re-cooled and sent into the room. In a crowded building such as departmentstore,theater, buisness office, etc., external air must be sucked at a rate sufficient for a maximum load which is too heavy for a nomal load and much energy is consumed wastefully. However, it is possible to save the re-heating or re-cooling energy very effectively while preventing pollution or degradation of room air by measuring CO_2 concentration in recovered air and controlling air suction rate in conjunction with measured results.

CONCLUSION

NDIR gas analyzer has various merits comparing with another method of analyzers and various demands for its applicatioin have been increasing. The single beam method CO_2 analyzer described in this paper has the features of high stability, compact and lightweight design. This will be the advantage for new applications especially in the use of inexpensive systems.

REFERENCES

Watanabe,A. (1979). NDIR Analyzer For Measuring Two Low Concentrated Components. ISA Proceedings of the 1979 symposium Newark, pp. 29-37.

Nakano,M. (1980). Single Beam Method Non Dispersive Infrared Gas Analyzer ISA Proceedings ,1980,pp.337-345.

Fig. 3. Ourlook of CO_2 Analzer

Copyright © IFAC Mathematical and Control Applications
Agriculture and Horticulture, Matsuyama, Japan 1991

GENERALIZED PREDICTIVE CONTROL OF AN
IRRIGATION CANAL REACH

V. M. Ruiz-Carmona* and K. Najim**

*Coordinación de Automatización, Instituto de Ingeniería, Universidad Nacional Autónoma
de México, AP.Po.70-474, Coyoacan 04510 México D.F. México

**Ecole Nationale Supérieure d'Ingénieurs de Génie Chimique, C.N.R.S. UA 192,
Greco Sarta, Chemin de la loge, 31078 Toulouse Cedex, FRANCE

Abstract. This paper describes an application of the generalized predictive control algorithm to an irrigation canal reach.

In 1984, the Engineering Institute of the National University of Mexico (II - UNAM) has begun to work in the development of a control system for the Mexican canal irrigation network. Initially a performance and stability study of the classical control systems used (downstream and upstream control) were made. As result of this, it was decided to use the "Downstream control" systems, in which the control variable is the upper gate of the canal reach and the water level at the beginning and/or the end of the canal is the controlled variable. The control objective is to maintain a desired flow rate and a water surface elevation. This leads to keeping the water level at the end of the canal close to a reference profile.

The equation of motion for water in an open canal is derived from momentum and mass balances for an elemental length. The model obtained is composed of two coupled non-linear partial differential equation systems called the Saint-Venant equations associated with non-linear boundary conditions. The simulation of this model was performed using the Preissman algorithm. This model is too complex to use for control purposes, so to represent the complex dynamics of the canal, linear discrete input-output models are derived.

The control algorithm is derived from plant output prediction over several steps on the assumption that beyond a certain horizon the control increments become constant.

The model parameters are estimated by a constant trace adaptation algorithm with the necessary features (prefiltering by the internal perturbations model, filtering by a selective bandpass filter, normalization, U/D factorization, dead zone, etc.) that enhance the robustness of the self-tuning control algorithm.

The obtained results illustrate the high performances and the excellent properties of the presented control scheme. Finally, some details on practical implementation are provided.

Keywords. Adaptive control; identification; irrigation canal; modelling; prediction.

INTRODUCTION

In the economy of many countries the irrigation of agricultural lands is an important factor. Surface irrigation usually requires an extensive canal system to transport water from storage reservoirs to farms. Because of variable weather conditions, the quantity of water required by irrigators changes with time. A control system that can improve the general operation of the canal system is very useful.

Good management of water in canals can make substantial savings in canal system operation and in new canal projects, because of the possible reduction of the canal cross sections (canal dimensions).

The main control objective is to maintain a desired flow rate and water surface elevation.

In this paper we shall be concerned with the modelling and the control of a single reach of an irrigation canal using the water levels near the gates as controlled variables. These are the most accessible measurement variables that describe the system dynamics.

The control of water flow between two sections or reaches of an irrigation canal is commonly accomplished with motor-driven gates.

For canal control purposes, open-loop control systems

were initially developed. These control systems require an a priori schedule of water demand. In this case, if a change in the demand occurs, it can induce either water loss or dissatisfaction of users (farmers) when there is insufficient water.

A pragmatic solution in the control of canal systems is the application of a feedback control that automatically adjusts the opening of gates in order to satisfy the control objective.

Two control systems are considered, the 'upstream control system' and the 'downstream control system' (Cunge and Woolhiser 1975) (Fig. 1). In the first system, the lower gate (B) regulates the water level at the upper end of the reach (Y_B). In the second system, the upper gate (A) controls the water level at the begining of the reach (Y_A), or at the end of the reach (Y_B), or an average of levels Y_A and Y_B.

The upstream system has been shown to have the same disadvantages as the open-loop control system. A good knowledge of the water needed is necessary and if the required quantity varies greatly this can lead to wastage or scarcity of water.

In the downstream control system, there are three cases: in the first, the output variable is the level Y_A and an increase in the canal banks is required; in the second, the level Y_B is the output variable and the gain margin of the control system is small; in the last case, named Bival, a weighted average of the water levels Y_A and Y_B is used as output variable (which means that a level in the middle of the reach is regulated). The last one has a great gain margin and the canal banks required are smaller than in the first case.

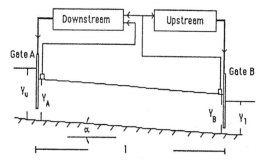

Fig. 1. Downstream and Upstream control systems.

In 1984 the Institute of Engineering of the National University of Mexico (UNAM) began a project for the design of a canal control system. The first step of this work was (España and Barrera, 1986) the development of a numerical simulator of a canal reach, used to verify the design of control structures, and the selection of the control system. The control system selected was the downstream one and it was seen that the water level Yb needs to be kept close to the reference profile in order to have the smallest possible canal banks and to satisfy the flow rate demanded by the canal network. The stability of the different downstream control systems was studied (España and Ruiz, 1987; Constantinides, Koojoori and Jacquot, 1976) and it was found that a modification of the bival control can improve the canal operation. In this control system, two different control structures are used for the control of water levels Y_B and Y_A. This control system has the same stability margins of the Bival control system using the Bival parameter design. It gives us the possibility of eliminating the static error at the water level Y_B. With this in mind, we used a proportional control at the water level Y_A and a proportional integral control for the water level Y_B (P-PI). In the operation of the P-PI control in a downstream control system (Fig. 2), it was seen that the selection of the control parameters was a very demanding job and a new parameter set was required if the canal operation points changed.

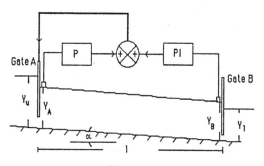

Fig. 2. P-PI control in a downstream control system

On-line determination of the optimal conditions requires excessive computer memory and time. Indeed, it is preferable to establish approximate on-line control policies which can be applied to practical situations and which can adapt themselves to varying conditions.

As a continuation of the work done in the II-UNAM we decided to study the feasibility of adaptive control strategies. Among the adaptive control algorithms developed in the last decade (Åström, 1987), the Generalized Predictive Control (GPC) approach (Clarke, Mohtadi and Tuffs, 1987) seems to be very attractive and robust. This control approach has been used to regulate the water level Y_B (output variable) in a downstream control system.

PLANT MODELS

The equations of motion for water flow in a rectangular open canal may be derived from momentum and mass balances for an elemental length of canal, making the following assumptions:

- The wave surface gradually varies and the vertical acceleration is small.
- Friction losses in unsteady flow are not significantly different from those in steady flow.
- Velocity distribution across the wetted area does not substantially affect the wave propagation.
- The wave movement can be considered as two-dimensional.
- The average slope of the channel bottom is so small that $\sin(\alpha)$ may be replaced by $\tan(\alpha)$, and $\cos(\alpha)$ by unity, where α is the angle made by the channel bottom with the horizontal (Fig. 1).

Momentum balance gives:
$$\partial v(x,t)/\partial t = - g \ \partial y(x,t)/\partial x - v(x,t) \ \partial v(x,t)/\partial x$$
$$- g \ (S_f - \alpha) \qquad (1)$$
Mass balance:
$$\partial y(x,t)/\partial t = - v(x,t) \ \partial y(x,t)/\partial x - y(x,t) \ \partial v(x,t)/\partial x \quad (2)$$

where $v(x,t)$ and $y(x,t)$ represent respectively the velocity and the water depth at location x $(0 \leq x \leq l)$ and time t $(0 \leq t \leq \infty)$; g the gravitational acceleration; S_f is the friction coefficient which is given by the empirical Manning equation (Chow 1973):
$$S_f = (n \ Q)^2 \ P^{-2} \ (P/A)^{4/3} \qquad (3)$$
with (for a rectangular canal):

P : $T + 2 \ y(x,t)$, the wetted perimeter.
A : $T \ y(x,t)$ the wetted area.
n : the Manning coeficient
Q : flow rate
T : canal width.
l : canal length.

The boundary conditions associated with the system are described by (Cunge and Woolhaiser 1975):

$$Q = \mu \ a \ \sqrt{H-a} \qquad \text{if } h \leq 2/3 \ H \ \text{ and } a \leq 2/3 \ H$$
$$Q = 0.385 \ \mu \ H^{3/2} \qquad \text{if } h \leq 2/3 \ H \ \text{ and } a > 2/3 \ H$$
$$Q = \mu \ h \ \sqrt{H-h} \qquad \text{if } h > 2/3 \ H \ \text{ and } a > h \qquad (4)$$
$$Q = \mu \ a \ \sqrt{H-h} \qquad \text{if } h > 2/3 \ H \ \text{ and } a \leq h$$

where:
$$\mu = b \ (2g \ / \ \gamma)^{1/2}$$
a : gate open.

428

H : upstream level
h : downstream level
b : width of the weir
γ : head loss coefficient

The model is simulated using the implicit method of finite differences named "Preissmann implicit method". For the non-linear terms it uses an approximation in power series where the terms of second and higher order are neglected. A full description and stability analysis of the method can be found in Liggett and Cunge (1975).

As continuation of the study made by Espana and Ruiz (1987), we decided to work with the rectangular canal reach which they considered. The physical parameters are given in table 1.

The complex canal reach dynamics, using a downstream control system for which the controlled variable is the water level Y_B, can be described by a single-input/ single-output model, which, in discrete time, can be approximated by a disturbed parametric linear model:

$$A(q^{-1}) y(k) = B(q^{-1}) u(k-d) + Dq(^{-1}) v(k-d_d)$$
$$+ [C(q^{-1})/\Delta(q^{-1})] e(t) \qquad (5)$$

where k, u(k), y(k), v(k) and d are respectively the discrete time, the input, the output, the measured disturbance and the time delay. {e(k)} is a zero mean disturbance sequence with finite variance. The degrees of the polynomials $A(q^{-1})$, $B(q^{-1})$, $C(q^{-1})$ and $D(q^{-1})$ are respectively n_a, n_b, n_c, and n_d. The polynomial $\Delta(q^{-1})$ is the differencing operator. It allows an integral action to be introduced into the control law (Tuffs and Clarke 1985).

GENERALIZED PREDICTIVE CONTROL

The Generalized Predictive Control is based on the minimization of the quadratic cost function (Clarke, Mohtadi and Tuffs, 1987):

$$J=E\left\{ \left[\sum_{j=Ns}^{Ny}(P\, y(k+j) - Q\, y_m(k+j))^2 + \lambda(k) \sum_{j=0}^{Nu-1} \Delta u(k+j)^2 \right] / \mu_k \right\} \qquad (6)$$

where :

- $E\{\ \}$ is the expectation operator conditioned on the set μ_k of data up to time k.

- Ny maximum prediction horizon.
- Ns minimum prediction horizon.
- Nu control horizon (Nu≤Ny) which defines the control scenario, $\Delta u(k+j)=v$ for $j \geq ch$.
- $\lambda(k)$ the control weighting sequence.
- $Q(z^{-1})$ polynomial of order nq.
- $P(z^{-1})$ an asymptotical stable monic polynomial of order np.

The reference sequence {$y_m(k)$} can be calculated as the output of a reference model:

$$A_m y_m(k) = B_m r(k) \qquad (7)$$

where {r(k)} is the set-point to be tracked by the system output.
To solve this minimization problem, the plant model needs to be reparametrized as a function of $[P(z^{-1})y(k)]$ and $[\Delta u(k)]$. Multiplying the plant model (5) by $\Delta(z^{-1})P(z^{-1})$, we obtain:

$$y_p(k) = (1-\alpha)\, y_p(k) + \beta\, \Delta u(k-d) + \gamma\, \Delta v(k-d_d) + \Gamma\, e(k) \qquad (8)$$

where
$$y_p(k) = P(z^{-1})\, y(k)$$
with
$$\alpha(z^{-1}) = A(z^{-1})\Delta(z^{-1}) \quad ; \quad \beta(z^{-1}) = P(z^{-1})B(z^{-1})$$
$$\gamma(z^{-1}) = P(z^{-1})D(z^{-1}) \quad ; \quad \Gamma(z^{-1}) = P(z^{-1})C(z^{-1})$$
$$n\alpha = na + n\Delta \ ; \quad n\beta = np + nb \ ; \quad n\gamma = np + nd \ ;$$
$$n\Gamma = np + nc$$

The generalized predictive controller involves the prediction of the auxiliary output $y_p(k+j)$. Using the "Concatenation property of the optimal predictor" (Goodwin and Sin, 1984; Favier, 1987), the j-step-ahead predictor of $y_p(k+j)$ can be written as follows:

$$y_{pk+j} = (1-\alpha)\, y_{pk+j} + \beta\, \Delta u(k-d+j) + \gamma\, \Delta v(k-d_d+j) + \widetilde{\Gamma}_j \omega(k) \qquad (9)$$

where
$$y_{pk+j} = \begin{cases} y_p(k+j/k) & j > 0 \\ \\ y_p(k+j) & j \leq 0 \end{cases}$$

$$\widetilde{\Gamma}_j = \begin{cases} \Gamma_j + \Gamma_{j+1} z^{-1} + \Gamma_{j+2} z^{-2} + ... + \Gamma_{n\Gamma} z^{-n\Gamma+j} & j \leq n\Gamma \\ \\ 0 & j > n\Gamma \end{cases}$$

$$\omega(k) = y_p(k) - y_p(k/k-1) = e(k)$$

The j-step ahead prediction error satisfies the following equation:

$$\alpha\, (y_p(k+j) - y_{pk+j}) = (1 + \Gamma_1 z^{-1} + \Gamma_2 z^{-2} + ... + \Gamma_{j-1} z^{-j+1})\, e(k+j).$$

The predictor equations has been decomposed by Favier (1987) into two terms; one is a function of the initial conditions and $\Delta u(k+m)$ for m = -1,-2,... ; the other is a function of the present and future of $\Delta u(k+m)$ for m = 0,1,... :

$$y_{pk+j} = \begin{cases} y_p^o{}_{k+j} + \sum_{i=d}^{j} m_{i-d}\, \Delta u(k+i-d) & \text{if } j \geq d \\ \\ y_p^o{}_{k+j} & \text{if } j < d \end{cases}$$

$$y_p^o{}_{k+j} = (1-\alpha)\, y_p^o{}_{k+j} + \beta\, \Delta u^o(k-d+j) + \gamma\, \Delta v^o(k-d_d+j) + \widetilde{\Gamma}_j\, \omega(k)$$

with
$$y_p^o{}_{k+j} = \begin{cases} y_p^o(k+j/k) & \text{if } j > 0 \\ \\ y_p(k+j) & \text{if } j \leq 0 \end{cases}$$

$$\Delta u^o(k+j) = \begin{cases} \Delta u(k+j) & \text{if } j < 0 \\ 0 & \text{if } j \geq 0 \end{cases}$$

$$v^o(k+j) = \begin{cases} v(k+j) & \text{if } j < 0 \\ v(k) & \text{if } j \geq 0 \end{cases}$$

As the future values of v(k) are required and are not available, we assume that they do not change in the future.

The coefficient m_i are computed using a recursive procedure developed by Favier (1987):

$$m_0 = \beta_0$$

$$m_i = \beta_i - \sum_{j=1}^{Min(n\alpha,i)} \alpha_j m_{i-j} \quad \text{for } i > 0$$

The quadratic cost function can be written in a vectorial form:

$$J = \| Y_p^o + M \Delta U - Q Y_m \|^2 + \lambda(k) \| \Delta U \|^2$$

where:

$$Y_p^o = [y_p^o{}_{k+Ns}, y_p^o{}_{k+Ns+1}, \ldots, y_p^o{}_{k+Ny}]^T$$
$$\dim Y_p^o = Ny-Ns+1$$
$$Y_m = [y_m(k+Ns), y_m(k+Ns+1), \ldots, y_m(k+Ny)]^T$$
$$\dim Y_m = Ny-Ns+1$$
$$\Delta U = [\Delta u(k), \Delta u(k+1), \ldots, \Delta u(k+Nu-1)]^T \quad \dim \Delta U = Nu$$

$$M = \begin{bmatrix} m_{Ns-d} \cdots m_0 \ 0 \ \cdots \cdots \ 0 \\ m_{Ns-d+1} \ \cdots \ m_0 \ 0 \ \cdots \ 0 \\ \cdots \cdots \cdots \cdots \cdots \cdots \\ m_{Nu-1} \ \cdots \cdots \cdots \ m_0 \\ \cdots \cdots \cdots \cdots \cdots \cdots \\ m_{Ny-d} \ \cdots \cdots \cdots \ m_{Ny-Nu-d+1} \end{bmatrix}$$
$$\dim M = Ny-Ns+1 \times Nu$$

The derivation of the control law from the cost function is given in Clarke, Mohtadi and Tuffs (1987). The resulting control law is given by:

$$\Delta U = (M^t M + \lambda(k) I)^{-1} M^t (Q Y_m - Y_p^o)$$

A recursive numerical stable algorithm for the inversion of the matrix $[M^t M + \lambda(k) I]$ developed by Favier (1987) was used to improve the numerical robustness of the control algorithm.

The adaptive GPC corresponds to the association of the GPC algorithm with a robust parameter identification algorithm.

The unknown process parameters are recursively identified and the control strategy is developed in a self-tuning manner by replacing the process parameters by their estimates.

The unmodeled dynamics are always present. It is necessary to use a normalizing signal (Ortega and Praly, 1986). This signal is given by:

$$\rho(t) = \mu \ \rho(t-1) + (1-\mu) \max \{ \rho_o ; \Phi^T(t-1) \ \Phi(t-1) \}$$

with $|\mu| < 1$, and ρ_o a positive constant parameter.

The regressor $\Phi(t)$ and the estimation error $\epsilon(t)$ are normalized as follows:

$$\Phi_n(t) = \Phi(t)/\sqrt{\rho(t)}, \ \epsilon_n(t) = \epsilon(t)/\sqrt{\rho(t)}.$$

To ensure the upper bound of the gain matrix P(t), the constant trace algorithm (Lozano, 1982; Irving and co-workers, 1979) has been used. In order to ensure numerical efficiency, it is advisable to use the U-D factorization $P(t) = U(t)D(t)U^T(t)$ (where U and D are an unitary upper triangular matrix and a diagonal matrix respectively) to update the covariance matrix (Bierman, 1977). It is straightforward to ensure a lower bound on the estimator gain matrix by monitoring (Ljung and Söderström, 1983) the elements of $D(t) = diag\{d_i(t)\}$, i.e., $d_i(t) = max\{d_\emptyset, d_i(t)\}$ where $d_\emptyset$ is a regularizing constant.

We also added a dead-zone in the identification algorithm. This means that whenever the prediction error is close to zero (when the available information is not likely to improve the parameter estimation) we turn off estimation.

When the system is over-parametrized or when the data supplying the estimation procedure are not exciting enough, a problem of stabilizability may occur. This problem can be avoided by an appropriate choice of the model order. A supervisory system has been developed to provide the adaptive control integrity. This system includes, among other procedures:

- An on-line time delay estimation using three candidate model structures excited by a white noise signal (De Keyser, 1986).
- An information measure (M'Saad, 1987) which indicates how much the recent information differs from the past data.
- A regularized normalized least squares algorithm with parameter projection (Ortega and Praly, 1986), etc.

All these modifications of the basic recursive least squares algorithm are introduced to improve the robustness properties of the adaptive controller with respect to a wide range of operating conditions.

EXPERIMANTAL RESULTS

In this section we shall be concerned with the preliminary results related to the adaptive predictive control of an irrigation canal.

Prior to on-line control on the canal, experimental studies were carried out to assess modelling and control requirements.

The preliminary studies were made to determine a suitable model order, time delay, filter and an appropriate sampling rate for control.

The sampling period selected is 110 seconds and for the antialiasing filter we have used a Butterword filter (Åström, 1984) with $\omega_o = 0.01$.

The aperture velocity of the gate is equal to 1 meter/minute.

The plant model is identified in open-loop, using a suitable PRBS. The parameter identification algorithm is intialized as follows:

$$nb = 1; \ na = 3; \ d = 4 \quad \Phi(0) = 0; \ \theta(0) = [1,0, \ldots, 0]^T;$$
$$P(0) = 10 \ I \ (I \text{ is the unit matrix of dimension 5.5});$$
$$d_0 = 0.01; \mu = 0.1 \ ; \rho_\emptyset = 0.001 \ ; \rho(0) = 0.001.$$

The control parameters are:

$$Nu = 1 \ ; \ Ny = 8 \ ; \ Ns = 4 \ ; \ \lambda = 0.1; \ P = 1; \ Q = 1 \ ;$$

The evolution of the water level Y_B and its reference profile are given in Fig. 3. Figures 4 and 5 show the behaviour of the system when there is a variation of +15 cm or -15 cm respectively in the aperture of the downstream gate (B).

Figure 6 show the performance of the P-PI regulator and the GPC in regulation operation. The downstream gate aperture variations are successively: +15 cm, -30 cm, +30 cm, -30 cm and +15 cm. In this figure the GPC can compensate the disturbance faster and with smaller output agitations.

These figures show that the generalized predictive control algorithm leads to much more satisfying results for the control of the water level Y_B than the control based on P and PI regulators. This is due to the GPC ability to compensate the non-linear character of the canal and to the adaptation loop that takes into account changes in the process dynamics.

CONCLUSION

In this paper the generalized predictive control algorithm associated with a robust parameter estimator was used to control a highly non-linear irrigation canal which is quite a difficult process, owing to the complex relationship between the variables of interest in the physical model.

The results show the high performance of this adaptive control algorithm over the P-PI regulator, in the case of tracking desired profiles and regulation operation.

Table 1 Physical characteristics of the rectangular canal reach

$\alpha = 0.0001$ °
$T = 3.048$ m.
$n = 0.0225$
$l = 1500$ m.
$g = 9.81$ m/s^2
$v(x,0) = 0.4109$ m/s.
$y(x,0) = 2.1336$ m.
$Y_u = 2.49936$ m.
$Y_l = 1.76784$ m.
$a_A = 0.7257$ m.
$a_B = 0.7253$ m.
$\mu = 6.0723$ m$^{3/2}$/s.

AKNOWLEDGEMENTS

V. Ruiz wishes to acknowledge the scholarship provided by the CONACYT (Mexico) during the period in which this work was done in France (ENSIGC-INPT, GRECO SARTA).

REFERENCES

Åström, K. J. and B. Wittenmark (1984). *Computer controlled systems: theory and design*. Prentice-Hall, New Jersey, U.S.A..

Åström, K. J. (1987). Adaptive feedback control. *Proceedings I.E.E.E.*, **75**, 185-217.

Bierman G. J. (1977). *Factorization methods for discrete sequential estimation*. Academic Press, New York, U.S.A..

Chow V. T. (1973). *Open-channel hydraulics*. McGraw-Hill. New York, U.S.A..

Clarke, D. W., C. Mohtadi, and P.S. Tuffs, (1987). Generalized predictive control - Part I. the basicalgorithm -Part II. Extensions and interpretations. *Automatica*, **23**, 137-164.

Constantinides C. T., T. Koojoori and R. G. Jacquot (1976). Stability of a feedback controlled distributed system by modal representation. *Int. J. Systems Sci.*, **7**, 989-1003.

Cunge, J.A., and D.A. Woolhiser (1975). Irrigation systems. In K. Mahmood, and U. Yevjevich, *Unsteady flow in open channels*, Vol II. WaterResources Publications, Colorado, U.S.A., pp. 509-514.

De Keyser R. M. C. (1986). Adaptive dead-time estimation. *Proc. of 2nd IFAC workshop on adaptive system in control and signal processing*. Lund, Sweden.

España V., M. and J. Barrera (1986). Estudio del control en simulacion de un canal de riego. Tech. report No. 5118 II-UNAM, Mexico.

España V., M. and V. M. Ruiz C. (1987). Control adaptable con modelo de regulacion de un canal de riego. Tech. report No.5142 II-UNAM, Mexico.

Favier, G. (1987). Self-tuning long range predictive controller. *Proc. of the 10th IFAC world congress*, Munich, F.R.G..

Goodwin G.C. and K.S. Sin (1984). *Adaptive filtering prediction and control*. Prentice-Hall. New York, U.S.A..

Irving, E., J.P. Barret, C. Charrossey, and J.P. Monville (1979). Improving power network stability and unit stress with adaptive generator control. *Automatica*, **15**, 31-46.

Liggett, J.A. and J.A. Cunge (1975). Numerical methods of solution to the unsteady flow equations. In K. Mahmood, and U. Yevjevich, *Unsteady flow in open channels*, Vol I. Water Resources Publications, Colorado, U.S.A., pp. 142-162.

Ljung, L., and T.Söderström (1983). *Theory and practice of recursive identification*. MIT press, Massachusetts, U.S.A..

Lozano L., R. (1982). Independent tracking and regulation adaptive control with forgetting factor. *Automatica*, **18**, 455-459.

M' Saad, M. (1987). Sur l'applicabilité de la commande adaptative. Thèse d'état, I.N.P. Grenoble, France.

Ortega, R., and L. Praly (1986). Robustesse des algorithmes de commande adaptative. In I. D. Landau and L. Dugard (Ed.), *Commande adaptative: Aspects pratiques et théoriques*. Masson, Paris.

Tuffs, P.S., and D. W.Clarke (1985). Self-tuning control of offset: a unified approach. *I.E.E. Proc.*, **132D**, 100-110.

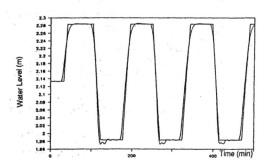

Fig. 3.a

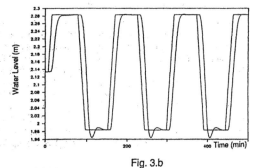

Fig. 3.b

Fig. 3. Evolution of the water level YB and its reference profile (a.- GPC; b.- P-PI).

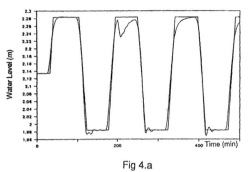

Fig 4.a

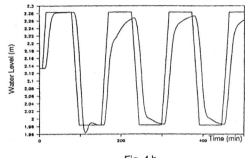

Fig. 4.b

Fig. 4. The influence of opening (15 cm.) the downstream gate on the water level YB, which tracks the reference profile (a.- GPC ; b.- P-PI).

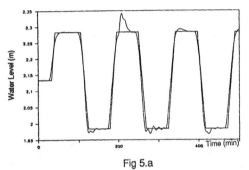

Fig 5.a

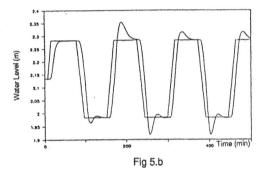

Fig 5.b

Fig. 5. The influence of closing (15 cm.) the downstream gate on the water level YB, which tracks the reference profile (a.- GPC ; b.- P-PI).

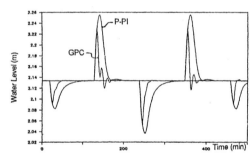

Fig. 6. The influence of closing and opening the downstream gate on the regulation of the water level YB.

432

AUTOMATION OF A DRIP IRRIGATION SYSTEM

A. Araya*, H. Ortíz*, E Van der Meer* and A. Torres**

*Departamento de Electrónica, Campus Norte, Universidad de Tarapacá, Arica, Chile
**Instituto de Agronomía, Campus Azapa, Universidad de Tarapacá, Arica, Chile

Abstract. The survey presents the most relevant aspects of automation -
of a drip-irrigation system. This technical devise is used in times and
through a low-cost personal computer or equivalent; in this case an Atari 800XL is used. The installation of the system is set to be carried -
out innational locations, particularly in Northern Chile. Implementa---
tion of the system and relevant hardware are discussed.

Keywords. Agriculture; drip-irrigation; computer control; computer -
applications; process control; software development.

INTRODUCTION

Technology has been growing steadily and   -
everyone has benefitted from it. With the
passing of time, it has become evident that
areas usually distant, have been interrela-
ting with each other in the pursuit of common goals.

In this context the engineering with their
skills can contribute to the development of
agriculture.

In this survey, causes for implementing au
tomation in irrigation networks are first
analysed. Then the software system develop
ment is describe, where problem definition,
requirements and objectives to be accomplis
hed are specified. Finally, a detailed implementation of the system is presented.

WHY AUTOMATION IN IRRIGATION NETWORKS

A number of distinct causes foster the im--
plementation of irrigation system's automation for various locations all over the   -
world.
Some of the causes are:

1. Practically, a better control and irriga
   tion dose is achieved.
2. Optimization of water requires for the -
   crops.
3. Low requirements of manpower.
4. Permits not only programmed irrigation,
   but also additional operations, such as
   fertirrigation, filter's cleansing, etc.
5. Unconvenient situations can be contro---
   lled, such as network failures, dry func
   tioning of pump, blocking of irrigation
   ducts.

DESCRIPTION OF SYSTEM CONSIDERED FOR AUTOMATION

Figure 1. shows system for automation. It
consists of a drip irrigation system loca--
ted at the Agronomy Institute of Campus Aza
pa. In this system, heads and hidraulic -
valves - as shown - will be autimized.

Valves required for automation are electrovalves having a selenoid that operates   -

through an electric pulse. This pulse   -
will be provided the output ports of the -
Atari 800XL computer.

The irrigation time for a specific crop is
determined by the "Irrigation Times Equa--
tion" as follows:

$$T_R = \frac{EB * Kc * PS * T}{nq * qg * 100} \qquad (1)$$

where,

EB: evaporation coefficient tray
Kc: standarized crop coefficient
PS: covering porcentage
T : inference area
nq: number of drip
qg: flow of drip

The farm is divided in five sectors named
a , b , c , d and e.
Sector a. Consists in olives and the irri
          gation is by micro-sprinkling.

Sector b. Consists in olives and the irri
          gation is by drip.

Sector c. Consists in tomatoes and the -
          irrigation is by drip.

Sector d. Consists in cucumbers and the -
          irrigation is by drip.

Sector e. Consists in flowers and the   -
          irrigation is by drip.

All the variables in equation (1) change -
according to the vegetative times to each
crop, just as indicated by standarized ta
bles.

SOFTWARE SYSTEM DEVELOPMENT

Definition of the problem.

Software capable of controlling hardware
system of a drip-irrigation system must be
designed.

Objectives.

The drip-irrigation system must execute -
the following objectives:

1. Control of irrigation valves in time pe
   riods.
2. Control of start/stop of motor-pump.

3. Control of filter's cleansing.
4. Delivery of relevant information.
5. Reporting when needed.

All the times sequences will be controlled by an intern clock.

The users will enter the time cycles for the drip-irrigation system functioning.

The maximum programming time is a week. After this time, if are not changed the specifications, system starts again with the former programming.

Software flow diagram.

Figure 2. shows the flow diagram implemented in BASIC language. The used variables are first sized and the ports A and B of the PIA (6520) are programmed as outs. After this the principal menu in screen will be presented, which consists of six options accessed from the keyboard. This options are:

a. clock programming
b. drip- irrigation time programming.
c. actual report
d. valves test
e. system run
f. system stop

Option a. Allows to enter the time (hours and minutes), setting the computer's internal clock (this time will not alter if other option is choosen).

Option b. There is two operations: Time programming and Filter's Cleansing.
The first one will ask about crop type and hours and minutes when will initiate the irrigation, tray evaporation and the week of vegetative period according to the crop.
The second operation will ask the beginning and term time of filter's cleansing.

Option c. Shows in the computer's screen a picture frame where appears crops type that were programmed and the variables of each them. The user can check if the programmed values were entered well.

Option d. This option shows that the hardware's micro-switch will be in off position. The aim of this test is to check through panel control the correct action of the valves.

Option e. Sets all the system in functioning. To go out from this option, strike the ESC key, returning to the principal menu.

Option f. This option stops all the drip - irrigation system.

IMPLEMENTATION OF AUTOMATION

The system implementation is showed in Fig. 3. and consists essentially of four parts:
1. Computing system
2. Control panel
3. Conditioning circuit
4. Electrovalves

Computer's system

For the automation of the drip-irrigation system, an Atari 800XL computer is used.

To control electrovalves, a peripheral interface adaptor integrated to this computer (PIA 6520) is used. This chip mainly consists of two ports of 8 bits each, being used as input or output, programmed - bit by bit as needed.

As operation deals with controlling automatically selenoid valves, it is necessary to generate output pulses through ports of 6520.

Pine designation of ports A and B.

Twelve electrovalves are to be controlled, of which five are head of the drip-irrigation system (including water pump) and sectors a,b,c,d and e will be controlled by the others seven (see Fig. 1). Thus, pine designation of ports A and B are the following:

PORT A : At this port, all output bits will be used (b0 - b7).
The configuration is as follows:

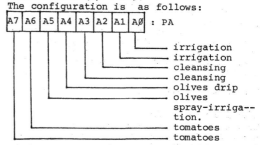

Bit description is as follows:

PA0 - PA1 : Bits representing electrovalves of head of irrigation system.

PA2 - PA3 : Bits representing electrovalves of head of filter cleansing system.

PA4   : Bit representing electrovalve of sector b, where olives are drip irrigated.

PA5   : Bit representing electrovalve of sector a, where olives are spray-irrigated.

PA6 - PA7 : Bits representing electrovalves of sector c, where tomato crop is irrigated.

PORT B : In this port only five bits will be used, b2, b3, b4, b5, b6. Three bits left cannot be utilized but in the hardware system of the Atari 800XL computer.

Configuration of five bits in use is the following:

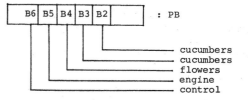

Bits decription is as follows:
PB2 - PB3 : These bits represent sector d electrovalves, where cucumbers

are irrigated.

PB4 : This bit represents electro--
valves of sector c, where flowers are irrigated.

PB5 : This bit represents pulse given
for starting and stopping motor
pump.

PB6 : This bit sends out pulse con--
trolling a three state gate.

Introduction of this control -
bit was necessary to solve the
problem of process start, because, when computer started,
all ports become outputs, sen
ding out a pulse to the elec--
trovalves. These remained -
open.

Control Panel

Control panel reports the state of electro
valves, whether open or closed, sector under irrigation, or filters operation.

Control panel was designed with 43 cm. -
length and 30 cm. width. Its irrigation -
network is represented over its surface, -
including heads system and the electrovalves are simulated by led's diods, which -
start and stop according to the user's pro
gramming.

Figure 4. shows control panel design.

Conditioning Circuit.

To increase current coming out through -
ports A and B of computer, and to produce
a pulse according to the computer electro--
valve interface and also to isolate the -
computing system from external influences,
a signal conditioning circuit was needed.
This consists basically of: Three-state ga
tes with buffer, microswitch, transistots,
transformer (220/24 VAC) and relays.

Figure 5 shows conditioning circuit design.

CONCLUSIONS

1. The drip-irrigation system is not complex and allows an easy manipulation.

2. Allows to accomplish crops in a zone whe
re water is little.

3. Automatic operation of irrigation sys--
tems has major functioning and economic
advantages.

4. Time automation is relative simple and
non expensive.

5. Time automation requires electric energy and fluctuations in the energy su--
ply will alter irrigation dose.

6. Due to the great technological advance,
using a personal computer, automation -
through times is feasible.

REFERENCES

CINADCO   (1984). El riego por goteo, Is
rael.

CORFO   (1989). Aplicación del riego -
por goteo, Chile.

Gurovich   (1985). Fundamento y diseño de
sistemas de riego, San Juan, -
Costa Rica, IICA.

Pisarro, F.(1987), Riegos localizados de -
alta frecuencia, Madrid prensa.

RAIN-BIRD (1989). Catálogo de programa--
ción de riego, E.E.U.U.

SEGUNDO CURSO INTERNACIONAL DE RIEGO LOCA
LIZADO. (1981). España.

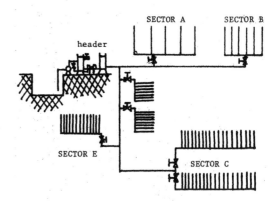

Fig. 1. System to be Automatized

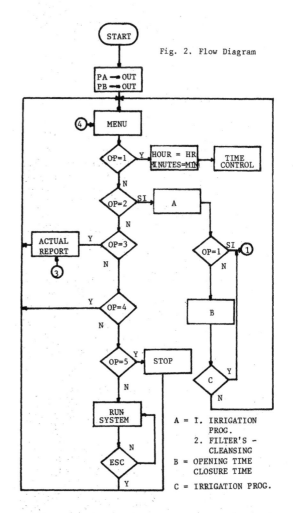

Fig. 2. Flow Diagram

A = 1. IRRIGATION
PROG.
2. FILTER'S -
CLEANSING

B = OPENING TIME
CLOSURE TIME

C = IRRIGATION PROG.

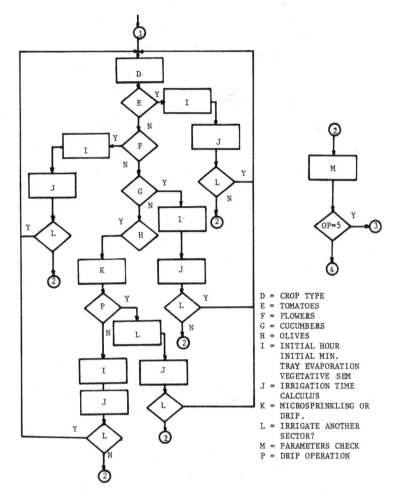

D = CROP TYPE
E = TOMATOES
F = FLOWERS
G = CUCUMBERS
H = OLIVES
I = INITIAL HOUR
    INITIAL MIN.
    TRAY EVAPORATION
    VEGETATIVE SEM
J = IRRIGATION TIME
    CALCULUS
K = MICROSPRINKLING OR
    DRIP.
L = IRRIGATE ANOTHER
    SECTOR?
M = PARAMETERS CHECK
P = DRIP OPERATION

Fig. 2. Flow Diagram

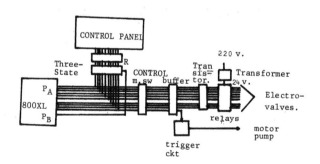

Fig. 3. Implemented System

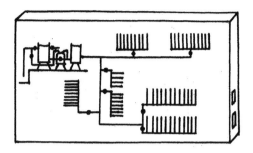

Fig. 4. Control Panel

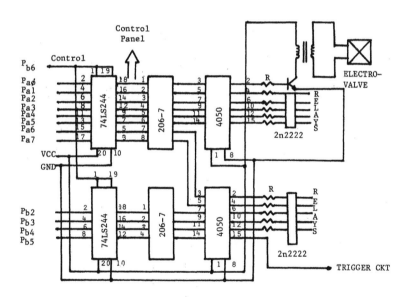

Fig. 5. Hardware System

A SHORT IRRIGATION PERIOD BY MEANS OF
CONTROL TECHNICS

B. de León Mojarro

Centro de Hidrociencias, Colegio de Postgraduados, 56230 Montecillo, Méx, México

Abstract. The current efforts to modernize mexican agriculture are aimed to increase
the agricultural yield of crops, with a strong participation of the farmers in the
operation and conservation of hydraulic infrastructure. At the same time, divulgation
of new technology will ameliorate the use of resources in order to make the irrigation
districts self-sufficient. To achieve this objective it is necessary to change the
traditional methods of water distribution (irrigation scheduling) and regulation methods of canals (manual operation), for a distribution "on-demand" and an automatic control of canals respectively. Mathematical simulation was used to study the operation
of a small irrigation district of 13,000 ha, which has a one week period of irrigation
scheduling and the regulation canal is manual. The results of this study show that by
changing current technics of control of canals and making transmission of information
more efficients is possible to reduce the duration of the irrigation program.

Keywords. Automatic control Canals; mathematical simulation canal; control technics
canals.

INTRODUCTION

In Mexico there are 6 millions ha of irrigated.
Nearly 50 per cent of the total agricultural produc
tion comes from this area under irrigation, and is
expected to increase to 65 per cent by the turn of
the century.

In order to increase the yield of irrigated agriculture it is required to modernize the operation
of hydraulic in such a way that permits the optimi
zation of the use of water that in Mexico is a
limiting factor for irrigated agriculture.

The operation of hydraulic infrastructure in irriga
tion districts must permit to reduce the water los
ses due to operation canals and ameliorate the
application of water to field with opportunity;
both aspects have a direct influence on growing
crops. To achieve these two objectives it is neces
sary to reduce the period of scheduling irrigation
and improve the regulation canals technics. The
development of canal regulation procedures will -
permit to simplify the process of transfering the
responsability of regulation and operation of ca -
nals to farmers, who should assume a dominant role
of control and management of water supply from the
irrigation district. By means of a simulation
mathematical model it was studied the operation of
a small irrigation district of 13,000 ha in which
the period of irrigation scheduling consist of one
week and the regulation canal is manual. Several
methods based on the linear system theory were tes
ted in this district. The results show that by
means of control technics and transmission of infor
mation it is possible to reduce the period of pro
gram irrigation to six hours. This reduction of
irrigation period permits to etablish crops with
high frequency of irrigation and to reduce the
losses of water due to operation canals.

REGULATION OF IRRIGATION CANALS

The regulation canals adapt the conveyance and
distribution water to spatial and temporal varia -
tion of irrigation needs of crops. In this way the
regulation canals is related to the crop water
requirements and the water delivery sheduling. Thus
when the water is in a strict rotation schedule
with a frequency, flow rate and duration fixed by
central policy it practices an upstream regulation
canal with manual gate operation (Burt, 1990).

In canals with side walls parallel to bottom (mexi
can canals), for delivery of water on demand it is
necessary to develop control algorithms which have
two fundamentals parts: the firts part represents
the water needs of crops and the farmers prevailing
practices of irrigation, which can be determined
by means of a module of PREDICTION OF IRRIGATION
WATER; the second part represents the propagation
of hydraulic perturbations along the canal reaches
and the set of operation of gates realized by the
ditchriders, which can be represented by modules
of CANAL and REGULATION respectively, Fig. 1.

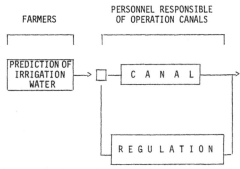

Fig. 1. The block diagram of the regulated system.

Prediction of Irrigation Water

In this work the prediction of irrigation water was based on the rotation scheduling elaborated by the personnel responsible of the operation canals.

Modeling of Flow in Canals

For regulating flow discharge it is necessary the use of a mathematical formulation of canals, which is described by Saint-Venant's equations (Pochat, 1981). These equations are not linears and their use in control algorithms does not results easy. However, it is possible to simplify these equations and obtain a modified model known as Hayami model (Verdier, 1986). The equation is:

$$\frac{\partial Q(x,t)}{\partial t} + \theta \frac{\partial Q(x,t)}{\partial x} + E \frac{\partial^2 Q(x,t)}{\partial x^2} = 0 \qquad (1)$$

Where:

$Q(x,t)$ = discharge
θ, E = constant coeficient representatives of canal.
x = distance along canal
t = time

Equation (1) allows for the Laplace transformation as follows:

$$\frac{Qs(P)}{Qe(P)} = F(\theta, E, P, x) \qquad (2)$$

Where:

$Qs(P)$ = output discharge in a pool canal
$Qe(P)$ = input discharge in a pool canal
P = Laplace variable

Equation (2) not always has an inverse Laplace transformation for all types of input $\{Qe(P)\}$; which makes difficult to obtain the searched fuction $Qs(t)$. Piquereau (1982) indicate that a solution to this problem, is using equation (2) to identify a second order system as follows:

$$\text{Min } C = \int_0^T \{Qe(P)F(\theta,E,P,x) - \frac{Qe(P)Wn^2}{Wn^2+2\delta WnP+P^2}\}^2 \, dt \quad (3)$$

Where:

C = performance criterion
Wn = natural frequency
δ = damping

For the case studied it is necessary the introduction of a time delay (γ) among $Qe(t)$ and $Qs(t)$ corresponding to propagation time of discharge within canal. Richard (1980) indicate that the transfert function of a delayed second order system is:

$$\frac{Qs(P)}{Qe(P)} = \frac{Wn \{EXP(-\gamma P)}{P^2 + 2\delta WnP + Wn^2} \qquad (4)$$

In practice the operation of gates is not continuous, because for elaborating control commands it needs the sampled expression of equation (4), which is:

$$F(Z) = Z^{-n} \frac{CZ^{-1} + DZ^{-2}}{1-AZ^{-1} + BZ^{-2}} \qquad (5)$$

Where:

Z = Z-transform
$n = \gamma/T$

T = sampling period
$C,D,A,B = f(\delta, Wn)$

Regulation Method

The block diagram of the regulated system is shown in Fig. 1. Guthikonda (1979) indicate that is general, the regulator is of the form:

$$R(Z) = \frac{\sum\limits_{i=0}^{m} \alpha i \, Z^{-1}}{1 + \sum\limits_{i=i}^{n} ai \, Z^{-1}} \qquad (6)$$

The regulator characteristics must be:

- The response of the closed-loop system corresponds to a second-order system with damping equal to one.

- For a stable system, the poles of closed-loop system must varying between (0,1).

- In order to eliminate the steady-state error it is necessary to introduce an integration in the regulator.

In this application we chose a sample period of one hour, due to the longitude of the pool canals, the time-delay results inferior to the sample period and in consequence will be considered as zero. Under this conditions De León (1986) found that the expression of the regulator for the different pools of the canal is:

$$R(Z) = \frac{1 - AZ^{-1}}{1 + (D-1) \, Z^{-1} + DZ^{-2}} \qquad (7)$$

The coeficients A and D are the parameters obtained in equation (5).

RESULTS

Characteristics of the Canal Studied

The method and concepts exposed in previous parts were applied in an irrigation district consisting of a reservoir, a main canal that supplies secondary and tertiary canals which contain turnouts to large individual fields, or to groups of fields.

The main canal is 25 km long with a slope of 20 cm/km. It has a trapezoidal cross-section shape and the bottom width is 5 m. Slope of the side walls is 1 on 1, with a maximum normal depth of 2.30 m and a design capacity of 10 m³/s.

Along the main canal are established 9 radial gates for obtaining a higher level of water to supply 11 secondary canals and turnouts, whose situation is shown in Table 1.

440

Table 1 Situation of Check-Drop Structure

Pool	Chech-drop structure	Distance (km)	Discharge (m³/s)
1	reservoir	0.00	10.00
2	gate 1	1.10	7.25
3	gate 2	3.10	4.28
4	gate 3	6.80	3.50
5	gate 4	9.95	2.35
6	gate 5	13.00	1.65
7	gate 6	15.00	0.36
8	gate 7	17.80	0.36
9	gate 8	20.30	0.36
	gate 9	22.00	0.36

The Hydraulic function and the regulation of the main canal were represented by means of a mathematical simulation model of the transient flux.

Break Irrigation Overnight

The interest in studying canal regulation methods is because of the possibility to correct fluctuations which are not considered in the hydraulic regime. These fluctuations have different origins, but the most frequent is the interruption of irrigated at night. The situation that we illustrate in this paper consist in interrupting irrigation at nigth and its continuation the following day, in pool 6 situated among gate 5 and gate 6 (Table 1), which is located 15 km downstream with a water supply of 1.65 m³/s. Figure 2 shows that 12 hours after interruptingt irrigation, the water level upstream of gate 6 reaches its maximum value (2.30 m), while water discharge in gate 5 (Fig. 3) reduces its value to the minimum necessary to supply the downstream turnouts (0.36 m³/s). Figure 2 also shows that 10 hours after resuming irrigation, the water level upstream of gate 6 reaches the minimum and does not permit lateral extractions for a period of six hours. This six-hour interruption is due to the fact that both events (interruption and resuming irrigation) are considered as strong perturbations not foreseen. Nevertheless, is important to note that even if the irrigation is not satisfied in a continuous way, the water losses are reduced by 100 per cent; the reduction of losses is possible due to a continuous evaluation of the hydraulic state of the canals.

Quantitative results. It is important to mention that using adequate data analysis facilities and efficient transmission of information in a frequent way (each hour) permits to reduce the duration of water delivery scheduling from one week to six hours. This continuous regulation permits an economy of water of 70,000 m³, which would be equal to 10 days of running a pumpling well with a discharge of 70 l/s. It is clear that the evaluation of hydraulic state of the canals realized each hour can not be done in a manual way, because in this case there are 72 actions taking place over a distance of 15 km at nigth.

CONCLUSIONS

Real time control offers possibilities to improve the operation of irrigation canals and to reduce the water losses. However, in the mexican irrigation districts the incorporation of control systems should be effectuated gradually, starting with main canals taking into account the experience and knowledge of the field man.

Finally the importance of simplicity of methods of regulation canals should not be overlooked and future capabilities of reliable water deliveries should be emphasized.

REFERENCES

Burt, Ch. (1990). Water delivery control workshop. Dept. of Agriculture Engineering. Cal. Poly, San Luis Obispo, California, U.S.A.

De León, M. B. (1986). Contribution a l'amelioration de la gestion des perimétres irrigués. Université de Montpellier, France.

Guthikonda, V. R. (1979). Complex digital control Systems. Van Nostrand Reinhold Company, New York, U.S.A.

Piquereaux, S. (1982). Gestion automatique des eaux d'étiage. Centre d'Etudes et de Recherche de Toulouse, France.

Pochat, R. (1981). Ecole d'été de mecanique de fluides. Hanoi.

Richard, C.D. (1980). Modern control system. Addison-Wesley publishing company. University of California, Davis, U.S.A.

Verdier, J. (1986). Information de la commande du transpport et de la distribution d'eau d'irrigation, Journées Regionales Européennes de la CIID, Espagne.

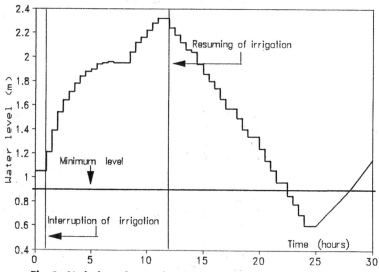

Fig. 2. Variation of water level upstream of gate 6.

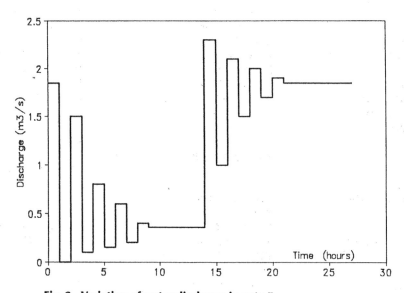

Fig. 3. Variation of water discharge in gate 5.

AUTHOR INDEX

KEYWORD INDEX

447

Printed and bound by CPI Group (UK) Ltd, Croydon, CR0 4YY

03/10/2024

01040323-0017